ELEMENTARY ANALYSIS

KENNETH O. MAY

Dover Publications, Inc.
Mineola, New York

Bibliographical Note

This Dover edition, first published in 2020, is an unabridged republication of the work as published in 1952 by John Wiley & Sons, Inc., New York. It was originally self-published by the author in 1950 under the title, *Analysis: A Freshman Course.*

Library of Congress Cataloging-in-Publication Data

Names: May, Kenneth O. (Kenneth Ownsworth), 1915-1977, author.
Title: Elementary analysis / Kenneth O. May.
Other titles: Author's self-published title: Analysis : a freshman course
Description: Dover edition | Mineola : Dover Publication, Inc., 2020. | An
  unabridged republication of the work originally published: New York : John
  Wiley & Sons, Inc., 1952. | Includes bibliographical references and index. |
  Summary: "This enthusiastically reviewed text, which assumes one year of
  high school algebra and one of plane geometry, provides a unified treatment
  of algebra, analytic geometry, trigonometry, and introductory calculus. Its
  well-organized and thoughtful presentation is fortified with many problems,
  for which solutions are provided. Readers at any level can profit from
  working their way through this book to a basic foundation in college-level
  mathematics"— Provided by publisher.
Identifiers: LCCN 2019038758 | ISBN 9780486842752 (trade paperback)
Subjects: LCSH: Mathematics. | Mathematical analysis.
Classification: LCC QA37 .M38 2020 | DDC 510—dc23
LC record available at https://lccn.loc.gov/2019038758

Manufactured in the United States by LSC Communications
84275401
www.doverpublications.com

2 4 6 8 10 9 7 5 3 1
2019

To John Van De Putte

The essence of mathematics lies in its freedom.

GEORGE CANTOR

Preface

The title of my book, *Elementary Analysis*, is intended to suggest a unified treatment of material usually labeled algebra, analytic geometry, trigonometry, and introductory calculus. The book is directed to those who are going to use mathematics in advanced courses or in science, engineering, or business. Although it assumes as a minimum only one year of high school algebra and one of plane geometry, there is plenty of material for a year's work by students with more extensive preparation. The book is designed for easy adaptation to classes with different backgrounds and to either introductory or terminal courses.

It was for my own students that I began to write this book. Although varying widely in background, aptitude, and intentions, they shared a desire both for enlightenment and for technical competence in mathematics. Those who planned to continue wanted to be prepared for a rigorous course in calculus. Those who did not expect to do further work in mathematics wanted to master the most useful skills and illuminating ideas that could be included in a year course. With these ideas in mind, we assigned students to sections meeting three, four, or five hours a week and looked for a text that would provide all students with a common background for further work. But in spite of the many excellent books available, it proved difficult to find one right for our purposes. Most of them started at too high a level or else did not go far enough. Others omitted important topics, lacked a unified approach, or took fundamental ideas for granted. Accordingly I embarked on this rash endeavor, writing and revising at first for my own classes, then later taking account of varying needs of other possible users.

Underlying the book is my conviction that rigor and plausibility, understanding and manipulative skill, and theory and practi-

ix

cal application are not antagonistic pairs but mutually reinforcing aspects of sound mathematical training. Important ideas are explicitly motivated, heuristic discussions and plausibility arguments are frequent and clearly distinguished, formal definitions and proofs are given in a form that will require no later correction, and over six thousand graded problems afford a wide choice for drill, application, treatment of optional topics, and creative effort. Exercises on applications were prepared by consulting textbooks in advanced mathematics and many other fields, including the physical sciences, biological sciences, engineering, economics, and business. Key ideas such as set, correspondence, number, function, inverse, graph, vector, transformation, and mathematical induction are introduced early and considered repeatedly. Many topics traditionally isolated in separate chapters are woven into the sections and exercises as part of the logical development and application of analysis. For example, arithmetic progressions appear as instances of linear functions, and mathematics of finance is developed in connection with various mathematical topics. Since the problems are indexed in detail, all material on any topic may easily be located.

A textbook writer is the beneficiary of so many sources that it is possible to acknowledge only the most obvious. Above all, I owe a debt of gratitude to Carleton College, which provides the atmosphere in which serious work is possible, and to my students, who are a constant source of stimulation. Professor John Dyer-Bennet of Purdue read several early versions and the final manuscript. The late C. H. Gingrich scrutinized it with the benefit of experience gained from four decades of teaching and a quarter century as editor of *Popular Astronomy*. My colleague, Professor Kenneth Wegner, examined the manuscript with meticulous care and suggested a non-denumerable number of changes and corrections. Professor Ruth Eliot of the Carleton English department was mercilessly helpful in matters of style. Several anonymous reviewers offered encouragement and helpful suggestions, and the publisher's editors attempted to correct what they tactfully called my "lapses" in spelling, grammar, and punctuation. It would be impossible to list individually all the suggestions and assistance contributed by colleagues and students, but I am no less grateful on that account. It appears that I can claim undivided credit

only for errors and weaknesses, which I hope the users of this first
edition will not hesitate to call to my attention.

KENNETH O. MAY

Northfield, Minnesota
March 1952

Note to the Student

Be sure to read the Introduction (Chapter 1) and especially Section 1.7, which contains advice on the best way to study this book. You may not be familiar with the numbering system used in this book, but you will soon find it very convenient. Double numbers indicate sections. For example, 3.12 and Ex. 3.12 indicate the twelfth section in Chapter 3 and the exercise immediately following it. Theorems, formal definitions, equations, figures, and problems are indicated by the number of the section in which they appear followed by a number indicating their position within the section. Thus *3.4.4 means theorem 4 in 3.4, and Ex. 3.8.2 means problem 2 in Ex. 3.8. The same system is used in the index, except that "Ex." is omitted there to save space.

K. O. M.

Note to the Teacher

In order to make possible rounded discussion of significant ideas, I have divided the book into relatively few sections. Each one is intended to be read and mastered as a unit, although perhaps in more or less than one assignment, depending on the class. Each section is followed by a graded exercise that begins with simple problems illustrating the section (numbered in bold-face italics), continues with various applications and developments (numbered in bold face), and ends with problems of some difficulty (marked with a dagger).

The pace of the book is initially slow. After an introduction on the nature of mathematics, Chapter 2 presents the simplest ideas of logic, largely in terms of elementary geometry, and Chapter 3 presents elementary algebra according to the pattern of the construction of the number system from the positive integers. Depending upon student ability and professorial interest, these chapters may be omitted and used for reference, assigned for outside reading, covered quickly with emphasis on troublesome topics, used for review, studied section by section with ample drill, or used as a framework for a rigorous axiomatic development of the number system. Beginning with Chapter 4, the usual material is organized around the study of elementary functions. As the book progresses, the pace quickens, and the student is expected to participate more actively in working out results.

A unified text cannot consist of independent units designed for arbitrary omission or study in any order. Nevertheless, I have written with flexibility as a prime goal. The graded exercises are a principal means of adapting the book to different classes and individuals within the same class. Choice of problem assignments may be used also to emphasize particular topics. For example, inequalities may be treated with unusual thoroughness by emphasizing section 4.12 and the problems on inequalities, which are

scattered throughout the book but can easily be located through
the index. Some sections may be omitted with little or no effect
on continuity. Such are those on the topics Arithmetic Progres-
sions (4.5), Determinants (4.10–4.11), Linear Inequalities (4.12),
Permutations and Combinations (6.4), Binomial Theorem (6.5),
Geometric Progressions (6.6), Geometric Series (6.7), Computa-
tions with Logarithms (7.4), Natural Logarithms and Change of
Base (7.5), Solution of Triangles (8.10–8.12), Families of Straight
Lines (9.4), Vectors (9.7), Complex Numbers (10.1–10.5), The
General Conic (11.4), Invariants (11.5), Geometric Properties of
Conics (11.6), Theory of Equations (12.1–12.4), and Number
Bases (12.8). Calculus topics may be omitted by leaving out 5.2,
5.4, 12.5–12.7, and the calculus problems in other sections. The
teacher will see other possibilities and may find it expedient to
vary the order. For example, Polynomials (Chapter 12) might
well be treated directly after Power Functions (Chapter 6) if the
student does not require an immediate acquaintance with logarithms and trigonometry. Where parts of the book are not assigned
because they cover topics already familiar to the class, it may still
be well to encourage the student to read them on his own in order
to relate his previous information to the new ideas he is acquiring.

I shall be very grateful to any teachers who will give me their
candid reaction to this book and their suggestions for making it
more useful.

K. O. M.

Contents

1

Introduction

What is it all about? Where is its place in my life? Why should
I study it? How should I study it? These questions, although
quite legitimate to ask about mathematics or any other subject,
cannot be completely answered in a brief introduction. We make
only a few comments that may help the student to answer them
for himself in the light of his own experience. Suggestions for
further reading are given at the end of the chapter.

1.1 Mathematics in our civilization

Mathematics, one of the oldest expressions of human culture,
has developed with the growth of civilization. Its roots go back
as far as those of language, art, and religion. The needs and aspi-
rations of society have stimulated its expansion, and science and
technology have provided the raw material for its theories. On the
other hand, mathematical discoveries have often made advances
in other fields possible. Modern mathematics and our industrial
civilization are so closely bound together that neither could exist
without the other. Professional mathematicians and specialists in
the numerous fields of applied mathematics are as essential to our
society as industrial workers, engineers, or doctors.

But mathematics is more than a reflection of the technical and
scientific needs of society. It is also an expression of human
imagination and creative ability. Mathematics has great beauty,
not merely the elegance of geometrical figures, although this is
one aspect of the matter, but the beauty of a logically connected
whole. The beauty of mathematics is a harmony of form, of
symmetries, of complexities simply related, of repeated and varied
patterns. It is no accident that a taste for mathematics is often
associated with a taste for the arts—especially for abstract art and
music. The impulses of the creative mathematician have much in
common with those of the creative artist, composer, or writer, and
the mathematics he creates has aesthetic value for the observer.

1

Mathematicians speak of a theorem as beautiful when it shows many relationships in a simple way. They speak of a proof as brilliant when it demonstrates gracefully and easily something that is difficult to see. They call a theory elegant when it brings together in balanced arrangement, and explains simply, a complex mass of details.

The student will find in this book many examples of the usefulness of mathematics in science, technology, and business, but he should be equally alert to the beauty of mathematics as one of mankind's greatest creative achievements.

1.2 Mathematics as a logical structure

Mathematics is sometimes called a logical structure, because its statements can be proved logically from a few assumptions. Mathematicians demand very high standards of logical precision and reject any theory that is not logically consistent. A mathematical theory may or may not be useful in a particular situation, but it is sound mathematics if it is consistent.

For example, the statement that $2 + 2 = 4$ is mathematically correct because it can be proved from the definitions of the symbols 2, 4, $+$, and $=$. The many situations in which this statement is appropriately applied include the classic example of the boy with two apples who acquires two more. But the statement $2 + 2 = 4$ does not apply to every situation in which we combine things. If we mix two quarts of water and two quarts of ethyl alcohol, we do not get four quarts of the resulting mixture. Instead we get about 3.86 quarts. Does this mean that in this case $2 + 2 = 3.86$? Of course not. It means that the usual notion of addition does not apply to this mixing process, a fact based upon the observations of chemists. Mathematics, however, is not concerned directly with such factual statements, but with abstract statements like $2 + 2 = 4$, which are true quite independently of any observations. Factual statements are true if they correspond to observations, whereas mathematical statements are true if they can be proved logically.

We have emphasized that mathematicians demand proof of statements. But it is impossible to prove everything, because we can prove one statement only by referring to another as justification. If we try to prove everything, we shall go on indefinitely or else prove each of two things by reference to the other. The result of this effort is therefore either an endless process or circular reasoning. Since we cannot prove everything, some statements called axioms must be accepted without proof. We are free to choose any axioms we wish as the foundations of a mathematical theory. This does not mean that it makes no difference what axioms we choose, for a poor choice may lead to contradictory, clumsy, uninteresting, or useless results. But the value of axioms can be fully understood only by working out their logical consequences; hence it is desirable in mathematics to try different axioms. Naturally, in a course in elementary analysis, we shall work with certain simple axioms, most of which are already familiar to the student.

It may be rather startling to learn that axioms are not "necessarily true." For many centuries the axioms of Euclidean geometry were accepted as "self-evident" and "obvious" truths that no sensible person should question. But during the last century mathematicians worked out geometries based on different assumptions. For example, a non-Euclidean geometry has been constructed in which one can draw, through a point, two different lines parallel to a given line. In another geometry, there are no parallel lines at all! Yet these geometries are just as logical as Euclidean geometry and equally consistent with our observations of space, as illustrated by the fact that Einstein used a non-Euclidean geometry in constructing the theory of relativity. The discovery of non-Euclidean geometry helped to free men from the unquestioning acceptance of traditional assumptions, and it encouraged the intellectual experimentation that has led and is still leading to important advances in the arts and sciences. In this book, however, the geometry is Euclidean.

If we cannot say that certain axioms are necessary truths, do we mean that we cannot know where we stand in mathematics? Should we hesitate to accept the geometry we learned in high school? Certainly not. We know just where we stand. We know that, if certain axioms of Euclid are accepted, we must logically accept the theorems that follow. Mathematics does not assert that certain things are true unconditionally, but that certain statements are true *if* certain other statements are true. Does it follow that mathematics is just a game that gives us no information? No, because it is very useful to know exactly what follows from assuming certain axioms. As we shall indicate in the next section, the construction of mathematical theories is very practical.

The logical structure of mathematics is considered further in Chapter 2, and there are many opportunities in the book for the student to develop his skill in logical thinking.

1.3 Mathematics as a tool of science

Mathematical theories are very useful in science, even though their correctness does not depend upon this fact. Since the scientist is trying to describe and to explain observable events, a logically constructed theory may be very helpful to him if it "fits" the situation he is studying, a question that may be answered by methods that are the subject of a special branch of applied mathematics—mathematical statistics. On the other hand, an illogical theory is of little use to the scientist because such a theory will give contradictory results. It will both fit and not fit the facts.

Since mathematics has developed in close touch with the sciences, it is quite natural that mathematicians have constructed, and are daily constructing, theories that are valuable for their applicability to the physical, biological, and social sciences. At the same time, in every scientific field there are those who specialize in the theoretical side and are concerned primarily with constructing special logical theories that "fit." To be successful they must work closely with other scientists who specialize in experiment and observation. They must also be particularly well acquainted with mathematics. In many fields this specialization is recognized by name, for example, there are mathematical physicists and experimental physicists, theoretical biologists and experimental biologists, mathematical economists and econometricians.

Although scientific method is too complicated for brief definition, its two main aspects may be described as the empirical (observation and experiment) and the mathematical (the construction of theories). An example of a very useful mathematical theory is analytical mechanics, the theory of the behavior of bodies under the action of forces. This theory, which is quite abstract and can be derived from a few axioms, including Newton's famous laws of motion, turns out to correspond to the facts in most situations. Hence it is very useful to the physicist, chemist, and engineer. But as is now widely known, Newtonian mechanics does not correspond to the behavior of bodies traveling at very high speeds, and a different theory, the mechanics of relativity, is more appropriate under such conditions. Another example is the mathematical theory of symbolic logic, built upon axioms about deduction. This theory is useful in solving complex problems, not only in science, but also in philosophy.

Although it is possible to develop mathematical theories without regard to their usefulness, it is important to keep applications in mind in order to understand "what it is all about." Pure and applied mathematics enrich and stimulate each other, and the combination of mathematical with factual knowledge is most fruitful for practical work. Mathematics is a theoretical, abstract subject, but it is at the same time eminently practical, if used with understanding and discretion.

This book contains examples of applications in many fields, including the social and natural sciences, engineering, and business. By consulting the index the student can find the applications in which he is particularly interested.

1.4 Mathematics as a language

We have been talking about mathematics as a structure composed of statements derived from axioms, but we have said nothing about the means by which these statements are communicated. Mathematicians have developed a special language that is very convenient for scientific thinking and for clear and concise expression of statements. Yet this language is not something entirely different from everyday verbal language; it is rather an extension or an addition to it.

The symbols of mathematics that stand for objects correspond to nouns in the verbal language. For example, x^2 stands for the number that results from squaring x. Mathematics similarly has symbols for relations and operations that correspond to verbs in everyday language. Thus "$6 = 6$" is a sentence in which "$=$" is the verb. Mathematics has rules of grammar and syntax, which must be followed in order to "make sense." An expression like "$6 =$" is an incomplete sentence. Moreover, the same logical rules are used in proving statements in both languages.

Although verbal language and mathematics have much in common as parts of our total means of communication, they have important differences. Mathematics is more precise and concise. It is a universal language, almost the same throughout the world. Finally, and perhaps most important, it is a highly specialized language. It has tremendous scope and power just because it is especially designed for use wherever precise statement and reasoning must be applied to a situation that can be represented symbolically. On the other hand, our everyday language is very well suited to conversation, exposition, and literature. Having developed spontaneously as a historical social product, it has further advantages of general familiarity, suggestiveness, and flexibility, in contrast to mathematics, which is created consciously for special purposes.

It is important to be able to translate from verbal language into mathematical language, and back again, in order to be able to use whichever is convenient for the problem in hand, and especially to be able to talk and write intelligibly about mathematics. There are many opportunities in this book to practice this translation.

1.5 Heuristic

We have indicated that mathematics may be looked upon as a logical structure, a tool of science, and a language. But we have said little about how people construct new mathematics and how they decide what mathematics to use in a given situation. Why are certain axioms chosen? What do they mean? How do we decide what theorems to try to prove? How can we find a proof of something we think may be true? How can we find a way to solve a given problem? These questions cannot be answered by the formal logical theory of mathematics. Imagination, creative thinking, insight, and intuition are involved in these matters. Of course insight and intuition are subjective faculties, which are different for each individual, and we cannot accept something just because someone says he "feels" that it is so; but "bright ideas" are very important when they are verified and supported by proof.

The word "heuristic" is used as a noun to refer to the art of original thinking, invention, and discovery. It is used as an adjective to describe thinking that has discovery as its purpose. Heuristic thinking involves "mulling things over," hitting upon new ideas, guessing at solutions to problems, and thinking of axioms that may lead to interesting results. Without being bound by precise logic, heuristic thinking proceeds in terms of what seems plausible. Hence it may lead to error unless it is backed by proof. But without heuristic thinking we should have very little mathematics, very little science, art, or literature.

Heuristic ability is highly prized but seldom discussed, perhaps because those who practice the art most effectively are often unconscious of their creative methods and uninterested in telling others about them. However, there is some literature on heuristic thinking (see the bibliography at the end of the chapter), and there are many opportunities throughout this book to study it in the most effective way—by practice.

1.6 Elementary analysis

Arithmetic, elementary algebra, and plane geometry may be compared to the finger exercises that are required before one can create music. The mathematics that the student has learned so far, although useful and interesting, is only a very small part, and not the most useful or interesting part, of mathematics.

Analysis is one of the broad fields of modern mathematics. It has developed in closest relation with science and may be described as "mathematics at work." It brings together arithmetic, algebra, and geometry as well as other subjects. In elementary analysis the student will for the first time be studying mathematics, rather than some isolated part of mathematics.

The function concept, which we introduce in Chapter 4, is a key idea in analysis. The text is organized around the functions that are basic in applications and future work. Chapter 2 presents some ideas of logic that are used throughout the course. Chapter 3 deals with numbers and the elementary operations of addition, subtraction, multiplication, and division. These chapters give the student a chance to review his arithmetic, algebra, and plane geometry in preparation for what follows.

1.7 How to study this book

In mastering mathematics, understanding and skill are more important than memory. Accordingly, systematic daily effort rather than "cramming" is the key to success. It is easy to think that the best way to do an assignment is to get it over with all at once. But two sessions of one hour are often more effective than one session of two hours. Mathematics is like playing an instrument. You have to practice regularly.

Each section of the book is designed to be studied as a unit. Begin by reading it over quickly to get the "drift." On the second reading, have plenty of paper on hand and *use* it. Check every statement by asking "Why?" Fill in omitted steps in proofs and computations. Draw large sketches to illustrate the discussion. Look up unknown words, using the index to this book and a good unabridged dictionary. Restate important ideas in your own words, and try to improve the proofs and calculations. Look for the author's errors. When you have worked through the section in this way, there may still be many points that are not clear, but you will be ready to begin the exercise.

Each exercise is designed to help you master the preceding section in relation to previous work and to give you opportunities to develop heuristic ability. The first few problems are of a rather routine character to provide "drill" on the definitions, theorems, and techniques of the section. You may find it helpful to refer to these even on your second reading of the section. Those problems whose mastery might be considered to represent a minimum acquaintance with the section are numbered in bold-face italics. You should certainly be sure that you can handle these before continuing. The remainder of the exercise invites you to develop further the ideas presented in the section and gives further examples illustrating the material, introducing various applications, presenting related topics not important enough to be included in the section, and anticipating future work. Problems that present some heuristic challenge are indicated by a dagger (†).

In doing problems, remember that written work is a form of communication and must, therefore, be understandable and convincing. A mere answer, without explanation of how it was found or a check to show that it is correct, is of little interest. A "proof" without reasons is not compelling. Your written work should be a clear record of your methods, reasons, and results, not merely some notes and computations. You should save it for review and for reference in connection with later exercises.

When you have "finished" the assignment it is time to look over the section again, do some additional problems, and review previous work. You should look at each problem, even though you may not be able to do them all. Some obscure points may now be cleared up, but you should make notes of those that remain and ask your instructor about them at the earliest opportunity. It is a good idea to look ahead occasionally to get a better idea of where you are going and to begin to get acquainted with material that is to be assigned.

On occasion you will find yourself "stuck." This may be a good time to review or just to browse through the text looking for ideas. But when you have made an energetic attempt and still seem to be making no progress, stop work. You are "in a rut," and you can't get a car out of a rut by racing the motor! Your best chance is to try again later with a new approach, and you will have time to do this if you have begun work on the assignment promptly. It is important to try hard because serious effort starts your mind working on the problem. It is important to quit work in order to give your mind a rest and perhaps to permit it to solve the problem for you while you are thinking consciously about something else. Of course you should seek help from your fellow students and teachers, but do not let them do your thinking for you if you wish to develop your own ability.

SUGGESTIONS FOR FURTHER READING

Among the very few books that discuss mathematics in both a broad and an elementary way are:

The Education of T. C. Mits, by Lillian R. Lieber, W. W. Norton and Co., New York, 1944.
Mathematics and the Imagination, by Edward Kasner and James Newman, Simon and Schuster, New York, 1940.

Somewhat deeper are:

Mathematics, Our Great Heritage, edited by William L. Schaaf, Harper and Brothers, New York, 1948.
An Introduction to Mathematics, by A. N. Whitehead, Oxford University Press, New York, 1948.

Books on the history of mathematics often give considerable insight into its nature and place in society. Good histories of elementary mathematics are:

A Short History of Mathematics, by Vera Sanford, Houghton Mifflin Co., Boston, 1930.
The River Mathematics, by Alfred Hooper, Henry Holt and Co., New York, 1945.

Biographies of mathematicians offer fascinating glimpses of mathematics as a human activity and show dramatically how closely related it is to science and philosophy. Among the best books in this field are:

Men of Mathematics, by E. T. Bell, Simon and Schuster, New York, 1937.
Makers of Mathematics, by Alfred Hooper, Random House, New York, 1948.
Whom the Gods Love, by Leopold Infeld, Whittlesey House, McGraw-Hill Book Co., New York, 1948.

Two very elementary books on the logical structure of mathematics are:

Mits, Wits, and Logic, by Lillian R. Lieber, W. W. Norton and Co., New York, 1947.
Non-Euclidian Geometry, by Lillian R. Lieber, Science Press, Lancaster, Pennsylvania, 1940.

Among elementary textbooks on mathematical logic might be mentioned:

Fundamentals of Symbolic Logic, by Alice Ambrose and Morris Lazerowitz, Rinehart and Co., New York, 1948.

More profound are:

An Introduction to Logic, by Alfred Tarski, Oxford University Press, New York, 1946.
Foundations of Logic and Mathematics, by Rudolph Carnap, University of Chicago Press, Chicago, 1939.
Principles of Mathematical Logic, by David Hilbert, Chelsea Publishing Co., New York, 1950.

To see the importance of mathematics as a tool of science, you need only page through advanced books and periodicals in any scientific field. The following books present mathematics in relation to its applications:

Mathematics for the Million, by Lancelot Hogben, W. W. Norton and Co., New York, 1940.
Mathematics, Queen and Servant of Science, by E. T. Bell, McGraw-Hill Book Co., New York, 1951.

An excellent and delightful book on heuristic is:

How to Solve It, by G. Polya, Princeton University Press, Princeton, 1945.

Interesting ideas on heuristic are to be found in:

An Essay on the Psychology of Invention in the Mathematical Field, by Jacques Hadamard, Princeton University Press, Princeton, 1949.

The following give helpful hints on study as well as on problem solving:

"How to Study Mathematics," by W. C. Arnold in the *American Mathematical Monthly*, December, 1940, p. 704.
How to Study—How to Solve, by H. M. Dadourian, Addison-Wesley Press, Cambridge, Mass., 1951.

A very informative article on careers in mathematics is:

"Professional Opportunities in Mathematics" in the *American Mathematical Monthly*, January, 1951, pp. 1 ff.

The tedium of numerical calculations is considerably reduced by using a slide rule or computing machine. Instruction books usually accompany these devices. An interesting book on the more powerful mathematical machines is:

Giant Brains—or Machines That Think, by C. Berkeley, John Wiley and Sons, New York, 1949.

The logic that is used in building mathematics is basically the same as that involved in all precise thinking. Usually we think more or less clearly without being conscious of what we are doing. But, in mathematics, "more or less" is not enough. We must be as clear as possible about the reasons for each statement we make, and this means being conscious of the logical way in which we are thinking. In this chapter we discuss some of the most elementary ideas of logic.

2.1 Propositions

A **proposition** is a statement or sentence. However, when we use the word "proposition" in mathematics, we have in mind that the statement is to be proved or consciously assumed without proof. Those assumed without proof are called **axioms;** those that are proved are called **theorems.** A typical axiom of mathematics is: "There is one and only one straight line through two points." A typical theorem is: "If two triangles are congruent, they have equal areas."

There are many statements that are not propositions in the mathematical sense. For example, "mathematicians cannot add" is a factual statement, which may or may not be true. We would not expect it to be assumed as an axiom or proved by reference to axioms. As we pointed out in 1.2, mathematics is not concerned with statements whose justification is based directly on facts of experience, but rather with theorems of an abstract character based on axioms. A mathematical proposition is always part of a logical structure.

A theorem in mathematics is always a conditional, not a categorical, statement. It asserts that *if* a certain statement is true, *then* another statement is true. The theorem we quoted above, "if two triangles are congruent, they have equal areas," does not state that any two triangles are congruent or that two particular triangles are congruent. It does not state that any two or a particular pair of triangles have equal areas. It asserts simply that, *if* two triangles are congruent, *then* they have equal areas. This theorem can be proved from the axioms of Euclid, but, if we took some other axioms, it might not be provable. Hence, when we state a theorem in mathematics, we have in mind that its truth depends on what has been assumed or proved before.

In a theorem the statement that follows "if" is called the **hypothesis.** The statement following "then" is called the **conclusion.** Thus a mathematical theorem takes the form: "If hypothesis, then conclusion." Suppose we let A and B stand for any two statements. Then a theorem can be written: "If A, then B." There are other ways of saying the same thing. In geometry we often write: "Hyp.: A; Con.: B." Still other forms are: "If A, B." "A implies B." "A, if B." "From A, it follows that B." When A implies B, it is said that "A is a **sufficient condition** for B." This expression is appropriate because if we know that A implies B, and we know that A is true, we can conclude that B is true. Thus knowing A is "sufficient" for knowing B. When A implies B, it is also said that "B is a **necessary condition** for A." This is appropriate, because, if A is true, B must be true. Hence B is "necessary" for the truth of A.

A very convenient and suggestive symbol is the arrow $\longrightarrow$, which means "implies." We can write "A implies B" as "$A \longrightarrow B$." The arrow emphasizes that B follows from A. If we know that $A \longrightarrow B$, and that A is true, we can conclude that B is true; but without this knowledge about A, we can draw no conclusion about B. The theorem $A \longrightarrow B$ says nothing about A and B taken separately. Moreover, as we shall see in 2.5, if $A \longrightarrow B$, it does *not* follow that $B \longrightarrow A$. From $A \longrightarrow B$ and B we cannot conclude that A is true.

It may happen that a theorem is not stated in the form "if A, then B." For example, the Pythagorean theorem is usually stated: "The square of the hypotenuse of a right triangle equals the sum of the squares of the other two sides." We may rewrite this theorem as follows: "If a triangle has a right angle, then the square of the side opposite the right angle is equal to the sum of the squares of the other two sides." This way of writing the theorem brings out the nature of the hypothesis.

EXERCISE 2.1

1. In each of the following propositions state the hypothesis and the conclusion separately:

(*a*) If two sides and the included angle of one triangle are equal respectively to two sides and the included angle of a second triangle, then the triangles are congruent.

(*b*) If the corresponding sides of two angles are parallel, the angles are equal. (*Solution.* Corresponding sides of two angles are parallel. The angles are equal.)

(*c*) If the corresponding sides of two angles are perpendicular, the angles are equal.

(*d*) If a straight line cuts two parallel lines, the opposite interior angles are equal.

(*e*) If two things are equal to the same thing, they are equal to each other.

(*f*) If two triangles are similar, their corresponding sides are proportional.

(*g*) If in a plane two lines are perpendicular to the same line, they are parallel.

(*h*) Two right triangles are similar, if an acute angle of one is equal to an acute angle of the other.

(*i*) If two circles intersect, their common chord is perpendicular to the line joining their centers.

(*j*) Circular arcs which subtend equal central angles are proportional to their radii. (*Solution.* Two circular arcs subtend equal central angles. The arcs are proportional to their radii.)

(*k*) If two planes intersect, the intersection is a straight line.

(*l*) If two planes are parallel to the same plane, they are parallel to each other.

(*m*) If a line is perpendicular to two intersecting lines, it is perpendicular to the plane of those lines.

(*n*) If a chord of a circle is bisected by a radius, it is perpendicular to it.

(*o*) If x is an even number, $5x$ is an even number.

(*p*) If $y = \sqrt{x}$, $y^2 = x$.

(*q*) If x is an odd number, so is x^2.

2. Restate the theorems in prob. 1 by using letters and writing in the form: Hyp.: .... Con.: .... (*Solution for e:* Hyp.: $a = b$, $c = b$. Con.: $a = c$.)

3. Restate the theorems in prob. 1 in as brief a form as possible, making use of the arrow to mean "implies." [*Solution for a:* (ABC and $A'B'C'$ are triangles with $AB = A'B'$, $BC = B'C'$, and $\angle B = \angle B'$) $\longrightarrow$ ($ABC \cong A'B'C'$).]

4. In the following propositions the hypothesis and conclusion are not clearly indicated. Restate so as to make them clear.

(*a*) A point on the perpendicular bisector of a segment is equidistant from the end points. (*Solution.* If a point lies on the perpendicular bisector of a segment, then it is equidistant from the end points.)

(b) A point on the bisector of an angle is equidistant from the sides of the angle.

(c) Parallel lines do not meet.

(d) An exterior angle of a triangle is equal to the sum of the opposite interior angles.

(e) A side of a triangle is less than the sum of the other two sides.

(f) Opposite sides of a parallelogram are equal. (*Solution.* If $ABCD$ is a parallelogram, then $AB = CD$ and $BC = DA$. Or: If a quadrilateral has its opposite sides parallel, then they are equal.)

(g) An angle that is inscribed in a semicircle is a right angle.

(h) A line perpendicular to a radius at the point where it meets the circumference of a circle is tangent to the circle.

(i) The medians of a triangle meet in a point. (*Solution.* If three segments are the medians of a triangle, then they meet in a point.)

(j) In regular polygons all sides are equal.

(k) The sum of the interior angles of any triangle equals two right angles.

(l) The area of a parallelogram is given by $A = bh$, where b is a side and h is the altitude on that side.

(m) The area of a triangle is $\frac{1}{2}bh$, where b is a side and h the altitude on that side.

(n) The area of a trapezoid is given by $A = \frac{1}{2}(b + b')h$, where b and b' are the lengths of the parallel sides and h is the altitude.

(o) The surface area and volume of a sphere of radius r are given by $S = 4\pi r^2$ and $V = \frac{4}{3}\pi r^3$.

(p) The lateral surface area and the volume of a right circular cone with altitude h, radius of base a, and slant height s are given by $S = \pi a s$ and $V = \frac{1}{3}\pi a^2 h$.

(q) One and only one plane passes through three non-collinear points in space.

(r) One and only one plane passes through a line and a point not on the line.

(s) One and only one plane contains both of two intersecting lines.

(t) A plane perpendicular to each of two intersecting planes is perpendicular to their intersection.

(u) There is one and only one plane perpendicular to two skew lines.

(v) A plane intersects a sphere in a circle.

(w) Any multiple of an even number is even.

(x) $a + b = b + a$ for any numbers a and b.

5. Reconsider the theorems in probs. 1 and 4. Draw a sketch to illustrate each one and make sure you can explain the meaning of all the terms that appear. Try to recall or reconstruct the proof of some of them. If necessary consult a geometry textbook.

6. Consider the proposition: "If the moon is made of green cheese, it contains calcium." (a) Is it true? (b) Is its hypothesis true? (c) Is its conclusion true? (d) Answer the same questions for "if the ocean is made of grape juice, it is not salty." (e) Similarly for "if the earth is flat, people do not fall off."

2.2 Definitions

If we are to understand a statement we must know the meaning of the words and other symbols that it contains. The meaning of an expression may be given by a definition. For example, the words "right triangle" mean "a triangle one of whose angles is a right angle." We could state this formally as follows: "A **right triangle** *is defined to be* a triangle one of whose angles is a right angle." As a second example, the symbol 6^2 means $6 \cdot 6$. More generally, for any number a, a^2 is defined to mean $a \cdot a$.

In a definition, the expression that is defined is usually written first and may be called the **defined term.** The expression that gives the meaning of the defined term is usually written second and is called the **defining term.** The definition gives the meaning of the defined term by stating that it is the same as that of the defining term. Thus $6^2 = 6 \cdot 6$ gives the meaning of 6^2 as $6 \cdot 6$. A definition is neither a theorem nor an axiom. It is simply a matter of convenience. We decide to invent a different way of expressing something, and we define the new way of expressing it in terms of the old way. We announce that we shall use a certain symbol, previously assigned no meaning, in a certain sense specified by the definition. Thus $6^2 = 6 \cdot 6$ cannot be proved, nor is it an assumption. It is just that we choose as a matter of convenience to write $6 \cdot 6$ in the briefer form 6^2. Definitions are thus in a sense arbitrary, but this does not mean that it is unimportant how we choose them. The definition $a^2 = a \cdot a$ is very useful. Good symbols and terms are very helpful in reasoning. You should always look for the motive behind a definition, but often it will be apparent only after you know and work with the definition for some time. Of course, unless you know exactly what is meant by a word or symbol you will not be able to handle it intelligently. Accordingly, *you should memorize definitions.*

Sometimes we are interested in defining a statement. For example, "X and Y are siblings" means "X and Y have the same parents." In such cases it is clear that each term of the definition implies the other. Thus if X and Y are siblings, X and Y have the same parents. Also if X and Y have the same parents, X and Y are siblings. This must be so, since the two statements have the same meaning. If we define a statement A by another statement B, we have $A \longrightarrow B$ and $B \longrightarrow A$ by definition, and we write "$A \longleftrightarrow B$ (Definition)." *The defined term is always on the left in this form.* The double-headed arrow indicates that each statement implies the other, and the word "definition" indicates that this is true by definition. Another way of stating "$A \longleftrightarrow B$ (Definition)" is: "If B, then *it is said that A.*" For example: "If x and y have the same parents, it is said that they are siblings." *Note that the defined term comes second in this form.*

In other cases we wish to define words or symbols that do not form a statement. For example, we wish to define "triangle" as "three non-collinear points and the segments joining them." In this case neither term is a statement, and we cannot speak of one implying the other. However, we can say that the defined term *is defined to be* the defining term or that the defining term *is called* the defined term. For example: "A parallelogram is defined to be a quadrilateral whose opposite sides are parallel." Or: "A quadrilateral whose opposite sides are parallel is called a parallelogram." Sometimes "is defined to be" or "is called" is replaced by "is." *Note that the defined term comes first in one form and second in the other.* Often we can use an equals sign to represent the relation between the defined term and the defining term. For example, we write: "$a^2 = a \cdot a$ (Definition)." *Here the defined term is always written on the left.* In this book defined mathematical terms are printed in bold-face type.

Often the word "definition" is used loosely for a statement that describes something. For example, a college education has been "defined" as "four years of subsidized irresponsibility." The statement is interesting, but hardly a definition in the mathematical sense. *A mathematical definition is not a statement of fact or opinion, but merely an agreement about the use of words and other symbols.*

It is evident from the foregoing discussion that a thing can be defined only in terms of something else, just as a statement can be proved only from another statement. We might like to define everything, but it is impossible. Either we would have to proceed in a circle, defining A in terms of B, B in terms of C, and then C in terms of A, or else we would start but never finish an endless series of definitions. An interesting experiment is to use the dictionary to try to find the meaning of a word. Look up the word, then each word that appears in its definition. Continuing in this way, we should soon find ourselves back where we started. The dictionary can tell us the meaning of a word only if we know the meanings of other words already. The notion that the dictionary really defines everything in an absolute sense is called the "fallacy of the dictionary."

Since we cannot define everything, we must take some basic things as undefined. These are called **undefined elements** or **basic concepts.** In geometry, the notions of "point" and "straight line" are usually taken as basic. The student may remember statements that a point is a thing with position but no magnitude, and that a line is the shortest distance between two points. But these are not genuine definitions, because "position," "dimension," "distance," and "shortest" are concepts that cannot be defined except in terms of "point" and "straight line." Such "definitions" are circular and are really informal discussions intended to explain ideas that are not actually defined. *A genuine definition must give the meaning of a thing in terms of something else previously defined or taken as undefined.* When certain symbols are taken as undefined, it does not mean that we "do not know what we are talking about." It means simply that we have to start with something as given.

A mathematical structure may be compared to a building. The basic, undefined elements are the bricks and other building materials. The axioms are the fundamental ways in which these materials can be combined. Logic is the technique of constructing the building so that it will hold together. The theorems are the parts of the building, constructed from the undefined ideas according to the axioms and held together by the cement of logical deduction. The meaning and function of a brick cannot be clear until one sees the buildings that can be built with bricks. In the same way the full significance of undefined elements is contained in the theory that is based on them. It is hard to define a "point," but Euclidean geometry gives us a great deal of information about "points." When a mathematical theory is developed without giving concrete meaning to its undefined elements, we call it **pure mathematics.** When the undefined ideas are given a definite interpretation, we call the resulting theory **applied mathematics.** Thus geometry may be considered an abstract theory based on axioms about undefined entities called "points," or it may be interpreted as a theory about ordinary space by associating the notion of "point" in the theory with a "very small" object, such as a dot on a piece of paper or the intersection of two strings.

Usually a pure theory gives rise to many applied theories, depending on the interpretation of its undefined elements. For example,
we shall interpret a "point" as a pair of numbers, and this interpretation is a very practical one. On the other hand, there are
often many different abstract theories that can be applied effectively to the same concrete situation.

EXERCISE 2.2

1. If we take the ideas of 1 and addition as basic, we can define 2 as $1 + 1$.
Write out definitions of the same sort for 3, 4, and 7.

2. Define a^3 as we defined a^2 in the section.

3. Define 3 in terms of 2, 1, and $+$.

4. We can define $2a$ to mean $a + a$. Define $3a$ in a similar way.

5. We can define **"complementary"** by saying: "Angles A and B are
complementary" is defined to mean "$A + B = 90°$." Give a similar definition for "supplementary."

6. Complete the following definitions (note that "is" stands for "is defined
to be"):

    (a) An equilateral triangle is a triangle in which.........................

    (b) A median of a triangle is a line....................................

    (c) The hypotenuse of a right triangle is.............................

    (d) A circle is the locus of all points.................................

    (e) A diameter of a circle is a line...................................

    (f) A rectangle is a quadrilateral....................................

    (g) A trapezoid is a quadrilateral...................................

    (h) A tetrahedron is a space figure consisting of.......................

    (i) An obtuse angle is an angle......................................

    (j) π is the ratio...

7. In the following definitions indicate the defined term and restate, using
a double arrow or the phrase "is defined to be."

    (a) A quadrilateral having its opposite sides parallel is called a **parallelo-
gram.**

    (b) When a triangle has all its angles less than 90° it is called **acute.**

    (c) A triangle is called **scalene** when no two of its sides are equal.

    (d) Sides opposite equal angles of two similar triangles are called **corre-
sponding sides.**

    (e) An **equilateral** triangle has all its sides equal.

    (f) If a triangle has two equal sides it is called **isosceles.**

    (g) The length of a perpendicular from a point to a line is called the **dis-
tance from the point to the line.**

    (h) A line that cuts two or more lines is called a **transversal** of these lines.

    (i) The distance from the vertex of a triangle to the opposite side is called
an **altitude** of the triangle.

(j) The **lateral area** of a cone is the area of its surface, not including the base.

(k) One integer is said to be **greater** than another if it comes later in the sequences of integers.

8. Define the following geometric terms by means of careful statements: (a) Congruent. (b) Equal angles. (c) Cone. (d) Oblique angle. (e) Similar triangles. (f) Obtuse angle. (g) Parallel lines. (h) Right angle. (i) Square. (j) Regular polygon. (k) Distance between two parallel lines. (l) Cube. (m) Exterior angle. (n) Perpendicular bisector. (o) Diagonal. (p) Diameter. (q) Concurrent lines. (r) Chord. (s) Secant. (t) A line perpendicular to a plane. (u) Prism. (v) Skew lines. (w) Cylinder. (x) Sphere. (y) Rhombus.

9. Define the following terms: (a) Even number. (b) Odd number. (c) Prime number. (d) Perfect square. (e) Multiple of four. (f) Proper fraction. (g) Lowest terms. (h) Binomial. (i) Exponent. (j) Least common multiple.

2.3 Proof

The word "proof" is used in ordinary conversation in different ways. For example, Snodgrass may offer to "prove" to Fignewton that he can "whip" him. On a rather more civilized level, a scientist may set out to "prove" the correctness of a physical hypothesis. In both cases the proof appeals to observation and experiment. This is not the kind of proof that we are discussing here. In mathematics, proof means deduction rather than verification, an appeal to axioms rather than to experience.

A **mathematical proof** consists of a sequence of statements. Each one must be justified by reference to axioms (including axioms of logic), previously proved theorems (including those of logic), definitions, previous statements in the proof, or a combination of these. The proof ends with a statement of the theorem to be proved. The statements in a proof are called **steps.** Each step is justified by something previously accepted. The person who is following the proof is compelled at every step either to deny something that has gone before or else to agree to take the next step. Since everything that has gone before has been justified in the same way, the reader must in the end either accept the proof or deny the axioms on which it is based. That is the meaning of a proof. If the axioms are accepted, the theorem must be also. When you read or listen to a proof it is *your* responsibility to be sure that each step really is justified. When constructing a proof, it is up to *you* to explain and to justify every step.

A proof is thus an argument, an argument that is thoroughly and genuinely convincing. It might seem that there would be no objective way to say whether a proof is good or bad, since what convinces one man may not convince another. This difficulty is avoided in mathematics by agreement in advance on "rules of proof." Certain axioms of logic are agreed upon, and proof proceeds in terms of these. In this way, objective standards of proof can be established. One rule of logic very frequently used in proofs is the principle that "if A implies B, and B implies C, then A implies C," or

*2.3.1 $(A \longrightarrow B)$ and $(B \longrightarrow C) \longrightarrow (A \longrightarrow C)$

The student will remember how he used this principle in geometry. He would start with the hypothesis of a theorem and show that it implied another statement. This in turn would imply a third statement. Then the rule would justify concluding that the hypothesis implied this third statement. Thus in a series of steps, if each step implies the following one, we may conclude that the first one implies the last. The subject of proof is too complicated to discuss here fully. In this book we try to be as careful as possible in our proofs, but our main interest is to become familiar with important ideas and methods rather than with the technical details of proof.

Proofs may take various forms. The most complete is the detailed **formal proof.** Such a proof consists in a list of numbered steps, each one accompanied by detailed reasons. A slightly less formal type of proof is the **informal proof,** written in literary style, with reasons merely indicated in some cases. Often we find a **sketch proof** that gives merely some key ideas and reasons upon which a formal proof could be based. Sometimes, also, discussions are labeled proofs that are really not proofs at all. Such **plausibility "proofs"** are merely arguments that suggest the possibility that the theorem may be true, or try to sway the reader to feel that the theorem is true without giving a genuine deductive proof that "holds water." Plausible arguments often help the reader to appreciate the meaning of a theorem. However, they may easily be given for false theorems, and are not substitutes for genuine proofs.

We shall give proofs in all these forms. In any proof, formal, informal, or sketch, steps or reasons may be omitted. It is essential that the student, when reading a proof, fill in for himself what is omitted. If an informal or sketch proof is given, the student should try to construct a formal proof. In addition to constructing proofs, we often discuss a theorem from the point of view of how a proof can be constructed by one who does not already know how. Such discussions will be labeled **heuristic discussions.**

In order to illustrate these ideas we consider the following theorem: "In an isosceles triangle, the angles opposite the equal sides are equal."

Plausibility "Proof." We recall first that an isosceles triangle is one that has two sides equal. We can see from a drawing that the angles opposite the equal sides are equal. We can fold the triangle down the center and test this. (Note that this discussion does not in any way prove the theorem, because we know from experience that what appears to be true may be false, and what is true in one case may be false in another.)

Heuristic Discussion. To help devise a proof we draw a figure (Fig. 2.3a). It does *appear* that the angles A and C are equal. But how to prove it? Measurement? No, because measurement is inaccurate at best. We want to prove that $\angle A = \angle C$ in *any* triangle in which $AB = BC$. Do we know any theorems or axioms about equal segments or equal angles? When are two angles equal?

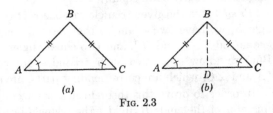

(a)   (b)

FIG. 2.3

They are certainly equal if they are corresponding parts of two congruent triangles. But here we have only one triangle. This suggests *making* two triangles by drawing a line through B as in Fig. 2.3b. We shall be able to show that angles A and C are equal if we can show that the triangles ABD and CBD are congruent. They have a side in common, BD, and two equal sides AB and BC. They would be congruent if the included angles ABD and CBD were equal. (Why?) Hence we draw BD as the bisector of the angle B. We are now ready to construct a formal proof.

Formal Proof. The theorem may be restated as follows:

Hyp.: *ABC* is a triangle with *AB* = *BC*

Con.: ∠*A* = ∠*C*

Proof.

1. Let the bisector of the angle *ABC* meet *AC* in *D*. (There is one and only one bisector of an angle, and any two non-parallel lines meet in a point.)

2. In triangles *ABD* and *CBD*, *AB* = *BC*. (By hypothesis.)

3. In triangles *ABD* and *CBD*, *BD* = *BD*. (Identically the same.)

4. In triangles *ABD* and *CBD*, ∠*ABD* = ∠*CBD*. (Step 1 and the definition of bisector.)

5. Triangles *ABD* and *CBD* are congruent. (Steps 2, 3, and 4, and the theorem: Two triangles are congruent if two sides and the included angle of one are equal respectively to two sides and the included angle of the other.)

6. ∠*A* and ∠*C* are corresponding parts of congruent triangles *ABD* and *CBD*. (Corresponding angles are defined as those opposite equal sides.)

7. ∠*A* = ∠*C*. (Step 6 and the theorem that corresponding parts of congruent triangles are equal.)

Informal Proof. If in the given triangle we construct the bisector of the angle *B*, we have two triangles that have a common side *BD*, two equal sides *AB* and *BC*, and two equal angles *ABD* and *CBD*. But we know that two such triangles are congruent. Hence ∠*A* and ∠*C*, which are corresponding parts, are equal.

Sketch. In order to prove this theorem it is sufficient to notice that the bisector of the angle *B* divides the triangle into two congruent triangles in which ∠*A* and ∠*C* are corresponding parts. The conclusion follows.

Notice that in the informal proof steps and reasons are omitted or merely indicated. Still the argument is convincing *if* the reader supplies these reasons and steps himself. In the sketch all that is given is a key idea or two. It is up to the reader to follow out what is hardly more than a hint to be sure that the proof really could be made step by step. We constructed the formal proof before giving an informal proof or sketch, because we are not justified in giving these briefer forms until we see whether the proof "goes." The only way to find this out is to try to give a complete proof.

Proofs are important in order that we may know for sure that a certain theorem is true if the axioms are accepted. It is impossible for every person who uses a mathematical theorem to prove it himself, but he must be confident that it can be proved and has been proved by someone. It is for this reason that mathematicians set very high standards of logical rigor, which they strive to maintain in connection with new mathematical research and reorganization of existing knowledge.

Proofs have another importance for the student. They are the best means of understanding the theorems. If you understand a proof, you will be able to apply the theorem. Moreover, in constructing proofs yourself you will develop problem-solving ability. It is not worth while to memorize proofs. The important thing is to understand the key ideas and to be able, by means of heuristic reasoning, to recall or reconstruct a proof if needed.

EXERCISE 2.3

1. Which of the following statements can be proved deductively from axioms of mathematics (including logic) and which require observation for their verification?

(a) John is taller than Mary.

(b) If Bill is taller than John, and John is taller than Mary, then Bill is taller than Mary.

(c) The statement "John is taller than Mary" is either true or false.

(d) Women tend to live longer than men.

(e) If you toss a penny in the air one thousand times, it will come down tails more often than heads.

(f) If you toss a "perfectly balanced" penny, the chances are fifty-fifty that it will come up heads.

(g) Mathematics is a logical structure composed of undefined elements, axioms, and theorems.

(h) There is no integer between zero and one.

(i) The first president of the United States was Abraham Lincoln.

(j) One-fourth equals 0.25.

(k) All equilateral triangles are isosceles.

(l) $9 + 6 = 15$.

(m) One apple and one apple are two apples.

(n) All men are males.

(o) All men grow beards.

2. Devise formal proofs for one or more of the theorems stated in Ex. 2.1. Accompany your formal proof by a heuristic discussion, an informal proof, and a sketch proof. If necessary, consult a geometry textbook, but avoid copying. Don't get any more help than you find absolutely necessary.

†3. Prove that the area of an isosceles triangle is given by $A = \frac{1}{2}b\sqrt{a^2 - (b^2/4)}$, where a is the length of the equal sides.

†4. Prove that the area of a triangle is given by $A = \sqrt{s(s - a)(s - b)(s - c)}$, where a, b, c are its sides and s is the semi-perimeter.

2.4 The negative of a statement

The negative of "the house has a green roof" is "it is false that the house has a green roof." The **negative of a statement** is simply the denial of that statement, and it is usually formed by negating the main verb. In the above example, the negative may be written "the house does not have a green roof." The negative should not be confused with other statements involving negative ideas. The negative of "the cat is white" is not "the cat is black" but "the cat is not white."

A mathematical proposition may be written in the forms: "A implies B." "If A, then B." Or "$A \longrightarrow B$." The main verb in the first form is "implies." In the second, it is not stated; but "if A, then B" is short for "B follows from A" or "A implies B." In the third form, the arrow stands for "implies." Accordingly, the negative of a proposition is formed by replacing "implies" by "does not imply," "follows" by "does not follow," or "$\longrightarrow$" by "$\longrightarrow\!\!\!|$." The negative of "$A \longrightarrow B$" is "$A \longrightarrow\!\!\!| B$." It should be noted that $A \longrightarrow\!\!\!| B$ does not involve the negative of A or B. Suppose we let not-A and not-B stand for the negatives of A and B. Then neither "not-$A \longrightarrow$ not-B," nor "not-$B \longrightarrow$ not-A," nor "$A \longrightarrow$ not-B" is the negative of "$A \longrightarrow B$."

To illustrate these ideas, we consider this rather trivial theorem: "If a triangle is equilateral, it is isosceles." Since an equilateral triangle is one with all three sides equal, and an isosceles triangle is one with two sides equal, the theorem is obviously true. We may write it: (ABC equilateral) $\longrightarrow$ (ABC isosceles). The negative is (ABC equilateral) $\longrightarrow\!\!\!|$ (ABC isosceles). This negative is false, since we take as an axiom

∗2.4.1   If A is true, not-A is false.

Now consider propositions involving the negatives of hypothesis and conclusion. The theorem (ABC not equilateral) $\longrightarrow$ (ABC

isosceles) is false, for a triangle could be both not equilateral and not isosceles. Another possibility, $(ABC$ equilateral) $\longrightarrow$ $(ABC$ not isosceles), is evidently false also, since an equilateral triangle cannot fail to be isosceles. Taking the negative of both hypothesis and conclusion, we have $(ABC$ not equilateral) $\longrightarrow$ $(ABC$ not isosceles), which is false since a triangle might be not equilateral and still isosceles. We see that from $A \longrightarrow B$ we cannot conclude not-$A \longrightarrow B$, $A \longrightarrow$ not-B, or not-$A \longrightarrow$ not-B. There may be special cases when more than one of these hold, but the first does not imply the others, since we have given an example when the first holds and the others do not. (We shall see in the next section that $A \longrightarrow B$ does imply not-$B \longrightarrow$ not-A.)

We have just proved a statement false by exhibiting a case where it does not hold. If we wish to prove that $A \nrightarrow B$, it is enough to cite one example in which A is true and B false, since $A \longrightarrow B$ means that B is always true when A is. Its negative denies this or, in other words, asserts that B is false in at least one case in which A is true. More generally, the negative of a statement that *all* of certain things have a given property is the statement that *some* (one or more) do not have it.

For example, suppose we wish to show that it is false that "for any number a, $a^2 = 2a$." It is sufficient to cite the example $a = 3$, since in this case a is a number, yet $a^2 = 9$ and $2a = 6$ so that $a^2 \neq 2a$. Thus we can prove a theorem false by citing a single "counter example." On the other hand, we cannot prove a theorem by citing any number of favorable special cases, unless the theorem refers to only a finite number of possibilities and we treat each one. The statement $a^2 = 2a$ is true for $a = 0$ and $a = 2$, but citing these in no way proves that "for any number a, $a^2 = 2a$."

EXERCISE 2.4

1. The negative of "$A \longrightarrow B$" could be defined formally as follows: $A \nrightarrow B \longleftrightarrow$ not $(A \longrightarrow B)$. (Definition.) Write a similar formal definition of $a \neq b$.

2. State the negative of each theorem in Ex. 2.1. Then state separately the negative of the hypothesis and conclusion of each one.

3. Give the negative of: (a) The cat is not white. (b) Men are not birds. (c) $2 \neq 3$. What logical principle is illustrated here?

4. State in words the negative of: (a) 6 is greater than 3. (*Answer.* 6 is not greater than 3, or 6 is less than or equal to 3.) (b) Triangle ABC is isosceles. (c) All men are mortal. (*Answer.* Some men are not mortal, or not all men are mortal.) (*d*) No circles are square. (*e*) All positive numbers are greater than 1. (*f*) Some women are good mathematicians. (*g*) No Americans are illiterate. (*h*) Two points determine a straight line. (*i*) For any number *a*, *a* + *a* = 2*a*. (*Answer.* *a* + *a* ≠ 2*a* for at least one value of *a*.) (*j*) Cement is not a virtue. (*k*) Temperance is a virtue. (*l*) Every point on the circumference of a circle is at the same distance from the center. (*m*) Mr. X is a conservative. (*n*) Carleton has a good chess team. (*o*) Mary is backward. (*p*) Mary is forward. (*q*) Mathematics is abstract. (*r*) Mathematics is practical.

5. Show that the following equations are *not* always true: (*a*) $3a \overset{?}{=} a^3$.

(*b*) $\dfrac{1}{\dfrac{1}{a} + \dfrac{1}{b}} \overset{?}{=} a + b$.   (*c*) $\dfrac{(a+b)}{(a-b)} \overset{?}{=} -1$.   (*d*) $(a+b)^2 \overset{?}{=} a^2 + b^2$.

2.5   The converse of a proposition

The **converse** of a proposition is the statement we get by interchanging hypothesis and conclusion. The converse of "if a triangle is equilateral, then it is isosceles" is "if a triangle is isosceles, then it is equilateral." The first proposition is true, but its converse is false. In general, therefore, if we know that a theorem is true we *cannot* conclude that the converse is true. From *A* implies *B* it does *not* follow that *B* implies *A*.

Of course it may happen that both a proposition and its converse are true. In this case we have both *A* implies *B* and *B* implies *A*. We say that "*A* is **equivalent** to *B*," "*A* implies *B* and **conversely**," "*A* **if and only if** *B*," or "*A* is a **necessary and sufficient condition** for *B*." Using the arrow for "implies," we write the converse of *A* $\longrightarrow$ *B* as *B* $\longrightarrow$ *A* or *A* $\longleftarrow$ *B*. If both *A* $\longrightarrow$ *B* and *A* $\longleftarrow$ *B*, we write *A* $\longleftrightarrow$ *B*. The expression "*A* is a necessary and sufficient condition for *B*" is consistent with the terminology explained in 2.1. There we said that "*A* is a necessary condition for *B*" means "*B* $\longrightarrow$ *A*," and that "*A* is a sufficient condition for *B*" means "*A* $\longrightarrow$ *B*." Hence "*A* $\longleftrightarrow$ *B*" means that *A* is both a necessary and a sufficient condition for *B*. The word "equivalent" is appropriate because *A* $\longleftrightarrow$ *B* means that *A* and *B* are both true or both false, that is, they have equivalent "truth values."

Definitions are examples of equivalence. When one statement is defined to mean the same as another, each implies the other, and we use the double arrow as indicated in 2.2. There are many other examples of equivalence. For example, both the following theorems are true: "If two sides of a triangle are equal, the opposite angles are equal." "If two angles of a triangle are equal, the opposite sides are equal." Each is the converse of the other. We could then write any of the following: "A necessary and sufficient condition that a triangle have two equal sides is that it have two equal angles." "In a triangle, if two sides are equal, then the two opposite angles are equal, and conversely." "In a triangle ABC, $(AB = BC) \longleftrightarrow (\angle A = \angle C)$."

In 2.4 we found that $A \longrightarrow B$ does not imply not-$A \longrightarrow$ not-B. However, it does imply not-$B \longrightarrow$ not-A. Consider, for example, this theorem: "If two triangles are similar, their corresponding sides are proportional." Interchanging hypothesis and conclusion and taking the negative of each, we have: "If two triangles do not have their sides proportional, they are not similar." This is true. It is a theorem of logic that: "If $A \longrightarrow B$, then not-$B \longrightarrow$ not-A." It can also be proved that the converse of this theorem is true. Hence we write

*2.5.1 $(A \longrightarrow B) \longleftrightarrow (\text{not-}B \longrightarrow \text{not-}A)$

This theorem is very helpful in proofs. If we find it difficult to prove $A \longrightarrow B$, we may start with not-B and try to prove not-$B \longrightarrow$ not-A. We "assume the contrary" of the conclusion and try to get a "contradiction," that is, derive the negative of the hypothesis. According to this theorem a statement that implies a false statement is false, since if $A \longrightarrow B$ and B is false, that is, not-B is true, then not-A is true and A is false. (Why?) In particular, if a statement implies its own negative it must be false, since a statement cannot be both true and false. Proofs based on these ideas are called **indirect proofs,** proofs by contradiction, or proofs by *reductio ad absurdum.*

EXERCISE 2.5

1. State the converse of each theorem in Ex. 2.1. Which are true? (*Hint.* In order to state the converse of a theorem it is essential to separate clearly the hypothesis and the conclusion. For example, the theorem "a diagonal of a parallelogram divides it into two congruent triangles" is not stated so that it is easy to interchange hypothesis and conclusion. It says that "*if* a quadrilateral is a parallelogram, *then* a diagonal separates it into two congruent triangles." The converse is now seen to be "*if* a diagonal separates a quadrilateral into two congruent triangles, *then* it is a parallelogram.")

2. In each of the following pairs of statements decide whether one implies the other, neither implies the other, or each implies the other:

(a) $ABCD$ is a parallelogram; $AB = CD$.

(b) Two lines are parallel; they are everywhere equidistant.

(c) $a = b$; $2a = 2b$.

(d) Triangles ABC and $A'B'C'$ are congruent; $AB = A'B'$, $AC = A'C'$, $BC = B'C'$.

(e) Triangles ABC and $A'B'C'$ are congruent; angles A, B, C equal angles A', B', C'.

(f) Triangle ABC is a right triangle with right angle at B; $\overline{AC}^2 = \overline{AB}^2 + \overline{BC}^2$.

(g) Triangles ABC and DEF have equal bases and altitudes; triangles ABC and DEF have equal areas.

(h) A is B's brother; B is A's brother.

(i) Axioms cannot be proved; axioms are false.

(j) Axioms cannot be proved; axioms are true.

(k) $s = t$; $s^2 = t^2$.

(l) Mathematics is theoretical; mathematics is impractical.

(m) The plane traveled at a constant speed of 400 mph for one hour; the plane covered 400 miles in one hour.

(n) If $ABCD$ is a trapezoid, $ABCD$ is a parallelogram.

3. The idea that "two negatives make a positive" may be expressed in our symbols by

$$\text{*2.5.2} \qquad \text{not-(not } A) \longleftrightarrow A$$

(a) Give examples of your own construction.

(†b) Use *2.5.2 to prove that (not-$A \longrightarrow B$) $\longrightarrow$ (not-$B \longrightarrow A$).

4. What is the error in the following reasoning? "Honest people are frank. Mr. Kilroy is frank. Hence Mr. Kilroy is honest." (*Hint.* Rewrite the reasoning as follows: "If X is honest, then X is frank. Mr. Kilroy is frank. Therefore Mr. Kilroy is honest.")

5. Why is the following reasoning correct? "Honest people are frank. Mr. Dooley is not frank. Hence Mr. Dooley is not honest."

6. If A is a necessary and sufficient condition for B, is B a necessary and sufficient condition for A? How about the case where A is a necessary but not a sufficient condition for B, and the case where A is a sufficient but not a necessary condition for B? Explain.

7. If A is logically equivalent to B and B is logically equivalent to C, is A logically equivalent to C? Why? What other relations can you recall for which a similar theorem is true?

†8. Devise a proof of the converse of the Pythagorean theorem. (*Hint.* Construct another triangle which is a right triangle and has the same legs.)

†9. Is it true that a statement that implies a true statement must be true? Explain and give examples. (*Hint.* Consider the following: $2 = 0 \longrightarrow 0 = 2 \longrightarrow 2 + 0 = 0 + 2 \longrightarrow 2 = 2!$)

†10. Use indirect proof to show that two distinct lines parallel to the same line are parallel to each other.

2.6 Logical fallacies

There are a few logical errors, or "fallacies," that are so common among mathematics students that we want to warn you particularly against them.

Circular reasoning is the error of assuming what is to be proved. Few people will indulge in obvious circular reasoning, yet it is easy to insert into an argument the thing to be proved in a disguised form. You must remember that the conclusion of the theorem is what you are trying to get by logical steps from the hypothesis. Obviously you will be able to get it if you insert it somewhere along the way! But such a "proof" means nothing, since anything could be "proved" by such juggling. In order to avoid this error, ask yourself: "What is given? What is to be proved?" Only the former may be used in the proof. The latter appears only as the final step.

The *notion that a theorem implies its converse* is a common error. We discussed this matter in 2.5. It often appears in proofs in the following way: It is required to prove that $A \longrightarrow B$. Instead it is proved that $B \longrightarrow A$, and the conclusion is then drawn that $A \longrightarrow B$. But it is possible that $B \longrightarrow A$ without $A \longrightarrow B$. To avoid this error, keep in mind the hypothesis and the conclusion, and remember to go from the first to the second.

A closely related error is *the fallacy that a statement that implies a true statement must be true* itself. People imagine that, if $A \longrightarrow B$ and B is true, then A must be true. But this is not so, since a true statement may be derived from a false statement, sometimes more easily than from a true one! It is a theorem of logic that a false statement implies any statement, true or false. Hence if you accept this fallacy, you will be able to "prove" that any false statement is true! We shall use this method to "show" that $2 = 4$. We begin by assuming that $2 = 4$, and show that it leads to a true statement. Now, if $2 = 4$, we can subtract 2 from each side to get $0 = 2$. Now we subtract 0 from the left side of the original equation and 2 from the right side, which is legitimate since equals may be subtracted from equals. The result is $2 = 2$. Since this is certainly correct, we "conclude" that our assumption $2 = 4$ must be true also!!?? The fallacy is in thinking that what implies a true statement must be true. In order to avoid this error, keep in mind that from $A \longrightarrow B$ and A you can conclude B, but from $A \longrightarrow B$ and B you *cannot* conclude A.

Finally we mention *the error of proof by special cases*, which we discussed in 2.4. Special cases may be helpful in the heuristic thinking that leads to the construction of a proof, and, if there is a suspicion that a theorem is false, special cases are the easiest way to verify this. However, a general theorem about all of a certain set of things can be proved only by showing that it holds for *every* case.

EXERCISE 2.6

1. Detect and discuss errors in the following:

(a) The triangle ABC is equilateral because it has two equal angles and we know that an equilateral triangle has two equal angles.

(b) My opponent in this campaign denies that he took bribes. Why, then, if his conscience is clear, did he fail to report this extra income in his tax return?

(c) I know that all Chinese are crooked, because a Chinese merchant once short-changed me and I never did get the money.

(d) If we are given $x^2 = 9$, we know that $x = 3$ because $3^2 = 9$, and if two numbers are equal their squares are equal.

(e) The theory of relativity is true because from it one can deduce results that correspond with observation and experiment. (*Note.* From Newtonian physics one can also deduce results that correspond to experience.)

(f) All odd numbers are prime, for example, 1, 3, 5, 7, 11, 13, $\cdots$.

(g) Socialists support public ownership. Mr. Jones supports public ownership. Therefore, Mr. Jones is a socialist.

(h) A student must study to deserve good grades. Jean studied. Therefore, she deserves good grades.

(i) In a certain triangle the sum of the squares of two sides equals the square of the third. Hence the triangle is a right triangle by the Pythagorean theorem.

2. Show that if you accept the fallacy that a theorem implies its converse you must accept the fallacy that a statement that implies a true statement must itself be true.

3. Construct or find in current books, magazines, or newspapers several examples of fallacious reasoning.

4. Discuss the following statement attributed to Huck Finn by Mark Twain: "Jim said bees wouldn't sting idiots; but I didn't believe that because I had tried them lots of times myself, and they wouldn't sting me."

5. Discuss the following: "I'm glad I don't like coffee because if I did I'd drink it, and I can't stand the stuff."

2.7 Sets and variables

We are going to introduce a concept that is sometimes considered to be the most basic in mathematics. It is so simple and so familiar that perhaps the student has never thought about it before. It is the concept of a set.

All our experience is in terms of things that are associated in classes, or groups, or collections, or aggregates, or sets. Even if we think of a single object isolated from all other things, we have thereby associated it in a special way with the set of all other things and also with the set that consists of itself alone. The idea of a set is so fundamental that it is difficult to define. On the other hand, it is a very useful notion in terms of which to define other concepts. Hence it is convenient to take it as an undefined idea. In this section we shall merely give a few examples of sets and explain what is meant by the word "variable" in mathematics.

The student body at Carleton College is a set whose members are those people who are registered at Carleton. Of course the student body is not the same thing as the students. When we speak of the student body we are thinking of the collection of students as a single entity. Another set is the class of all presidents of the United States, past and present. We can list the members of this set. A third example is the set of living former presidents. This set contains just one member, as of the year 1951. Finally, we mention the set of all women who have been president. This set contains no members.

In each of these examples we have a certain collection that we call a **set**. We have a rule that tells us whether any object is in the set or not. An object that satisfies the condition for being in the set is called a **member** of the set and is said to belong to it. The set is said to **contain** its members, or to consist of its members. We give further examples.

The set of all sentences in this book. In order to be a member of this set an object must be a sentence and must appear in this book. The previous sentence is a member. Examples of objects that do not belong to this set are a house, a number, any sentence that does not appear in this book.

The set of digits 0, 1, 2, 3, 4, 5, 6, 7, 8, 9. In order to be in this set an object must be one of the objects listed. All other things are not members.

The set of Americans who are not college students. All members of this set must be members of the set of Americans and also not members of the set of college students. Things that fail to have these two properties are not in the set.

The set of all points on a line segment. This set consists of all those points and only those points that lie on the given segment. There are an infinite number of objects in this set, that is, no matter how large a number we mention, there are more than this many points in the set.

The important thing about all these examples is that when we speak of a set we have in mind a rule that enables us to say for any object whether or not it is in the set, that is, whether or not it has the property possessed by members of the set and no other things. If the set has only a few members, we may define it by simply listing them. If it consists of a large or even infinite number of things, it is sufficient to have some rule for telling whether an object belongs or not. Finally, the set may consist of only one or even of no objects at all. This is true if just one or no object satisfies the condition for membership.

In this chapter we have been talking about a set of statements— the propositions of mathematics. We have discussed the rules for deriving some of its members (theorems) from others (axioms). The axioms are members of the set of statements accepted without proof. The theorems make up the set of statements that are proved. We made a number of statements that are true for any propositions. For instance, we said that if one proposition implies a second, and the second implies the third, then the first implies the third. We found it convenient to write this as follows: If $A \longrightarrow B$ and $B \longrightarrow C$, then $A \longrightarrow C$. What is the meaning of the letters A, B, and C in this expression? Certainly one letter cannot imply another! The statement is nonsense if we interpret it in that way. What we have in mind is that, if A, B, C are replaced by any statements, the foregoing formula becomes a true statement. We expressed this by saying that A, B, and C "stand for" statements. Thus A, B, and C are symbols that "stand for" any member of the class of statements in which we are interested. We have also made statements of the following kind: $s^2 = s \cdot s$. Here we have in mind that this becomes a true statement if s is replaced by any number. The letter s "stands for" any member of the set of all numbers. What we mean by saying that it "stands for" any number is that it can be replaced by any number.

This idea of using a letter or other symbol to stand for any member of a set of objects is of fundamental importance. It means that we can write very briefly a theorem about all the members of a certain set of objects. All we have to do is to make the statement in terms of a symbol that stands for any member of the set. Thus we write: "$(A \longrightarrow B) \longrightarrow (\text{not-}B \longrightarrow \text{not-}A)$." If we want to be sure that the meaning of A and B is not misunderstood, we may write: "If A and B are statements, $(A \longrightarrow B) \longrightarrow (\text{not-}B \longrightarrow \text{not-}A)$." Similarly we write: "If s is a number, $2s = s + s$." Such forms are easier to grasp, remember, and apply than the same propositions stated in words.

A symbol used to stand for an unspecified member of a set is called a **variable**. The set of objects for which the variable stands is called the **range** of the variable. When we say that a variable stands for an unspecified member of a set of objects we mean that it can be replaced by any one of these objects. Each member of the range is called a **value** of the variable. When we replace a variable by one of the objects in its range, we say that the variable "takes" this value. A variable whose range consists of a single object is called a **constant**.

For example, suppose we let v stand for any digit in the set $\{1, 2, 3, 4\}$. Then the range of v is the set $\{1, 2, 3, 4\}$, and the values of v are 1, 2, 3, and 4. If we replace v by 2, we say that v takes the value 2. Again, let X stand for any student at Carleton. Then the range of this variable is the student body. Any particular student is a value of X. We can make general statements about X. For example, we can say "X is a human being." This is true for all values of X; that is, if we replace X by the name of any student at Carleton, we get a true statement.

EXERCISE 2.7

1. List the members of the following sets: (a) The even integers between 1 and 9. (b) The initial letters in the words in (a) above. (c) The weekdays. (d) The numbers x such that $2x = 1$. (e) Cities in the United States with populations of over 5,000,000. (f) The social sciences. (g) The natural sciences.

2. Name three sets that have just one member each.

3. Name three properties that are possessed by no objects. (*Note.* Any of these properties defines a set with no members. Mathematicians refer to any such set as the **null set.** There is essentially just one null set, a notion that was not appreciated by the soda jerker who in reply to a request for "a chocolate milk shake without any chocolate syrup" said: "Sorry, we have no chocolate syrup. You'll have to take it without vanilla.")

4. Name three sets with an infinite number of members.

5. In the following statements indicate the range of each variable and give an example of a value of the variable.

(*a*) If ABC is a triangle, the sum of the interior angles is $180°$. (*A, B,* and *C* alone stand for points, but the variable here is the symbol ABC.)

(*b*) If *A, B,* and *C* are triangles and *A* is congruent to *B* and *B* is congruent to *C*, then *A* is congruent to *C*. (*Answer.* The range of *A* is the set of all triangles. Any particular triangle is a value of *A.* The same applies to *B* and *C.*)

(*c*) For all numerical values of A, $3A = A + A + A$.

(*d*) $S^3 = S \cdot S^2$.

(*e*) $B \longrightarrow B$.

(*f*) If X is a Minnesotan, X lives in the Middle West. (*Answer.* The range of X is the set of all people. Any person is a value.)

(*g*) If Y does not study, Y will have regrets.

(*h*) $a + b = b + a$.

6. Go over this chapter, picking out all the cases where variables were used and identifying the range in each case.

7. Often words are used as variables. For example, if we say that a proposition implies itself, we are using "a proposition" as a variable standing for any proposition. This statement has the same meaning as 5(*e*) above, and "a proposition" and "itself" play the role of *B*. In the following statements identify the variables, replace them by letters, and indicate their ranges:

(*a*) A proposition always takes this form: "Statement implies statement."

(*b*) The negative of the negative of a proposition is equivalent to the proposition.

(*c*) If a proposition implies its own negative it is false.

(*d*) If twice a number is 4, the number must be 2. (*Answer.* $2x = 4 \longrightarrow x = 2$.)

(*e*) A proposition does not imply its converse.

(*f*) The statement that one assertion is a necessary and sufficient condition for another is often abbreviated by writing that the first is a N. and S. condition for the second.

(*g*) If two lines are parallel, they do not meet.

(*h*) If Socrates is a man, Socrates is mortal.

(*i*) If two triangles are equilateral, they are similar.

†**8.** Prove that at least two people in New York City have identical initials.

9. When Lincoln freed the slaves what sets of people did he abolish? Did he abolish the members of these sets? (Note the importance of distinguishing between a set and its members.)

10. When Germany became a republic in 1918, what set of one member was abolished? What became of the member of this set?

11. The statement "*a* is a member of *A*" is expressed in symbols by writing "*a* ∈ *A*." (∈ is read "is a member of.") Its negative is written "*a* ∉ *A*." (∉ is read "is not a member of.") Indicate which of the following statements are true and which are false: (*a*) 1 ∈ {1, 2, 3}. (*b*) 3 ∈ [the even integers]. (*c*) John ∈ [men's names]. (*d*) Square *ABCD* ∈ [rectangles]. (*e*) A circle *O* ∈ [polygons]. (*f*) *a* ∈ {*c*, *b*, *a*, 1}. (*g*) Loyalty ∈ [virtues]. (*h*) An educated man ∈ [illiterates]. (*i*) regular polygon ∈ [symmetric figures].

12. Letting △ stand for "triangles," read the following in words: (*a*) *X* ∈ [△]. (*b*) *X* ∈ [right △]. (*c*) *X* ∈ [isosceles △]. (*d*) *X* ∉ [isosceles △].

13. Which of the following are true and which false? (*a*) *X* ∈ [△] ⟶ *X* has three sides. (*b*) *ABCD* ∈ [parallelograms] ⟶ *AB* = *CD*. (*c*) *X* ∈ [equilateral △] ⟶ *X* ∈ [isosceles △]. (*d*) *ABC* ∉ [right △] ⟶ $\overline{AB}^2 \neq \overline{BC}^2 + \overline{CA}^2$.

†**14.** If two sets *A* and *B* are such that every member of *A* is a member of *B*, we say that "*A* is a **subset** of *B*," "*A* is **included** in *B*," or "*B* **includes** *A*," and we write "*A* ⊆ *B*." (*a*) Is it true that for any set *A*, *A* ⊆ *A*? (*b*) Does *A* ⊆ *B* ⟶ *B* ⊆ *A*? (*c*) If *T* is the set of all triangles, *R* the right triangles, *I* the isosceles triangles, and *E* the equilateral triangles, does *R* ⊆ *T*? *T* ⊆ *R*? *R* ⊆ *E*? *I* ⊆ *E*? (*d*) The null set is considered to be a subset of every set. How is this consistent with the definition? (*e*) Express in words and justify: *A* ⊆ *B* ⟷ [*x* ∈ *A* ⟶ *x* ∈ *B*]. (*f*) List several subsets of each set in prob. 1.

15. Write down the set of all different subsets of each of the following sets: (*a*) {1, 2, 3}. [*Answer.* {Null set, {1}, {2}, {3}, {1, 2}, {1, 3}, {2, 3}, {1, 2, 3}}.] (*b*) {*a*, *b*, *c*}. (*c*) {1, 2}. (*d*) {*A*, *B*, *C*, *D*}.

†**16.** When two sets have the same members, they are said to be **equal** and we write *A* = *B*. Prove that (*a*) *A* = *B* ⟷ (*A* ⊆ *B* and *B* ⊆ *A*). (*b*) (*A* ≠ *B* and *A* ⊆ *B*) ⟶ *B* ⊈ *A*.

†**17.** Interpret and prove *A* ⊆ *B* and *B* ⊆ *C* ⟶ *A* ⊆ *C*.

†**18.** Two sets are called **disjoint** if they have no members in common. In symbols: *A* and *B* are disjoint ⟷ *x* ∈ *A* ⟶ *x* ∉ *B* and *x* ∈ *B* ⟶ *x* ∉ *A*. List several pairs of disjoint subsets of the set of all Americans. List some pairs that are not disjoint.

3

Numbers and
Elementary Operations

Our purpose in this chapter is not primarily to teach the student to do arithmetic and algebraic manipulations, but rather to increase his understanding of these operations. Modern high-speed computing devices have reduced the importance of personal skill in "figuring," but it is now more important than ever to understand these operations in order to be able to compute intelligently and to appreciate more advanced mathematics and its applications. Accordingly, we review the fundamentals of arithmetic and algebra in the course of indicating the pattern by which the number system is built up from the integers and by which the rules of algebra are derived from a few axioms.

We could assume some of the propositions of the chapter and derive the others from them, or we could prove all of them from the axioms of logic and the theory of sets. But this would take too much time and divert us from the main purpose of the book. Hence we give only a few proofs in order to familiarize the student with important ideas and to show how the rules of algebra could be proved with the same logical precision as the theorems of geometry.

The entire book provides opportunities to use the theorems of this chapter, and the student should develop the habit of justifying his algebra by reference to them. The student who is "weak" in arithmetic or elementary algebra should do as many as needed of the drill problems, but he should find himself making more rapid progress when he has a knowledge of the reasoning behind what he is doing.

3.1 Counting

Everybody knows what 3 means. But it is not easy to give an exact definition. Suppose we were trying to put across the idea of 3 to someone who did not know our language. We might proceed by pointing to various collections of things such as the following set of asterisks: * * *. Each time we would say, "Three." We would be careful to pick out sets with the same number of objects. We might also point to some sets with more or fewer members, at the same time saying, "Not three." We would be explaining 3 in terms of the idea of a set.

What did we mean by saying that we must pick sets with the same number of members? How would we check to see if two sets had this common property? We could check off the members of one set against those of the other, taking out one member from each set to form pairs. If all members paired off, we would say that the sets had the "same number." We now describe precisely this method of comparing sets. Suppose we have two sets and that we can form a set of pairs of objects, each pair being formed by choosing one member from each set, so that all the members of the sets are used and none is used twice. Then we say that we have paired off the sets or established a **one-to-one correspondence** between them. Such sets are called **equivalent.**[1]

Consider the following two sets: $\{a, b, d, k\}$ and $\{R, S, T, U\}$. We can form the following set of pairs: $\{\{a, R\}, \{b, S\}, \{k, T\}, \{d, U\}\}$. Each pair in the new set is composed of a member from one set and a member from the other set, all the members of the two original sets are used, and no member appears in more than one pair. Hence we have established a one-to-one correspondence between these two sets, and they are equivalent. Of course other correspondences, or pairings, could be established. But all that is required for equivalence is that *some* one-to-one correspondence be possible.

We can now say what we mean by the number 3. It is a symbol associated with sets that are equivalent to the following set: $\{a, b, c\}$. Similarly, 1 is associated with sets equivalent to the set $\{a\}$, and 2 is associated with sets equivalent to the set $\{a, b\}$. In this way we could give precise meaning to the symbols 1, 2, 3, 4, 5, 6, 7, 8, 9, 10, 11 $\cdots$. But this sequence of integers is itself a set, the set of **positive integers** or "natural numbers." Let us consider some of its properties. Notice first of all that it has a first member. However, there is no last member because we can always continue the sequence by writing another number. We

---

[1] This use of "equivalent" is quite distinct from the use in 2.5.

express this by saying that it is an infinite set. Notice also that the members of this set are in a definite order. If we pick out any two different positive integers we can always say which one comes first, or at the left of the second. The one at the left is called "smaller." Moreover, each number has a "next" number or successor, which immediately follows it, and each positive integer, except one, has an immediately preceding positive integer.

These properties of the positive integers enable us to use them to determine the number of objects in any finite set. We simply compare the given set with a part of the infinite set of integers above. We pair off the elements of the set with part of the set of integers in order. When the set is exhausted, the last integer used in the pairing is the number of elements in the set. This pairing process is "counting." We pair off the positive integers in the definite order 1, 2, 3, $\cdots$, but the count of the set is the same no matter in what order we count its elements. The process works because the set $\{1, 2, 3\}$ is equivalent to the set $\{a, b, c\}$; the set $\{1, 2, 3, 4, 5\}$ is equivalent to the set $\{a, b, c, d, e\}$, etc. If we cut off the integers at any point, the last integer remaining is just the number of positive integers in the remaining set. Thus the positive integers form a sequence that can be used as a standard for comparing the size of sets. Instead of comparing two sets by pairing them off directly, it is easier to count them both and see which is larger. Counting means, of course, pairing them off with the positive integers.

EXERCISE 3.1

1. Show that the following two sets are equivalent by pairing off directly: $(x, *, 1, 17, \$), (2, †, !, /, ?)$.

2. Pair off each of the above sets with positive integers and thus show that they are equivalent.

3. Establish a different pairing (one-to-one correspondence) between the sets in prob. 1.

4. What is the meaning of "4"?

5. Write the successor of each of the following numbers: 27; 29; 99; 314; 298; 119; 999; four thousand and ninety-nine.

6. How many different ways can the following two sets be paired? (a, b, c) (s, t, u). Write out these possibilities.

†*7.* It is evident from the way we defined the positive integers that other symbols could be used in place of the familiar decimal notation. Thus the Romans used in place of 1, 2, 3, 4, $\cdots$ 10 $\cdots$ the symbols I, II, III, IV, $\cdots$ X $\cdots$. The Greeks used letters of their alphabet. Some modern computing

machines work best when numbers are expressed by using just two symbols, 0 and 1. Then the first few positive integers are written 1, 10, 11, 100, 101, 110, 111, 1000. Write the next few positive integers in this notation. Formulate a rule for writing the successor of any integer in this system. (See 12.8.)

†8. It is possible to apply our definition of equivalent sets to infinite as well as finite sets. Of course we cannot write down all the pairs of members of two infinite sets, but we say that two infinite sets are **equivalent** if we can state a rule that pairs the members in a one-to-one correspondence. Using this method, we can say that the set of all the integers 1, 2, 3, ⋯ is equivalent to the set of even integers 2, 4, 6, ⋯ because each number in the first can be paired with its double in the second. How about the set of all the even integers and the set of all the odd integers 1, 3, 5, 7, ⋯? [*Note.* This illustrates the fact that, for infinite sets, the whole may not be greater than each of its parts. The even integers are part of the integers (in fact, they seem to be just half of them), yet there are the "same number" in each set!]

†9. Prove that a pair of persons exists in New York who have the same number of hairs on their heads.

†10. Writing $A \sim B$ for "A is equivalent to B," where A and B are sets, show that (*a*) $A = B \longrightarrow A \sim B$. (*b*) $A \sim B \not\longrightarrow A = B$. (See Ex. 2.7.16.)

3.2 Numbers and points

The relation between positive integers and sets was discussed in the last section. Another intepretation of positive integers is in terms of points on a line. This geometric interpretation is very useful because it can be extended to all numbers. Suppose we imagine a straight line upon which we designate a reference point called the **origin.** We agree on a segment of fixed length called the **unit length.** Then we mark off points to the right of the origin by laying off distances equal to the unit length, twice the unit length, etc. The result is sketched in Fig. 3.2.1. Such a line

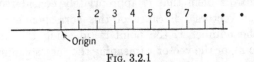

Fig. 3.2.1

with origin, unit length, and direction of measurement is called an **axis,** or **directed line.** It is obvious that the set of points at distances 1 unit, 2 units, etc., to the right of the origin is a set that is equivalent to the set of positive integers, that is, there is a one-to-one correspondence between these points and the positive integers. The number corresponding to a point is called the **coordinate** of the point. Of course there are points to which no integer corresponds, but, as we shall see in 3.13, there is a one-to-one correspondence between all points on the line and all real numbers.

There is a second way to visualize the correspondence we have described. For each number we take an arrow with its initial point at the origin and its terminal point at the point corresponding to the number. Thus the unit length is thought of as an arrow one unit long pointing to the right, and for each positive integer there is a corresponding arrow whose length is found by laying out the unit arrow the appropriate number of times. An arrow used in this way to indicate a length measured in a certain direction is called a **vector.** The number 1 corresponds to a vector of unit length, called the **unit vector.** In Fig. 3.2.2 several vectors are

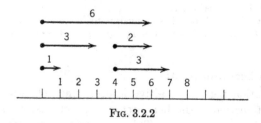

FIG. 3.2.2

drawn so as to show their relation to the points of the axis and the coordinates of these points. The sets of numbers, points, and vectors are in one-to-one correspondence, each vector having a length equal to the corresponding number.

This is very convenient, because it allows us to visualize geometrically any problem involving numbers and to translate into numbers any problem that is appropriately considered in terms of points or vectors. Because of this correspondence we may speak of the number x, the point x (meaning the point corresponding to x), or the vector x (meaning the corresponding vector).

EXERCISE 3.2

1. Sketch the points x, where $x = 1, 3, 7$.

2. Sketch the vectors x, where $x = 2, 1, 8$.

3. Lay out vectors with initial and terminal points as follows: origin and 4, 1 and 6, 2 and 5, 2 and 7, 3 and 8.

4. List several different pairs of initial and terminal points that yield the vector of length 4 pointing to the right. Sketch. (Vectors with the same

length and direction are considered equivalent even though they may be in
different positions. Hence all these pairs determine the same vector.)

5. Explain carefully what is meant by saying that there is a one-to-one
correspondence between the positive integers and (*a*) the set of points at integral distances to the right of the origin, (*b*) the set of vectors with integral
lengths pointing to the right.

3.3 Equality and inequality

Perhaps the most obvious property of numbers [1] is that any two
of them are either equal or unequal, but not both. We may
interpret equality geometrically by saying that equal numbers
correspond to the same point or to the same vector. As we know,
the same number may be represented by different symbols. For
example, 6 and VI stand for the same number. When two numbers
a and *b* are equal, we write $a = b$ and call this statement an
equation in which *a* is the **left member** and *b* is the **right
member.** (The members are called **sides** also.) In the contrary
case, we write $a \neq b$. Here *a* and *b* are variables that stand for
any numbers, that is, the range of each is the set of all numbers.
The principle that we stated above may be formulated briefly:

*3.3.1   Either $a = b$, or $a \neq b$, but not both

(Dichotomy Law)

An obvious property of equality of numbers is that any number
equals itself:

*3.3.2 $a \equiv a$ (Reflexive Property of Equality)

In *3.3.2 we used the symbol "$\equiv$" to indicate that $a = a$ for
all values of *a*, that is, when *a* is replaced by any number. The
sign "$\equiv$" is called the **sign of identity,** and an equation true for
all values of the variables that appear is called an **identity.** Of
course "all values" means all values in a certain range, and usually
this range is the set of values for which the equality has meaning.
Thus $a^2 = a \cdot a$ does not hold if we replace *a* by "a shaggy dog,"
but we write "$a^2 \equiv a \cdot a$" as an abbreviation for "if *a* is any
number, $a^2 = a \cdot a$." An equation that is true only for some

[1] We are thinking in terms of positive integers, but this statement, like
many others in the chapter, is true of all numbers. In such cases we use the
word "number" without qualification.

values of the variables is called a **conditional equation.** For such equations we use the usual equality sign.

Another property of equality is:

*3.3.3 $$a = b \longleftrightarrow b = a$$

(Symmetric Property of Equality)

The foregoing properties of equality are so "obvious" that the student has probably never considered them, even though he uses them in his daily thinking. We now mention a property that the student remembers from plane geometry in the form: "Two things that are equal to the same thing are equal to each other." We state it in symbols:

*3.3.4 $$a = b \text{ and } b = c \longrightarrow a = c$$

(Transitive Property of Equality)

The transitive property is used repeatedly in proofs and manipulations. Often we show that $A = B$ and $B = C$ in order to conclude that $A = C$. The work is usually abbreviated in this form:

(1)
$$A = B$$
$$= C$$

This form indicates that we state $A = B$, then $B = C$, and hence $A = C$. For example: $2 + 3 \cdot 10 = 2 + 30 = 32$ is a brief way of writing $2 + 3 \cdot 10 = 2 + 30, 2 + 30 = 32$, and, therefore, $2 + 3 \cdot 10 = 32$. Often the number of steps is greater than two, and the computation or proof takes the form

(2)
$$A = B$$
$$= C$$
$$= D$$
$$\vdots$$
$$= Y$$
$$= Z$$

This form is short for $A = B$, $B = C$, $C = D$, $\cdots$ $Y = Z$, and hence $A = Z$.

We have disposed of the case where two numbers are equal. What if they are unequal? We mentioned in 3.1 that when two integers are not equal one of them is less than the other, that is, it comes first in the sequence 1, 2, 3, 4, $\cdots$. The expression "$a < b$" means "a is less than b" in this sense. This can also be written $b > a$, which in words is "b is greater than a." The inequalities $a < b$ and $c < d$ are said to be alike in sense, while $a < b$ and $c > d$ are in opposite sense.

When $a \neq b$, there are two possibilities, either $a < b$ or $a > b$, depending upon which comes first in the sequence of integers. Hence we can improve upon the dichotomy law by writing:

*3.3.5   One and only one of the following holds:

$$a < b, a = b, a > b \quad \text{(Trichotomy Law)}$$

The student can easily assure himself that inequality is not reflexive or symmetric, but he already is acquainted with this idea: "If a is less than b, and b is less than c, then a is less than c." That is,

*3.3.6 $a < b$ and $b < c \longrightarrow a < c$ (Transitive Law)

The geometric interpretation of these relations is valuable. When two expressions are equal, they correspond to the same point or the same vector. When $a < b$, point a lies to the left of point b; when $a = b$, the points coincide; when $a > b$, point a lies to the right of point b. The geometric interpretation of the transitive law for inequality is that, if one point lies to the left of a second, and the second lies to the left of a third, then the first lies to the left of the third.

When we wish to state that $a < b$ or $a = b$, without saying which, we write $a \leq b$, which is read "a is less than or equal to b." Similarly we write $a \geq b$, meaning "a is greater than or equal to b." A statement like $a \leq b$ or $a \geq b$ represents two alternative possibilities, the equality and inequality, and it is true if *either one* is true.

EXERCISE 3.3

1. Restate in words the numbered theorems of this section.
2. Construct numerical examples to illustrate each theorem. Sketch and give the geometrical interpretation.

3. State which of the following are true and which false. Justify your decisions and illustrate with sketches.

(a) XI = 11	(f) $1,000,000 \geq 17$	(k) $21 > 23$
(b) $9 + 2 \neq 11$	(g) $7 > 9$	(l) not-$(5 \geq 4)$
(c) $3 > 3$	(h) $19 \leq 1$	(m) $11 < 3$
(d) $4 \leq 2$	(i) not-$(1 > 2)$	(n) $15 \leq 17$
(e) $9 > 9$	(j) $8 \leq 8$	

4. The negative of $a > b$ is $a \leq b$. State the negatives of:

(a) $a < b$	(e) $a > b$ or $a < b$
(b) $a \geq b$	(f) $a \neq b$
(c) $a \leq b$	(†g) $a \leq b$ and $a \geq b$
(d) $a = b$	

5. Show that $a \geq b$ and $a \leq b \longrightarrow a = b$.

6. Consider the relation of being a sibling, where we define "X is a sibling of Y" to mean "X and Y have the same parents." (a) Is this relation reflexive, i.e., is X a sibling of X for all X? (b) Is it symmetric, i.e., does it follow that Y is a sibling of X, if X is a sibling of Y? (c) Is it transitive?

7. Answer the same questions for the relation of brotherhood. (*Note.* Define "X is a brother of Y" to mean "X is male and X and Y have the same parents.")

†8. A relation that is reflexive, symmetric, transitive is called an **equivalence relation.** Which of the following are equivalence relations? (a) "$\leq$." (b) "Is older than." (c) "Is parallel to." (d) "Lives in the same state as."

†9. Suggest a way of telling which of two sets is bigger by direct comparison without counting.

†10. Formulate a rule by which for any two integers a and b, written in decimal notation, we can tell whether $a < b$, $a = b$, or $a > b$.

†11. Prove:

*3.3.7 $a > b$ and $b > c \longrightarrow a > c$

*3.3.8 $a \leq b$ and $b \leq c \longrightarrow a \leq c$

*3.3.9 $a \geq b$ and $b \geq c \longrightarrow a \geq c$

†12. Is the relation "$\subseteq$" for sets reflexive, symmetric, and transitive? How about the relations "$\sim$" and "$=$" for sets?

3.4 Addition

Suppose we have two sets: $\{a, b, c, d\}$ and $\{r, s, t\}$. The set $\{a, b, c, d, r, s, t\}$ consists of all the elements that are members of either set. The number of the first set is 4, that of the second 3, and that of the combined set is 7. We say that "four plus three equals seven." This simple example suggests the way in which we might define the addition of positive integers in terms of the "addition" of sets. Suppose we have two sets, one with m members and one with n members, and with no members in common. We form a set out of all members of both sets and define $m + n$ to mean the number of members in this set. To add two positive integers m and n, we pick out any two sets with m and n members and count the combined set. It would make no difference what sets with m and n members we chose, nor in what order we counted the resulting set.

In this way, for any two positive integers a and b we have a unique **sum** $a + b$. The numbers to be added, a and b, are called **terms** of the sum. When we say that the sum is unique we mean that there is only one answer. Another way of saying this is:

*3.4.1 $a = c \text{ and } b = d \ \longrightarrow \ a + b = c + d$

(Uniqueness of Addition)

Since the sum $a + b$ is found by counting a set made up by combining one set with a members and one set with b members, it seems reasonable that the sum $b + a$ is the same number, because the combined set is identical for the two sums. We describe this by writing:

*3.4.2 $a + b \equiv b + a$ (Commutative Law of Addition)

We use the identity sign as before in order to indicate that the equality holds for all numerical values of a and b.

We can always visualize an addition of positive integers in terms of the sequence $1, 2, 3, 4, 5, \cdots$. For example, we visualize the addition of 4 and 3 as follows:

$$\overbrace{1, 2, 3, 4,}^{4} \ \overbrace{1, 2, 3}^{3}$$
$$1, 2, 3, 4, 5, 6, 7$$

Here we take as our set of four objects the set $\{1, 2, 3, 4\}$ and as our set of three objects the set $\{1, 2, 3\}$. We form a set by laying these out in a row, and then we count. In this way adding may be reduced to counting, and no doubt it originated in counting, men using their fingers in place of $1, 2, 3, \cdots$. This interpretation of addition in terms of sets and counting enables us to add any two positive integers, but it would be laborious for large ones. Additions are actually done by means of mechanical rules, which depend upon knowing by heart the sums of any two numbers less than 10 and following a definite procedure involving adding the digits column by column and "carrying" when necessary. The exact procedure is, of course, valid only for numbers written in the decimal system, but it can be fully justified from the fundamental properties we are developing in this chapter. (See 12.8.)

A very important interpretation can be given to addition in terms of vectors. For example, if we place the vector 4 with its initial point at the origin and then place the vector 3 with its initial point at the terminal point of the 4 vector, the terminal point of the 3 vector will be at 7. If we now draw a vector from the initial point of the first to the terminal point of the second, we have the vector 7. The sums $4 + 3$ and $3 + 4$ are shown in Fig.

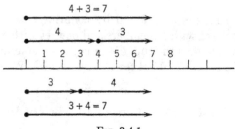

FIG. 3.4.1

3.4.1. This interpretation has many applications, and it has the advantage that it can be applied without change when the numbers are not integers.

So far we have talked only about the addition of two numbers. Addition of three or more can always be considered in terms of adding two of them, then another, etc. For example, we may add 2 and 3 and then 4 to the result. The student is familiar with the fact that the same number results from adding 2 and the sum of 3 and 4. We express this equality by writing $(2 + 3) + 4 = 2 + (3 + 4)$. The parentheses are used to indicate the terms that are to be added in connection with each plus sign. The left side means that the numbers $(2 + 3)$ and 4 are to be added. The right side indicates that the numbers 2 and $(3 + 4)$ are to be added. We illustrate with vectors in Fig. 3.4.2. These remarks would apply if 2, 3, and 4 were replaced by any numbers, that is,

*3.4.3
$$(a + b) + c \equiv a + (b + c)$$

(Associative Law of Addition)

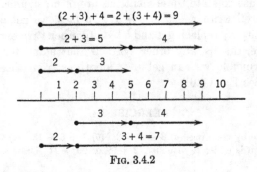

Fig. 3.4.2

A sum of terms is sometimes called a **summation** or a **multinomial.** When there are just two terms as in $a + b$ or $(x + y) + z$, it is called a **binomial.** When there are three as in $a + b + c$ or $x + y + z$, it is called a **trinomial.** When one term stands alone, it is sometimes called a **monomial.**

The commutative and associative laws of addition may be taken as axioms or deduced from the properties of sets as suggested in the exercise. They enable us to rearrange the terms in any sum according to convenience. Since $(a + b) + c \equiv a + (b + c)$, it is customary to write $a + b + c$, and similarly for other multinomials.

The law of uniqueness of addition (*3.4.1) tells us what happens when we add equals to equals. What happens in the case of unequals? If we add 3 to both sides of the inequality $6 > 4$ we get $9 > 7$. This is easy to understand in terms of our vector interpretation of addition. The point 6 lies to the right of the point 4. If we move three units further to the right of each point, the resulting points are in the same relative position. In terms of sets, if one set is bigger than another, and the same number of objects are added to each, the resulting sets have the same relation of inequality. In symbols,

*3.4.4
$$a < b \longrightarrow a + c < b + c$$

(Monotonic Law of Addition)

We stated the principles of this section in terms of variables. It must be remembered that a statement like $a + b \equiv b + a$ means that if a and b are replaced by any numbers the expressions $a + b$ and $b + a$ become equal numbers. Obviously, the particular letters we use to state these identities are of no significance. We might as well write $x + y \equiv y + x$. In fact, we get an equally valid identity by replacing a and b by any other variables with the same range, that is, any other symbols that stand for numbers. By this principle we can get new identities from those already stated. (See Ex. 3.4.6.)

EXERCISE 3.4

1. Prove by counting: (a) $3 + 5 = 8$. (b) $7 + 6 = 13$. (c) $9 + 7 = 16$.

2. Which is easier to visualize, $2 + 99$ or $99 + 2$? Draw a vector diagram.

3. What would be the best way to add mentally 17, 18, and 2? Which of the following represents your choice: $(17 + 18) + 2$ or $17 + (18 + 2)$? What numbers substituted for what variables in what identity give the equality of these two expressions?

4. State in words each of the theorems in this section, giving the hypotheses explicitly.

5. Illustrate each of the theorems in this section by giving the variables numerical values and drawing vector diagrams.

6. Rewrite *3.4.2 and *3.4.3 with a, b, c replaced by:

(a) 2, 3, 14	(f) $r + s, t, u$	(k) $r + s, t + u, v + w$
(b) u, v, w	(g) AB, AC, AD	(l) $2s, RS, a + b$
(c) $a + b, c, 3d$	(h) $r + s, t + u, v$	(m) $2x, a + b, 2z$
(d) Rr, rr', s	(i) $a + b, c, d + e$	(n) $2b + 3c, d, d + e$
(e) P, W, W, M	(j) $x, y + z, w + x$	

Solution for (f): *3.4.2: $(r + s) + t \equiv t + (r + s)$.
     *3.4.3: $\{(r + s) + t\} + u \equiv (r + s) + (t + u)$.

In cases like this, where parentheses appear within others, it is best to use different symbols of grouping. The most common are **parentheses** ( ), **brackets** [ ], **braces** { }, and the **vinculum** $\overline{}$. Thus $(b + c) + a \equiv [b + c] + a \equiv \{b + c\} + a \equiv \overline{b + c} + a$.

7. Justify the following by citing theorems and indicating the substitutions to be made:

(a) $(a + c) + d \equiv d + (a + c)$. (*Solution.* $a + b \equiv b + a$ with $a + c, d$ for a, b.)

(b) $(a + b) + (c + d) \equiv [(a + b) + c] + d$.

(c) $(a + b) + c \equiv (a + c) + b$. (Three steps.)

8. You know this rule: "The same number may be added to both sides of an equation." How does it follow from the uniqueness of addition? (*Hint.* See *3.3.2 as well as *3.4.1.)

9. You have $357 = 300 + 50 + 7$ and $428 = 400 + 20 + 8$. Hence $357 + 428 = 300 + 50 + 7 + 400 + 20 + 8$, and these terms can be arranged in any convenient way. To add mentally, you can proceed from left to right with the successive totals: 357, 757, 777, 785. This amounts to transferring a certain number of units at a time from one number to the other. You could also add hundreds, then tens, etc., to get the successive totals: 300, 700, 750, 770, 777, 785. Still another order (the one adopted in doing additions by hand) is to add units, then tens, then hundreds. What laws assure that the sum is the same regardless of the method? Experiment with the three orders and use the one by which you find it easiest to do the following additions mentally. (Note that the first method requires less memory and may be visualized as moving along the axis from one number by adding successive parts of the other.)

(a) $436 + 220$ (d) $199 + 821 + 347$ (g) $1948 + 1066$
(b) $597 + 243 + 19$ (e) $399 + 46 + 114$ (h) $1492 + 1766 + 17$
(c) $27 + 44 + 82 + 125$ (f) $1496 + 831$ (i) $32{,}875 + 4121 + 107$

Answers. (a) 656. (c) 278. (e) 559. (g) 3014. (i) 37,103.

10. A **magic square** consists of the first n^2 positive integers written in a square of n rows and columns in such a way that the sums of the numbers in each row, column, or diagonal are the same. Test the following to see whether they are magic squares:

(a)
8	1	6
3	5	7
4	9	2

(b)
18	1	24	7	15
5	23	6	14	17
22	10	13	16	4
9	12	20	3	21
11	19	2	25	8

(c)
46	55	44	19	58	9	22	7
43	18	47	56	21	6	59	10
54	45	20	41	12	57	8	23
17	42	53	48	5	24	11	60
52	3	32	13	40	61	34	25
31	16	49	4	33	28	37	62
2	51	14	29	64	39	26	35
15	30	1	50	27	36	63	38

(*Note.* Magic squares and other unusual arrays have interested many mathematicians. The article on magic squares in the *Encyclopaedia Britannica* gives the history of the subject and other examples.)

†**11.** The following game involves more than the ability to add. Select two numbers n and N; the second much larger than the first. Choose two teams (possibly of one member each). The first player selects any number x, such that $0 < x < n$. The other side adds to this any number satisfying the same conditions. Each team does this in turn. The team to reach exactly N is the winner. Play the game with various choices of n and N and then see whether you can work out a correct strategy—that is, a rule which will tell you in every case what number to add in order to have the best chance of winning.

†**12.** Solve problem E751 on p. 38 of the *Mathematical Monthly* for 1947, and prove that your solution is the only one.

13. Prove and illustrate with numerical examples:

*3.4.5 $a < b$ and $c < d \longrightarrow a + c < b + d$

Solution:

(1)	$a < b$	(Hypothesis)
(2)	$a + c < b + c$	(Monotonic Law of Addition)
(3)	$c < d$	(?)
(4)	$b + c < b + d$	(Monotonic Law and ?)
(5)	$a + c < b + d$	(Steps 2 and 4 and ?)

*3.4.6 $a > b \longrightarrow a + c > b + c$

*3.4.7 $a \leq b \longrightarrow a + c \leq b + c$

*3.4.8 $a \geq b \longrightarrow a + c \geq b + c$

†14. If A and B are sets, $A \cup B$ (called the **logical sum**) means the set consisting of those elements that are in either A or B, i.e.,

$$X \in (A \cup B) \longleftrightarrow X \in A \text{ or } X \in B$$

(We use "or" to mean "and/or," i.e., "in A or B" means "in A or in B or in both A and B.")

Find $A \cup B$, where A and B are:

(*a*) $A = \{1, 2, 3\}$, $B = \{2, 3, 4\}$.
(*b*) $A = $ men, $B = $ women.
(*c*) $A = $ the even integers, $B = $ the odd integers.
(*d*) $A = \{a, b, 1, c, 2\}$, $B = \{d, e, 3, f, 4\}$.
(*e*) $A = $ the chairs in the classroom, $B = $ the tables in the classroom.

Answers. (*a*) $\{1, 2, 3, 4\}$. (*c*) The integers.

†15. Show that:

(*a*) $A \cup B = B \cup A$.
(*b*) $A \cup (B \cup C) = (A \cup B) \cup C$.
(*c*) $A \cup A = A$.

Hint. Show that an element is in the left member if and only if it is in the right.

†16. The **sum of two integers** may be defined as follows. Let A and B be any disjoint sets, with m and n members, respectively. Then $m + n$ is the number of members in $A \cup B$. From this and prob. 15 prove *3.4.2 and *3.4.3.

†17. If A and B are sets, $A \cap B$ (called the **logical product**) means the set consisting of those elements that are in *both* A and B, i.e.,

$$X \in (A \cap B) \longleftrightarrow X \in A \text{ and } X \in B$$

Show that A and B are disjoint if and only if $A \cap B = $ the null set.

†18. Find $A \cap B$ for each part of prob. 14 and for: (a) A = numbers which are perfect squares, B = even numbers. (b) A = American citizens, B = Christians. (c) A = isosceles triangles, B = right triangles. (d) A = equilateral triangles, B = right triangles. [*Answers.* (14a) $\{2, 3\}$. (14c) Null set. (a) $\{4, 16, 36, \cdots \}$. (c) isosceles right triangles.]

†19. Show that:

(a) $A \cap B = B \cap A$.

(b) $A \cap (B \cap C) = (A \cap B) \cap C$.

(c) $A \cap A = A$.

3.5 Multiplication

Why does two times three equal six? A reasonable answer is: "Because two times three means three plus three, and that is six." Multiplication can thus be described in terms of addition. We can define the **product** of two positive integers m and n as follows: $mn = n + n + n + n \cdots$ (m terms). For example, $3n = n + n + n$ and $4n = n + n + n + n$. The numbers that are to be multiplied are called **factors** of the product. The product is called a **multiple** of either factor. Also either factor is said to be the **coefficient** of the other. In ab, a is the coefficient of b, and b is the coefficient of a. Since $(3)(6) = 18$, we say that 3 and 6 are factors of 18, and 18 is a multiple of 3 and of 6. The product mn may be written $m \times n$, $m \cdot n$ or $(m)(n)$. The last form is convenient when both factors are numbers. The definition gives a unique answer for the product of any two numbers, that is,

*3.5.1 $a = c$ and $b = d \longrightarrow ab = cd$

(Uniqueness of Multiplication)

Also there is only one multiple of a number that equals a given number. Thus if $ac = bc$, we must have $a = b$.

*3.5.2 $ac = bc$ and $c \neq 0 \longrightarrow a = b$

(Cancellation Law)

It is easy to give a geometric interpretation of multiplication by referring to the vector interpretation of addition. Since $(3)(4) = 4 + 4 + 4$, it is the vector found by laying out the vector 4 three times. But 4×3 is by definition $3 + 3 + 3 + 3$ or 12, the same result as we got above for 3 × 4. (See Fig. 3.5.) The student is aware that this equality of 3 × 4 and 4 × 3 is not coincidental.

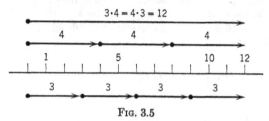

FIG. 3.5

For any two numbers:

*3.5.3 $ab \equiv ba$

(Commutative Law of Multiplication)

The definition of multiplication did not refer to more than two numbers. However, just as in the case of addition, when more than two numbers are involved a product is interpreted in terms of successive multiplications. For example, we may multiply 3 by 4 and the result by 2. We write this $2 \cdot (4 \cdot 3)$. We get 12 and then 24. Or we could multiply 4 by 2 and then multiply 3 by this result. We write this $(2 \cdot 4) \cdot 3$. The final result is the same either way. That is, $2 \cdot (4 \cdot 3) = (2 \cdot 4) \cdot 3$. This illustrates the general principle that the result of multiplying three numbers is independent of the way in which they are associated, that is,

*3.5.4 $(ab)c \equiv a(bc)$

(Associative Law of Multiplication)

The commutative and associative laws may be taken as axioms or proved from the properties of sets. They enable us to rearrange the factors in a product in any convenient way. Since $(ab)c \equiv a(bc)$, we write abc without indicating the way in which the product is to be found. The same remarks apply to more than three factors. When expressions are rewritten by the commutative and associative laws of addition and multiplication we may speak of "rearrangement" to avoid having to refer to all four laws on each occasion.

When the factors in a product are the same, we write the factor with a raised number, called an **exponent,** after it to indicate the number of times the factor appears. For instance, $a \cdot a \cdot a$ is written a^3. In general,

(1) $$a^n = a \cdot a \cdot a \cdot a \cdots \text{ to } n \text{ factors}$$

We call a^n the **nth power** of a. Exponents will be considered in 6. Meanwhile this definition suffices for simple manipulations, such as:

(2) $$a^2 \cdot a^3 \equiv (a \cdot a)(a \cdot a \cdot a)$$

(3) $$\equiv a \cdot a \cdot a \cdot a \cdot a \equiv a^5$$

We found in 3.4 that equals added to unequals yield unequals in the same sense. A similar property holds for multiplication. If unequals are multiplied by a *positive* number, the vector interpretation is that each is laid out the same number of times. The one that started at the right finishes there also. Thus multiplication is monotonic, that is,

3.5.5 $a < b \longrightarrow ca < cb$ (Monotonic Law of Multiplication for Positive Numbers)

The student should note that we state this law here *only* for positive numbers. Unlike the other theorems stated so far, it requires modification before it can be stated for all numbers. Hence we have not starred its number. (See *3.10.14 for the general law.)

We close this section with a special property of the integer 1, namely,

*3.5.6 $(1)a \equiv a$

EXERCISE 3.5

1. Find the following products by addition and make vector sketches: (a) (5)(3) and (3)(5). (b) (2)(5) and (5)(2). (c) (3)(6). (d) (9)(3). (e) 3^2.

2. What would be the most convenient way to multiply 2, 76, and 5? Which theorems justify the rearrangement? Explain what substitutions in which identities give each of these steps:

(1) $(2 \cdot 76) \cdot 5 = (76 \cdot 2) \cdot 5$ ($ab \equiv ba$ with ? and ? substituted for a and b)
(2) $= 76 \cdot (2 \cdot 5)$

3. Suggest an easy way to multiply 47 by 16. (*Hint.* $16 = 2 \cdot 2 \cdot 2 \cdot 2$.)

4. Find the following mentally: (*a*) (2)(23)(4)(5). (*b*) (4)(128)(25). (*c*) (73)(32). (*d*) (24)(47). [Solution to (*d*): (24)(47) = $3 \cdot 2 \cdot 2 \cdot 2 \cdot 47$ = 1128.] (*e*) (67)(72). (*f*) (33)(72). (*g*) (19)(27). (*h*) (17)(134).

5. Give numerical examples to illustrate each starred theorem in the section and draw sketches where convenient.

6. State in words the commutative and associative laws of multiplication.

7. Rewrite *3.5.3 and *3.5.4 with *a*, *b*, and *c* replaced by:

(*a*) 2, *c*, *d*   (*e*) *U, S, A*
(*b*) *a*, *b*, *c* + *d*   (*f*) *ch, e, ss*
(*c*) (*a* + *b* + *c*), *S, T*   (*g*) (*a* + *b*), (*a* + *b*), (*c* + *d*)
(*d*) *S, S, U*   (*h*) *x*, (*x* + *y*), *yz*

8. Justify each of the following identities by citing theorems and indicating the substitution used:

(*a*) $(r + s) \cdot t \equiv t \cdot (r + s)$ [Solution: $a \cdot b \equiv b \cdot a$ with *a* and *b* replaced by (*r* + *s*) and *t*.]
(*b*) $(a \cdot b)(c \cdot d) \equiv (c \cdot d)(a \cdot b)$
(*c*) $2 \cdot (3 \cdot 5) \equiv (2 \cdot 3) \cdot 5$
(*d*) $(a \cdot b) \cdot (c \cdot d) \equiv [(a \cdot b) \cdot c] \cdot d$
(*e*) $(ab \cdot c) \cdot d \equiv (c \cdot d) \cdot ab$ (Two steps)

9. How does the rule "both sides of an equation may be multiplied by the same number" follow from *3.5.1?

10. Justify each of the following steps in the proof that $(ab)(cd) \equiv (ac)(bd)$ by reference to a principle of this section:

(*a*) $(ab)(cd) \equiv [(ab)c]d$ (*3.5.4 with *ab, c, d* for *a, b, c*) [1]
(*b*) $\equiv [a(bc)]d$
(*c*) $\equiv [a(cb)]d$
(*d*) $\equiv [(ac)b]d$
(*e*) $\equiv (ac)(bd)$

11. In the following, rearrange in order to simplify. Justify in detail each step by citing an identity and indicating what substitutions are made in order to apply it.

(*a*) $(3)[(a)(2)] \equiv (3)[(2)(a)]$ ($ab \equiv ba$ with *a*, 2 for *a*, *b*)
                 $\equiv (3 \cdot 2)a$ [$a(bc) \equiv (ab)c$ with 3, 2, *a* for *a*, *b*, *c*]
                 $\equiv 6a$

(*b*) $3 \cdot (4x)$ (*e*) $(6a)8$ (*h*) $(xy)(bx)$
(*c*) $a(ba)$ (*f*) $2(a) \cdot 4$ (*i*) $(ST)[S(tS)]$
(*d*) $(2x)(3x)$ (*g*) $(ab)(ba)$ (*j*) $A(Bc)A$

Answers. (*c*) a^2b. (*e*) 48*a*. (*g*) a^2b^2. (*i*) S^3Tt.

12. The previous exercises were for the purpose of becoming fully acquainted with the commutative and associative laws. Henceforward, we

---

[1] In order to save space we cite theorems by number. The student should write them out, however, since the numbers mean nothing apart from this book.

take for granted that any product may be rearranged as to order and associa-
tion. Simplify the following:

(a) $(ab)(2)4a^2$

(b) $(a + b)(c + d)(a + b)$

(c) $xabxybca$

(d) $(4ac)(abc)$

(e) $P(1 + r)P(1 + r)^3$

(f) $xyz(xyz)$

(g) $(x + y + z)(ab)(x + y + z)$

(h) $(1 + r)(1 + s)(1 + r)(1 + t)^2$

Answers. (a) $8a^3b$. (c) $a^2b^2cx^2y$. (e) $P^2(1 + r)^4$. (g) $ab(x + y + z)^2$.

13. In prob. 12 what is the coefficient of:

(a) b in (a)

(b) xyz in (f)

(c) $(a + b)^2$ in (b)

(d) 4 in (d)

(e) $(1 + s)$ in (h)

(f) $(1 + r)^3$ in (e)

(g) ab in (g)

(h) b in (g)

Answers. (a) $8a^3$. (c) $(c + d)$. (e) $(1 + r)^2(1 + t)^2$. (g) $(x + y + z)^2$.

14. Do the following practice exercises in arithmetic: (a) Evaluate
$(83476)(9)(8)(7)(6)$. (b) Multiply 827 by all possible different numbers containing the digits 5, 6, and 0 just once each. (c) Find $(10)(9)(8)(7)(6)(5)(4)(3)(2)$.
(See *6.4.2.) (d) Compute the square of the previous result. (This is a long
computation, but if you want practice this is one way to get it!) (e)
$(123456789)(9) = ?$ (f) $(123456789)(63) = ?$ (g) $(123456789)(81) = ?$ (h)
Compute 25! defined by *6.4.2. [*Answers.* (a) 252,431,424. (c) 3,628,800.
(d) 13,168,189,440,000. (e) 1,111,111,101. (g) 9,999,999,909. (h) 15,511,210,-
043,330,985,984,000,000.]

15. Show that: (a) $2a = 2c \longrightarrow a = c$. (b) $ax = xy \longrightarrow a = y$.

16. What is the converse of *3.5.2? Is it true?

†**17.** Show that:

*3.5.7 $0 < a < b$ and $0 < c < d \longrightarrow ac < bd$

†**18.** If A and B are disjoint sets, $A \times B$ (called the **combinatorial product**)
is defined as the set of all pairs of elements formed by choosing one member of
the pair from A and one from B. Thus $\{1, 2\} \times \{0, 3\} = \{(1, 0), (1, 3),$
$(2, 0), (2, 3)\}$. Find:

(a) $\{1, 3\} \times \{2, 4\}$

(b) $\{1, 2\} \times \{4, 3\}$

(c) $\{1, 2\} \times \{a\}$

(d) $\{d\} \times \{a, b, c\}$

(e) $\{1, 2, 3\} \times \{4, 5, 6\}$

†**19.** The **product of two integers** may be defined as follows. Let A
and B be any two disjoint sets with m and n members. Then mn is defined to
be the number of members in $A \times B$. In the examples of prob. 18 check this
definition against your knowledge of arithmetic. For example, the number
of members in $\{1, 2\}$ is 2 and in $\{0, 3\}$ is 2, whereas the number of members
in $\{1, 2\} \times \{0, 3\}$ is 4. This checks with $2 \cdot 2 = 4$.

†**20.** Show that: (a) $A \times B \sim B \times A$. (b) $A \times (B \times C) \sim (A \times B) \times C$.

†**21.** From prob. 20 prove *3.5.3 and *3.5.4.

3.6 The distributive law

We have seen that addition and multiplication follow similar
commutative, associative, and monotonic laws. Now we come to
a law that involves both multiplication and addition. Suppose

we add 3 and 2, then multiply the result by 5. The addition gives 5, and the following multiplication gives 25. In symbols, $5(3 + 2) = 5(5) = 25$. Suppose, instead, we multiply 5 times 3, then 5 times 2, and then add. We get 15 and 10, whose sum is 25. In briefer form, $5(3) + 5(2) = 15 + 10 = 25$. It appears that we get the same answer by adding first and then multiplying as we would by multiplying and then adding. This property of numbers is quite general, that is,

*3.6.1 $a(b + c) \equiv ab + ac$ (Distributive Law)

We say that *multiplication is distributive over addition*, because we can "distribute" the multiplication in $a \cdot (b + c)$ over the two terms to be added, $a \cdot b + a \cdot c$.

Addition is not distributive over multiplication, that is, $a + (b \cdot c) \not\equiv (a + b) \cdot (a + c)$. This fact requires us to be careful with parentheses when both addition and multiplication are involved. When parentheses are omitted, it is agreed that multiplication takes priority, that is, multiplications are to be performed first. For instance, $6 + 2 \cdot 5$ means $6 + (2 \cdot 5)$ not $(6 + 2) \cdot 5$. In this way we write $a + bc$, meaning $a + (b \cdot c)$. Similarly $a \cdot b + c \cdot d$ means $(ab) + (cd)$ and not $a(b + c)d$. Also $a + b \cdot c + d$ means $a + (bc) + d$ and not $(a + b) \cdot (c + d)$. This convention saves writing parentheses, but when any doubt is possible parentheses should be used. Forms like $6 + 2 \times 5$ should be avoided in favor of $6 + (2 \cdot 5)$ or $6 + 2(5)$.

The distributive law is the justification for all the rules about parentheses that involve "removing parentheses," "multiplying out," "taking out a common factor," etc. The student should strive to use it instead of these rules of thumb. It enables us to rewrite a sum as a product or a product as a sum. Writing a sum as a product is called **factoring.** Writing a product as a sum is called **expanding.** Depending upon the circumstances, it may be convenient to expand or to factor an expression. Suppose we have $rst + rsW$. We note that rs appears in both terms. Hence we can use the distributive law with rs, t, and W substituted for a, b, and c. We rewrite $rst + rsW$ as $rs(t + W)$, and say that we have factored it. On the other hand, if we rewrite $rs(t + W)$ as $rst + rsW$, we say that we have expanded it. In a sum to which the distributive law applies, the factor that appears in both terms is called the **common factor,** and such terms are called **similar terms.**

Thus rst and rsW are similar terms in $rst + rsW$ and rs is the common factor. We "took out the common factor" when we factored the expression into $rs(t + W)$.

Since the commutative, associative, and distributive laws hold when the letters are replaced by any numbers, we can get other true identities by replacing the letters by any variables standing for numbers. We have only to be careful to replace each variable by the *same* expression *throughout*. Of course, we could always go back to the fundamental laws, but it is easier to work out many identities to fit different situations. We list here some of these:

*3.6.2 $$(b + c)a \equiv ba + ca$$

This theorem is almost like the distributive law. It indicates that we can "multiply out" from the right as well as from the left. It is easy to prove. The commutative law asserts that $(b + c)a \equiv a(b + c)$, the distributive law immediately gives $a(b + c) \equiv ab + ac$, and the commutative law permits us to rewrite this as $ba + ca$. It is this law that justifies combining similar terms with numerical coefficients. For example, $8x + 4x \equiv (8 + 4)x \equiv 12x$.

*3.6.3 $$a(b + c + d + \cdots) \equiv ab + ac + ad + \cdots$$

(Generalized Distributive Law)

*3.6.4 $$(a + b)(c + d) \equiv ac + ad + bc + bd$$

This identity suggests a convenient way of multiplying sums. Start with the first term of the first sum and multiply it in turn by the terms of the second sum. Then do the same with the second. The process may be diagramed as follows:

(1)
$$(a + b)(c + d) = ac + ad + bc + bd$$

The rule may be generalized to sums of more than two terms. We multiply each term of the first sum by each term of the second in succession, that is,

*3.6.5 $(a + b + c + \cdots)(e + f + g + \cdots) \equiv ae + af + ag + \cdots$
$$+ be + bf + bg + \cdots + ce + cf + cg + \cdots$$

The proofs of *3.6.3 and *3.6.5 depend on the principle of finite induction and so will be postponed to 6.2. An important special case of *3.6.4 occurs when the two sums are the same. Then we have:

*3.6.6 $$(a + b)^2 \equiv a^2 + 2ab + b^2$$

This identity, which is probably familiar to the student, can be illustrated in many ways. For example, we may interpret $(a + b)^2$ as the area of a square of side $(a + b)$. (See Fig. 3.6.) Then a^2

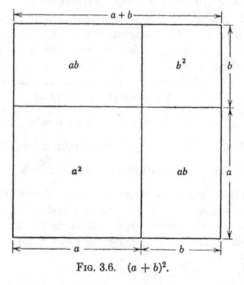

Fig. 3.6. $(a + b)^2$.

and b^2 are areas of squares of side a and b, and $2ab$ is the area of two rectangles of base a and height b.

*3.6.7 $(a + b + c + \cdots)^2 \equiv a^2 + b^2 + c^2 + \cdots$
$$+ 2ab + 2ac + \cdots + 2bc + \cdots$$

This theorem states that the square of a sum equals the sum of the squares of the terms and twice the products of each term by every term that follows it. It follows from *3.6.5 by letting the two factors be the same.

EXERCISE 3.6

1. State the distributive law in words.

2. Rewrite the distributive law and *3.6.2 with a, b, c replaced by:

(a) r, s, t	(e) 2, 2, 20	(i) $2a$, b, 1
(b) 2, 7, 3	(f) $3v$, v, v	(j) a, $x + y$, 1
(c) $(a + b)$, c, d	(g) $a + b + c$, 1, c	(k) $m + n$, c, 1
(d) 24, $3a$, $5c$	(h) 1, s, x	(l) mn, $n + m$, mn

3. Rewrite *3.6.3 and *3.6.4 with a, b, c, d replaced by:

(a) 1, r, s, t	(d) a, b, 1, d	(g) $a + c$, b, $a + b$, d
(b) $3cd$, x, c, 2	(e) $2s$, $3m$, $5t$, $9s$	(h) 20, 8, 50, 7
(c) $a + b$, R, S, T	(f) c, $a + b$, e, f	(i) 10, a, 20, b

4. Rewrite *3.6.6 with a, b replaced by:

(a) c, y	(e) x, 2	(i) $a + b$, 2	(m) $2xy$, 1
(b) $2x$, y	(f) y, 3	(j) st, 7	(n) 3, $5st$
(c) s, $3t$	(g) s, 20	(k) uv, xw	
(d) x, 1	(h) $3t$, $5s$	(l) $x\sqrt{2}$, 1	

5. Justify the following identities by reference to appropriate theorems. Indicate the required substitutions.

(a) $c(s + 2t) = cs + c(2t)$

(b) $ab(r + uv) = (ab)r + (ab)(uv)$

(c) $(3 + 4c)2 = 6 + 8c$

(d) $3c + 4c + 8c = 15c$

(e) $110^2 = 100^2 + 2000 + 100$

(f) $4xy + bxy = (4 + b)xy$

Solution for part a. *3.6.1 with a, b, c replaced by c, s, $2t$.

6. Make a diagram like (1) for $(a + b + c)(e + f + g)$. Draw a square illustrating $(a + b + c)^2$. (See Fig. 3.6.)

7. Do the following multiplications by using the principles of this section. See how much of the work you can do mentally.

(a) 3(29)	(e) 8(2123)	(i) (28)(43)	(m) (52)(88)
(b) 6(73)	(f) 3(9165)	(j) (64)(72)	(n) (39)(67)
(c) 7(94)	(g) 5(1875)	(k) (19)(36)	(o) (128)(32)
(d) 4(1321)	(h) (37)(29)	(l) (105)(23)	(p) (83)(286)

(*Note.* The purpose of this problem is to become familiar with the distributive law. The following methods are similar to those used by "lightning calculators." In part a, think "$3 \cdot 20 = 60$, plus $3 \cdot 9 = 27$, gives 87." Only the 60 and 27 should be remembered. In part d, double twice or multiply from the left to get successively 4000, 5200, 5280, 5284. Here each product is added to the previous total. In part h, visualize the product as $(30 + 7)(20 + 9)$. Then 600, 870, 1010, 1073 are successive totals found by applying *3.6.4 and adding each product to the previous total.)

8. Find the following by using *3.6.6 and *3.6.7, and check by the usual method of multiplication:

(a) $(73)^2$ (d) $(28)^2$ (g) $(246)^2$ (j) $(904)^2$
(b) $(34)^2$ (e) $(13)^2$ (h) $(121)^2$ (k) $(1001)^2$
(c) $(67)^2$ (f) $(101)^2$ (i) $(325)^2$ (l) $(5020)^2$

(*Note.* To do these mentally, add each term in the product to the previous total and remember only this. Thus $73^2 = (70 + 3)^2 = 70^2 + 2\cdot70\cdot3 + 3^2 = 4900 + 420 + 9$; but in working without paper think 4900, find 420 and add to get 5320, then add 9 to get 5329.)

9. Use the identities of this and previous sections to expand the following, citing identities and indicating substitutions:

(a) $2(3 + a)$ (l) $(a + b)(a + b + 2)$
(b) $3a(1 + c)$ (m) $(a + b)(a + b + 1)$
(c) $ab(R + ST)$ (n) $(x + y + z)(x + y + 1)$
(d) $(x + 5)(x + 2)$ (o) $(r + 2x)(r + 2x + c)$
(e) $(x + 3)(x + 1)$ (p) $(s + uv)(s + a + b)$
(f) $(a + b)(a + c)$ (q) $(2x + 3y)^2$
(g) $(c + x)(3 + x)$ (r) $(a + b + 2)^2$
(h) $(1 + x)xy$ (s) $(3a + 2b + 1)^2$
(i) $2a(x + 4y + z)$ (t) $(4s + 3t + 5)(s + t + 2)$
(j) $(a + 2b)^2$ (u) $(2t + y)(a + 2c + d + 6)$
(k) $(2c + 1)^2$ (v) $(a_1+b_1+c_1+d_1)(a_2+b_2+c_2+d_2)$

Answers. (a) $6 + 2a$. (c) $abR + abST$. (e) $x^2 + 4x + 3$. (g) $x^2 + (3 + c)x + 3c$. (i) $2ax + 8ay + 2az$. (k) $4c^2 + 4c + 1$. (m) $a^2 + 2ab + b^2 + a + b$. (o) $r^2 + 4rx + 4x^2 + cr + 2cx$. (q) $4x^2 + 12xy + 9y^2$. (s) $9a^2 + 4b^2 + 1 + 12ab + 6a + 4b$. (u) $2at + 4ct + 2dt + 12t + ay + 2cy + dy + 6y$.

10. Use the identities of this and previous sections to factor the following, citing theorems as in prob. 9:

(a) $6 + 2a$ (m) $9x^2 + 12x + 4$
(b) $a + a^2$ (n) $2a + 2b + sa + sb$
(c) $2q + 4pq$ (o) $st + ut + s^2 + us$
(d) $x^2 + x$ (p) $abx + acx + 2ac + 2ab$
(e) $x + 6xz + x^2$ (q) $4 + s^2 + t^2 + 4s + 4t + 2st$
(f) $s^2 + 7s + 10$ (r) $1 + a^2 + 2a + 2b + 2ab + b^2$
(g) $a + 2a$ (s) $x^2 + 2xy + y^2 + 2x + 2y$
(h) $as + 2sta + Ua$ (t) $4s^2 + 4xs + 2x + 4s + x^2$
(i) $x^2 + 2xy + y^2$ (u) $2x + 2y + ax + by + bx + ay$
(j) $s^2 + 2s + 1$ (v) $4r^2(1-r)^2+4r(1-r)^3+(1-r)^4$
(k) $t^2 + 4t + 4$ (w) $6x^2 + 9ax + 15x + 18sx$
(l) $9x^2 + 6x + 1$

Answers. (a) $2(3 + a)$. (c) $2q(1 + 2p)$. (e) $x(x + 6z + 1)$. (g) $3a$. (i) $(x + y)^2$. (k) $(t + 2)^2$. (m) $(3x + 2)^2$. (o) $(s + u)(s + t)$. (q) $(2 + s + t)^2$. (s) $(x + y)(x + y + 2)$. (u) $(x + y)(2 + a + b)$.

11. Simplify, giving reasons for each step:

(a) $(2x + 5a) + (x + 2a + 1)$ (b) $(2y + 5a + 7) + (a + 4y + 1)$

(c) $(s + t) + (t + 3)$ (e) $(x + y) + (y + z) + (z + x)$
(d) $(ab + c + 2x) + (5c + 2ab + x)$

Solution for part a:

$$(2x + 5a) + (x + 2a + 1) \equiv (2x + x) + (5a + 2a) + 1 \quad \text{(Why?)}$$
$$\equiv 3x + 7a + 1 \quad (*3.6.2)$$

†12. Write out formal proofs of *3.6.2, *3.6.4, and *3.6.6, using only previously stated identities.

†13. Prove *3.6.3 for the special case $a(b + c + d) \equiv ab + ac + ad$.

†14. Prove from *3.6.6, without using *3.6.7, that

$$(a + b + c)^2 \equiv a^2 + b^2 + c^2 + 2ab + 2ac + 2bc$$

†15. Use the notions of set theory (Ex. 3.4 and Ex. 3.5) to prove the distributive law.

16. Show that $(a + b) \equiv a + b$ by using *3.5.6 and the distributive law.

3.7 Subtraction and division

We say that $10 - 7 = 3$ because $7 + 3 = 10$. More generally, $a - b$ means the number which must be added to b to get a, that is,

*3.7.1 $x = a - b \;\longleftrightarrow\; b + x = a$ (Definition of Subtraction)

We call $a - b$ the **difference** of a and b or "a minus b." The first number is called the **minuend**; the second is called the **subtrahend**. We say that we have subtracted b from a. It may happen that there is no positive integer equal to $a - b$. For example, $2 - 6$ does not exist among the numbers treated so far. However, when $a - b$ exists, it is unique; that is, there is only one number x such that $b + x = a$.

*3.7.2 $a = b$ and $c = d \;\longrightarrow\; a - c = b - d$

A subtraction may be conveniently represented by vectors, as is done for $6 - 2 = 4$ in Fig. 3.7. The vector $6 - 2$ is the vector

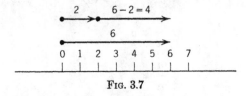

Fɪɢ. 3.7

with initial point 2 and terminal point 6, and it gives the distance from 2 to 6. More generally, if the vector $a - b$ is placed with its

initial point at b, its terminal point lies at a, and we call $a - b$ the
directed distance from b to a. Note that the diagram of $4 = 6$
-2 is the same as that for $2 + 4 = 6$.

The relation between addition and subtraction suggests calling
subtraction the inverse of addition. The word "inverse" refers
to this property: if we add and then subtract or subtract and then
add the same number, we return to the starting point. The two
operations "cancel each other out." Another way of stating it is:

*3.7.3 $(a + b) - b \equiv a \equiv (a - b) + b$

Subtractions are carried out by rules that are based on the
inverse relation between addition and subtraction. We do not
learn subtraction tables, but we make use of addition tables
"backwards."

Division is the inverse of multiplication in the same sense that
subtraction is the inverse of addition. If someone asks why
$12 \div 3 = 4$, a legitimate answer is "because $3 \cdot 4 = 12$." Thus
our definition of division is like that of subtraction, with $+$ and $-$
replaced by $\cdot$ and $\div$. The **quotient** $a \div b$ is defined to be the
number that, when multiplied by b, yields a, that is,

*3.7.4 $x = a \div b \longleftrightarrow bx = a$ (Definition of Division)

The quotient $a \div b$ is also described as "a divided by b" or "a
over b." It is written as $\dfrac{a}{b}$, $a:b$, or a/b. In a/b, a is called the
numerator or **dividend,** b is called the **denominator** or **divisor,**
and the whole expression is called a **ratio,** the "ratio of a to b."

It follows from the definition that:

*3.7.5 $\dfrac{ba}{b} \equiv a \equiv b\left(\dfrac{a}{b}\right)$

*3.7.6 $\dfrac{a}{1} \equiv a$

*3.7.7 $\dfrac{a}{a} \equiv 1$

The first is similar to *3.7.3. The second follows from $1 \cdot a = a$.
The third is implied by $a \cdot 1 = a$.

Just as subtraction did not always have an answer, so now division is not always possible. When a/b is an integer, we say that a is **divisible** by b or that b is a **factor** of a. But it is easy to find examples where there is no integer which satisfies the definition. Thus $2/3$ is not an integer, since there is no positive integer x such that $3x = 2$. When a/b does exist, it is unique, that is, there is only one number x such that $bx = a$.

*3.7.8
$$a = b \text{ and } c = d \neq 0 \longrightarrow \frac{a}{c} = \frac{b}{d}$$

The condition $d \neq 0$ will be explained in 3.8.

There are many different ratios that equal the same integer. For example, $3/1 = 6/2 = 51/17$. In fact $ca/c \equiv a$ for any c. On the other hand, if a ratio a/b equals an integer, the denominator must be a factor of the numerator, for in that case $a/b = x$ and $bx = a$. If $a/b = c/d$, we have, after multiplying both sides by bd and using *3.7.5, $ad = bc$. Conversely, if $ad = bc$, then $a/b = c/d$. (Why?) Hence:

*3.7.9
$$\frac{a}{b} = \frac{c}{d} \longleftrightarrow ad = bc$$

From this it follows at once (why?) that:

*3.7.10
$$\frac{ca}{cb} \equiv \frac{a}{b} \quad (c \neq 0)$$

EXERCISE 3.7

1. Restate each theorem in this section in words and also in letters, using different ones from those in the text.

2. Draw vector diagrams of: (a) $10 - 3$. (b) $2 - 1$. (c) $23 - 14$. (d) $102 - 56$. (e) $52 - 6$.

3. Justify *3.7.3 in your own words.

4. Diagram $12/4 = 3$ by diagraming $4 \cdot 3 = 12$. Why is this proper? Draw similar diagrams for $6/2$, $28/7$, $39/13$, $72/9$.

5. Which of the following are true and which false? Justify your answer by reference to *3.7.1 and *3.7.4.

(a) $2(3/2) = 4/3$
(b) $7 - 11 = 4$
(c) $66/22 \neq 3$
(d) $17/1 = 16$
(e) $23/24 \neq 3/4$

(f) $43 - 18 = 25$
(g) $3 + (4 - 3) \neq 4$
(h) $5(999/5) \neq 999$
(i) $42 - 6 = 36$
(j) $42/36 = 6$

(k) $795/795 = 1$

(l) $42/7 \neq 12/2$

(m) $2 - 2 = 1$

(n) $a/a = 1$

(o) $b - b \neq 1$

(p) $1918 - 1945 = 1945 - 1918$

(q) $51/17 = 3/1$

(r) $2 + 4 = 8 - 2$

(s) $2x - x \neq x$

(t) $2x/x = 2$

(u) $2a/2a \neq 2$

(v) $3a/3b = a/b$

(w) $36/37 = 6/7$

(x) $ab/cb = b$

Answers. (a) F. (c) F. (e) T. (g) F. (i) T. (k) T. (m) F. (o) T.

6. Practice long division by testing whether the following ratios equal integers: (a) 681/227. (b) 383/178. (c) 3533/871. (d) 396/22. (e) 5938/359. (f) 10831/843. [*Answers.* (a) Yes. (c) No. (e) No.]

7. State the rules for subtraction and division of integers in decimal form and illustrate by the following examples:

(a) $3801029 - 2557331$

(b) $901033000 - 82341234$

(c) $10009818/582$

(d) $506518353/893$

Answers. (a) 1243698. (c) 17199.

8. To **factor** an integer is to write it as a product of other integers. It is said to be completely factored if it is written as the product of prime integers. (A **prime** is an integer greater than 1 that has no factors except itself and 1.) To factor a number, we may try to divide by the primes in order, the first few being 2, 3, 5, 7, 11, 13, $\cdots$, or we may recognize it as the product of other numbers. Factor:

(a) 99   (e) 41   (i) 81   (m) 117   (q) 427

(b) 136 (f) 102 (j) 97   (n) 360   (r) 243

(c) 51   (g) 37   (k) 122 (o) 715   (s) 301

(d) 450 (h) 435 (l) 495 (p) 2001 (t) 1711

Answers. (a) $3^2 \cdot 11$. (c) $3 \cdot 17$. (e) Prime. (g) Prime. (i) 3^4. (k) $2 \cdot 61$. (m) $3^2 \cdot 13$. (o) $5 \cdot 11 \cdot 13$. (q) $7 \cdot 61$. (s) $7 \cdot 43$.

9. A number was called **perfect** by the ancient Greeks if it equaled the sum of all its factors, including 1 but not itself. Thus $6 = 1 + 2 + 3$ and $6 = 1 \cdot 2 \cdot 3$. Show that 28 and 496 are perfect. (*Note.* Superstition has attached mystical significance to these numbers, but among mathematicians the term has no such connotations. There is only one other perfect number less than 10,000. See *Mathematical Recreations and Essays*, by W. W. R. Ball.)

10. Two numbers are called **amicable** if each is equal to the sum of all the factors of the other. Show that 220 and 284 are amicable.

11. If a number is divisible by nine, so is the sum of its digits. Check this for the first few multiples of nine. Devise a heuristic explanation.

12. Why is it that multiplication by 5 can be performed by dividing by 2?

13. Deal three equal piles of cards in a row. Take two cards from each end pile and put in the center pile. Then take from this pile as many cards as there are in either of the others. Show that there are always six cards left in the center pile. (*Hint.* Let x be the original number in each pile.)

†**14.** Prove that subtraction is unique. (*Hint.* Assume that x and x' both satisfy $a + x = b$.)

†**15.** Prove that division is unique.

†16. Prove *3.7.5.

†17. Construct a careful proof of *3.7.9. (*Hint.* In proving $ad = bc$ ⟶ $a/b = c/d$, let $a/b = x$ and $c/d = y$. Then $a = bx$ and $c = dy$. From these, it follows that $x = y$.)

†18. Prove that $a - (b + c) \equiv (a - b) - c$.

†19. Show that subtraction and division are not commutative, i.e., $a - b \not\equiv b - a$ and $a/b \not\equiv b/a$.

†20. Is the relation "is divisible by" an equivalence relation? Justify your answer.

21. Explain the trick called "Even and Odd" described on p. 174 of *Numerology*, by E. T. Bell (Baltimore, 1933).

†22. Write for subtraction an identity analogous to *3.7.10. Is it true?

3.8 Zero

We introduced the positive integers in terms of sets but did not assign any number to the set that contains no elements, the null set. (See 2.7.) We now assign the symbol 0, called **zero,** to this set. It may seem rather artificial to talk about a set with no members, but there are many situations when this idea is very practical. Consider, for example, a man who has a deposit of ten dollars in the bank. If he draws a check for ten dollars, he has no money in the bank. (We neglect the fact that the bank will probably charge his account a few cents for cashing the check!) It would not be correct to say that he has no account, even though he has no money in his account. We want to be able to say that he has an account but that the balance is zero! Thus we visualize the null set as still a set but with no members.

We explained addition of positive integers in terms of adding sets. Thus to add m and n, we take any set of m objects and any set of n different objects and count the set consisting of all members of both sets. (See 3.4.) This notion of addition can be applied when one of the sets is the null set, and it is evident that, if we combine the null set with any set, the new set is identical with the old, that is,

*3.8.1 $a + 0 \equiv 0 + a \equiv a$

This identity holds for all a. In particular, when $a = 0$ we have $0 + 0 = 0$. We see also that $a - a = 0$ for any a. For by the definition of subtraction, $a - a = x$ means $a + x = a$, and 0 is just such an x, since $a + 0 = a$. Hence:

*3.8.2 $a - a \equiv 0$

Zero has a geometrical interpretation that is familiar to the student. The origin on an axis is made to correspond to the number 0. Then the vector corresponding to 0 is one with both its initial and terminal points at the origin. It is a vector of zero length. With this interpretation of 0, the geometrical picture of addition is the same whether or not zero is involved.

How should we interpret multiplication when one of the factors is zero? Consider first $a \cdot 0$. Consistent with the vector interpretation of multiplication, we find this by laying out the zero vector a times. But this evidently leaves us still at the origin. This suggests $a \cdot 0 = 0$. If multiplication is still to be commutative for 0 with the positive integers, we must have $0 \cdot a = 0$ also. Hence:

*3.8.3 $0 \cdot a \equiv a \cdot 0 \equiv 0$ (Definition)

This can be taken as a definition or derived as suggested in the exercise. A special case is $0 \cdot 0 = 0$. Thus, if either or both of two factors is zero, the product is zero. The converse is true also, that is, if the product of two numbers is zero, one or both must be zero. It is easy to see that this is true when a and b are positive integers or zero, for, if neither a nor b is zero, the product is a positive integer and hence not zero.

*3.8.4 $ab = 0 \longleftrightarrow a = 0$ or $b = 0$

We now turn to the properties of 0 with respect to division. First of all, we note that if b is zero, a/b is undefined for the following reasons: If $a \neq 0$, there is no number x, such that $x = a/0$. For by the definition of division (*3.7.4), no matter what number x we take, $0 \cdot x = 0$ and hence cannot equal a. On the other hand, $0/0$ could equal any number according to the definition of division, since for any x we have $0 \cdot x = 0$. Thus $a/0$ yields either no answer or an indeterminate result, and we say that *division by zero is undefined*. When dealing with variables that stand for numbers, it is important to see that no variable that appears as a divisor stands for zero. If this should happen the expression would be undefined. On the other hand, there is no objection to 0 in the numerator of a division. In fact, since $b \cdot 0 = 0$,

*3.8.5 $0/b \equiv 0$ for $b \neq 0$

We now have a number system consisting of the positive integers and zero. In accordance with the interpretation of 0 in terms of sets and vectors, we place 0 at the left of the sequence of positive integers to get the sequence $0, 1, 2, 3 \cdots$. We say that any positive integer is greater than zero, that is, $a > 0$ or $0 < a$ for all positive a. It can be shown that all the theorems so far stated in this chapter hold when the variables are replaced by any members of the set $0, 1, 2, 3, \cdots$, except that a denominator must not be zero.

EXERCISE 3.8

1. Give examples of the way 0 is used to talk about the following: (*a*) Scores in an examination. (*b*) Depth of water in a tank. (*c*) Speed of a car. (*d*) Temperature. (*e*) Height above sea level. (*f*) Number of chapel cuts. (*g*) Number of faulty products turned out by a machine. (*h*) Number of mistakes in a computation. (*i*) Error in a measurement. (*j*) Debts. (*k*) Other situations of your own construction.

2. Explain in terms of the definition of subtraction why $3 - 3 = 0$.

3. If $3x = 0$, what conclusion can you draw about x? Why?

4. If $xy = 0$, what conclusion can you draw about x and y? Why?

5. If $xy \neq 0$, what conclusion can you draw about x and y? Why?

6. Which of the following are true and which false? Give reasons.

(*a*) $6 \cdot 0 \neq 1$
(*b*) $0/2 = 0$
(*c*) $18/0 = 17$
(*d*) $5/0 = 5$
(*e*) $0/1 \neq 1$
(*f*) $1/1 = 0$

(*g*) $1 - 1 = 0$
(*h*) $(4-4) + (2-2) = 0$
(*i*) $3/3 = 0/0$
(*j*) $1/1,000,000 \neq 0$
(*k*) $1/0 = 100\%$
(*l*) $2/0 = 200\%$

(*m*) $0 \div 20 = 20$
(*n*) $2a/0 \neq 2a$
(*o*) $a/b = 0 \cdot a/0 \cdot b$
(*p*) $0/0 \neq 1$

Answers. (*a*) T. (*c*) F. (*e*) T. (*g*) T. (*i*) F. (*k*) F. (*m*) F. (*o*) F.

7. Simplify the following, giving reasons:

(*a*) $(2 + 0)(a + b)$
(*b*) $a + b - c + c - b$
(*c*) $(st + u) - st + u$

(*d*) $uv - (vu)$
(*e*) $(r + s + t)(2a - 2a)$
(*f*) $1 - a - 1 - a$

Answers. (*a*) $2(a + b)$. (*c*) $2u$. (*e*) 0.

8. If $a \neq 0$, is there a number b such that $a/b = 0$? Explain.

9. What idea of 2.5 was used in justifying *3.8.4?

†*10.* Use the ideas of set theory to show that, if θ is the null set, then:

(*a*) $A \cup \theta = \theta \cup A = A$
(*b*) $A \cap \theta = \theta \cap A = \theta$
(*c*) $A \times \theta = \theta \times A = \theta$

†*11.* Use prob. 10 to prove *3.8.1 and *3.8.3.

3.9   Negative integers

With the numbers so far defined, the left half of the axis is quite empty. It is natural to label points toward the left as we have already done toward the right. Then for each point on the right there is a corresponding point on the left at the same distance from the origin. Designating by $-m$ the point at the left corresponding to the positive integer m, the axis appears as in Fig. 3.9.

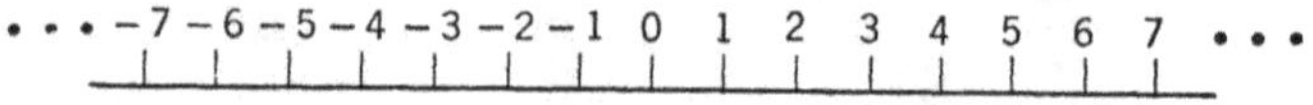

$$\cdots \bullet \bullet -7 -6 -5 -4 -3 -2 -1 \quad 0 \quad 1 \quad 2 \quad 3 \quad 4 \quad 5 \quad 6 \quad 7 \bullet \bullet \bullet$$

Fig. 3.9

We thus invent a new number corresponding to every positive number m. We call it "minus m" or "negative m" and write it $-m$. The corresponding vector has its initial point at the origin and its terminal point at $-m$. It is a vector of length m pointing to the left.

We call $-1, -2, -3, -4, \cdots$ the **negative integers.** We speak of any member of this set as being **negative,** and any member of the set $1, 2, 3, \cdots$ as being **positive.** The set consisting of the positive and negative integers and zero is called **the integers.** We now have a row of numbers extending in both directions without end: $\cdots -5, -4, -3, -2, -1, 0, 1, 2, 3 \cdots$. As before, we say that two expressions are equal if and only if they stand for the same number, and we say that one number is greater than another if it is at the right in the above row. Thus $1 > -5$ because 1 lies to the right both in the row of numbers and on the axis. Since any positive number is greater than zero, we write $a > 0$ for "a is positive" and $a < 0$ for "a is negative."

Thus the non-zero numbers exist in pairs represented by points at equal distances from the origin. We use the minus sign in front of *either* number to indicate the *other* number in the pair. For example, -6 means the negative number that goes with 6, and $-(-6)$ means the positive number that goes with -6. The number 0 does not have a different negative, but we agree that $-0 = 0$; and this is consistent with our geometrical interpretation of numbers. (Why?) Then:

*3.9.1 $-(-a) \equiv a$

Since the minus sign before a number indicates the other number in the pair, it is clear that $-a$ is negative if a is positive and positive if a is negative, that is,

*3.9.2
$$a > 0 \longleftrightarrow -a < 0$$
$$a < 0 \longleftrightarrow -a > 0$$

Also, if two numbers are equal, their negatives are equal, that is,

*3.9.3 $$a = b \longleftrightarrow -a = -b$$

It is often convenient to have a symbol that represents the positive member of the pair $\{a, -a\}$. This number is called the **absolute value** of a and is designated by $|a|$. Its formal definition is:

*3.9.4
$$|a| = a \text{ if } a \geq 0 \qquad \text{(Definition)}$$
$$= -a \text{ if } a < 0$$

The geometric interpretation of $|a|$ is the **distance** (undirected) of the point a from the origin; a is interpreted as the **directed distance** of the point a from the origin. Note that by definition the absolute value of zero is zero. According to the definition, $|7| = 7$, $|-2| = 2$, and $|0| = 0$. We cannot tell whether $|a|$ is a or $-a$ unless we know whether $a \geq 0$ or $a < 0$, but we do know that $|a|$ is always positive or zero, that is,

*3.9.5 $$|a| \geq 0$$

Suppose we have two numbers a and b with $a < b$. If we take the negatives of both, the number that was at the right will now be at the left, since taking the negative of a vector means swinging it around to the other side of the origin. Thus:

*3.9.6 $$a < b \longrightarrow -a > -b$$

EXERCISE 3.9

1. What is the negative of each of the following? (*a*) 1. (*b*) -3. (*c*) -8. (*d*) 0. (*e*) $-(-1)$. (*f*) $-(144)$. (*g*) $-(-a)$. (*h*) $-a$.

2. State in words and interpret geometrically:

(a) $-6 < 0$ (d) $3 > -2$ (g) $a \leq c$ (j) $0 > -1$
(b) $7 > 0$ (e) $a < -b$ (h) $-(-3) > -3$ (k) $0 < -a$
(c) $-a > 0$ (f) $-c \geq d$ (i) $-3 < 3$ (l) $2a < a$

Answers. (a) -6 is negative; -6 lies to the left of the origin. (e) a is less than $-b$; a lies to the left of $-b$.

3. In each part of prob. 2 in which letters appear, suggest values of the letters that make the statement true and others that make it false.

4. Which of the following are true and which false? Give reasons.

(a) $-2 > 0$ (h) $-1 \leq 0$ (o) $2 < -13$
(b) $-(-5) \leq 5$ (i) $1 < -5$ (p) $a \geq a$
(c) $1 > -1$ (j) $-3 < -4$ (q) $a < a$
(d) $14 < -100$ (k) $0 < -6$ (r) $-8 < 0$
(e) $0 < 16$ (l) $932 > -1000$ (s) $-27 > -1$
(f) $-(-(-1)) < 0$ (m) $-23 < -1$
(g) $7 > -7$ (n) $-(-(-(-3))) > 0$

(t) $|x| = 0 \longrightarrow x = 0$ (w) $|a| = |b| \longrightarrow a = b$
(u) $|x| = 1 \longrightarrow x = 1$ (x) $|a| > |-a|$
(v) $|y| = 2 \longrightarrow y = 2 \text{ or } y = -2$

Answers. (a) F. (c) T. (e) T. (g) T. (i) F. (k) F. (m) T. (o) F. (q) F. (s) F. (u) F. (w) F.

5. Diagram a and $-a$ for: (a) $a = 4$. (b) $a = -6$. (c) $a = -3$. (d) $a = 10$.

6. Illustrate *3.9.1 and *3.9.2 for $a = 2, 0, -3, 4, -6$.

7. Draw diagrams illustrating the following examples of *3.9.6:

(a) $2 < 5$ and $-2 > -5$ (c) $-4 < -1$ and $-(-4) > -(-1)$
(b) $-3 < 2$ and $-(-3) > -2$ (d) $-1 < 6$ and $-(-1) > -6$

†**8.** Prove that $|a| = |-a|$. (*Hint.* Consider $a > 0$, $a = 0$, and $a < 0$.)

3.10 The integers

We have extended the number system to include all the positive and negative integers and zero, but we have not said what is meant by the sum, difference, product, or quotient of two numbers when negatives are involved. We shall do this in such a way as to preserve the usual vector interpretations and so that the theorems stated for positive integers and zero hold for all integers.

We define the **sum** $a + b$, when a and b are any integers, according to the rule that is already familiar to the student. If both a and b are positive, $a + b$ is the sum of the absolute values, which in this case are just the numbers themselves. If both a and b are negative, we add the absolute values and prefix a minus sign. Thus $(-3) + (-2) = -(3 + 2) = -5$, by definition. If a and

b are of different sign (that is, one positive and the other negative), we subtract the smaller absolute value from the larger and prefix the sign of the number with the larger absolute value. Thus $(-2) + 3 = 3 - 2 = 1$ and $(-3) + 2 = -(3 - 2) = -1$. If the absolute values are equal, the sum is zero. As before, we define the sum of any number and zero to be the number itself.

This definition enables us to compute any sum in terms of just positive numbers and zero, since the definition is in terms of absolute values. It can be shown that the commutative and associative laws and other theorems about addition hold for all integers. Therefore, in dealing with variables representing numbers, we do not need to consider whether they stand for positive or for negative numbers until we come to numerical computation. It is easy to see by examples that the vector interpretation of addition is always the same regardless of the signs of the numbers involved. The vector $a + b$ is always found by laying out the vector b with its initial point on the terminal point of a. We illustrate for various cases in Fig. 3.10.1.

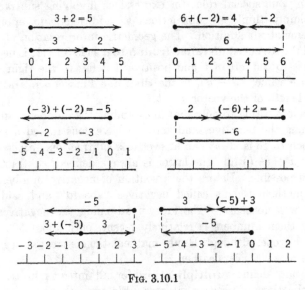

FIG. 3.10.1

Having defined addition for integers, we can extend the definition of subtraction given in 3.7. By $a - b$ we still mean the number x such that $b + x = a$. It is not hard to show from our

definition of addition of integers that this number is the same as
$a + (-b)$, that is,

*3.10.1 $a + (-b) \equiv a - b$

A special case of this identity is obtained when $a = b$. Then
we have $a + (-a) = a - a = 0$. Thus any number and its negative add to zero, that is,

*3.10.2 $a + (-a) \equiv 0$

The last two identities justify the use of the minus sign to
designate both subtraction and the negative of a number. Since
$a + (-b)$ is always the same as $a - b$, and since $-b$ is the same
as $0 - b$, the two uses of the minus sign do not lead to contradictory
results. These identities establish also a relation between addition
and subtraction that enables us to consider any subtraction as an
addition. Thus we think of $a - b$ as the sum of a and $-b$ rather
than as a difference. This point of view saves us the trouble of
working out special rules for expressions involving subtractions.
We simply think of subtractions as additions and apply the
theorems about addition. The geometric interpretation of $a - b$
is still the **directed distance from b to a** as in 3.7. It is negative
if a lies to the left of b and positive if it lies to the right. The
absolute value, $| a - b |$, is the **distance between a and b.** It
is the length of the vector $a - b$.

It is interesting to note that in extending the number system by
adjoining the negatives and zero we have constructed a system
in which there is always an answer to a subtraction. Since $a - b$
equals $a + (-b)$ and the latter is always defined, subtraction is
always possible. Before the invention of negative numbers and
zero, mathematicians called negatives "absurd" and said that
$6 - 8$ was nonsense. The idea of adjoining the negatives and
calling them numbers is relatively recent, only about 300 years
old in Europe, although Hindu mathematicians treated negatives
as numbers 900 years earlier.

We now define **multiplication** for all integers in terms of
multiplications of positive integers. We agree that ab is to mean
the product of the absolute values of a and b if they have the same
signs (both positive or both negative), and the negative of this
product if they have different signs. If either factor is zero, the

product is defined as zero. Thus $(6)(5) = 30$ and $(-6)(-5)$ $= |-6| \cdot |-5| = 6 \cdot 5 = 30$ but $(6)(-5) = -30$ and $(-6)(5)$ $= -30$, and $(-6)0 = 0$. This definition enables us to compute any product in terms of multiplications of non-negative numbers. With this and the previous definitions of addition, it can be shown that the integers obey all the theorems of this chapter, and in particular the commutative, associative, and distributive laws of multiplication.

From the definition of multiplication it follows that:

*3.10.3 $\qquad\qquad\qquad (-a)(-b) \equiv ab$

From *3.10.3,

*3.10.4 $\qquad\quad (-1)(-1) \equiv 1$

*3.10.5 $\qquad\quad (-a)b \equiv a(-b) = -ab$

*3.10.6 $\qquad\quad (-1)a \equiv -a$

*3.10.7 $\qquad\quad (-1)^n = 1$ when n is even

$\qquad\qquad\qquad\quad = -1$ when n is odd

By "*n is **even**" we mean that n is a multiple of 2, and by "*n is **odd**" we mean that n is not even. These identities, rather than rules of thumb, should be used to simplify expressions involving minus signs. In case of doubt, minus signs should be replaced by multiplication by (-1) according to *3.10.6.

Having defined multiplication for integers, we can apply the definition of division given in 3.7. By a/b we still mean the number x such that $bx = a$. Then it can be proved that:

*3.10.8 $\qquad\qquad\qquad \dfrac{a}{b} \equiv \dfrac{-a}{-b}$

*3.10.9 $\qquad\qquad\qquad \dfrac{-a}{b} \equiv \dfrac{a}{-b} \equiv -\dfrac{a}{b}$

Thus any division of integers can be rewritten so that it involves only a division of positive integers. In fact, a/b equals $|a|/|b|$ if a and b have the same sign, and $-|a|/|b|$ if they have different signs. For example, $6/2 = (-6)/(-2) = 3$, but $(-6)/2$ $= 6/(-2) = -(6/2) = -3$.

The following identities can be derived from principles already stated:

*3.10.10 $a(b - c) \equiv ab - ac$

*3.10.11 $(a + b)(a - b) \equiv a^2 - b^2$

*3.10.12 $(a - b)^2 \equiv a^2 - 2ab + b^2$

*3.10.13 $-(a + b + c + \cdots + z)$

$$\equiv (-a) + (-b) + (-c) \cdots + (-z)$$

The last one justifies all the rules of thumb about "removing parentheses" preceded by minus signs.

We now give a more general theorem to replace the monotonic law stated in 3.5 for positive numbers only.

*3.10.14
$$c > 0 \text{ and } a < b \longrightarrow ca < cb$$
$$c < 0 \text{ and } a < b \longrightarrow ca > cb$$

This means that multiplying an inequality by a *positive* number does not change the sense of the inequality, whereas multiplication by a *negative* number does change the sense. To illustrate the theorem we note that $2 < 5$ and $3 \cdot 2 < 3 \cdot 5$, but $(-3)2 > (-3)5$. Also $-6 < -1$ and $-12 < -2$, but $12 > 2$. Finally, $-2 < 3$, $-6 < 9$, but $6 > -9$.

EXERCISE 3.10

1. Carry out the following additions, explaining in each case the application of the definition of addition of integers:

(a) $3 + 4$ (e) $2 + (-2)$ (i) $245 + (-826)$
(b) $(-5) + (-2)$ (f) $(-1) + (-1)$ (j) $(-4) + 0$
(c) $2 + (-3)$ (g) $(-3) + (-12)$ (k) $(-3x) + (-4x)$
(d) $(-4) + 5$ (h) $0 + (-20)$ (l) $2ax + (-4ax)$

Answers. (a) 7. (c) -1. (e) 0. (g) -15. (i) -581. (k) $-7x$.

2. Diagram each addition in prob. 1.

3. Illustrate the commutative law of addition by comparing vector diagrams of the sums in prob. 1 with those of the same sums in opposite order.

4. Illustrate the associative law of addition by computing $(-5) + [3 + (-1)]$ and $[(-5) + 3] + (-1)$ and comparing their vector diagrams.

5. Rewrite the following differences as sums:

(a) $4 - 7$	(d) $-1 - 9$	(g) $3x - 2y$	(j) $17 - 0$
(b) $2 - 2$	(e) $5 - (-3)$	(h) $a - b$	(k) $-7a - 4a$
(c) $0 - 5$	(f) $-4 - (-6)$	(i) $a - (-b)$	(l) $0 - 2z$

Answers. (a) $4 + (-7)$. (e) $5 + 3$. (k) $-7a + (-4a)$.

6. Diagram $6 - 2 = 4$ as in Fig. 3.7 and compare with the diagram of $6 + (-2)$ in Fig. 3.10.1. Draw two such diagrams for each numerical part in prob. 5.

7. Find the directed distance from the first to the second point in each of the following pairs of points:

(a) (1), (2)	(d) (3), (7)	(g) $(-7), (-3)$	(j) $(-4), (-4)$
(b) (2), (1)	(e) $(-4), (2)$	(h) $(-2), (-3)$	
(c) $(2), (-3)$	(f) $(5), (-1)$	(i) $(5), (-5)$	

Answers. (a) $d = 2 - 1 = 1$. (c) -5 as in Fig. 3.10.2. (e) 6. (g) 4.

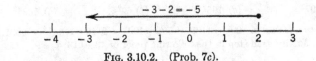

Fig. 3.10.2.   (Prob. 7c).

8. Find the distance in each case in prob. 7.

9. If the temperature changes from $-10°F$ to $36°F$ what is the change in temperature? Write a formula for the change in temperature when the initial and final temperatures are T_1 and T_2.

10. Carry out the following multiplications, explaining in each case the application of the definition of multiplication for integers:

(a) $2(-5)$	(d) $(-59987)(0)(154)$	(g) $[(-2)(1)][(5)(-3)]$
(b) $(-5)(-5)$	(e) $(-2)(-3)$	(h) $[(-1)(-4)][(2)(-3)]$
(c) $(-1)[(-1)(-1)]$	(f) $(-2)(16)$	

Answers. (a) -10. (c) -1. (e) 6. (g) 30.

11. Draw a vector diagram to illustrate $2(-5) = -10$.

12. Carry out the following divisions, and justify your results by reference to the definition.

(a) $2/(-1)$	(c) $0/(-1)$	(e) $32/8$	(g) $(4x)/(-2x)$
(b) $(-6)/3$	(d) $(-14)/(-7)$	(f) $-9/3$	(h) $81ab/9b$

Answers. (a) -2 since $(-1)(-2) = 2$. (c) 0 since $0 (-1) = 0$.

13. Use *3.10.3 to *3.10.9 to simplify:

(a) $(-a)(-b)(-c)$	(d) $(3a)(-2a)(-4)$	(g) $(-2)(-1)^2$
(b) $(-2)(-b)(-1)(-1)$	(e) $(-x)(-3y)(-5x)$	(h) $(-1)^{2n+1}$
(c) $(2x)(-4x)$	(f) $(-1)^5$	(i) $(-1)^{2n}$

Answers. (a) $-abc$. (c) $-8x^2$. (e) $-15x^2y$. (g) -2. (i) 1.

14. Illustrate the identities *3.10.10 to *3.10.13 by replacing the variables by numbers and computing both sides.

15. Rewrite *3.10.10 through *3.10.13, with a, b, c replaced by:

(a) $2, x, y$ (c) $-1, a, b$ (e) $RS, 3T, U$ (g) $3x, -2y, -4$
(b) $1, x, y$ (d) $2x, b, d$ (f) $2x, 3y, 1$ (h) $8, 100, 2$

16. Use *3.10.10 to compute:

(a) $3(98)$ (b) $4(97)$ (c) $5(196)$ (d) $(60)(49)$ (e) $(32)(99)$

Solution to part a. $3(98) = 3(100 - 2) = 300 - 6 = 294.$

17. Use *3.10.11 to compute:

(a) $98^2 - 97^2$ (c) $47^2 - 45^2$ (e) $102^2 - 98^2$
(b) $17^2 - 16^2$ (d) $152^2 - 137^2$ (f) $345^2 - 216^2$

Solution to part a. $98^2 - 97^2 = (98 + 97)(98 - 97) = 195.$

18. Make diagrams for *3.10.11 and *3.10.12 similar to Fig. 3.6.

19. Use *3.10.12 to compute:

(a) 99^2 (c) 93^2 (e) 48^2 (g) 995^2
(b) 97^2 (d) 195^2 (f) 69^2 (h) 1025^2

Solution to part a. $99^2 = (100 - 1)^2 = 100^2 - 2 \cdot 100 \cdot 1 + 1^2 = 9801.$

20. Justify each step in the following proof of *3.10.13:

(1) $-(a + b + c + \cdots) \equiv (-1)(a + b + c + \cdots)$
(2) $\equiv (-1)a + (-1)b + (-1)c + \cdots$
(3) $\equiv (-a) + (-b) + (-c) + \cdots$

21. Expand and simplify:

(a) $-(a - b)$
(b) $-(-B - A)$
(c) $-[s - (-3)]$
(d) $-[t - (1 - x)]$
(e) $-[u - (a - b)]$
(f) $-[5 - (rs + 3)]$
(g) $-2a[b - (3 - b)]$
(h) $-[-5 + 6s - (rt + 2)]$
(i) $-(2 - 3y + 7c)$
(j) $-4s(b + c - a)$
(k) $-[b - (a - b) - a]$
(l) $(2s - 45) - (2 - s)$
(m) $a(b - c) - c(a - b) - b(c - a)$
(n) $(x - y) - (y - z) - (z - x)$
(o) $(1 - x) - 2(3 - x) - 4(x - 1)$
(p) $-x - y - z - 3$
(q) $1 - c(a + b) + ac$
(r) $-(a + b) - 2(a - b)$

Answers. (a) $b - a$. (c) $-3 - s$. (e) $-u + a - b$. (g) $2a(3 - 2b)$. (i) $3y - 2 - 7c$. (k) $2(a - b)$. (m) $2a(b - c)$. (o) $-(3x + 1)$. (q) $1 - bc$.

22. In the following expressions factor sums and expand products, justifying each step:

(a) $2(3 - x)$
(b) $-3(x - 2)$
(c) $(2 - a)^2$
(d) $2ab - 4ac$
(e) $3a - 4ac$
(f) $3st - 2sv$
(g) $-21x - 7y$
(h) $2x - x + x^2$
(i) $4x^2 - 12x + 9$
(j) $9y^2 - 6y + 1$
(k) $ax - 4ax + axt - 6axt + axs$
(l) $1 - 2x + x^2$
(m) $9a^2b^2 - 12ab + 4$
(n) $(2s - 1)(2s + 1)$
(o) $(s + 2t)(s - 2t)$
(p) $(R + Su)(R - Su)$

(q) $(ax + b)(b - ax)$

(r) $9 - y^2$

(s) $16x^2 - y^2$

(t) $4a^2 - 25b^2$

(u) $2x^2 - 2y^2$

(v) $4 - (x - 1)^2$

(w) $(x - 3cd)^2$

(x) $(a^2 + 1)(a^2 - 1)$

Answers. (a) $6 - 2x$ (*3.10.10 with 2, 3, x for a, b, c). (c) $4 - 4a + a^2$ (*3.10.12 with 2, a for a, b). (i) $(2x - 3)^2$ (*3.10.12 with $2x$, 3 for a, b). (o) $s^2 - 4t^2$ (*3.10.11 with s, $2t$ for a, b).

23. Factor:

(a) $a^2 + b^2 - c^2 + 2ab$ (b) $a^4 + a^2b^2 + b^4$ (c) $9x^4 + 4y^4 + 3x^2y^2$

Suggestion for part b. Add and subtract a^2b^2 and consider the result as the difference of two squares.

24. Expand, justifying each step:

(a) $5A(16A - 3)$

(b) $-2[s(3a - 4) - t(b - a)]$

(c) $(-a - b)(a + b)$

(d) $(a + 2s)(b + c)(b - c)$

(e) $(a + 2b + c)[(a + 2b) - c]$

(f) $(2x + y - 1)(2x + y + 1)$

(g) $(a + b + c)(a - b - c)$

(h) $(a + 4t - c)(a - 4t + c)$

(i) $(2s - t + u)(2s + t - u)$

(j) $(a - b + c)(b + a - c)$

(k) $(a - b)(a^2 + ab + b^2)$

(l) $(a + b)(a^2 - ab + b^2)$

(m) $(x^2 - 2x - 1)(3 - x + x^3 - 2x^7)$

(n) $(1 + x + 3x^2 + 3x^3 + 4x^4)(x^2 - 1)$

Answers. (a) $80A^2 - 15A$. (c) $-(a^2 + 2ab + b^2)$. (e) $(a + 2b)^2 - c^2$. (g) $a^2 - (b + c)^2$. (i) $4s^2 - (t - a)^2$. (k) $a^3 - b^3$.

25. Factor:

(a) $16b^2 - x^2 - 2xy - y^2$

(b) $x^2 + y^2 + 2xy - 1$

(c) $2ab - c^2 + a^2 + b^2$

(d) $\dfrac{x^2}{9} + \dfrac{y^2}{-4}$

(e) $a^2x^2 - b^2$

(f) $ar - bt + at - br$

(g) $(x + y)^2 - (x - y)^2$

(h) $(2s - t)^2 - (t + 2s)^2$

(i) $(a - b)(a + b) + (a - b)^2$

(†j) $a^3 - b^3$

(k) $x^2 + y^2 + z^2 - 2xy - 2xz + 2yz$

(l) $4s^2 + 4xt + 4t^2 + x^2 - 8st - 4sx$

Answers. (a) $(4b + x + y)(4b - x - y)$. (c) $(a + b + c)(a + b - c)$. (e) $(ax + b)(ax - b)$. (g) $4xy$. (i) $2a(a - b)$. (k) $(x - y - z)^2$.

† 26. Prove *3.10.10 to *3.10.12 from previously stated theorems. (*Suggestion.* *3.10.1 is particularly useful.)

† 27. Illustrate *3.10.14 by sketching on an axis both sides of the following inequalities and the results of multiplying both sides as indicated:

(a) $1 < 6$ by 2

(b) $-2 < -1$ by 3

(c) $-3 < -2$ by -2

(d) $-3 < 2$ by -3

(e) $4 < 5$ by -4

(f) $-5 < -2$ by -1

† 28. Derive *3.10.4 to *3.10.6.

† 29. Prove *3.10.3.

30. If the area of a circle is given by πr^2, where r is the radius, show that the area between two concentric circles of radius R and r is given by $\pi(R + r)(R - r)$.

31. Show that:

$$(x_1{}^2 + y_1{}^2 + z_1{}^2)(x_2{}^2 + y_2{}^2 + z_2{}^2) - (x_1x_2 + y_1y_2 + z_1z_2)^2$$
$$\equiv (x_1y_2 - x_2y_1)^2 + (y_1z_2 - y_2z_1)^2 + (z_1x_2 - z_2x_1)^2$$

32. Find the error in the following reasoning. Let $A = B$. Then $AA = AB$ and $AA - AB = 0$. This means that $A(A - B) = 0$. Dividing both sides by $A - B$, we find $A = 0$! Hence any number whatever is shown to be equal to zero!?

33. Construct along two straight edges scales of the integers with the same unit length and show how to use them to perform additions and subtractions.

34. Show that if $x > 1$, then $x - 1 > 0$, $x + 2 > 3$, $x - 4 > -3$, $2x > 2$, $-x < -1$, and $x > 0$.

35. Show that $a < b \longleftrightarrow c - a > c - b$.

36. Prove the following:

(a) $ab > 0 \longleftrightarrow a/b > 0$

(b) $ab > 0 \longleftrightarrow$ ($a > 0$ and $b > 0$) or ($a < 0$ and $b < 0$)

(c) $a < b$ and $c < d \longrightarrow\!\!\!/ \;\; a - c < b - d$

(d) $a < b$ and $c < d \longrightarrow a - d < b - c$

†37. Show that:

*3.10.15 $0 \leq a < b$ and $0 \leq c < d \longrightarrow ac < bd$

(*Hint.* $ac \leq bc$ and $bc < bd$.)

†38. Prove *3.10.14 by considering special cases.

39. Find the error in the following. We know that $ad = bc$ implies $a/b = c/d$. Also, if $a > b$, then $c > d$ because of this proportion. Now let $a = d = 2$ and $b = c = -2$. Then $ad = bc$ and $a > b$. Hence $-2 > 2!!?$

40. Interpret geometrically and sketch:

(a) $x - y = 2$	(f) $x - 3 = 4$	(k) $\lvert 3 - 6 \rvert = 3$
(b) $\lvert x - y \rvert = 2$	(g) $x + 2 = 5$	(l) $6 - 3 = 3$
(c) $y - x = -2$	(h) $\lvert x - a \rvert = 6$	(m) $3 - 6 = -3$
(d) $y - x = 2$	(i) $\lvert -2 \rvert = 2$	(n) $\lvert 0 - 8 \rvert = 8$
(e) $\lvert x - 2 \rvert = 1$	(j) $\lvert 6 - 3 \rvert = 3$	(o) $5 - 7 = -2$

Solutions. (a) The point x lies 2 units to the right of the point y. (e) The point x lies at a distance of 1 from the point 2. It may be either to the right or to the left. (g) x is a distance 2 to the left of 5.

†41. If AB represents the directed distance from A to B: (a) Show that $AB = -BA$. (b) Explain why $\lvert AB \rvert$ represents the undirected distance and show that $\lvert AB \rvert = \lvert BA \rvert$. (c) If A, B, and C are any three points on a line show that $AB + BC = AC$. (d) Under what conditions will $\lvert AB \rvert + \lvert BC \rvert = \lvert AC \rvert$?

42. Prove:

*3.10.16 $|a - b| \equiv |b - a|$

*3.10.17 $|a + b| \leq |a| + |b|$

*3.10.18 $|a + b| \geq |a| - |b|$

*3.10.19 $|ab| \equiv |a| \cdot |b|$

*3.10.20 $|a/b| \equiv |a| / |b|$

43. Under what condition is each of the following true?

(a) $|x| = -x$ (d) $x - 1 = |1 - x|$ (g) $|x - y| = y - x$

(b) $|2x| = 2x$ (e) $|1 - x| = 1 - x$ (h) $|xy| = xy$

(c) $|x + 1| = x + 1$ (f) $|x - y| = x - y$ (i) $|xy| = -xy$

Answers. (a) $x \leq 0$. (c) $x + 1 \geq 0$. (e) $x \leq 1$. (g) $x \leq y$.

†**44.** Suppose that x and y are numbers and that $\max(x, y)$ and $\min(x, y)$ stand for the larger and the smaller of the two (or either one if they are equal). Show that:

(4)
$$\max(x, y) = \tfrac{1}{2}(x + y + |x - y|)$$
$$\min(x, y) = \tfrac{1}{2}(x + y - |x - y|)$$

†**45.** A set is called an **integral domain** if operations of addition and multiplication are defined so that the following propositions hold:

(I) For any two elements of the set there are a unique sum and a unique product, which are in the set.

(II) Addition and multiplication are commutative and associative.

(III) Multiplication is distributive over addition.

(IV) There are elements e and e' in the set such that for any element a in the set

$$e + a = a + e = a \quad \text{and} \quad e'a = ae' = a$$

(V) For any element a in the set there is an element $\bar{a}$ in the set such that $a + \bar{a} = \bar{a} + a = e$.

(VI) For $c \neq 0$ and any elements a, b in the set

$$ac = bc \longrightarrow a = b$$

(a) Show from the properties of the integers that they form an integral domain. (*Hint.* $e = 0$ and $e' = 1$.)

(b) Starting from (I)–(VI) as axioms, derive *3.8.3 and *3.10.3. [Actually all the identities can be derived from (I)–(VI).]

(c) Show that the following sets are *not* integral domains: the positive integers, the even integers, the odd integers.

3.11 The rationals

In the last section we extended the number system to include the positive and the negative integers and zero. The sum, product, or difference of two of these numbers is again such a number, but the quotient of two integers is not necessarily an integer. Hence we must either consider such an expression as 2/3 to be meaningless or extend to it the idea of number. To carry out this extension we define equality, inequality, addition, subtraction, multiplication, and division for expressions of the form a/b, with a and b integers and $b \neq 0$. The new numbers are called **rationals.** They include the integers, satisfy the same basic algebraic laws, and have a similar geometric interpretation.

In 3.7 we found that when a/b and c/d are integers, $a/b = c/d$ if and only if $ad = bc$. We take this as the definition of **equality for rationals,** that is,

*3.11.1 $\dfrac{a}{b} = \dfrac{c}{d} \longleftrightarrow ad = bc$ (Definition of Equality)

Thus $2/3 = 4/6$ by definition, since $2 \cdot 6 = 3 \cdot 4$. Among the rationals are those whose denominators are 1. We identify $a/1$ with the integer a. (Compare *3.7.6.) With this understanding the rationals include the integers as special cases. It follows immediately that *3.7.10 ($ca/cb \equiv a/b$) holds for all integral a, b, and $c \neq 0$.

A statement of equality between two fractions, that is, expressions of the form a/b, is sometimes called a **proportion.** In this language, $a/b = c/d$ is called a proportion between the ratios a/b and c/d. The proportion may be written $a{:}b{:}{:}c{:}d$ and is read "a is to b as c is to d." The numbers b and c are called **means;** a and d are called **extremes.** The definition of equality for rationals becomes in the proportion language: "The product of the means equals the product of the extremes." It is also sometimes said that a, b, c, d are in proportion. When several ratios are equal, say $a/b = c/d = e/f$, it is said that the numerators are proportional to the denominators, that is, a, c, and e are **proportional** to b, d, and f. The usual language in terms of the equality of fractions is preferable, especially for computations, and it is advisable to translate proportionality statements into the more familiar and tractable equations involving fractions.

We now define the **sum of two rationals** with the same denominator in conformity with the usual rule: "Add numerators over the common denominator." That is,

*3.11.2 $\dfrac{a}{c} + \dfrac{b}{c} \equiv \dfrac{a+b}{c}$ (Definition of Addition)

With this definition we can find the sum of *any* two rationals. For by *3.7.10, $a/b + c/d = ad/bd + cb/bd$. Now the two fractions have the same denominator and *3.11.2 yields:

*3.11.3 $\dfrac{a}{b} + \dfrac{c}{d} \equiv \dfrac{ad+bc}{bd}$

We define **multiplication of rationals** according to the familiar rule: "Multiply numerators and denominators." That is,

*3.11.4 $\left(\dfrac{a}{b}\right)\left(\dfrac{c}{d}\right) \equiv \dfrac{ac}{bd}$ (Definition of Multiplication)

Subtraction and division are now defined as the inverses of addition and multiplication. (See 3.7.) We define inequality of rationals in the next section. Then it can be shown that the rationals obey all the numbered theorems of this chapter. This means that the variables in these theorems may be thought of as standing for any rationals. In particular, the definitions of this section become true statements when the variables are replaced by any rationals.

The most useful relation in dealing with fractions is the identity $ca/cb \equiv a/b$ (*3.7.10). Used judiciously, it can save much trouble. For example, in order to simplify

(1) $\dfrac{1}{\dfrac{1}{a} + \dfrac{1}{b}}$

we multiply numerator and denominator by ab as follows:

(2) $\dfrac{1}{\dfrac{1}{a} + \dfrac{1}{b}} \equiv \dfrac{ab(1)}{ab\left(\dfrac{1}{a} + \dfrac{1}{b}\right)}$

(3)
$$\equiv \frac{ab}{\dfrac{ab}{a} + \dfrac{ab}{b}}$$

(4)
$$\equiv \frac{ab}{b + a}$$

Note that we used the distributive law to get from (2) to (3), but that we used *3.7.10 to write (2) and to get from (3) to (4). The final result can be written in a single step if the procedure is understood. This method is preferable to adding terms in the denominator and then "inverting."

Besides extending old identities to the new numbers, we can state theorems that would not have had meaning before. For example, we can deal with numbers of the form $1/b$, where b is any integer or any rational. The number $1/b$ is called the **reciprocal** of b. Its fundamental property is

*3.11.5
$$\left(\frac{1}{b}\right) b \equiv b\left(\frac{1}{b}\right) \equiv 1$$

This theorem states that the product of any number and its reciprocal is 1. Other identities are:

*3.11.6
$$\left(\frac{a}{b}\right)\left(\frac{b}{a}\right) \equiv 1$$

*3.11.7
$$\frac{1}{\dfrac{a}{b}} \equiv \frac{b}{a}$$

*3.11.8
$$\frac{a}{b} \div \frac{c}{d} \equiv \frac{ad}{bc}$$

*3.11.9
$$\frac{a}{b} \equiv \left(\frac{1}{b}\right) a$$

The first two of the foregoing identities are really different ways of saying that the reciprocal of a/b is b/a. The third is the familiar rule: "To divide one fraction by another, invert the divisor and multiply." The last means that we may consider division by a number to be the same as multiplication by its reciprocal. This is similar to thinking of a subtraction as addition of the negative.

When adding fractions the student should keep in mind the fundamental definition for fractions with equal denominators. By appropriate multiplications of numerators and denominators he should rewrite all fractions with equal denominators and then use this definition. Of course *3.11.2 can be extended to any number of fractions with equal denominators, that is,

*3.11.10
$$\frac{a}{e} + \frac{b}{e} + \frac{c}{e} + \cdots \equiv \frac{a + b + c + \cdots}{e}$$

EXERCISE 3.11

1. From *3.11.1 show that:

(a) $2/3 = -2/(-3)$ (d) $ab/a \equiv b$ (g) $2x/4y \equiv x/2y$
(b) $-7/2 = 7/(-2)$ (e) $ba/a \equiv b$ (h) $-as/st \equiv a/(-t)$
(c) $100/75 = 12/9$ (f) $abc/ac \equiv b$ (i) $(x-y)/(y-x) \equiv -1$

2. From *3.11.1 show that *3.7.10 holds for any integral a, b, c.

3. Translate from the proportion language into the equation language: (a) 2:3::4:6. (b) x is to y as 2 is to 3. (c) The numbers 5, 6, 10, and 12 are in proportion. (d) R is the fourth proportional to a, b, and c. (e) The ratio of a to b is 3. (f) 1, 4, and -1 are proportional to 2, 8, and -2. (g) x, $2a$, and y are proportional to $-x$, $-2a$, and $-y$. (h) r, s, and t are proportional to 3, -1, and 2. [*Answers.* (a) $2/3 = 4/6$. (c) $5/6 = 10/12$. (e) $a/b = 3$. (g) $x/(-x) = 2a/(-2a) = y/(-y)$.]

4. Rewrite *3.11.2 and *3.11.4, with a, b, c, d replaced by (a) 2, 3, 4, 1. (b) 2, a, -5, s. (c) s, t, $a + b$, a. (d) $2a$, $b + c$, $3b$, $c - b$.

5. Perform the following additions, justifying each step:

(a) $2/3 + 5/3$ (e) $5/4 - 7/3$ (i) $2 + 1/3$
(b) $2/3 - 4/3$ (f) $15/(-2) + 5/2$ (j) $2/5 - 3 + 3/2$
(c) $8/5 - 3/5$ (g) $-4/5 - 1/6$ (k) $ab/c - b/c$
(d) $1/3 + 1/2$ (h) $-9/2 + 1/4$ (l) $2x/3 + x/2 + 1/3$

Answers. (a) 7/3. (c) 1. (e) $-13/12$. (g) $-29/30$. (i) 7/3. (k) $b(a - 1)/c$.

6. Show that $1/a + b \not\equiv 1/(a + b)$.

7. Perform the following multiplications and divisions, justifying each step:

(a) $5(2/15)$ (f) $(ab/s)(1/s)$ (k) $(2/3)/(-1/2)$
(b) $(2/5)(3/4)$ (g) $(2/3)/6$ (l) $(1/7)(2/9)(21/2)$
(c) $(2/3)(2/3)$ (h) $(9/8)(2/7)$ (m) $(6/5)(1/3)(9/4)$
(d) $(3/2)^2$ (i) $(5/4)(18/14)$ (n) $(a/b)/c$
(e) $(-1/2)(2/3)$ (j) $(c/ab)/c$ (o) $(1/2)/[(4/3)(-5/4)]$

Answers. (a) 2/3. (c) 4/9. (e) $-1/3$. (g) 1/9. (i) 45/28. (k) $-4/3$. (m) 9/10. (o) $-3/10$.

8. Find the reciprocals of:

(a) 3
(b) −3
(c) −0.1
(d) 3/7

(e) −14/11
(f) 27/10
(g) st/b
(h) $(a + b)/2$

(i) $(a + b)/(b − a)$
(j) $(a/b)/(c/d)$

9. Complete the following by replacing x by the proper number:

(a) $2/3 = x/6$ (c) $4/7 = 12/x$ (e) $7/5 = x/21$ (g) $x/2 = 3/7$
(b) $1/9 = 4/x$ (d) $7/3 = x/9$ (f) $21/12 = x/60$ (h) $1/x = 5/6$

Answers. (a) 4. (c) 21. (e) 147/5. (g) 6/7.

10. In order to add fractions by using the definition, it is necessary to rewrite them with a common denominator. Any common denominator will do, but it is neater to use the smallest number that is a multiple of all the denominators. For example,

$$(5) \qquad \frac{5}{6} + \frac{3}{2} = \frac{5}{6} + \frac{9}{6} = \frac{14}{6} = \frac{7}{3}$$

We note that 6 is a multiple of 2. Hence we take it as common denominator and have only to multiply numerator and denominator of the second fraction by 3. In order to find the *least common denominator* we can factor each of the denominators and then build up the common denominator by multiplying together all different prime factors that appear. A factor that appears more than once in any denominator is included in the common denominator the *greatest* number of times it appears in *any one* denominator. For example, in $2/3 + 5/6 + 1/7 + 2/9$, the factored denominators are 3, 2·3, 7, and 3·3. The number 3 appears in three denominators, but the greatest number of times it appears in any one is 2. Hence the least common denominator is $2·3^2·7 = 126$. We multiply numerator and denominator of each fraction by the factors that *do not* appear in its denominator:

$$(6) \qquad \frac{2}{3} + \frac{5}{6} + \frac{1}{7} + \frac{2}{9} = \frac{(42)2}{(42)3} + \frac{(21)5}{(21)6} + \frac{(18)1}{(18)7} + \frac{(14)2}{(14)9}$$

$$(7) \qquad = \frac{84 + 105 + 18 + 28}{126} \qquad (*3.11.10)$$

Do the following additions:

(a) $2/3 + 1/6 − 3/5$
(b) $1/9 + 1/4 + 7/2$
(c) $14/3 − 11/10 + 3/11$
(d) $7/8 + 1/2 − 3/4$
(e) $3 + 2/11 − 1/2$

(f) $2/2a + 1/2ab$
(g) $3/2 + 1/x + 1/3$
(h) $a/b + 3/2$
(i) $5/8 + 3(2/3) + (4/7)(5/2)$
(j) $5/12 − 1/18 + 4/15$

Answers. (a) 7/30. (c) 1267/330. (e) 59/22. (g) $(11x + 6)/6x$.

11. Sometimes an integer plus a fraction is written without a plus sign, e.g., $2\frac{1}{2}$ means $2 + \frac{1}{2}$. Such expressions are called "mixed numbers." Except for very simple cases, mixed numbers should be avoided and the number written as a single fraction or decimal. Write the following as single fractions:

(a) $3\frac{1}{2}$ (d) $16\frac{1}{4}$ (g) $1\frac{7}{16}$ (j) $-8\frac{2}{3}$

(b) $5\frac{2}{3}$ (e) $8\frac{2}{5}$ (h) $9\frac{4}{15} + \frac{2}{3}$ (k) $14\frac{3}{32}$

(c) $-10\frac{1}{2}$ (f) $5\frac{3}{8}$ (i) $2\frac{1}{2} - 5\frac{1}{3}$ (l) $5\frac{2}{9}$

Answers. (a) 7/2. (c) $-21/2$. (e) 42/5. (g) 23/16. (i) $-17/6$.

12. Expand and simplify, giving reasons:

(a) $22/33$

(b) $-2/(-3)$

(c) $81/27$

(d) $-63/(-28)$

(e) $a(x-y)/b(x-y)$

(f) $(a-b)/(b-a)$

(g) $2(a^2-b^2)/(a-b)$

(h) $1/a + 1/b$

(i) $1/a + 1/b + 1/c$

(j) $a/b - 1$

(k) $x/y + 2/y^2$

(l) $(a/2)(2/ab)(4/b)$

(m) $(1/s)(st + 2s)$

(n) $\left(\dfrac{1}{x}+\dfrac{1}{y}\right)\left(\dfrac{1}{x}-\dfrac{1}{y}\right)$

(o) $\left(\dfrac{1}{x}+2\right)^2$

(p) $\left(1-\dfrac{x}{2}+\dfrac{y}{3}\right)^2$

(q) $a/b + c/b^2 + d/3$

(r) $(a+b+c)/3a + (b+c)/3$

(s) $abc(x^2 + 20x + 100)/(xb + 10b)$

(t) $\left(\dfrac{x^2}{y^2}-\dfrac{y^2}{x^2}\right)\bigg/\left(x+\dfrac{y^2}{x}\right)$

(u) $(1/2)(1/3 + 3/2)$

(v) $(5/6)(1/5 + 2/3)$

(w) $\left(-\dfrac{1}{2}+\dfrac{13}{5}\right)\bigg/\left(1+\dfrac{2}{5}\right)$

(x) $\left(\dfrac{1}{3}+\dfrac{4}{7}-\dfrac{1}{2}\right)\bigg/\left(2-\dfrac{1}{6}\right)$

(y) $8(y-2)^2/(y^2-4)$

(z) $(cy^2 + cx^2 + 2axy + ay^2 + 2cxy + ax^2)/(a+c)$

Answers. (a) 2/3. (c) 3. (e) a/b. (g) $2(a+b)$. (i) $(ab + bc + ca)/abc$. (k) $(xy + 2)/y^2$. (m) $t + 2$. (o) $(1 + 4x + 4x^2)/x^2$. (q) $(3ab + 3c + db^2)/3b^2$. (s) $ac(x + 10)$. (u) 11/12. (w) 3/2. (y) $8(y - 2)/(y + 2)$.

13. Cite the theorems or definitions that justify the steps in the following proof of *3.11.5, and construct similar proofs of *3.11.6 through *3.11.9.

(8) $$\left(\frac{1}{b}\right)b \equiv \left(\frac{1}{b}\right)\left(\frac{b}{1}\right)$$

(9) $$\equiv \frac{1\cdot b}{b\cdot 1}$$

(10) $$\equiv \frac{b}{b}$$

(11) $$\equiv 1$$

14. In $a/b = c/d$, d is called the **fourth proportional** to a, b, and c. Find the fourth proportional to 3, 8, and 10.

15. In $a/x = x/d$, d is called the **third proportional** to a and x, and x is called the **mean proportional** to a and d. Find the third proportional to 2 and 4 and the mean proportional to 6 and 2/3. (*Answers.* 8 and 2.)

16. Prove that $\dfrac{E}{R + (r/2)} \equiv \dfrac{2E}{r + 2R}$.

17. If $a/b = c/d$, show that:

(a) $b/a = d/c$ (Called "inversion" in the proportion language.)
(b) $a/c = b/d$ (Called "alternation.")
(c) $(a + b)/b = (c + d)/d$ (Called "composition.")
(d) $(a - b)/b = (c - d)/d$ (Called "division.")
(e) $(a + b)/(a - b) = (c + d)/(c - d)$ (Called "composition and division.")

18. Show that if $a/b = c/d = e/f$, then:

$$\frac{a + c + e}{b + d + f} = \frac{a}{b}$$

Restate this result in the proportion language.

19. Prove that:

$$\frac{a + b}{c + d} \not\equiv \frac{a}{c} + \frac{b}{d}$$

20. Consider the following argument: To compute $(a^2 - b^2)/(a - b)$ we note that a^2 over a is a, b^2 over b is b, and minus over minus is plus. Hence the result is $a + b$. Is the answer correct? Is the argument correct? (See 2.6 again for a discussion of the fallacy that a statement is valid if it leads to a true result.)

21. What theorem about fractions is equivalent to the rule of thumb that factors common to numerator and denominator may be canceled? Is it true that any term that appears in both terms of a fraction may be canceled? Illustrate your answer by discussing the following:

(a) $\dfrac{\cancel{a} + 1}{\cancel{a}} \overset{?}{=} \dfrac{1 + 1}{1} = 2??$ (c) $\dfrac{2a\cancel{x} + 2\cancel{x} + 2a}{2\cancel{x}} \overset{?}{=} a + a = 2a??$

(b) $\dfrac{\cancel{a} - \cancel{b}}{\cancel{a} + \cancel{b}} \overset{?}{=} -1??$ (d) $\dfrac{3\cancel{a}b + \cancel{a}}{\cancel{a}} \overset{?}{=} 3b??$

†**22.** Show that rationals with denominators equal to 1 behave under addition and multiplication exactly like the corresponding integers.

†**23.** A set is called a **field** if it has the properties I–V of an integral domain (Ex. 3.10.45) and also:

(VI′) For every element a there is an element a' such that $aa' = a'a = e'$.

(a) Show that the rationals form a field.
(b) Show that the integers do not form a field.
(c) Show that any field is an integral domain, but not conversely.

†**24.** If the field properties are taken as axioms for rationals, all the identities of this chapter can be derived from them. Prove in this way *3.11.1, *3.11.2, *3.11.3.

†25. Show that equality as defined by *3.11.1 is an equivalence relation. (Note that this equality is not the same as identity since two fractions can be equal without being identical, i.e., $a/b = c/d \; \not\leftrightarrow \; a = c$ and $b = d$.)

3.12 Interpretation of rationals

Any rational may be written as the quotient of two positive integers, preceded or not by a minus sign. (See *3.10.8 and *3.10.9.) If the numerator and denominator have a common factor (other than 1), we may eliminate it by applying *3.7.10. In this way, we can rewrite any rational in the form m/n or $-(m/n)$, where m and n are positive integers with no common factor other than 1. Such a fraction is said to be in **lowest terms.** Those of the form m/n we call **positive,** and those of the form $-(m/n)$ we call **negative.**

This argument shows that any rational is equal to some fraction in lowest terms. On the other hand, no two fractions in lowest terms can be equal unless they are identical. To prove this, we assume $m/n = m'/n'$. (Read "m-prime" and "n-prime.") From the definition of equality, $mn' = nm'$. Since the two members are equal, they must have the same factors. But m and n have no common factors, and m' and n' have no common factors. Hence the factors of m and m' must be identical. It follows that $m = m'$ and that $n = n'$. The set of fractions in lowest terms thus includes all the essentially different rational numbers in the sense that every quotient of integers equals one of them, and no two of them are equal. (That each rational can be represented in many ways as the quotient of two integers is no more surprising than that each integer can be represented in many ways as the difference of two positive integers.)

We now wish to give a geometric picture of rationals that fits our interpretation of integers. Since $n(1/n) = 1$, we may consider $1/n$ the vector that yields 1 when laid out n times. Such a vector may be found by dividing the segment between the origin and the point 1 into n parts. We diagram $1/5$ in Fig. 3.12a. Since m/n can be thought of as the product $m(1/n)$, we may interpret m/n as a vector found by laying out the vector $1/n$ m times. The ratio $7/4$ is sketched in Fig. 3.12b. An alternative interpretation of m/n follows directly from the definition of division. Since $n(m/n) = m$, m/n is a vector that gives m when laid out n times. We find m/n by "dividing" the vector m into n parts. The fraction $7/4$ is sketched this way in Fig. 3.12c.

We interpret negative rationals according to the identity $-(m/n)$ $= (-1)(m/n)$, where the factor -1 reverses the direction of the vector. Hence $-(m/n)$ is a vector of length m/n pointing to the left. The point $-(m/n)$ and the point m/n are at the same

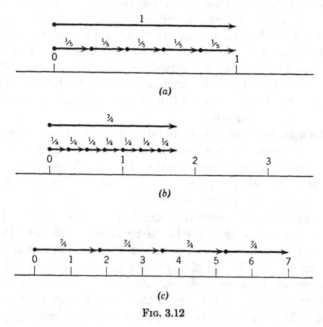

Fig. 3.12

distance from the origin, but in opposite directions. As in the case of integers, we call the positive number of this pair the absolute value of either one. (Compare 3.9.) With this interpretation of negative rationals, we can locate a point corresponding to every rational.

Since the rationals are thought of as laid out on a line, it is natural to extend to them the idea of order that we established for integers. We took $a < b$ to mean that the point a lies to the left of the point b. Another way of saying this is that there is a positive number x such that $a + x = b$, or that $b - a$ is positive. With this notion of inequality, any negative number is less than any positive number, the greater of two positives is the one with the greater absolute value, and the greater of two negatives is that with the smaller absolute value. We now carry over these ideas to the rationals. First we define inequality for positive rationals. Since we wish $m/n < m'/n'$ to mean $m'/n' - m/n > 0$, we note that $m'/n' - m/n = (m'n - n'm)/nn'$. Since all variables stand for positive integers, the last fraction is positive or negative according to whether $m'n - n'm$ is positive or negative. This suggests:

*3.12.1 $$m/n < m'/n' \longleftrightarrow mn' < nm'$$

(Definition of Inequality for Positive Rationals)

The definition holds whether or not the rationals are in lowest terms. It could be extended to the case $m = 0$, but it is easier to agree that any positive rational is to be considered greater than 0.

The foregoing definition will not do for negatives. For example, $2/(-3)$ lies to the left of $1/2$; hence we would want to have $2/(-3) < 1/2$. But we do not have $2 \cdot 2 < (-3) \cdot 1$, which is what *3.12.1 would require. Since the application of *3.12.1 to negative rationals would not correspond to our ideas of order, we make use of the concepts that apply to integers. When two rationals have opposite signs we say that the negative one is less. Of two negative rationals, we say that the smaller is the one with the larger absolute value. With this understanding, the geometric interpretation of inequality and the theorems about inequalities apply to all rationals.

The points corresponding to rational numbers are rather thickly clustered on the axis. Suppose that $a < b$, where a and b are rationals. Then $c = (a + b)/2$ is rational, and

(1) $$c = \frac{a + b}{2} = a + \frac{(b - a)}{2} > a \qquad \text{(Why?)}$$

(2) $$= b - \frac{(b - a)}{2} < b \qquad \text{(Why?)}$$

Hence $a < c < b$, and the point c lies in the segment joining a and b. Evidently, we can always find a rational between any two given rationals. We could continue by finding a rational d between a and c, then a rational e between a and d, etc., without limit. In this way we could find a rational as close to a as we like. Geometrically this means that any interval on the axis contains an unlimited (infinite) number of rational points, and that we can find a rational point as close as we like to a given rational point. This property is described by saying that the rational numbers are a **dense** set. Since the rationals are so tightly packed on the line, it might seem that they would include all points on the line. But this is not so! On the contrary, the rational points are only an insignificant portion of all the points on the line, as we shall see in the next section.

EXERCISE 3.12

1. Reduce the following fractions to lowest terms:

(a) 120/80	(d) 42/112	(g) 99/(−33)	(j) 2010/1005
(b) 18/48	(e) 1568/126	(h) 0/27	(k) −82/(−12)
(c) 8/(−4)	(f) (−6)/8	(i) 0/(−1)	(l) 117/39

Answers. (a) $120/80 = 3 \cdot 40/2 \cdot 40 = 3/2$. (c) −2. (e) 112/9. (g) −3. (i) 0. (k) 41/6.

2. Sketch the rationals in prob. 1.

3. Sketch as in Figs. 3.4.1, 3.5, and 3.7:

(a) $3/2 + 2/3 = 13/6$	(e) $3.27 = 3 + 2/10 + 7/100$
(b) $5/6 − 1/3 = 1/2$	(f) $−1.75 = −1 − 0.7 − 0.05$
(c) $4(2/3) = 8/3$	(g) $10.25 = 10 + 0.2 + 0.05$
(d) $3.2 = 3 + 2/10$	(h) $3.2 = 4 − 0.8$

4. Explain how m/n can be found from m and n by ruler and compass. Illustrate by finding 3/4 in two ways.

5. Indicate which of the following are true and give reasons:

(a) $2/3 < 3/4$?	(g) $15/4 > −1/6$?	(m) $0.99 < 3498/3497$?
(b) $−3/2 < 1/2$?	(h) $12/7 > 1/(−2)$?	(n) $1/10 < 1/15$?
(c) $−9/8 < −1$?	(i) $−2/3 > −1/3$?	(o) $−0.5 > −0.02$?
(d) $0.8 \leq 4/5$?	(j) $−1/3 < −2/(−5)$?	(p) $(−0.5)(−0.5) > 0$?
(e) $−1/2 \leq 1/2$?	(k) $38/39 < 40/100$?	(q) $−7/(−17) < 0$?
(f) $−2 = −4/(−2)$?	(l) $2/10 \geq 0.21$?	(r) $281/987 < 1$?

Answers. (a) True since $8 < 9$. (c) True since $9/8 > 1$. (e) True since any negative is less than any positive. (m) True since $0.99 < 1$ and $3498/3497 > 1$.

6. Justify the inequalities (1) and (2) in the section.

7. Prove that $(a + b)/2$ is the midpoint of the segment from a to b by proving that:

$$\left| \frac{a + b}{2} - a \right| = \left| \frac{a + b}{2} - b \right|$$

8. Find a rational point within a distance 0.0000001 of the point 2.3. Write a formula that gives a rational point within a distance d of a given rational point a.

9. The student is assumed to be familiar with the *decimal* way of writing numbers. Thus:

$$24905.3902 = 2(10,000) + 4(1000) + 9(100) + 0(10) + 5$$
$$+ 3/10 + 9/100 + 0/1000 + 2/10,000$$
$$= 2 \times 10^4 + 4 \times 10^3 + 9 \times 10^2 + 5 + 3 \times 10^{-1}$$
$$+ 9 \times 10^{-2} + 2 \times 10^{-4}$$

Here the negative exponents indicate the product of the corresponding number of tens in the denominator. Write out the following decimals in the forms above and give a graphical interpretation:

(a) 235   (c) 0.235   (e) −21.39   (g) 1.481
(b) 23.5   (d) 2.35   (f) 461.3   (h) 894.01003

Answers. (a) $2 \times 10^2 + 3 \times 10 + 5$. (c) $2 \times 10^{-1} + 3 \times 10^{-2} + 5 \times 10^{-3}$.
10. Compute:

(a) 34.284 − 26.12 − 15.4   (d) 6.518 + 4.2813 − 35.880
(b) 182.9 + 0.003 + 5.109   (e) 10.0000 − 8.3819
(c) 193.218 − 264.50   (f) 43.00591 − 0.004889 + 500.1

Answers. (a) −7.236. (c) −71.282. (e) 1.6181.
11. State the rule for placing the decimal point in a product and compute:

(a) (235.08)(51.1)   (c) (189)(0.09132)
(b) (0.00189)(0.0399)   (d) (0.426)(0.20001)

Answers. (a) 12012.588. (c) 17.25948.
12. Find the products of the numbers appearing in each part of prob. 10. [*Answers.* (a) 13790.670432. (c) −51106.161.]
13. Any terminating decimal is equal to a quotient of integers, the denominator being some power of ten. Thus $2.98 = 298/100$, where the second member is found by multiplying and dividing by 100. Write the following decimals as fractions:

(a) 1.1   (c) 0.112   (e) 0.01234   (g) 1.00503
(b) 9.632   (d) 374.11   (f) 196.00001   (h) −52.8007

Answers. (a) 11/10. (c) 112/1000. (e) 1234/100,000.
14. State a rule for dealing with the decimal point in division and find to four decimal places:

(a) (437.1)/(27.990)   (c) (2886.52)/(0.0863)
(b) (0.01829)/(0.000364)   (d) (0.4897)/(0.00608)

Answers. (a) 15.6163. (c) 33,447.5087.
15. Find the reciprocal to five decimal places of each number in prob. 13. (Check by multiplication.)
16. If a fraction can be written so that its denominator is some power of ten, it can be expressed as a terminating decimal. This can be done by long

division, which will come out even, or by multiplying numerator and denominator so as to get a power of ten in the denominator. Thus $1/8 = 125/125 \cdot 8 = 125/1000 = 0.125$. Use both methods to find the decimal equivalents of the following:

(a) 3/8	(c) 7/20	(e) 73/80	(g) 1/32
(b) 9/4	(d) 1/16	(f) 1/125	(h) 3/64

Answers. (a) 0.375. (c) 0.35. (e) 0.9125.

17. In order for the foregoing process to be possible, it is necessary that the denominator have only 2's and 5's as prime factors. (Why?) Otherwise long division will not come out even, but will eventually begin to repeat. It is customary to write dots over the group of digits which repeats. Thus $5/3 = 1.\dot{6}$ means that division of 5 by 3 yields an unlimited number of sixes. Sometimes this is written $1.666 \cdots$. Such an expression is called a repeating decimal. Find as repeating decimals:

(a) 10/3	(c) 1/7	(e) 29/26	(g) 215/211
(b) 1/6	(d) 1/9	(f) 3/17	(h) 1/37

Answers. (a) $3.\dot{3}$. (c) $0.\dot{1}4285\dot{7}$. (e) $1.1\dot{1}5384\dot{6}$.

18. Two numbers can be compared in various ways. The most common ways are to find their ratio or their difference. If $a/b = x$, we say that a is x **times as big** as b. If $a - b = y$, we say that a is y **more than** b or that b is y **less than** a. Compare the following pairs of numbers in these two ways, letting each in turn play the role of a:

(a) 12, 14	(c) 0.000013, 0.00003	(e) 100.28, 98.28
(b) 200, 204	(d) 1,000,000, 1,000,003	(f) 2.28, 0.28

Answers. (a) 12 is 2 less than 14, 12 is 6/7 as big as 14. (c) 0.00003 is 0.000017 more than 0.000013, 0.00003 is about 2.3 times as big as 0.000013.

19. Numbers, especially proper fractions, are often expressed in **percentage**. This notation is defined as follows:

$$x\% = x/100 \quad \text{(Definition)}$$

Thus 2% ("two per cent") means 2/100, (1/2)% means 1/200, and 500% means $500/100 = 5$. Rewrite the following percentages as rationals in fractional and decimal form:

(a) (2/3)%	(d) 2.37%	(g) 0.05%	(j) −600%
(b) 8%	(e) 1/8%	(h) 0.2%	(k) 0%
(c) 387%	(f) −2%	(i) 1000%	(l) −100%

Answers. (a) $2/300 = 0.00\dot{6}$. (c) 3.87. (e) $1/800 = 0.00125$.

20. Express the following as percentages:

(a) 0.03	(e) 27/63	(i) −21/6	(m) 7.5
(b) 8	(f) 0.012	(j) 39/14	(n) −5/4
(c) 0.001	(g) 0.37	(k) 10	(o) 1/100
(d) 7/50	(h) −3.154	(l) 2/3	(p) 5000

Answers. (a) 3%. (c) 0.1%. (e) (300/7)%. (g) 37%. (i) −350%.

21. The expression "a is $b\%$ of c" means "$a/c = b\% = b/100$." Translate the following into equations:

(a) R is 2.1% of T.

(b) x is 0.31% of y.

(c) AB is $(1/2)\%$ of C.

(d) Q is -21% of U.

(e) P_1 is 543% of P_2.

(f) A if 50% of B.

(g) x is 20% of 15.

(h) y is 3.8% of 1451.

Answers. (a) $R = 0.021T$. (c) $AB = 0.005C$. (e) $P_1 = 5.43P_2$.

22. Thirty club members vote for a resolution and seventy-three vote against it. What percentage is in favor of the resolution? (*Answer.* About 29%.)

23. The *efficiency of a machine* is defined as the ratio of power output to power input. Calculate the per cent efficiency of a machine which uses 5800 watts of electricity to produce 4983 watts of mechanical power.

24. In a series of experiments on muscle efficiency it was found that a man at rest consumed 2397 calories per day. When he did 546 calories of work pedaling a bicycle, he required 5120 calories. What is the efficiency of this man's body as a muscle machine? (*Answer.* 20%.)

25. The *concentration* of a solution is sometimes measured by the ratio of the weight of the solute (dissolved substance) to the total weight of the solution. If 5 grams of salt are dissolved in 75 grams of water, what is the percentage concentration?

26. If 17 out of 1231 tires have to be discarded because of faults in manufacture, what is the percentage of scrap? (*Answer.* 1.4%.)

27. If a family with a yearly income of $3600 spends $55 a month on rent, what percentage of income goes for this purpose?

28. A sample of wheat flour contains 12% protein and 15% moisture. What percentage of the actual flour (without moisture) is protein?

29. A merchant gives a stamp with each 10-cent purchase. The customer pastes the stamps in a book with 25 stamps to the page. When 60 pages are full the book may be redeemed for $3. What is the percentage of saving?

30. Suppose that of 100 people with severe colds 60 are treated with an anti-histamine drug. Of those treated, 13 show immediate improvement; of those not treated, 9 show immediate improvement. Express the results of the experiment in percentages. Can you draw any conclusions?

31. In a differential count of white blood cells, the following figures were found: leucocytes: 4800 per cubic millimeter; lymphocytes: 1640 per cubic millimeter; and other white cells: 670 per cubic millimeter. Calculate the percentage of each in the total blood count. (*Answers.* 67.5%, 23.1%, 9.4%.)

32. The *specific gravity* of a substance is the ratio of its weight to that of an equal volume of water at 4 °C. If the weight of 1 cubic foot of water at 4 °C is 62.4 pounds, what is the specific gravity of a certain aluminum alloy 2.1 cubic feet of which weigh 340 pounds.

33. The **per cent change** in a quantity is defined as the ratio in percentage of the change to the original value. Thus if a variable changes from 2 to 3, the per cent change is $(3 - 2)/2 = 50\%$. Find the per cent change when old and new values of a variable are as indicated: (a) 3, 4. (b) 1, 2. (c) 5, 6. (d) 101, 116. (e) 2, 1. (f) 1, 0.5. (g) 8, -1. [*Answers.* (a) 33.$\dot{3}\%$. (c) 20%.]

34. If the price of coffee goes up from 50 to 55 cents, what is the per cent increase? If it returns to 50 cents, what is the per cent decrease?

35. Prove:

*3.12.2
$$0 < \frac{1}{m} < \frac{1}{n} \longleftrightarrow m > n > 0$$

3.13   Real numbers

We have seen that a point on the axis corresponds to each rational number. But is there a rational number corresponding to each point? We shall answer this question by locating a particular point and showing that it does not correspond to a rational. We construct a square of side 1 and locate a point by laying out its diagonal from the origin. (See Fig. 3.13.1.) Letting

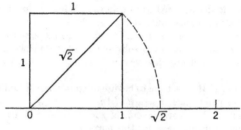

FIG. 3.13.1

x stand for the distance of this point from the origin, we know from the Pythagorean theorem that $x^2 = 1^2 + 1^2 = 2$ or $x = \sqrt{2}$. We have located a point that corresponds to $\sqrt{2}$, but we shall prove that $\sqrt{2}$ is not a rational number, that is,

*3.13.1   There is no rational number m/n,
                  such that $(m/n)^2 = 2$.

Heuristic Discussion. We are going to prove the theorem by assuming the contrary and showing that this leads to a contradiction. We shall assume that there *are* integers m and n, such that $(m/n)^2 = 2$. Since any fraction can be reduced to lowest terms, we can assume that m and n have no factor in common except 1. Then we shall find that the assumption $(m/n)^2 = 2$ implies that m and n do have the factor 2 in common. Hence we must either accept *3.13.1 or deny that all fractions can be reduced to lowest terms.

Proof of *3.13.1. If there is a rational m/n in lowest terms, such that $(m/n)^2 = 2$, it follows that $m^2 = 2n^2$. (Why?) Since the right member of this equation is divisible by 2, m^2 must also be divisible by 2. It follows that m itself is divisible by 2. (Otherwise m would be of the form $2x + 1$ and we would have $m^2 = 4x^2 + 4x + 1$, which is not divisible by 2.) Hence $m = 2p$, where p is some integer. Substituting in $m^2 = 2n^2$ and simplifying, we find $n^2 = 2p^2$. Now by exactly the same argument n must be divisible by 2. Thus we have a contradiction of our original assumption, and the theorem must be true.

By similar methods it can be proved that if a is not a perfect square, $\sqrt{a}$ is not a rational. Hence, if we limit ourselves to rational numbers we cannot solve equations of the type $x^2 = a$. Nor are the rationals sufficient to deal with higher roots (cube roots, fourth roots, etc.), logarithms, or trigonometry. The familiar π is not a rational number, and there are countless other examples. Evidently there are infinitely many points on the axis that do not correspond to rational numbers, and we need to extend the number system further. This is done by defining new numbers corresponding to the points that are not rational. These new numbers are called **irrational.** Together with the rationals, they make up the **real number system.** For every real number there is a corresponding point on the axis, and for every point a corresponding real number. We now call the axis the **axis of reals.** As before, the number corresponding to a point is called its **coordinate.** Since the correspondence is one-to-one, we speak of the point x or the number x. It can be shown that the real numbers obey the same algebraic laws as the rationals and are subject to the same geometric interpretations. Throughout this book, the word "number" means "real number" unless the contrary is stated.

To justify the statements made in the previous paragraph, we would have to give precise definitions of real numbers and the operations upon them. This task is much more difficult for irrational numbers than for the previous extensions that we have outlined. Accordingly, we give only a few indications of the numerical and geometrical interpretation of real numbers.

Since any finite decimal represents a rational number (see Ex. 3.12.13), we must resort to infinite decimals to represent irrationals. By an **infinite decimal** we mean an integer followed by a decimal point and a string of digits that continues without end. If N is the integer, such a decimal will look like this:

(1) $$N.a_1a_2a_3a_4 \cdots a_n \cdots$$

where a_1 represents the tenths digit, a_2 the hundredths digit, a_3 the digit in the third place, etc. The digit in the nth place is represented by a_n, and the dots indicate that the digits continue indefinitely. It may happen that after a certain place the decimal consists of the repetition of a single digit or group of digits. Examples are

(2) $$4.233333333 \cdots$$

(3) $$1.891212121212 \cdots$$

(4) $$10.2300000000000 \cdots$$

(5) $$10.22999999999999 \cdots$$

We shall show in 6.7 that decimals of this kind, which are called **repeating decimals,** equal rational numbers. Those that end in zeros are usually called **terminating** and are obviously rational. Those that end in 9's can always be written as terminating decimals. Thus the decimal in (5) equals 10.23 and has the same value as the decimal in (4). Hence, in order to avoid having two decimals for the same number we exclude decimals that end in 9's. If the decimal is not repeating, it represents an irrational number.

In order to find the point corresponding to an infinite decimal, we consider the decimals formed by cutting it off at one decimal place, two decimal places, etc. For example,

(6) $$\pi = 3.1415926545 \cdots$$

Evidently π lies between 3.1 and 3.2. (Why?) Also it must lie between 3.14 and 3.15. Continuing, it must lie between each of the following pairs of numbers: 3.141 and 3.142, 3.1415 and 3.1416, 3.14159 and 3.14160, 3.141592 and 3.141593, etc. Each pair of numbers determines a segment on the real axis. Each segment lies within the previous one and is $\frac{1}{10}$ as long. These segments are shown schematically in Fig. 3.13.2, each line showing a portion of the previous one enlarged 10 times. It appears plausible that, when the process is continued indefinitely, there is just one point that is common to all these segments, and this point is the point corresponding to π. A similar process could be carried out for any infinite decimal. The nth segment would have a length of $1/10^n$. The nature of the process makes us feel certain that there is a definite point corresponding to each infinite decimal, but it does not enable us to locate the point *exactly*. We must be content

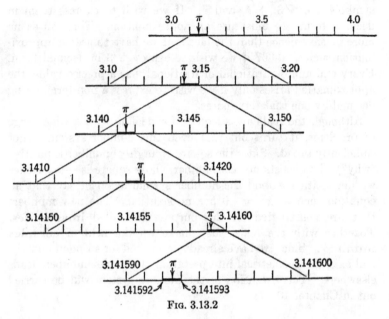

FIG. 3.13.2

with an approximation, but we can make the approximation as close as we find practical or desirable.

In a similar way we can find an infinite decimal corresponding to any point. Of course, if measurement is involved, we shall be able to find only as many decimal places as our instruments allow. But if the point (number) is defined mathematically, we can find as many decimal places as we wish. As an example, consider $x = \sqrt{2}$. We are looking for a number x such that $x^2 = 2$. Since $(1.4)^2 = 1.96$ and $(1.5)^2 = 2.25$, x lies between 1.4 and 1.5. By trying numbers involving one more decimal place, we find that it lies between 1.41 and 1.42. We can continue in this way to find pairs of decimals that "bracket" $\sqrt{2}$ to as many decimal places as time, patience, and interest permit. Knowing the first n decimal places in $\sqrt{2}$, we could always find the next one by trying numbers involving one more decimal place. In this way the digit in any given decimal position could be found.

The discussion above shows that in practice we may treat symbols representing irrational numbers just as we would those representing rationals. We can represent irrationals precisely by symbols like $\sqrt{3}$, $\sqrt{5}$, and π. If we wish to express them in decimal form, we must use an approximation. This causes no more inconvenience than the fact that we have to use an approximation such as 0.667 if we wish to express 2/3 in decimal form. Every real number, rational or irrational, has a precise value, the approximation arises only if we wish to express a non-terminating decimal by one that terminates.

Although the reals are adequate for dealing with a wide range of problems, it turns out that even this number system is not sufficiently broad. Since the square of any real number is positive (why?), there exists no real number whose square is -9. Hence we must either extend the number system again or say that an equation such as $x^2 = -9$ has no solution. The new numbers that are constructed for this purpose are called **imaginaries.** Together with the reals, they make up the field of **complex numbers.** Since we have already accounted for all points on the real axis, the geometrical interpretation of complex numbers leads elsewhere. This extension of the number system will be carried out in Chapter 10.

EXERCISE 3.13

1. Show how to construct by ruler and compass the square root of any integer.

2. Assuming that, if m^2 is divisible by a prime number p, m is also divisible by p, show that $\sqrt{3}$ is irrational.

3. Try to prove that $\sqrt{4}$ is irrational by the method of the section, and explain why the proof breaks down.

4. Prove that $\sqrt{5}$ is irrational.

5. Give a geometrical interpretation of the following statements, where the variables stand for real numbers:

(a) $a < b$ (g) $a - b = 3$
(b) $b > 2$ (h) $b - a < 2$
(c) $a > b$ (i) $a + 4 = b$
(d) $a \leq b$ (j) $a - b > 2$
(e) $a > b > c$ (k) $a \leq 0$
(f) $-1 < 0 < 1$ (l) $3.6 < 3.65 < 3.7$

(m) $-2.8 < -2.74 < -2.7$	(s) $|a| \le |b|$
(n) $3 < \pi < 4$	(t) $|x - 4| < 2$
(o) $3.1 < \pi < 3.2$	(u) $|x - 3| > 5$
(p) $3.14 < \pi < 3.15$	(v) $|x + 1| \le 3$
(q) $|3 - 2| < |4 - 1|$	(w) $|2x - 1| < 10$
(r) $|b - a| > 0$	(x) $2.3 < x < 2.4$

Answers. (a) a lies to the left of b. (c) a lies to the right of b. (e) b lies between a and c, where a is to the right of c. (g) a is 3 to the right of b. (r) b does not coincide with a. (t) x lies within a distance 2 of 4.

6. Make a drawing like Fig. 3.13.2 for the first five decimal places in the following repeating decimals:

(a) $0.\dot{3} = 0.333\cdots$ (d) $1.\dot{2}\dot{8}$ (g) $-2.6\dot{3}\dot{7}$

(b) $1.2999\cdots$ (e) $10.5\dot{1}\dot{8}$ (h) $3.15\dot{1}$

(c) $9.\dot{9}$ (f) $3.14\dot{1}$

7. Explain with the aid of a diagram how you would go about finding the decimal corresponding to a given point by measurement, provided indefinite accuracy were possible.

8. Find the closest numbers to three decimal places that bracket $\sqrt{2}$. Draw a diagram showing the intervals $1.4 < x < 1.5$, $1.41 < x < 1.42$, and the one you have found. Continue by finding two more bracketing pairs and so determine $\sqrt{2}$ to four decimal places.

9. Find the first five intervals that bracket $\sqrt{3}$ to units, tenths, etc. Draw a sketch. (Partial solution: $1 < \sqrt{3} < 2$ since $1^2 = 1$ and $2^2 = 4$. Moreover $\sqrt{3}$ appears to be closer to 2. (Why?) By calculating 1.5^2, 1.6^2, etc., we find that $1.7 < \sqrt{3} < 1.8$. Since 1.7 seems closer, we try 1.71^2, 1.72^2, etc., and find $1.73 < \sqrt{3} < 1.74$.)

10. Find the first three intervals that bracket $\sqrt[4]{10}$. (By $y = \sqrt[4]{x}$, we mean $x = y^4$.) Draw a sketch.

†**11.** Prove that $\sqrt[3]{3}$ is not rational.

†**12.** Suggest a simple way to show that $\pi \ne 22/7$. (By "simple" we mean involving only high school geometry.)

13. Recall or look up the discussion of π in elementary geometry. Explain how it would be possible to compute as close an approximation as required. (This has been done by more efficient methods to over 2035 decimal places as of the year 1950.)

†**14.** Suggest a rule by which you can tell which of two infinite decimals is larger. (Remember that, although we cannot write down all the digits in an infinite decimal, we always have a method of finding the digit in any given position.)

15. Give an intuitive explanation of why $0.\dot{9} = 1$.

16. Explain why the decimals in (5) and (4) represent the same point.

†**17.** Are the following equivalence relations?

(a) "Is the reciprocal of."   (c) "Does not equal."

(b) "Is divisible by."   (d) "Is as strong as."

18. What was the hypothesis of *3.4.1 when it was originally stated? How has this hypothesis been extended? What is it now in the light of this section? Discuss the other theorems of this chapter in this light.

19. What is the converse of *3.4.1. Is it true?

†**20.** If addition were not associative, would it be possible to prove *3.6.6? Justify your answer, and give examples of theorems that do and do not depend on associativity.

†**21.** It can be shown that the real numbers (defined as infinite decimals or otherwise equivalently) form a field. What bearing does this have on the statements in this section? Starting from the field properties, derive as many of the theorems of this chapter as you can.

†**22.** Prove *3.3.5, *3.3.6, *3.4.4, *3.10.14 from the assumption that among the reals there is a set P with the following properties:

(A) $x \in P$ and $y \in P \longrightarrow x + y \in P$
(B) $x \in P$ and $y \in P \longrightarrow xy \in P$
(C) One and only one of the following hold: $x \in P$, $-x \in P$, $x = 0$;

and from the following definition of $a < b$:

$$a < b \longleftrightarrow b - a \in P \qquad \text{(Definition)}$$

23. Let A and B stand for actions, and let AB stand for the action of performing A and then B. Let $A = B$ mean that the actions have the same result. In which of the following is multiplication commutative?

(a) A = putting on socks; B = putting on shoes.
(b) A = putting on coat; B = putting on hat.
(c) A = drying dishes; B = washing dishes.
(d) A = walking 10 miles North; B = walking 10 miles East.

†**24.** Suppose $A(x)$ stands for x^2 and $B(x)$ stands for $2x$, and $AB(x)$ means $A[B(x)]$ and $BA(x)$ means $B[A(x)]$. Does $AB(x) = BA(x)$? Answer the same question for the following: (a) $A(x) = x + 1$, $B(x) = x - 2$. (b) $A(x) = x + 1$, $B(x) = 2x$.

3.14 Approximation

Mathematics is an exact science in the sense that it is based on precise logical thinking, but approximation is an inevitable part of much mathematical work. Numbers that arise from measurement cannot be more accurate than the precision of our senses and measuring instruments. Irrational numbers and rationals equal to infinite repeating decimals cannot be written exactly as terminating decimals. Moreover, it may be convenient to work with an approximation involving fewer digits than a number given exactly to a large number of decimal places.

If a number A is an approximation to a number T we write $T \doteq A$ and say "T equals A approximately." Thus $\pi \neq 3.1416$, but we write $\pi \doteq 3.1416$ because this is the closest approximation to four decimal places. The difference between the true value and the approximation is called the **error** and is defined by

*3.14.1 $E = A - T$ (Definition of Error)

Note that E is *the approximation minus the true value* and hence is positive or negative according as the approximation is too big or too small. Thus if 23.2 is an approximation to 23.7, the error is -0.5.

If an approximation is to be of much use, we must have some idea of the maximum size of the error. We call $\bar{E}$ the **possible error** if it is the smallest number that we know is not exceeded by $|E|$. For example, if 6.3154 is the best approximation of four decimal places to a number, we know that the number lies between 6.31535 and 6.31545. Hence $|E| < 0.00005$, and 0.00005 is the smallest number with this property. We say that a number is **known to a certain decimal position** if the possible error is 5 in the next position. Thus, if 3 is the last accurately known digit in 7.22387, the possible error is 0.0005; and, if 4 is the last known digit in 28456, the possible error is 50.

There is no point in writing uncertain digits, since they do not reduce the possible error. In addition, they may give a false impression of accuracy. Accordingly, it is a universal convention to *write only those digits that are known* so that the last digit written indicates the possible error. This convention cannot always be followed if an integer ends in zeros that are not exact, since in the ordinary way of writing integers final zeros cannot be omitted without changing the meaning of the number. However, this difficulty is avoided if the number is written in the following way, which has many other advantages also. We move the decimal point to the right of the first non-zero digit and compensate by multiplying by some power of ten. Thus:

(1) $2385 = 2.385 \times 10^3$

(2) $23.85 = 2.385 \times 10$

(3) $0.2385 = 2.385 \times 10^{-1}$

(4) $0.002385 = 2.385 \times 10^{-3}$

The negative powers indicate division by powers of ten, that is, $10^{-1} = 1/10$, $10^{-3} = 1/10^3$, and in general $10^{-n} = 1/10^n$. This way of writing numbers is called the **scientific notation,** and it is evident that final zeros are no longer necessary unless we wish to indicate that they are accurately known. Thus $x \doteq 2.3000 \times 10^4$ indicates that x is nearer to 23,000 than to any other integer, whereas $x \doteq 2.3 \times 10^4$ means that x is 23,000 to the nearest thousand.

The process of replacing a number by one with fewer digits is called **rounding off.** It may be done because final digits are uncertain or simply because a less accurate approximation is more convenient. The convention for rounding off is to eliminate digits from the right and to leave the last remaining digit unchanged or increase it by one, according as the digits eliminated represent less or more than 5 in the next position. If they represent *exactly* 5, it is customary to choose the *even* possibility. It follows that the possible error of the rounded-off number is always 5 in the first neglected position. If the last digit is even, the error is less than or equal to this quantity. Otherwise it is always less than the possible error. This means that the rounded-off value is always the most accurate approximation with the same number of digits, or the even possibility if there are two equally accurate approximations. According to this convention rounding two digits off the numbers 4.8135, 8.2949, 2.9150, and 7.3181 yields 4.81, 8.29, 2.92, and 7.32.

Frequently the size of the error is less important than the relation between the error and the true value. The **relative error** is the ratio of the error to the true value, that is,

*3.14.2 $R = \dfrac{E}{T}$ (Definition of Relative Error)

It may be expressed in percentage as $100R\%$ and is then called the **per cent error.** When E is small, E/T and E/A are nearly the same since $E/T = (E/A)(A/T)$, and the second factor is nearly 1. Since T is usually unknown, E/A is often used as an approximation for R. The **possible relative error** is defined as $\bar{E}/|T|$, where $\bar{E}$ is the possible error. It is the smallest number that is not exceeded by the absolute value of the relative error. The possible relative error in $\pi \doteq 3.1416$ is approximately $0.00005/3.1416$. We say "approximately" because we have replaced T by A in the denominator.

A useful concept in dealing with relative error is order of magnitude. When a number is written in scientific notation, that is, in the form $N \times 10^n$, where $1 \leq N < 10$, its **order of magnitude** is said to be 10^n. Thus the order of magnitude of 1283 is 1000 and of 0.031 is $10^{-2} = 1/100$.

Another helpful notion is that of significant digits. The **significant digits** in a number are all its digits except initial zeros. By initial zeros we mean those not preceded by non-zero digits. We assume the convention that only known digits are written. For example, all digits in 30082 and 3.100×10^2 are significant, whereas the zeros in 0.0015 are not significant. The relative accuracy of a number is conveniently indicated by the number of significant digits because, if there are n significant digits in a number, the order of magnitude of its possible relative error is $1/10^n$ or $1/10^{n+1}$. For example, there are five significant digits in 3.1416, and its possible relative error is approximately 1.6×10^{-5}, but the possible relative error in 9.12 is $0.005/9.12 \doteq 5 \times 10^{-4}$.

Usually computations begin with numbers whose accuracy is known, and we are interested in estimating the error in the result. We have the following two theorems:

*3.14.3   The error of a sum is the sum of the errors.

*3.14.4   The relative error of a product is approximately the sum of the relative errors of the factors.

These theorems mean that it is convenient to consider the error when addition and subtraction are involved and relative error when multiplication and division are involved. There is no simple relationship for the error of a product or the relative error of a sum.

Proof of *3.14.3 for Two Terms.   Let T_1 and T_2, A_1 and A_2, E_1 and E_2 be the true values, approximations, and errors. Then the error of the sum is $E = (A_1 + A_2) - (T_1 + T_2) = E_1 + E_2$. (Why?)

Proof of *3.14.4 for Two Factors.   The error in the product is

(1) $E = A_1 A_2 - T_1 T_2$

(2) $= (T_1 + E_1)(T_2 + E_2) - T_1 T_2$ (Why?)

(3) $= E_1 T_2 + E_2 T_1 - E_1 E_2$ (Why?)

Hence the relative error is:

(4) $$R = \frac{E}{T} = \frac{E_1}{T_1} + \frac{E_2}{T_2} + \frac{E_1}{T_1}\frac{E_2}{T_2} \qquad \text{(Why?)}$$

(5) $$\doteq \frac{E_1}{T_1} + \frac{E_2}{T_2} = R_1 + R_2 \qquad \text{(Why?)}$$

The approximate equality is suggested by the fact that, if the relative errors are small, their product, which is the last term in (4), is insignificant. For example, if $R_1 = 0.01$ and $R_2 = 0.03$, $R_1 R_2 = 0.0003$.

It follows from these theorems and *3.10.17 that the possible error in a sum is the sum of the possible errors, and the possible relative error in a product is approximately the sum of the possible relative errors. However, because of the cancellation of errors, which may be positive or negative, we may expect the actual error of the result to be smaller. If several numbers are added, errors of the same order of magnitude tend to cancel, so that the error in a sum may be expected to be of the same order of magnitude as the largest error in the terms. This means that the *number of accurate decimal places in a sum is the same as the number in the least accurate term.* Similar remarks apply to the cancellation of relative errors when several numbers are multiplied. The relative error in the result may be expected to be of the same order of magnitude as the largest relative error in the factors. This means that *the number of significant digits in a product is given by the smallest number of significant digits in the factors.* Results should be rounded off according to these ideas.

The last paragraph concerned what we may "expect" about the accuracy of results. Precise statement and proof involve the theory of probability and statistics.

EXERCISE 3.14

1. In each of the following pairs of numbers the first is the approximation and the second is the true value. Find the error in each case.

(a) 23.3; 23.7   (e) 0.67; 2/3   (i) 1.830; 1.832
(b) 23.9; 23.7   (f) 10; 10.3   (j) 2,000,000; 2,000,001
(c) 100; 98   (g) 10; 10.7   (k) 9.9341; 9.9350
(d) 832; 1000   (h) 1830; 1832   (l) 68.001; 67.001

(m) 2; 1   (p) -4.1; -4.06 (s) -0.001; -0.01

(n) 1; 2   (q) 1; -1 (t) 3.84771; 4

(o) 0.0002; 0.0001   (r) 392.2; 291.5   (u) $-a, a$

Answers. (a) -0.4. (c) 2. (e) 1/300. (g) -0.7. (i) -0.002. (k) -0.0009. (m) 1. (o) 0.0001. (q) 2. (s) 0.009. (u) $-2a$.

2. Find the relative errors in prob. 1.

Answers. (a) $-4/237$. (c) 2/98. (e) 1/200. (g) $-7/107$. (i) $-1/916$. (k) $-9/99350$. (m) 1. (o) 1. (q) -2. (s) -0.9. (u) -2.

3. Write the following numbers in scientific notation:

(a) 82.4   (e) 0.8411   (i) 0.11893   (m) 0.00008

(b) 115   (f) 0.037   (j) 0.00074   (n) -0.000119

(c) 4999   (g) -0.9 (k) 6,000,000   (o) 0.10002

(d) 5/4   (h) 0.0091   (l) 5 billion   (p) 114.0005

Answers. (a) 8.24×10. (c) 4.999×10^3. (e) 8.411×10^{-1}. (g) -9×10^{-1}. (i) 1.1893×10^{-1}. (k) 6×10^6 on the assumption that the zeros are uncertain. (m) 8×10^{-5}. (o) 1.0002×10^{-1}.

4. Rewrite the following without scientific notation:

(a) 1.83×10^2 (c) 4.5586×10^{-4} (e) 2.475×10^{-1}

(b) 9.999×10^{-2} (d) 8×10^9 (f) 8.000×10^{-3}

Answers. (a) 183. (c) 0.00045586. (e) 0.2475.

5. How many significant digits are there in each number in prob. 3?

Answers. (a) 3. (c) 4. (e) 4. (g) 1. (i) 5. (k) 1 to 7, depending on accuracy. (m) 1. (o) 5.

6. How many decimal places are there in each number in prob. 3, that is, what is the position of the last known digit?

Answers. (a) 1. (c) Accurate to units. (e) 4. (g) Same as a. (i) 5. (k) Accurate to nearest million if zeros unknown. (m) 5. (o) 5.

†**7.** Is the number of decimal places a function of the number of significant digits. Is the number of significant digits a function of the decimal places? Illustrate your answer with numbers from previous problems. (*Answer.* No, the two are independent.)

†**8.** Answer the same question for numbers written in scientific notation. If your answer is yes, find the relation.

9. Round off each number in prob. 3 to one less decimal place.

Answers. (a) 82. (c) 5000. (e) 0.841. (g) -1. (i) 0.1189. (k) 10,000,000 if the zeros are uncertain. (m) 0.0001. (o) 0.1000.

10. Round off the following to three significant digits: (a) 8.24500. (b) 8.23500. (c) 1835. (e) 0.07995. (f) 0.1465. [*Answers.* (a) 8.24. (c) 1840. (f) 0.146.]

11. Calculate in decimal form each of the relative errors found in prob. 2 and round off to the nearest two significant digits. (This requires doing the computations to three significant digits first. Why?) [*Answers.* (a) -0.017. (c) 0.020. (e) 0.0050. (g) -0.065. (i) -0.0011. (k) -0.000091. (m) 1.0.]

12. Express each relative error in prob. 11 in percentage. [*Answers.* (a) −1.7%. (c) 2.0%. (e) 0.50%. (g) −6.5%. (i) −0.11%. (k) −0.0091%. (m) 100%. Note that 0.11% means 11/100 of 1%.]

13. What is the possible error if accuracy is to the nearest (a) inch, (b) foot, (c) 2 feet, (d) mile, (e) 30 seconds, (f) unit, (g) $\frac{1}{10}$, (h) million, (i) dollar, (j) cent, (k) dime? [*Answers.* (a) $\frac{1}{2}$ in. (c) 1 ft. (e) 15 sec. (g) $1/20 = 0.05$. (i) 50 cents. (k) nickel.]

14. What is the possible error in each number in prob. 3? [*Answers.* (a) 0.05. (c) 0.5. (e) 0.00005. (g) 0.05. (i) 0.000005. (k) 500,000 if zeros are not known. (m) 0.000005. (o) 0.000005.]

15. Supposing that the numbers in prob. 3 have been rounded off from other numbers, write inequalities that must be satisfied by the unknown original numbers. [*Answers.* (a) $82.35 \leq x \leq 82.45$. (c) $4998.5 < x < 4999.5$. (e) $0.84105 < x < 0.84115$. (g) $-0.95 < x < -0.85$. (i) $0.118925 < x < 0.118935$. (k) $5,500,000 \leq x \leq 6,500,000$. (m) $0.000075 \leq x \leq 0.000085$.]

16. Find to one significant digit the possible relative error in each number in prob. 3. [*Answers.* (a) 6×10^{-4}. (c) 1×10^{-4}. (e) 6×10^{-5}. (g) 6×10^{-2}. (i) 5×10^{-5}. (k) 8×10^{-2} if zeros uncertain. (m) 6×10^{-2}. (o) 5×10^{-5}.]

17. Compare the order of magnitude of the possible relative error with the number of significant digits in each number in prob. 3 and so check the statement made in the section.

18. What is the order of magnitude of each number in prob. 4? [*Answers.* (a) 10^2. (c) 10^{-4}. (e) 10^{-1}.]

19. Calculate the possible relative error and its order of magnitude for each of the following numbers assumed accurate to two digits: (a) 1.3. (b) 28. (c) 50. (d) 0.51. (e) 8.3. (f) 7.2. (g) 900.

†**20.** What determines whether the possible relative error is 10^{-n} or $10^{-(n+1)}$? Prove the statement in the section and your more precise result.

21. In each part of prob. 1 find E/A and compare it with the exact relative error E/T.

22. If $x \doteq 3.149$, what is the possible error in using E/A in place of E/T?

23. What is the possible percentage of error in using 22/7 as an approximation to π?

24. Criticize the following statement: "The metal is one-sixth gold. It weighs 100 ounces. Hence there are 16.6667 ounces of gold in it."

25. Criticize the following statement: "The foreign trade of Poldavia is $1,000,000 annually, of which the United States has two-thirds. Hence our foreign trade with Poldavia amounts to $666,666.66."

26. The number of molecules in one mole of any substance is a constant equal to 6.023×10^{23} and is called Avogadro's number. If molecules could be counted by a person at the rate of 1 every 10 sec, roughly how many thousand years would it take two billion people to count the molecules in a mole?

27. What is the order of magnitude of each possible error found in prob. 14. Compare with the position of the last accurate digit.

†**28.** Justify the assertion that, if a number has n decimal places, the order of magnitude of the possible error is $10^{-(n+1)}$ and, if the last accurate digit is in the mth place before the decimal point, the order of magnitude of the possible error is 10^{m-2}. Could the same statement be made about the error?

29. The following ten numbers were selected by opening *Tables of Random Sampling Numbers*, by Kendall and Smith, in a haphazard way and reading off digits: 6.675, 7.989, 5.592, 3.759, 3.431, 4.320, 4.558, 2.545, 4.436, 9.265. Find the exact sum. (a) Round off each number to two decimal places and compare the resulting sum with the original sum rounded to two places. Check by finding the errors due to rounding and adding. Add the first five exact numbers to the rounded-off values of the rest and compare with the true sum. (b) Do the same for the numbers rounded off to one decimal place. (c) The same for the numbers rounded to the units position. (*Answer to a.* 52.570; 52.57.)

30. The results were unusually close in prob. 29. Carry out the same experiment with the following numbers, which were read off from the table immediately after those found above: (a) 11.26, 63.45, 45.76, 50.59, 77.46. 34.66, 82.69, 99.26, 74.29, 75.16. (b) 17.87, 23.91, 42.45, 56.18, 1.46, 93.13, 74.89, 24.64, 25.75, 92.84.

31. Show that the error in $-x$ is the negative of the error in x. How does this justify applying the rule for additions to subtractions?

32. Find the exact result in each case below. Then round off numbers to one less digit, compute the result, round off and compare with the true value rounded off. Finally, round off the numbers to one significant digit, compute and round off the result, and compare with the rounded-off true value.

(a) $(7.55)(1.02)(1.77)$

(b) $(1.04)(33.9)(3.36)$

(c) $(4.27)(5.76)/(0.235)$

(d) $(561)(13.2)/989$

Answer for a. 13.630770; 14; 14; 20; 10.

†**33.** Show that the relative error in $1/A$ is approximately the negative of the relative error in A. How does this justify the application to division of the rule for multiplication?

†**34.** Under what conditions are the following approximations good? Find a formula for the error and relative error in each case.

(a) $(1 + a)(1 + b) \doteq 1 + a + b$

(b) $1/(1 - a) \doteq 1 + a$

(c) $(1 + a)^2 \doteq 1 + 2a$

(d) $a + b \doteq a$

35. The ideas of this section are very useful in checking computations by approximation. Thus to check $(0.563781)(2.911207) = 1.641283193667$ or to compute it approximately, we find $(0.6)(3) = 1.8$, which is of the same order of magnitude, and $(0.56)(2.9) = 1.624$, which is the same to two significant digits. Ordinarily, we can expect the result to be accurate to the same number of digits or decimal places as the numbers, but practical certainty is achieved by keeping one more digit or decimal place and rounding off the result. Calculate the following approximately as indicated:

(a) Order of magnitude of $(0.113924)(152)(3.826)$.

(b) Same for $(3,008,123)(1.9)(0.009934)(0.4001)$.

(c) One significant digit in $(34.992)(27)(0.001)$.

(d) Two significant digits in each of the previous.
(e) Order of magnitude of 1.301 + 28.99 − 13.2 − 0.0051.
(f) Same for 2.99 + 100.2 − 0.0059 and for 10.0139 + 1.0244 − 1.04.
(g) One decimal place in parts e and f.

Answers. (a) 10. (c) 9×10^{-1}. (e) 10.

36. The following approximation is used in aerial photography:

$$\frac{(H \pm h)^2 - H^2}{H} \doteq \pm 2h$$

Find the per cent error if $h/H = 0.01$.

4

Linear Functions

Linear functions are those of the form $y = mx + b$. They are basic to the study of all other functions and have themselves a tremendous variety of applications. The name "linear" is associated with the fact that these functions can be represented by straight lines. In the first section of this chapter we study formulas of the type $y = mx$. In 4.2 we introduce the basic ideas of graphing. We define the word "function" and discuss the general linear function in 4.3. The remainder of the chapter is devoted to related topics of algebra and geometry.

4.1 Direct variation

The following table represents the cost in cents of several quantities of 3-cent postage stamps:

x	y
0	0
1	3
2	6
3	9
4	12

We have labeled x the column giving the number of stamps and y the column giving the cost. Obviously each number in the cost column is three times the corresponding number in the quantity column. The correspondence between these two sets of numbers may be summed up by writing the equation $y = 3x$. Here x and y are variables, standing respectively for numbers in the first and second columns. The meaning of the equation is that if any value of x is substituted for x, the corresponding value of y will be given by $y = 3x$. A formula of this kind has advantages over a table. It includes all the possibilities, whereas the table can give only a few pairs of values. We can get the cost of any number of stamps by substitution in the equation. For example, $x = 15$ gives $y = 3 \cdot 15 = 45$. (Of course, we could substitute a fractional value for x, but this would have no meaning because the range of x is the positive integers.)

There are a very large number of situations in which two sets of numbers correspond in this way, that is, they are paired off so that for every number in one set multiplication by a fixed number gives a corresponding number in the other set. If x and y are variables standing for the numbers in the sets and m is a constant, the correspondence can be stated in the equation $y = mx$. When $y = mx$, it is said that y **varies directly** as x (or y is directly proportional to x), and m is the **constant of proportionality.** (Sometimes the relation is written $y \propto x$.)

The use of the word "proportional" is appropriate for the following reason. If x_1 (read "x-sub-one") is a particular value of x, and y_1 the corresponding value of y, then $y_1 = mx_1$. If x_2 and y_2 are another pair of corresponding values, $y_2 = mx_2$. Then:

(1) $$\frac{y_1}{y_2} = \frac{x_1}{x_2} \qquad \text{(Why?)}$$

or

(2) $$\frac{y_1}{x_1} = \frac{y_2}{x_2} \qquad \text{(Why?)}$$

Thus corresponding values of x and y are proportional, or the y's are proportional to the x's. (See 3.11 and Ex. 3.11.17.)

A well-known example of direct variation is the relation between time and distance when a body moves with a constant speed. The equation is $d = vt$, where d is the *distance*, t the *time*, and v the *speed*. Here the speed is the constant of proportionality.

It may happen that we know that y varies directly as x, but that we do not know the constant of proportionality. In this case we know that $y = mx$ for *some* m without knowing the value of m. If we are also given one pair of values of x and y, we can easily find m. We have only to substitute these values in the equation. Then $m = y/x$ by the definition of division. We know, for example, that the *mass* of a homogeneous body varies directly as its *volume*. We have $M = DV$, where M is the mass, V the volume. The constant of proportionality is called the *density* and depends upon the substance. Now suppose we are given a quantity of aluminum and find that its volume is 2 cubic centimeters and its mass 5.4 grams. Then we have $5.4 = D \cdot 2$. Hence $D = 2.7$ grams per cubic centimeter. Now we can write the formula $M = 2.7V$ and use it to calculate other values of M corresponding to different volumes.

EXERCISE 4.1

1. Given $y = (\frac{1}{2})x$, make up a table showing the pairs of corresponding values of x and y for $x = 0, 1, 2, 7, 10, \frac{1}{2}, \frac{1}{4}, -1, -10, -1.6$.

2. An airplane is traveling at a constant speed of 400 mph. Write the equation relating the distance traveled and the time. Find the distances covered in 2 hours and in 1 hour, 40 minutes.

3. An automobile goes 111 miles in 3 hours. Assuming constant speed, how far will it go in 5 hours? (*Answer.* 185 miles.)

4. Explain (1) and (2) in the section.

5. The wage of a worker paid by the piece is proportional to his output. The constant of proportionality is called the *piece rate*. A certain worker gets $8.75 for a day in which he turns out 125 units. Let W be the wage, Q the output, and r the piece rate. Find r, and calculate what he would earn for 200 units. (*Answer.* $14.)

6. Find the density of a body that weighs 535 grams and has a volume of 45 cubic centimeters.

7. If the density of water is 62.4 pounds per cubic foot, what volume will weigh 100 pounds? (*Answer.* 125/78 cubic feet.)

8. The *atomic weight* A of an element is proportional to the weight W of one of its atoms, and the constant of proportionality is such that the atomic weight of oxygen is 16. Write a formula for A in terms of W_0, the weight of an atom of oxygen.

9. The heat required to melt a substance varies directly as its mass, the constant of proportionality being called the *heat of fusion*. Find the heat of fusion of ice if it requires 1.52×10^4 calories to melt 1.91×10^2 grams. (*Answer.* 7.96×10 calories per gram.)

10. An unknown number and 6 are the means in a proportion in which the extremes are 7 and 5. What is the number? (See 3.11.)

11. The *pressure* P of a gas of constant volume is proportional to the absolute temperature, T. (a) Write the relationship between P and T. (b) Suppose that pressure in a tire is 30 in the morning when the temperature is 50° Fahrenheit. What will it be after the heat of the day and friction on the road have raised the temperature in the tire to 90°? (*Note.* 50° and 90° Fahrenheit are 283 and 305 in the absolute temperature scale. Assume constant volume.)

12. The *specific heat* of a substance is the constant of proportionality between heat absorbed by a unit mass and the temperature change. How many calories are required to heat one gram of asbestos from 60°C to 80°C if its specific heat is 0.195 calorie per gram per degree Centigrade.

13. The area of a circle A varies directly as the square of the radius r. The constant of proportionality is π, and $A = \pi r^2$. In the following, indicate the constant of proportionality and write the equation. Assume that variables not mentioned are constants.

(a) The area of a circle varies directly as the square of the diameter.

(b) The circumference of a circle varies directly as the diameter.

(c) The circumference of a circle is proportional to the radius.

(d) The area of a triangle is proportional to the product of its base and altitude.

(e) The value of a load of coal is proportional to the amount.

(f) The number of fingers in a room is proportional to the number of people in the room.

(g) The area of a parallelogram is proportional to the base.

(h) For a man being paid by the hour the wage is proportional to the number of hours worked.

(i) The annual interest on a debt varies directly as the debt.

(j) Force is proportional to the acceleration produced. (Newton's second law.)

(k) The moment or torque exerted by a force tending to cause a rotation is proportional to the length of the lever arm.

(l) Momentum is proportional to velocity.

14. Two pipes are used to fill a tank. The large one will fill it in 7 hours, the small one in 24 hours. How long will it take if both are used? (*Hint.* Find the rate of flow of each pipe in tanks per hour, then the rate together.)

15. The *magnifying power of a telescope* is given by $M = F/f$, where F and f are the focal lengths of the objective and eyepiece lenses. If the focal length of the objective is 3, what must be the focal length of the eyepiece in order to have a magnifying power of 10? (*Answer.* 0.3.)

†**16.** A motorcycle courier is continuously riding back and forth between two columns approaching each other. One column moves at 2 mph, the other at $2\frac{1}{2}$ mph. The motorcycle travels at an average of 40 mph and consumes a gallon for every 30 miles. If the columns are 20 miles apart, how many gallons of gas will the courier use before they come together? (*Note.* There are two types of direct variation here: gas-distance and distance-time.)

17. When a variable varies directly as the product of several variables, it is said to **vary jointly**. For example, if $y = mxzw$, y varies jointly as x, z, and w. If y varies jointly as S and T, and $y = 10$ when $S = 3$ and $T = \frac{1}{2}$, find y when $S = 2$ and $T = 5$. (*Answer.* 200/3.)

18. When a map is drawn to scale, distances on the map are directly proportional to the corresponding distances of the ground, and the constant of proportionality is called the *scale*. If the scale of a map is 1:25,000:

(a) What ground distance is represented by one inch on the map?

(b) What length on the map indicates one mile?

(c) Will a map with scale 1:65,000 give more or less detail?

19. A car travels 5 miles at 30 mph and then 5 miles at 40 mph. What is its average speed? (*Answer.* 34.3.)

20. A textbook in human biology states that an average man requires about 17 calories per pound of body weight per day or 2500 calories per day. What has the author taken as the weight of an "average man"?

21. Frequently particular values of a variable are indicated by placing numbers or other symbols below and to the right as we do in equations (1) and (2). Such symbols are called **subscripts**. The symbol "x_n" is read "x-sub n."

(a) Read x_1, x_2, x_3, x_0, x_r, x_s, x_{150}, y_5, y_m.

(b) If the range of y is the presidents of the United States, suggest the interpretation of y_n.

22. A friend who is unacquainted with mathematics writes you that he is puzzled by a "mathematical" passage on page 718 of the August 1950 issue of the *Sewanee Review*. Write an explanation for your friend.

4.2 Graphing pairs of numbers

It would be good to have a geometrical picture of the correspondence discussed in the last section. In Chapter 3 we indicated a way to interpret single numbers as points on a line. But here we are concerned with pairs of numbers, each y being paired with a corresponding x so that $y = mx$. An obvious idea is to set up two axes, one for x and one for y. For each value of x there is a point on the x-axis, and for each corresponding value of y a point on the y-axis. We have only to indicate in some way which points correspond. This might be done by drawing the axes one above the other and indicating by an arrow the y corresponding to a given x. We have done this for $y = 2x$ in Fig. 4.2.1.

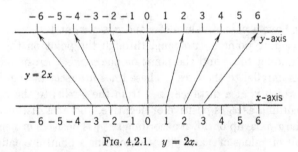

FIG. 4.2.1. $y = 2x$.

A very suggestive and powerful method is based on drawing the axes perpendicular, with their origins coincident. Then, instead of joining corresponding points, we draw lines from these points parallel to the axes. These lines meet in points that represent the pairs of numbers. We sketch $y = 2x$ in Fig. 4.2.2. It is customary to draw the x-axis horizontal and the y-axis vertical, with the positive directions on the axes being right and up. The point of intersection is called the **origin.** The axes are called **coordinate axes,** and such a scheme is called a **rectangular coordinate system** or coordinate plane.

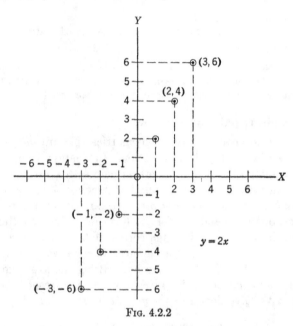

FIG. 4.2.2

If we have a pair of values of x and y, say x_1 and y_1, we find the corresponding point by drawing, through the point on the x-axis corresponding to x_1 and the point on the y-axis corresponding to y_1, lines parallel to the axes. These lines intersect at the desired point, which is at a distance $|x_1|$ from the y-axis, to the right or left according as x_1 is positive or negative, and at a distance $|y_1|$ from the x-axis, up or down according as y_1 is positive or negative. The pair of values of x and y that determine a point are called the **coordinates** of the point. They may be described as the directed distances of the point from the axes. The point is written (x, y). In Fig. 4.2.2 we so label the points given by $x = 2$, $y = 4$, and $x = -3$, $y = -6$. The value of x is called the **x-coordinate** or abscissa; the value of y is called the **y-coordinate** or ordinate.

Evidently there is just one point determined by any pair of values (x, y), since the lines meet in one and only one point. Conversely, for any point there is just one pair of values (x, y), since the point has unique directed distances from the axes. Therefore, there is a one-to-one correspondence between points in the plane and ordered pairs of numbers, just as there is a one-to-one correspondence between points on an axis and single numbers. We speak of *ordered* pairs because it makes a difference which of the two numbers is considered x and which y. The point (a, b) is not the same as the point (b, a) unless a and b are equal. The correspondence between ordered pairs of numbers and points is one of the foundations of analytic geometry. It means that we can think in terms of the geometrical concept of point or the corresponding algebraic concept of an ordered pair of numbers. Accordingly we speak of the point (x, y), meaning either the ordered pair of numbers or the corresponding point.

The axes divide the plane into four regions, numbered in counterclockwise fashion (Fig. 4.2.3). They are called the first, second, third, and fourth **quadrants.** In each quadrant x and y have characteristic signs, as indicated on the figure. On the y-axis, $x = 0$. On the x-axis, $y = 0$. At the origin, both x and y are zero.

We may think of the coordinates of a point as vectors. We get to a point (x, y) by laying out a vector x along the x-axis and then

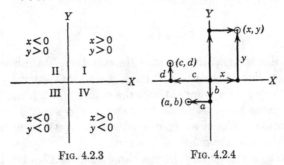

FIG. 4.2.3   FIG. 4.2.4

a vector y, parallel to the y-axis. Or we could go first along the y-axis and then parallel to the x-axis. The four vectors form a rectangle, as indicated in Fig. 4.2.4.

Returning to Fig. 4.2.2, we note that the plotted points appear to lie on a straight line through the origin. If they actually do, the correspondence $y = 2x$ can be visualized as a straight line, which can be located by any two of its points. The y corresponding to any x can then be found by drawing a vertical through the point x on the x-axis and seeing where it meets the line. (See Fig. 4.2.5.) We prove in 4.7 that all the points that satisfy the equation $y = mx$ lie on a straight line through the origin and, conversely, that every

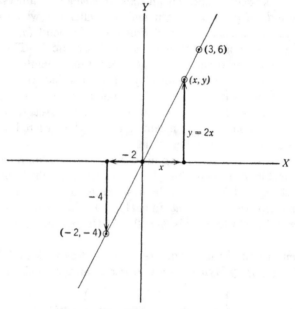

FIG. 4.2.5. $y = 2x$.

point on this straight line satisfies the equation. In other words, the straight line consists of just those points that satisfy $y = mx$. When we say that a point satisfies the equation, we mean that, if its coordinates are substituted for x and y, $y = mx$ will become a true equation. We speak of $y = mx$ as the *equation of the line*, and of the line as the *graph* of $y = mx$.

We use coordinates and graphs throughout the book. Indeed, this is one of the most important techniques in analysis. We mention, therefore, some conventions that the student should observe in making graphs:

1. Draw the x-axis horizontal and the y-axis vertical.
2. Take positive directions right and up.

3. Label the axes on the positive sides of the origin.

4. Indicate the unit length on each axis by labeling a few points.

5. Use the same unit length on each axis. (There are situations where different scales are desirable, but we use the same scales unless they are otherwise indicated.)

6. Circle plotted points and label key points by writing near by their coordinates in the form (x, y).

7. Make your graphs large and clear, avoiding too much detail.

EXERCISE 4.2

1. Draw a coordinate system and label the following points:

(a) $(0, 0)$, $(3, 0)$, $(0, 5)$, $(4, 5)$, $(-2, 3)$, $(-3, -4)$, $(4, -1)$

(b) $(0, -4)$, $(-1, 0)$, $(6, 2)$, $(2, 6)$, $(1, 1)$, $(4, -2)$, $(-2, 4)$

2. Graph $y = \frac{1}{2}x$ by plotting two points other than the origin and drawing a straight line through them.

3. Indicate, without plotting, the quadrant in which each of the following lies: $(1, -3)$, $(-2, -5)$, $(1, 1)$, $(-3, 4)$, $(-4, -10)$, $(1000, -1)$, $(-49, 13)$.

4. Make a drawing like Fig. 4.2.1 for the equation $y = \frac{1}{2}x$.

5. Graph:

(a) $y = 6x$ (d) $y = -x$ (g) $2y = 5x$ (j) $y = 1.7x$

(b) $y = -2x$ (e) $y = 0 \cdot x$ (h) $-3y = x$ (k) $8y = -5x$

(c) $y = x$ (f) $y = (1/10)x$ (i) $50x = y$ (l) $y = 0.2x$

6. Graph the equations of Ex. 4.1.2, 3, 5, 6, 7, and 9.

7. Money *value v* is proportional to the *quantity q* of a commodity. The constant of proportionality is called the *price p*. (a) Write the equation. (b) Graph on the same coordinate plane the lines corresponding to $p = 2$ and $p = 3$. (c) What is the meaning of the vertical distance between these lines for any particular value of q? (Take the q-axis horizontal, v-axis vertical.)

8. Plot $y = 1.35x$ by using the origin and the point for which $x = 4$. Use a large unit length so as to be as accurate as possible. Then estimate from the graph the y that corresponds to $x = 2.1$ and compare this estimate with the calculated value. Similarly for $x = 3.21$.

9. Graph $y = 2x$,where the range of x is the positive integers. [*Answer.* Discrete points $(0, 0)$, $(1, 2)$, $(2, 4)$, $(-1, -2)$, etc.]

†*10.* Prove that $y = x$ is the bisector of an angle between the x-axis and the y-axis. [*Hint.* This requires proving that for every point (x, y) on the bisector and for no other points we have $y = x$. What theorem about bisectors do you remember?]

†*11.* Similarly show that $y = -x$ is the equation of the other bisector.

†*12.* Let A be the set $\{1, 2, 3\}$ and $B = \{-1, -2, -3\}$. Find $A \times B$ and plot its members.

13. Indicate the set of points that satisfy: (a) $y < x$. (b) $y > x$. (c) $y > -2x$.

4.3   The function concept

The student is undoubtedly acquainted with the two most common scales for measuring temperatures, the Centigrade and the Fahrenheit. For every temperature given in degrees Centigrade there is a corresponding temperature in degrees Fahrenheit. For example, the freezing point of water is 0°C and 32°F. There is a simple rule that gives this correspondence and tells us how to find the Fahrenheit temperature when given the Centigrade temperature. It may be stated: "Multiply by 9/5 and add 32." If we let x stand for temperatures in degrees Centigrade and y for those in degrees Fahrenheit, we have:

$$(1) \qquad\qquad y = \frac{9}{5}x + 32$$

From this rule, stated in words or in the formula, we could construct a table showing the values of y corresponding to any set of values of x. Here x and y are variables standing for temperatures in the two scales, and we have a correspondence between the values of these variables, such that for every value of x there is a value y given by formula (1). A correspondence represented by an equation of this type is called a linear function.

More generally, suppose we have two variables x and y, and suppose that for each value of x there is a corresponding value of y. Then this correspondence is called a **function.** The variable x is called the **independent variable,** and the variable y is called the **dependent variable.** It is said that y is a function of x. The word "function" is applied also to the rule or formula that states the correspondence. Thus we say that $y = (9/5)x + 32$, or just $(9/5)x + 32$, is a function of x. Sometimes also the dependent variable itself is referred to as "the function" and its values as "values of the function." The function, that is, the correspondence, may be given by any rule that tells us what value of y corresponds to each value of x. If the correspondence can be expressed in the form $y = mx + b$, where m and b are constants, it is called a **linear function.**

In order to appreciate the meaning of the foregoing definition, the student should recall the meanings of the words "variable," "value," and "constant" given in 2.7. When we say that x and y are variables, this means that x and y stand for objects in two sets, the x-set and the y-set (sets of numbers representing temperatures in our example). A value of x or y means an object in one of the sets (temperature in our example). Therefore, when we say that for each value of x there is a corresponding value of y, we mean that for each object in the x-set there is a corresponding object in the y-set. The rule that gives the correspondence by telling what y-object corresponds to each x-object is the function. In this course, the sets are usually sets of numbers, so that x and y take numerical values. But the function idea can be extended to sets of any objects whatsoever.

The term linear function is used to describe functions of the type $y = mx + b$, because the graph of any such function is a straight line. (See 4.7.) In order to graph a linear function, we need to locate only two points. (Why?) An easy one to find is the point for which $x = 0$. We have $y = m \cdot 0 + b = b$. Hence $(0, b)$ is one point. It is called the **y-intercept** of the line. (The term "y-intercept" is used also to refer to the y-coordinate of this point and to the segment joining it to the origin.) One more point is sufficient, but to avoid error it is advisable to plot a third. We illustrate in Fig. 4.3 for the function $y = -2x + 3$. Letting $x = 0$, we have $y = 3$. For $x = 3$, we have $y = (-2)3 + 3 = -3$.

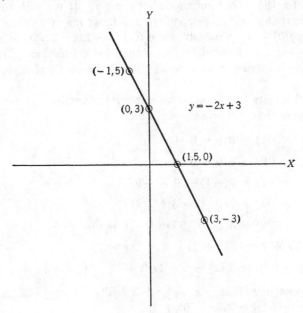

FIG. 4.3

A third point is $(-1, 5)$. We plot the three points, draw the straight line, and write the equation near it.

Direct variation, which we discussed in the last two sections, is clearly a special kind of linear function, for $y = mx$ is of the form $y = mx + b$, with $b = 0$. The point $(0, b)$ becomes $(0, 0)$, and the line goes through the origin.

In talking about a function given by a formula such as $mx + b$, it is awkward to have to write it out repeatedly. Instead we adopt a very important notation: a letter, followed by the independent variable in parentheses, for example, $f(x)$ or $g(x)$, which are read "f of x" and "g of x." This notation does *not* mean f times x. The letter f does not stand for a number. It stands for the function or rule that gives y as a function of x. Thus, if $y = -2x + 3$, we may write $f(x) = -2x + 3$ and then write $f(x)$ in place of $-2x + 3$ afterwards. The letter x in parentheses indicates the independent variable and means that $f(x)$ stands for a rule or formula involving x. The full value of this functional notation will become clear as the student sees its many uses. Not only can we save ourselves rewriting formulas, but we can also indicate briefly the y that corresponds to any x. Thus, if $f(x) = -2x + 3$, then $y = f(x)$ stands for the y that corresponds to any x. If we wish to write the fact that $y = 3$ corresponds to $x = 0$, we may do so briefly by writing $f(0) = 3$. Similarly, since $f(x) = -2x + 3$, $f(3) = -2(3) + 3 = -3$. In general, if $f(x)$ stands for a function (a rule or formula) involving x, then $f(a)$ indicates the result of replacing x by a.

We illustrate this notation with several examples.

Suppose that $f(x) = 3x - 6$. Then

$$f(1) = 3 \cdot 1 - 6 = -3$$
$$f(0) = 3 \cdot 0 - 6 = -6$$
$$f(-1) = 3(-1) - 6 = -9$$
$$f(s) = 3s - 6 = 3(s - 2)$$
$$f(2k) = 3(2k) - 6 = 6k - 6 = 6(k - 1)$$
$$f(x + 1) = 3(x + 1) - 6 = 3(x - 1)$$
$$f[f(x)] = 3f(x) - 6 = 3(3x - 6) - 6 = 3(3x - 8)$$

Suppose now that $g(x) = 2x + 3$. Then $f[g(x)] = 3(2x + 3) - 6$ and $g[f(x)] = 2(3x - 6) + 3$.

EXERCISE 4.3

1. In equation (1) which is the independent and which the dependent variable? Why do we say that it is in the form $y = mx + b$?

2. Put the following functions in the form $y = mx + b$, and indicate the values of m and b in each case. Graph:

(a) $y = 4x + 3$ (f) $y = x/2 - 5 + x/3$

(b) $y = 4 - 3x$ (g) $y = -3(x - 2)$

(c) $y = x - 2$ (h) $2y = 6(x - 1 + y/6)$

(d) $y = 3$ (i) $4x + y = -3(1 - 2y)$

(e) $y = -2x + 3 + 4x$

Answers. (a) $m = 4$, $b = 3$. (c) $m = 1$, $b = -2$. (e) $m = 2$, $b = 3$. (g) $m = -3$, $b = 6$.

3. Suppose that $f(x)$ represents the function (1). Find $f(0)$, $f(5)$, $f(-20)$, and graph the function $y = f(x)$.

4. Find $f(3)$, where $f(x)$ stands successively for each function in prob. 2. [*Answers.* (a) 15. (c) 1. (e) 9. (g) -3. (i) 3.]

5. Suppose $f(x) = 6x + 1$. Find:

(a) $f(1)$ (e) $f(10)$ (i) $f(bc)$ (m) $f(x_2 - x_1)$

(b) $f(0)$ (f) $f(s)$ (j) $f[f(x)]$ (n) $f(5x + 3)$

(c) $f(-3)$ (g) $f(10t)$ (k) $f(x_1)$ (o) $f(x - 1)$

(d) $f(\frac{1}{6})$ (h) $f(-\frac{1}{2})$ (l) $f(x_2) - f(x_1)$ (p) $f(a + b + c)$

Answers. (a) 7. (c) -17. (e) 61. (g) $60t + 1$. (j) $36x + 7$.

6. Suppose that $f(x) = x^2$ and $g(x) = 2x + 1$. Find:

(a) $f(x) + g(x)$ (e) $f(2) - f(1)$ (i) $g(x + h) - g(x)$

(b) $f(x) - g(x)$ (f) $1 - f(x^2)$ (j) $f[g(x)]$

(c) $f(x)g(x)$ (g) $f(x + h)$ (k) $g[f(x)]$

(d) $f(x)/g(t)$ (h) $f(x + h) - f(x)$ (l) $f[-g(x)]$

Answers. (a) $(x + 1)^2$. (c) $2x^3 + x^2$. (e) 3. (g) $(x + h)^2$. (i) $2h$.

7. Graph on the same coordinate plane $f(x) = 2x + 1$ and $g(x) = x + 3$. Find and interpret graphically:

(a) $f(0)$ (c) $g(0) - f(0)$ (e) $f(2) - g(2)$ (g) $f(4) - f(2)$

(b) $g(0)$ (d) $f(1) - g(1)$ (f) $f(3) - g(3)$ (h) $g(1) - g(0)$

Answers. (a) $f(0) = 1$, which means that the graph of $y = f(x)$ crosses the y-axis 1 unit above the origin. (c) $g(0) - f(0) = 2$, which means that along the line $x = 0$ the graph $y = g(x)$ is 2 units above $y = f(x)$.

8. Sometimes when $y = f(x)$, we write $y(x)$ for $f(x)$, thus using the letter y to stand for the function as well as for the dependent variable. Suppose $y = y(x) = 1 - x$. Find: (a) $y(0)$. (b) $y(1)$. (c) $y(-4)$. (d) $y(t)$. (e) $y(R)$. (f) $y(-x)$. [*Answers.* (a) 1. (c) 5. (e) $1 - R$.]

9. The following are functions that are not linear:

(a) The day on which Easter comes is a function of the year.

(b) The area of a circle is a function of the radius.

(c) The speed of an automobile is a function of the amount of gas fed to the motor.

Justify these statements in terms of the definition of function, and give other examples.

10. In a certain retail store, sales-force salaries W are related to sales income S by the linear function $W = 0.04S + 5000$. What percentage of sales income goes to salaries when $S = \$100,000$? When sales are $\$50,000$? What conclusion do you draw?

11. If P dollars are left in the bank at *simple interest* rate r, the interest after a time t is Prt.

(a) Show that the total amount on deposit is $A = P(1 + rt)$.

(b) Graph A for $P = 10$, $r = 5\%$.

(c) Assuming that interest is paid only once a year in a single lump instead of continuously as we assumed above, graph A as a function of t.

12. On the Réaumur temperature scale the freezing and boiling points of water are $0°$ and $80°$, whereas they are $0°$ and $100°$ on the Centigrade scale and $32°$ and $212°$ on the Fahrenheit scale. Find formulas that give the Réaumur temperature r corresponding to a Centigrade temperature x and the corresponding Fahrenheit temperature y.

13. If $f(x) = 2x - 3$, show that:

(a) $f(x) + 1 \neq f(x + 1)$ (c) $f(x + y) \neq f(x) + f(y)$

(b) $f(2x) \neq 2f(x)$ (d) $f(x^2) \neq [f(x)]^2$

14. If y is a function of u, $y = f(u)$, and u is a function of x, $u = g(x)$, then y is a function of x (why?) and is given by $y = f[g(x)]$. In this case y is called a **composite function** of x. Find y as a function of x if:

(a) $y = 3u, u = 2x - 1$ (c) $y = 7u - 1, u = 1 - x$

(b) $y = 1 - u, u = 1 - 5x$ (d) $y = 1 - 4u, u = 1 - 4x$

15. Show that if y is a linear function of u, and u is a linear function of x, then y is a linear function of x.

16. The functional notation is by no means confined to functions in which the variables are numbers. For each of the following functions identify the range of the independent and dependent variables and find the values indicated:

(a) $f(x) =$ the world's champion boxer in the year x. What is $f(1950)$? $f(1940)$? $f(1930)$?

(b) $T(n) = 2n + 1$ is an odd number. Find $T(1)$, $T(2)$, $T(3)$.

(c) $T(x) =$ If x is a man, then x is mortal. Find $T(\text{Socrates})$, $T(\text{Arthur})$, $T(\text{a dog})$.

(d) $P(m) = (y = mx$ is a linear function). What are $P(1)$, $P(2)$, $P(-9)$, $P(a)$?

(e) $L(s) =$ the length of s. Find $L(\text{segment from } x = 1 \text{ to } x = 2$ on the axis).

†17. Find $f(x)$ if: (a) $f(a + b) = a + 1 + b$.   (b) $f(1 - x) = x - 1$.

18. If $f(x) = 1/x$, show that $f(1/a) = 1/f(a)$.

19. If $f(x) = x + 1$, show that $f(1/a) \neq 1/f(a)$.

20. If $f(x) = mx + b$, under what conditions does $f(x_1 + x_2) = f(x_1) + f(x_2)$?

†21. The **change** or **increment** in $y = f(x)$ when x changes from x_1 to x_2 is $f(x_2) - f(x_1)$. If two variables x and y are so related that the increment in y is directly proportional to the increment in x, show that y is a linear function of x.

22. Plot the set of points for which $y = 3 + 2x$ and $x = 1, 0, 2, 3$, or 5.

†23. Graph: (a) $y = |\, 2x - 1 \,|$.   (b) $y = |\, x + 3 \,|$.   (c) $y = 1 + |\, x - 1 \,|$.

†24. The **signum function** is defined as follows:

$$\text{sg } x = 1 \text{ when } x > 0$$
$$= -1 \text{ when } x < 0$$
$$= 0 \text{ when } x = 0$$

(a) Graph $y = \text{sg } x$.

(b) Show that $|\, x \,| = (\text{sg } x)x$.

(c) Graph $y = (\text{sg } x)2x + 1$.

(d) Graph $y = (\text{sg } x)(2x + 1)$.

4.4 Linear equations

Conditional equations of the type $mx + b = 0$, where m and b are constants with $m \neq 0$, are called **linear equations** in one unknown. The variable x is called the **unknown**. A conditional equation poses a problem to find the value of the unknown that makes the equation true. Such a value is said to satisfy the equation and is called a **root** or solution. It is also called a **zero** of $mx + b$.

It is easy to see that a linear equation in one unknown has a root. If we substitute $x = -b/m$ in the left member of

(1) $$mx + b = 0$$

we find:

(2) $$m(-b/m) + b = (-b) + b \quad \text{(Why?)}$$

(3) $$= 0 \quad \text{(Why?)}$$

Hence $-b/m$ is a solution of (1).

The algebraic problem of solving an equation of this type is always the same. Collect all terms involving x on one side of the equation and the rest on the other side; then divide both sides by the coefficient of x. In general terms, we write (1) as $mx = -b$ and then divide both sides by m.

Suppose, for example, that we wish (for some unaccountable reason) to find a number that added to its double is equal to 3, less one-half the sum of the number and 2. We let x stand for the unknown number. Then the number added to its double is $x + 2x$, and half the sum of the number and 2 is $(\frac{1}{2})(x + 2)$. The required condition is then

(4) $\qquad x + 2x = 3 - \dfrac{1}{2}(x + 2)$

(5) $\qquad\qquad 3x = 3 - \dfrac{x}{2} - 1 \qquad$ (Distributive Law)

(6) $\qquad 3x + \dfrac{x}{2} = 3 - 1 \qquad \left(\text{Adding } \dfrac{x}{2} \text{ to both sides}\right)$

(7) $\qquad\qquad \dfrac{7}{2}x = 2 \qquad$ (Distributive Law)

(8) $\qquad\qquad x = \dfrac{4}{7} \qquad \left(\text{Multiplying both sides by } \dfrac{2}{7}\right)$

We give the solution in somewhat greater detail than would be required in writing, but all the steps are necessary, even when some are carried out mentally. In solving equations, the student should be prepared to justify every step in terms of the theorems of Chapter 3.

In proceeding from equation to equation in the solution, we make two types of changes. The first type involves the use of identities to replace one expression by an identically equal expression. For example, in (5) we used the distributive law to replace $x + 2x = x(1) + x(2)$ by $x(1 + 2) = 3x$. The second type of change involves adding the same number to both sides of the equation or multiplying both sides by the same number. This procedure is justified by the uniqueness of addition and multiplication stated in *3.4.1 and *3.5.1. For either type of change the resulting equation follows from the original equation. That is, if the first equation is true for some value of the variable, so is the second. In the case of steps using identities, this is obvious since there has been no change in value. In the case of multiplication or addition applied to both sides, we have a new equation whose sides have values different from the old. But if the first equation is true for some value of the variable, so is the second. Thus each step in the solution follows from the previous one.

More than that, all the steps could be reversed. This is obvious for those based on identities. For the others, if we have proceeded by addition we can return by adding the negative. If we have proceeded by multiplication, we can return by multiplying by the reciprocal, *provided we did not multiply by zero.* Obviously, we cannot multiply by $1/0$ or, what is the same thing, divide by zero, since this operation is undefined. With this caution about multiplying or dividing by zero, every equation implies the following equation, and conversely. They have the same roots and are called **equivalent.** Thus a solution proceeds from one equivalent equation to another until one gives the roots we seek. The caution about multiplication by zero is important. If we multiply or divide by an expression containing the unknown (an expression that may be zero), we may lose roots or get additional ones (called **extraneous**) that do not satisfy the original equation.

So far we have been talking about methods of *finding* a root of an equation. This is not the same thing as *proving* that a particular value of the unknown is actually a root. To prove this we must substitute in the original equation to see whether the supposed root does actually make it a true statement. In order to avoid the error of assuming what we are trying to prove, it is desirable to substitute in each side separately, then see whether the two sides yield the same number. The process is called checking. We illustrate the process of checking the supposed root of equation (4). Replacing x by $4/7$ in the two members of (4), we have:

$$x + 2x = \frac{4}{7} + 2\left(\frac{4}{7}\right) = \frac{4}{7} + \frac{8}{7} = \frac{12}{7}$$

$$3 - \frac{1}{2}(x + 2) = 3 - \frac{1}{2}\left(\frac{4}{7} + 2\right) = 3 - \frac{1}{2}\left(\frac{18}{7}\right) = \frac{21}{7} - \frac{9}{7} = \frac{12}{7}$$

Solutions may sometimes be found by methods other than algebraic manipulation, for example, guesswork or mechanical means. Certain methods always lead to correct solutions if no errors are made, but the proof lies in checking.

Our study of linear functions suggests a simple geometric interpretation of the solution of a linear equation. The left member of $mx + b = 0$ is a linear function of x. The equation $y = mx + b$ graphs as a straight line. The value of x that makes $y = 0$ is just the solution of $mx + b = 0$. But this is the x-coordinate of the point where the line crosses the x-axis. This point is called the **x-intercept** of the line. (Sometimes "x-intercept" is used to refer to the x-coordinate of the point or to the segment joining it to the origin.) Hence to solve $mx + b = 0$ is the same as to find the x-intercept of the line $y = mx + b$. This fact can be used to find approximate solutions graphically.

EXERCISE 4.4

1. Solve for x the following linear equations, citing a reason for each step, and check all answers:

(a) $3x - 8 = 0$

(b) $5 - 10x = 0$

(c) $x - 6 = 2x + 1$

(d) $-x - bc = 2x$

(e) $0 = 5x - 100/9$

(f) $2x - 1 - 2(x + 1) = 15x/4$

(g) $-4x - c = -(3x - 1)$

(h) $2.4 - 9.6x = 1$

(i) $(4/3) - (3x/2)$
　　$= 1 - 2(x + 0.5) + 4(1 - x)$

(j) $ax - bx = c$

(k) $(2 + x)/(3 + x) = 2/3$

(l) $(2 + x)/(3 + x) = 1$

(m) $(2 + x)/(3 + x) = 2$

(n) $4/x = 3/5$

2. Put each of the equations in prob. 1 in the form $mx + b = 0$, graph the left member, and compare the algebraic solution with the intercept.

3. Justify (4)–(8) in the section.

4. Why did we specify $m \neq 0$ in $mx + b = 0$? What conditions must be put on a, b, and c in part j of prob. 1 so that there is a solution?

5. Solve the following for x and check:

(a) $wx + t = 0$

(b) $ab - cx = ax$

(c) $3x - 3c = bx + 4 - cx$

(d) $x/(x + sx) = 2s - a$

(e) $(x/D) - (x/E) = 1$

(f) $y = (9/5)x + 32$

(g) $ax + by + c = 0$

(h) $u = x - x_0$

(i) $(x/2) - (2x/3) = 1/6$

(j) $(1/x) - (2/3x) = 7$

Answers. (a) $-T/w$. (c) $(4 + 3c)/(3 - b + c)$. (e) $DE/(E - D)$.

6. Solve for each variable in turn:

(a) $I = 2E/(R + 2r)$

(b) $N = 1 \left/ \left(\dfrac{1}{s} + \dfrac{1}{k} \right) \right.$

7. The following equations are taken from the fields indicated. Solve each one for each variable for which it is linear.

(a) $E = \frac{1}{2}mv^2$　　(Physics)

(b) $w = kbE/R^2$　　(Astronomy)

(c) $A = P(1 + i)^n$　　(Finance)

(d) $F = C + 2 - P$　　(Chemistry)

(e) $M = mgl^3/4sa^3b$　　(Engineering)

(f) $C = Krr'/(r - r')$ (Electricity)

(g) $1/F = 1/f_1 + 1/f_2$ (Optics)

(h) $k\bar{u} = \bar{x} - x_0$ (Statistics)

(i) $\pi = up - q$ (Economics)

(j) $v/V = \Delta/\lambda$ (Sound)

(k) $1/R = (\cos^2 \theta)/p_1 + (\sin^2 \theta)/p_2$ (Differential Geometry)

(l) $m = E/2(1 + d)$ (Physics)

(m) $w = hv - w_0$ (Quantum Mechanics)

(n) $v = 2dgr^2/9n$ (Physics)

(o) $x/(x + y + 1) = y'$ (Differential Equations)

(p) $2/(R + L) = d/D$ (Astronomy)

(q) $G_d = G_m(1 - P)$ (Mining Engineering)

(r) $F = (1 - i)(1 - s)(1 - r)$ (Petroleum Engineering)

(s) $2d = b(b - 1)/n$ (Genetics)

8. Find a number such that: (a) 2/3 of it is 18. (b) 9/8 of it is 36. (c) Its reciprocal is −2/3. (d) 37% of it is 98. (e) 2% of it is 41.3. (f) 99% of it is 149.
Answers. (a) 27. (c) −3/2. (e) 2065.

9. Find a number that added to its double is equal to 3 less than one-half the sum of the number and 2. (Compare with the problem in the section and notice that "A less B" means $A - B$, "A less than B" means $B - A$, and "A is less than B" means $A < B$.)

10. Without graphing, find the x-intercepts of the following straight lines: (a) $y = -2x + 3$. (b) $y = 3.17x + 28$. (c) $2y - 3x = 1$. (d) $1 = 2x + 8y$. [*Answers.* (a) 3/2. (c) −1/3.]

11. The fat man at the circus unexpectedly lost one-third of his weight. By careful dieting he managed to regain 10 pounds. Then he found some pills that increased his weight by 20%. His final weight was 500. What was his original weight? (*Answer.* 610.)

12. A young man became a great success as a radio singer. He earned a large income the first year. His agent took 10%, income taxes took 40% of the remainder, and his living expenses rose by 1000% so that he did not save any money. If he formerly lived on $2000 a year, what was his new income?

13. A wealthy old gentleman gave his eldest son one-third of his estate when he reached 21. The next son got one-third of the remainder, and the youngest son one-third of what was left. When someone protested that this was unfair, the father replied: "The young rascal still got $120,000!" How much was the original estate? (*Answer.* $810,000.)

14. The following procedure will enable you to tell a number that has been chosen by another person. Require him to subtract one, multiply the result by 4, add 5 to this, double this result, and tell you the final number. Work out the rule by which you can quickly tell the number he chose.

15. A person's age is 25 plus half his age. How old is he?

16. Find three consecutive even numbers whose sum is 204.

17. If each person in the world absorbs from his food 1.8×10^{-10} micrograms of plutonium per year from each atomic bomb that has been exploded in the world, and if a person cannot safely absorb more than 0.07 microgram per year, how many bombs must be exploded to endanger the world's population by food contamination. (*Answer.* 4×10^8.)

18. Two grades of nuts sell at 69 and 52 cents a half pound. A mixture of two parts of the first to one of the second sells for 60 cents. Is this the proper price? What should the price be? (You should be able to answer the first question without calculation.)

19. A soldier hears a bullet "crack" overhead, and one second later he hears the "thump" of the weapon that fired it. If the speed of sound is 1100 feet per second and such bullets travel at 3000 feet per second, how far away is the weapon? (*Answer.* About 600 yards.)

20. The radiator of a car is full of a mixture of equal parts of water and anti-freeze. The owner wishes to have the mixture 75% anti-freeze. He drains out part of the mixture and replaces it with pure anti-freeze. What part?

21. How many grams of 8% salt solution are required to yield 1.7 grams of salt? (*Answer.* About 21.)

22. (a) How much solvent must be added to a 60% solution to reduce it to 50%? (*Answer.* 20% of the original.)

(b) How much solvent must be added to an 83% solution to reduce it to 62.5%.

(c) How much solute must be added to increase the concentration of a solution from 18% to 40%.

23. The following genetical formula, known as Dahlberg's formula, gives the probability that certain types of individuals have parents who were first cousins: $k = C(1 + 15q)/16q$. Solve for q. [*Answer.* $C/(16k - 15C)$.]

24. Solve the following for y' in terms of the other variables:

(a) $xy' + 3xy^2 + 2y' = 0$ (c) $2axy' - 3byy' - y = 0$

(b) $6x + 8yy' = 2$ (d) $zy - 2zyy' = y'$

Answers. (a) $-3xy^2/(x + 2)$. (c) $y/(2ax - 3by)$.

25. Solve for k/h: $2a^2k - 3x^2yk + 3xyk + 5k - 2h = 0$.

26. Solve for T and simplify: $2T\pi r = \pi r^2 hpq + (\frac{1}{3})\pi r^3 pq$.

†**27.** For what value of m is there no x satisfying $x + a = m(x + b)$? Why?

†**28.** If $F \cos \theta + f - w = 0$, $R - F \sin \theta = 0$, and $mR = f$, show that $mF \sin \theta = w - F \cos \theta$.

†**29.** Show that $\dfrac{1}{x} = \dfrac{1}{x} + 1 \longrightarrow x = 0$ but not conversely, so that 0 is an extraneous root of this equation.

†**30.** Prove that a linear equation has *only* one root. (*Hint.* Assume that there are two and prove them equal.)

4.5 Arithmetic progressions

In this section we study linear functions in which the independent variable takes only positive integral values. The values of the function arranged in the same order as the corresponding positive integers form what is called an arithmetic progression. Consider the function $y = 2x + 1$. Its values for the first few integral values of x are shown in the table:

$$x \quad 1, \quad 2, \quad 3, \quad 4, \quad 5, \quad 6, \quad \cdots$$

$$y \quad 3, \quad 5, \quad 7, \quad 9, \quad 11, \quad 13, \quad \cdots$$

Note that the position of each y is given by the corresponding x. Thus the first value of y is 3 and the fifth is 11. Also the difference between any two successive values of y is the same, namely, 2. We say that the numbers 3, 5, 7, 9, $\cdots$ form an arithmetic progression with first term 3 and common difference 2.

In order to give a precise general definition of an arithmetic progression, we first introduce the idea of **sequence**. Suppose that y is a single-valued function of n, defined for all positive integral values of n. We designate by y_n (read "y– sub n") the value of y corresponding to n. The set of values $y_1, y_2, y_3, y_4, \cdots$ $y_{n-1}, y_n, y_{n+1}, \cdots$ is called a sequence. The individual values of y are called **terms** of the sequence, and y_n is called the **general term** or nth term. We use n in place of x for the independent variable and y_n in place of $f(x)$ in order to emphasize that the independent variable takes only positive integral values.

The positive integers themselves form a sequence; 1, 2, 3, $\cdots$. The first term is $y_1 = 1$, the second term is $y_2 = 2$, etc. The nth term is n itself, that is, $y_n = n$. There are an unlimited number of different types of sequences. Any set of objects arranged in correspondence with the positive integers will do. For example, $-1, 1, -1, 1, -1, \cdots$ are the first few terms of the sequence whose nth term is $(-1)^n$. Also 2, 2, 2, $\cdots$ is the beginning of the sequence whose terms are all 2. In order to define a sequence, it is sufficient to know the function that gives the nth term, in other words, a rule that tells what each term will be. This rule may be a formula that tells directly each value of y_n, or it may be a rule that tells how to get from each term to the following one. The entire sequence may be indicated conveniently by writing the formula for the nth term in braces, for example, $\{y_n\}$. The example with which we began this section is the sequence 3, 5, 7, $\cdots$. The first term is $y_1 = 3$. The nth term is $y_n = 2n + 1$. The entire sequence may be represented by $\{2n + 1\}$. We noted that the difference between two successive terms is 2, that is, $y_2 - y_1 = 2$, $y_3 - y_2 = 2$, etc. In general, $y_{n+1} = y_n + 2$. We take this feature as the defining feature of arithmetic progressions.

*4.5.1 An **arithmetic progression** is a sequence of numbers $\{y_n\}$, such that the difference between any two successive terms is a constant: $y_{n+1} = y_n + d$. (Definition.)

Suppose we have an arithmetic progression with **first term** y_1 and **common difference** d. Then the above definition gives $y_2 = y_1 + d$, $y_3 = y_2 + d = y_1 + 2d$. $y_4 = y_1 + 3d$, $y_5 = y_1 + 4d$, etc. Noting that the number of d's added is one less than the number of the term, we conjecture (that is, guess) that this holds in all cases and write:

*4.5.2 The nth term of an arithmetic progression with common difference d is given by

$$y_n = y_1 + (n-1)d$$

This formula enables us to find any term directly, without using the definition to go on from term to term. Thus in the sequence $\{2n+1\}$, $y_{30} = 3 + 29\cdot2 = 61$. The formula *4.5.2 shows that the nth term of an arithmetic progression is a linear function of n. Conversely, any linear function defines an arithmetic progression, for, if $y_n = dn + e$, it is easy to show that $y_{n+1} - y_n = d$, and $y_1 = d + e$.

It is often of interest to calculate the sum of the first n terms of an arithmetic progression. For example, in statistics we run across sums like

(1) $\qquad 1 + 3 + 5 + 7 + 9 + 11 + \cdots + (2n-1)$

We note that $1 + 3 = 4 = 2^2$, $1 + 3 + 5 = 9 = 3^2$, and $1 + 3 + 5 + 7 = 16 = 4^2$. It seems reasonable to conjecture that the sum of the first n odd integers is just n^2, a neat and practical result, if true. We shall find heuristically a formula that gives the sum of the first n terms of any arithmetic progression and, incidentally, the justification of our guess about the odd integers.

Let s_n stand for the **sum of the first n terms of a sequence:**

(2) $\qquad s_n = y_1 + y_2 + y_3 + \cdots + y_{n-1} + y_n$

For an arithmetic progression, this becomes:

(3) $\quad s_n = y_1 + (y_1 + d) + (y_1 + 2d) + \cdots$
$$+ (y_n - 2d) + (y_n - d) + y_n$$

We now rewrite this sum in reverse order:

(4) $\quad s_n = y_n + (y_n - d) + (y_n - 2d) + \cdots$
$$+ (y_1 + 2d) + (y_1 + d) + y_1$$

It appears that pairs of corresponding terms in the two sums add to $y_1 + y_n$. Since there are n terms, adding (3) and (4) apparently gives

(5) $\qquad\qquad 2s_n = n(y_1 + y_n)$

and we have

*4.5.3 The sum of the first n terms of an arithmetic progression is given by

$$s_n = \frac{n}{2}(y_1 + y_n)$$

The formula is easy to remember since it gives the sum as one half the product of the number of terms and the sum of the first and nth term. (For a proof see Ex. 6.2.3.)

In Fig. 4.5 we sketch the first few values of $y_n = y_1 + (n - 1)d$. We have not drawn the line upon which the circled points lie

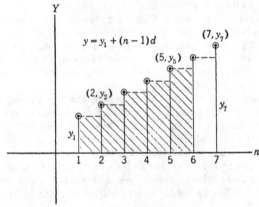

FIG. 4.5

because we are interested only in the values of y corresponding to the integral values of n. Each y_n is represented by the coordinate of its circled point and also by the area of a rectangle of base 1 and height y_n. Hence s_n is the sum of the first n of these rectangles, or the total area under the step-like dotted line. We have shaded s_5.

EXERCISE 4.5

1. Write the formula for the nth term of each of the arithmetic progressions whose first term and common difference are:

(a) 0, -2	(c) 2, 1	(e) 75, 20	(g) $-3, 0.2$
(b) 3, 4.5	(d) $\frac{1}{2}, \frac{1}{3}$	(f) 6, 4.8	(h) $-1.7, -0.03$

Answers. (a) $-2(n-1)$. (c) $n+1$. (e) $55+20n$. (g) $-3.2+0.2n$.

2. Find the 21st term and the sum of the first 17 terms of each progression in prob. 1.

Answers. (a) $-40, -272$. (c) 22, 170.

3. Write out the first five terms, the 20th term, and the sum of the first ten terms of each of the following arithmetic progressions:

(a) $\{3n\}$	(d) $\{-1 - n/2\}$	(g) $\{14n - 13\}$
(b) $\{-n\}$	(e) $\{n/6\}$	(h) $\{-0.25n + 0.8\}$
(c) $\{1 + 3n\}$	(f) $\{2n + 2\}$	

Answers. (a) 3, 6, 9, 12, 15; 60; 165. (c) 4, 7, 10, 13, 16; 61; 175.

4. Sketch several of the progressions in the foregoing problems and shade some s_n. (Note that areas below the x-axis have to be considered negative.)

5. A printer charges $5 for the first one hundred posters. Each additional hundred costs 50 cents less than the preceding until a flat rate of $2 per hundred is reached. (a) What is the first hundred for which $2 is charged? (b) How much would 1000 cost? [*Answers.* (a) 7. (b) $30.50.]

6. A company wishes to establish a schedule of regular salary increases. It wants to have the annual salary rise from a minimum of $2800 to a maximum of $3600 in 10 years. If increases are given every two years, what will be the annual increase? Suppose instead that the company decides to give a raise every year. What should it be? Which system will cost the company more and by how much per employee?

7. A progressive income tax is set so that the first thousand dollars is taxed 1%, the second thousand 2%, and each successive thousand 1% more than the preceding one. Thus the tax on $3000 would be $60. (a) Find the tax on $90,000. (b) Find the tax on $94,300. (*Suggestion.* Consider the sequence of payments on successive thousands.)

8. Pipe is piled so that each length rests on the depression formed by two lengths in the layer below and each layer has one less length than the one below. If a pile is 8 rows high, how many lengths are in the bottom layer? In the whole pile?

9. A clock strikes the hours and also once on the quarter, twice on the half, and three times at a quarter to the hour. How many times does it strike in 24 hours? (*Answer.* 300.)

10. If $45 is deposited at simple interest of 2% per year what will be on deposit after 1 year, 2 years, 40 years, n years? (See Ex. 4.3.11.)

11. Show that when P dollars are deposited at simple interest rate r payable annually, the amounts in the account at the ends of successive periods form an arithmetic progression. What is the common difference? (*Answer.* A_n = $P + Prn$, where n is the number of years.)

12. Suppose a man deposits $100 in postal savings at the end of each year for 20 years. Each deposit earns 2% simple interest. How much will he have at the end of the 20th year? (*Suggestion.* Start at the end of the period. His last deposit has just been made; hence it amounts to just $100. The previous year's deposit is now worth $102.)

13. A sequence of equal periodic payments or deposits is called an *annuity*. Find the value at rate of simple interest r of an annuity of n payments of R each at the moment of making the nth payment. [*Answer.* S_n = $(Rn/2)(2 + (n-1)r)$.]

14. In formulas *4.5.2 and *4.5.3 there are five variables: y_1, y_n, s_n, d, and n. Given any three, we can find the other two by manipulating the formulas. For example, if we have $y_1 = 4$, $d = -2$, and $n = 6$, then $y_6 = 4 + 5(-2)$ = -6, and $s_n = (6/2)[4 + (-6)] = 3(-2) = -6$. In the following find the missing variables:

(a) $n = 8$, $y_1 = 3$, $d = 4$

(b) $n = 7$, $y_1 = 2$, $d = -1$

(c) $y_1 = 5$, $y_6 = -5$

(d) $y_1 = -3$, $y_{10} = 20$

(e) $y_1 = 2$, $d = 0.5$, $n = 4$

(f) $d = 1.5$, $y_9 = 20$

(g) $y_1 = 4$, $y_n = 25$, $d = 3$

(h) $y_1 = -3$, $y_n = 21$, $s_n = 45$

(i) $y_n = -20$, $y_1 = 5$, $s_n = -75$

(j) $y_8 = 15$, $s_8 = 80$

(k) $s_9 = -50$, $y_1 = 4$

(l) $y_1 = -8$, $s_{11} = 30$

(m) $d = 3$, $s_5 = 40$

(n) $d = -1$, $s_{10} = 14$

(†o) $y_1 = 5$, $d = -1$, $s_n = 5$

(†p) $d = 2$, $y_n = 8$, $s_n = 20$

Answers. (a) $y_8 = 31$, $s_8 = 136$. (c) $d = -2$, $s_6 = 0$. (e) $y_4 = 3.5$, $s_4 = 11$. (g) $n = 8$, $s_8 = 116$. (i) $n = 10$, $d = -25/9$. (k) $y_9 = -136/9$, $d = -43/18$. (m) $y_1 = 2$, $y_5 = 14$. (o) $n = 1$, $y_n = 1$ or $n = 10$, $y_n = -4$.

15. Solve *4.5.2 for y_1, n, and d. Solve *4.5.3 for n, y_1, and y_n.

16. Use *4.5.3 to prove the following:

*4.5.4 $\quad 1 + 3 + 5 + 7 + 9 + \cdots + (2n - 1) = n^2$

*4.5.5 $\quad\quad 2 + 4 + 6 + 8 + 10 + \cdots + 2n = n(n + 1)$

*4.5.6 $\quad\quad 1 + 2 + 3 + 4 + 5 + \cdots + n = \dfrac{n(n + 1)}{2}$

17. Finding the $n - 2$ terms that lie between y_1 and y_n in an arithmetic progression is called inserting $n - 2$ **arithmetic means** between y_1 and y_n. Note that the number of means inserted is two less than the total number of terms in the resulting arithmetic progression. Insert: (a) Four arithmetic means between 5 and 15. (b) Six arithmetic means between -1 and 13. (c) Two arithmetic means between 4 and 6. (d) Twenty means between 3 and 50. (e) Three means between -8 and 17. (f) One mean between 9/2 and 11/2. [*Answers.* (a) 7, 9, 11, 13. (c) 14/3, 16/3. (e) -1.75, 4.50, 10.75.]

18. When just one mean is inserted between two numbers it is called **the arithmetic mean.** Show that the arithmetic mean of a and b is given by $\bar{x} = (a + b)/2$.

19. If x_1, x_2, $\cdots$ x_n are n numbers, their arithmetic mean or **average** is by definition:

$$(6) \qquad \bar{x} = \frac{x_1 + x_2 + \cdots + x_n}{n}$$

Use *4.5.4,5,6 to show that: (a) The average of the first n odd integers is n. (b) The average of the first n even integers is $n + 1$. (c) The average of the first n integers is $(n + 1)/2$.

20. Show that the average of the first n terms of any arithmetic progression is the average of the first and last terms. Use this result to restate *4.5.3 in words.

21. Prove the statement in the section that any linear function defines an arithmetic progression whose common difference is the coefficient of the variable.

22. Write the first five terms of the following sequences and indicate which are arithmetic progressions:

    (a) $\{n^2\}$ (d) $\{n(n - 1)\}$ (g) $\{(-\tfrac{1}{2})^n\}$
    (b) $\{4n + 1\}$ (e) $\{n^3\}$ (h) $\{1/n\}$
    (c) $\{n^2 - 1\}$ (f) $\{2^n\}$ (i) $\{1/2n\}$

    (j) $\{n + (n - 1)(n - 2)(n - 3)(n - 4)\}$
    (k) $\{n^2 + (n - 1)(n - 2)(n - 3)\}$
    (l) $\{2n + 3 + (n - 1)(n - 2)(n - 3)(n - 4)\}$

Answers. (a) 1, 4, 9, 16, 25. (c) 0, 3, 8, 15, 24. (e) 1, 8, 27, 64, 125. (g) $-\tfrac{1}{2}$, $\tfrac{1}{4}$, $-\tfrac{1}{8}$, $\tfrac{1}{16}$, $-\tfrac{1}{32}$. (i) 1, $\tfrac{1}{4}$, $\tfrac{1}{6}$, $\tfrac{1}{8}$, $\tfrac{1}{10}$.

23. A relation that gives the general term of a sequence in terms of preceding terms is called a **recurrence relation.** The equation in *4.5.1 is such a relation for an arithmetic progression. In each case below, the first term or terms of a sequence and a recurrence relation are given. Find the next three terms:

    (a) $y_1 = 1$, $y_n = 2y_{n-1}$ (d) $y_1 = a$, $ny_n = ay_{n-1}$
    (b) $y_1 = 1$, $y_n = y_{n-1} + y^2_{n-1}$ (e) $y_1 = 1$, $y_{n+1} = (1 + \sqrt{y_n})^2$
    (c) $y_1 = 4$, $y_n = (-1)^n y_{n-1}$ (f) $y_1 = 10$, $y_n = (y_{n-1} + y_{n-2})/2$

Answers. (a) 2, 4, 8. (c) 4, -4, -4. (e) 4, 9, 16.

24. The numbers defined by $y_1 = 0$, $y_2 = 1$, and $y_n = y_{n-1} + y_{n-2}$ are called Fibonacci numbers. Calculate the first 8.

25. A sequence of numbers whose reciprocals form an arithmetic progression is called a **harmonic progression.** Show that the following are harmonic progressions:

    (a) 1, $\tfrac{1}{2}$, $\tfrac{1}{3}$, $\cdots$, $1/n$, $\cdots$ (d) $\{2/(3 - 2n)\}$
    (b) 2, $\tfrac{4}{3}$, 1, $\tfrac{4}{5}$, $\tfrac{2}{3}$, $\cdots$ (e) $\{1/(1 + n)\}$
    (c) 3, 2, $\tfrac{3}{2}$, $\tfrac{6}{5}$, 1, $\cdots$ (f) $\{n/(n^2 + 3n)\}$

26. What is the formula for the general term of the harmonic progression that begins with the terms shown in part *b* in the preceding problem? Part *c*?

27. The terms between two terms of a harmonic progression are called **harmonic means.** Insert: (*a*) Two harmonic means between $\frac{1}{6}$ and $\frac{1}{2}$. (*b*) Three harmonic means between $\frac{1}{6}$ and $\frac{1}{2}$. (*c*) Three between *a* and *b*. (*d*) One between *a* and *b*. (This is called the **harmonic mean** of *a* and *b*.) *Answers.* (*a*) $\frac{1}{4}$, $\frac{1}{3}$. (*d*) $2ab/(a+b)$.

28. Numbers that are the sums of arithmetic progressions with integral differences have interesting geometric representations. For example, the sums of the first *n* terms of the progression of positive integers ($d = 1$) are given by the following arrangement of dots in triangles:

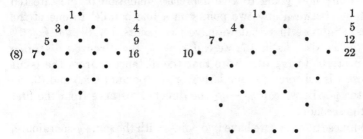

(7)

The terms of the progression appear at the left (equal to the number of dots in the line), and the sums of terms appear at the right (the number of dots in the triangle down to that row). Such numbers may be called "triangular." For $d = 2$ and $d = 3$, we get the "square" numbers and the "pentagonal" numbers, which may be sketched as follows:

(8)

Add additional rows of dots to each of the diagrams above. Construct a similar schema for $d = 4$ and suggest a name for these numbers. Write formulas for the *n*th pentagonal and hexagonal numbers.

4.6 Analytic geometry

We have been using geometry to illustrate arithmetic and algebra, but we could equally well use algebra to study geometry. We have interpreted a pair of numbers as a point and a linear function as a straight line, but it would be just as reasonable to interpret a point as a pair of numbers and any geometric figure as the set of its points, which are in turn looked upon as pairs of numbers. Consider, for example, the function $y = -2x + 3$. We used the straight line (Fig. 4.3) to show the relation between x and y, but we could just as well use the equation $y = -2x + 3$ to study the geometrical properties of the straight line.

When geometry is studied in this way it is called **analytic geometry.** We set up a coordinate system and so establish a one-to-one correspondence between points and pairs of coordinates. Then a geometrical figure (or locus, or curve) is considered as the set of points (pairs of coordinates) of which it is composed. If we have an equation that is satisfied by these coordinates and by no others, we call it the **equation of the locus** or the equation of the curve. We call the locus the graph of the equation. The **graph of an equation** is defined as *the set of points whose coordinates satisfy the equation.* (See **2.7.**) Since we think of a point as just an ordered pair of numbers, we speak of a point as satisfying an equation, meaning, of course, that its coordinates do so. Thus the graph of an equation is the set of points that satisfy it, and a point lies on a graph *if and only if* it satisfies the equation. Sometimes we speak of the **graph of a function,** $f(x)$, meaning the graph of $y = f(x)$.

We are now going to give a precise definition of the directed distance between any two points on a line parallel to one of the axes. Suppose first that we have two points, $(x_1, 0)$ and $(x_2, 0)$, on the x-axis. As in 3.7, we call $x_2 - x_1$ the **directed distance** from $(x_1, 0)$ to $(x_2, 0)$. Note that the distance is from the point whose x is subtracted. Similarly, for two points $(0, y_1)$ and $(0, y_2)$ on the y-axis, we call $y_2 - y_1$ the directed distance from the first to the second.

Suppose now that we have two points with the same y-coordinate, (x_1, y_0) and (x_2, y_0). Again we call $x_2 - x_1$ the **directed distance** from the first to the second. Similarly, $y_2 - y_1$ is called the directed distance from (x_0, y_1) to (x_0, y_2). The absolute value of the directed distance is called the **distance** between the two points, as in Chapter 3. Typical points and the vectors representing the directed distances are sketched in Fig. 4.6.1. This convention means that we regard the plane as covered by lines parallel to the axes. Along each of these lines we locate points, lay out vectors, and measure distances just as we did on the axes in Chapter 3. The coordinates of a point are vectors or directed distances

measured along these lines. On ordinary graph paper a large
number of these lines are printed to facilitate location of points.

At this stage of our discussion it would be natural to ask where
the coordinate axes should be placed. Of course there is nothing

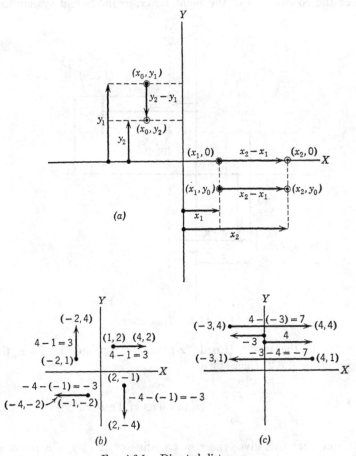

FIG. 4.6.1. Directed distances.

to stop us from drawing them where we wish, but it may be more
convenient to locate them in some particular way. For example,
it turns out that in studying the circle it is best to take the origin
at the center. In order to answer the question, we have to study
what happens when we move the axes about. In this section we

limit ourselves to **translation,** that is, moving the axes parallel to themselves without rotation. What effect will it have on the coordinates of a point if we move the axes from an original position XOY to a new position $X'O'Y'$ with origin at the point (x_0, y_0)? Let the coordinates of the point be (x, y) in the old system and

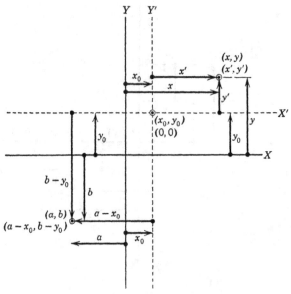

Fig. 4.6.2.   Translation.

(x', y') in the new. (Read "x-prime," "y-prime.") Then the figure (Fig. 4.6.2) suggests:

*4.6.1
$$x' = x - x_0$$
$$y' = y - y_0$$
(Equations of Translation)

Note that this gives the new coordinates (x', y') of a point as the old coordinates (x, y) minus the coordinates of the new origin (x_0, y_0) with reference to the old axes. The formulas follow from our discussion of directed distance, since x' is by definition the directed distance from the point (x_0, y) to the point (x, y), and this is always $x - x_0$. Similarly for the y-coordinates. The equations can of course be solved for the old coordinates to give $x = x' + x_0$, $y = y' + y_0$.

Suppose we have axes XOY and another set $X'O'Y'$ parallel to the first with origin at $(3, -2)$ (Fig. 4.6.3). Consider the line whose equation is $y = -2x + 4$. What is its equation in the new coordinates x' and y'? Since every $x = x' + 3$ and every $y = y' - 2$, the set of points whose old coordinates satisfy $y = -2x + 4$

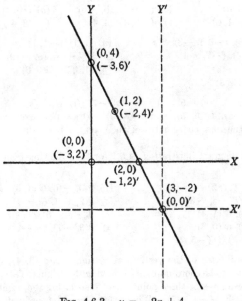

FIG. 4.6.3. $y = -2x + 4$.

is just the set of points whose new coordinates satisfy $(y' - 2) = -2(x' + 3) + 4$. Hence the new equation of the *same* line is $y' = -2x'$.

When we make a translation, we replace the old variables by new ones. We look at the same point or the same graph in new terms, in terms of the new variables. Such a change may be described as a *transformation* of variables or a change of frame of reference. In the example above, the line $y = -2x + 4$ became $y' = -2x'$ in the new variables. Evidently we could similarly eliminate the constant term in any linear function by translating the origin to some point on its graph (why?) and so reduce any linear function to a direct variation. This illustrates an important fact, namely, that frequently a judicious change of variables simplifies a problem.

EXERCISE 4.6

1. Find the directed distance from the first to the second point in each of the following pairs of points, and sketch the corresponding vectors in a coordinate plane:

(a) $(3, 0)$, $(8, 0)$ (d) $(0, -3)$, $(0, 6)$ (g) $(4, 1)$, $(-3, 1)$
(b) $(8, 0)$, $(3, 0)$ (e) $(0, -1)$, $(0, -6)$ (h) $(10, 5)$, $(-8, 5)$
(c) $(4, 0)$, $(-1, 0)$ (f) $(-3, 4)$, $(4, 4)$ (i) $(8, 4)$, $(6, 4)$

(j) $(0.5, -3)$, $(-0.5, -3)$ (n) $(1, 1)$, $(-1, 1)$
(k) $(-2, -4)$, $(-7, -4)$ (o) $(5, -18)$, $(5, -19)$
(l) $(-4, 3)$, $(5, 3)$ (p) $(1.3, 2)$, $(1.3, 8)$
(m) $(2, 3)$, $(2.5, 3)$

Answers. (a) 5. (c) -5. (e) -5. (g) -7. (i) -2. (k) -5. (m) 0.5.

2. Plot $(2, 8)$, $(-5, 5)$, and $(-3, 2)$. Draw new axes with center at $(1, 4)$. Find the new coordinates of these points from the graph and from *4.6.1.

3. Draw a diagram illustrating *4.6.1 for the following points Q and new origins O':

(a) $Q(6, 3)$, $O'(1, 2)$ (f) $Q(0, -6)$, $O'(4, 6)$
(b) $Q(4, 5)$, $O'(-5, -3)$ (g) $Q(2, 4)$, $O'(3, -1)$
(c) $Q(-2, 3)$, $O'(1, 4)$ (h) $Q(-3, -5)$, $O'(-3, 5)$
(d) $Q(4, -2)$, $O'(0, 3)$ (i) $Q(3, -7)$, $O'(-1, -10)$
(e) $Q(-3, -7)$, $O'(-5, 2)$

4. The coordinates of the vertices of a triangle are $(3, 4)$, $(-1, 5)$, and $(4, -1)$. What are the coordinates after moving the origin to $(-1, 5)$? Sketch.

5. The coordinates of three points after translating the origin to $(2, -1)$ are $(0, 0)$, $(5, -3)$, $(-1, -7)$. What were the old coordinates? [*Answer.* $(2, -1)$, $(7, -4)$, $(1, -8)$.]

6. The new and old coordinates of a point are $(3, -1)$ and $(-5, 2)$. Where is the new center of coordinates? Sketch old and new axes.

7. For each of the following equations, sketch the line and the new axes with origin as indicated. Find the equation in the new coordinates.

(a) $y = 3x - 2$, $(1, 1)$ (e) $y = -2$, $(0, -2)$
(b) $y = 5$, $(3, -1)$ (f) $y = -2x + 3$, $(1.5, 0)$
(c) $y = -x + 7$, $(0, 7)$ (g) $y = mx + b$, $(0, b)$
(d) $2y = 4x + 6$, $(-1, -3)$ (h) $y = mx + b$, $(-b/a, 0)$

Answers. (a) $y' = 3x'$. (c) $y' = -x'$. (e) $y' = 0$. (g) $y' = mx'$.

8. For the points $(2, 3)$, $(-1, 5)$, $(6, 9)$, and $(3, 5)$:

(a) Find the average $\bar{x}$ of the x's and the average $\bar{y}$ of the y's.
(b) Plot $(\bar{x}, \bar{y})$.
(c) Shift the origin to $(\bar{x}, \bar{y})$ and compute the coordinates of the original points in the new system.
(d) Calculate the averages of the new x's and y's.
(†e) Suggest and prove a theorem suggested by the result in part d.

9. The zero point (origin) of the Absolute temperature scale is at $-273\,°\text{C}$. Letting x and x' stand for temperatures Centigrade and Absolute, describe the relation between them as a translation and express x' in terms of x.

10. If A, B, C, and D are points on a straight line and AB, BC, CD, and DA represent directed distances, show that $AB + BC + CD + DA = 0$.

11. A point P is said to **divide the segment P_1P_2 in the ratio** r_1/r_2 if P is on the line through P_1 and P_2 and if $P_1P/PP_2 = r_1/r_2$. It is here assumed that a positive direction has been chosen on the line so that P_1P and PP_2 are directed distances. However, the ratio P_1P/PP_2 is not changed by changing the positive direction on the line. (Why?) Show by using the similar triangles in Fig. 4.6.4 that:

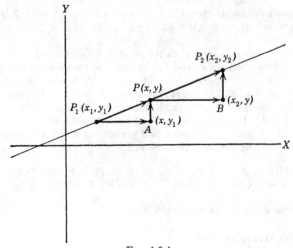

FIG. 4.6.4

*4.6.2 The point $P(x, y)$ that divides the segment joining $P_1(x_1, y_1)$ and $P_2(x_2, y_2)$ in the ratio r_1/r_2 is given by

$$x = \frac{r_1 x_2 + r_2 x_1}{r_1 + r_2}$$

$$y = \frac{r_1 y_2 + r_2 y_1}{r_1 + r_2}$$

12. Show that if P is the midpoint of P_1P_2, then $r_1/r_2 = 1$ and the coordinates of P are $[\frac{1}{2}(x_1 + x_2), \frac{1}{2}(y_1 + y_2)]$.

13. Find and sketch the midpoints of the following segments:

(a) $(3, -7)$, $(1, 5)$ (c) $(5, 0)$, $(-2, 0)$ (e) $(a - b, 0)$, $(b - a, 0)$

(b) $(-5, 2)$, $(5, -1)$ (d) $(-4, 0)$, $(0, 3)$ (f) $(c - 1, 0)$, $(0, 1 - c)$

14. Find and sketch the point that divides each of the following segments in the given ratio. The first point is to be taken as P_1. Note that when the ratio is an integer, it may be considered to have a denominator 1:

(a) (2, 4), (8, 4), 3/1   (f) (4, s), (6, $s - 2$), 1

(b) (0, 0), (8, 6), 2/3   (g) (r, s), ($r + 2$, $s - 2$), 0.5

(c) (−1, 2), (3, 5), −1/3   (h) (−21, 2), (1, 4), 5/3

(d) (−2, 0), (4, 0), −3/4   (i) (1, 3), (8, 20), 100

(e) (2, 8), (−2, 9), 1   (j) (1, 3), (8, 20), 0.01

Answers. (a) (13/2, 4). (c) (−3, 1/2). (e) (0, 17/2). (g) [(3r + 2)/3, (3s − 2)/3]. (i) (801/101, 2003/101).

15. Sketch the points that divide the segment from (2, 1) to (5, 9) in the following ratios: (a) −0.9. (b) −0.1. (c) 0. (d) 1. (e) 3. (f) 100. (g) −110. (h) −2. (i) −1.1.

16. What happens to r_1/r_2 when the point of division is: (a) At P_1? (b) At P_2? (c) Halfway between P_1 and P_2? (d) Outside the segment on the side of P_1? (e) Outside the segment on the side of P_2? (f) Near P_2 but outside the segment? (g) Near P_1 but outside the segment? (h) Near P_2 but inside the segment? (i) Very far away in either direction? [*Answers.* (a) 0. (c) 1. (d) Between 0 and −1. (e) Less than −1. (i) Nearly −1.]

17. In each triple of points below find the ratio in which each point divides the segment joining the other two. Sketch.

(a) (0, 1), (−1, 3), (−5, 11)   (c) (1, 3), (−2, −6), (0, 0)

(b) (−3, −6.5), (8, −1), (10, 0)   (d) (−1, 7), (0, 5), (2.5, 0)

18. Rewrite *4.6.2 in terms of $r_1/r_2 = r$.

4.7 The straight line

Our purpose in this section is to state the basic theorems on the straight line, but we begin with special cases.

*4.7.1   The equations of the x-axis and y-axis are respectively $y = 0$ and $x = 0$.

This is easy to see, for a point lies on the x-axis if and only if its y-coordinate is zero. Similarly, a point lies on the y-axis if and only if its x-coordinate is zero. Since the axes are set up as straight lines, we can show that a certain locus is a straight line if we can show that it can be made to coincide with one of the axes. We use this idea to prove the following theorems:

*4.7.2   The equation of a line parallel to the x-axis is $y = b$, where (0, b) is the y-intercept.

*4.7.3 The equation of a line parallel to the y-axis is $x = a$, where $(a, 0)$ is the x-intercept.

Proof of *4.7.2. If we translate the axes to the point $(0, b)$, the new coordinates are given by $x = x'$ and $y = y' + b$. (See Fig. 4.7.1.) The equation $y = b$ becomes $y' + b = b$ or $y' = 0$, the equation of the new x-axis by *4.7.1. Thus $y = b$ is the equation

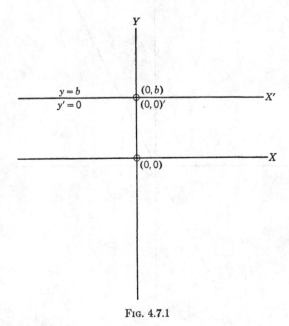

FIG. 4.7.1

of the indicated line since its locus coincides with the new x-axis, which is parallel to the old x-axis and at a distance b from it.

We now state the fundamental theorem upon which the theory of the straight line rests:

*4.7.4 The locus of $y = mx + b$ is a straight line.

Proof. Since the constant term may be eliminated by translating the axes (see 4.6 and Ex. 4.6.7), it is sufficient to show that the locus of $y = mx$ is a straight line. Evidently $(0, 0)$ is a point on the locus. Let $P_1(x_1, y_1)$ be on the locus so that $y_1/x_1 = m$. If $P_2(x_2, y_2)$ is a point lying on the straight line OP_1, than $\triangle AOP_1$

$\smile \triangle BOP_2$.   (See Fig. 4.7.2.)   Hence $y_2/x_2 = y_1/x_1$, and P_2
satisfies the equation.   Conversely, if a point $P_2(x_2, y_2)$ satisfies
$y = mx$, then $y_2/x_2 = y_1/x_1$.   It follows that the triangles are
similar, that $\angle AOP_1 = \angle AOP_2$, and that P_2 lies on the line OP_1.

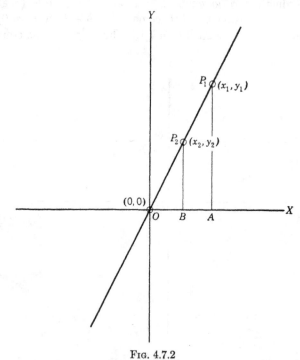

FIG. 4.7.2

Hence a point lies on the straight line OP_1 if and only if it satisfies
the equation, and the theorem is proved.   (An alternative proof
is given in 9.1.)

From *4.7.4 it follows that:

*4.7.5   Every non-vertical line has an equation of the form
$y = mx + b$.

Proof. Since the line is not parallel to the y-axis, it must meet
it in a point. Let this point be $(0, c)$, and let (d, e) be another
point on the given line. Then $y = (e - c)x/d + c$ is the equation
of the given line, since it is the equation of *some* straight line (by
*4.7.4), it is satisfied by $(0, c)$ and (d, e), and there is only one
line through two points. But the equation $y = (e - c)x/d + c$ is

of the form $y = mx + b$. Hence the theorem is proved. We note also that $(0, b)$ is the y-intercept of the line.

Heuristic Discussion. The student may wonder how we came to pick out the equation $y = (e - c)x/d + c$. We reasoned somewhat as follows. The given line must be specified in some way. Since two points determine a straight line, let us consider that it is specified by two points. (Certainly if the line is given, we can locate any number of points on it.) We know that $y = mx + b$ represents *some* straight line for every value of m and b. Now, we ask, can we determine m and b so that it is the equation of the given line? It appears that we can, because a point lies on the locus if its coordinates satisfy the equation. If we substitute the coordinates of two points in $y = mx + b$, we find two equations from which we may be able to find m and b. This is easiest to do if we choose a point where one of the coordinates is zero. Now we notice the provision that the line is not parallel to the y-axis. This means that it must meet it at some point, say $(0, c)$. Substituting in $y = mx + b$, we find that $b = c$. If the line is to pass through some other point (d, e), we must have $e = md + c$ or $m = (e - c)/d$. Thus we get the equation $y = (e - c)x/d + c$, which is easily shown to be the equation of the given line. It should be noted that the provision about not being parallel to the y-axis is essential. A line parallel to the y-axis does not have an equation of the form $y = mx + b$.

From the foregoing discussion it follows that:

*4.7.6 For any two points (x_1, y_1) and (x_2, y_2) on the straight line $y = mx + b$:

$$m = \frac{y_2 - y_1}{x_2 - x_1}$$

This theorem means that the ratio of the difference of y's to the difference of the corresponding x's is the same for any two points on a straight line. It is proved as follows:

(1) $y_2 = mx_2 + b$ (Why?)

(2) $y_1 = mx_1 + b$ (Why?)

(3) $y_2 - y_1 = m(x_2 - x_1)$ (Why?)

(4) $m = \dfrac{y_2 - y_1}{x_2 - x_1}$ (Why?)

The constant ratio of the difference of y's to the difference of x's is called the slope of the line. Its definition is:

*4.7.7   The **slope** of a straight line is defined by

(5)
$$\frac{y_2 - y_1}{x_2 - x_1}$$

where (x_1, y_1) and (x_2, y_2) are the coordinates of any two points on the line.

We may now summarize our results in the following theorem:

*4.7.8   The locus of $y = mx + b$ is a straight line with y-intercept $(0, b)$ and slope m.

A line parallel to the y-axis does not have a slope. (Why?) Any other line has a slope that is independent of the choice of points in (5). The slope is sometimes called the *gradient* by engineers. If we think of the straight line as a road, the slope or gradient is the distance climbed divided by the horizontal distance moved. Evidently the slope is positive or negative according to whether the line inclines up or down as we move from left to right. (See Fig. 4.7.3.) Also we may think of the slope as the rate at

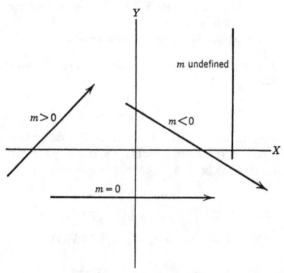

FIG. 4.7.3. Slope.

which y is changing as x changes, since it equals the change in y divided by the change in x.

EXERCISE 4.7

1. Graph the following special cases of $x = a$: (a) $a = 2$. (b) $a = -3$. (c) $a = 0$.

2. Graph the following special cases of $y = b$: (a) $b = 0$. (b) $b = -17/2$. (c) $b = 2.7$.

3. Graph the following special cases of $y = mx + b$:

(a) $m = 2, b = -5$
(b) $m = 0, b = 1$

(c) $m = 1, b = 0$
(d) $m = -2, b = -1$

4. Find the slopes of the lines through the following pairs of points. (Check by sketching.)

(a) $(2, 3), (5, 1)$
(b) $(-1, -5), (0, 3)$
(c) $(0, -2), (-5, 6)$
(d) $(18, -4), (-6, -3/2)$
(e) $(1/2, 7), (-1/3, -2)$

(f) $(14, 15), (11, -1)$
(g) $(0, b), (a, 0)$
(h) $(m, -m), (n, -n)$
(i) $(r, s), (-r, -s)$
(j) $(2 - r, 3), (r - 1, s - 2)$

Answers. (a) $-2/3$. (c) $-8/5$. (e) $54/5$. (g) $-b/a$. (i) s/r.

5. Explain each step (1)–(4) in the text.

6. Explain why $x = a$ is not a special case of $y = mx + b$, whereas $y = b$ is.

7. Prove *4.7.3 as we did *4.7.2.

8. Construct a proof of *4.7.5 with the line determined by $(0, d)$ and (e, c). Why can we not take the x-intercept for our second point?

9. Explain the difference between a line having no slope and one having a zero slope.

10. If we take two points such that $x_2 - x_1 = 1$, the slope is just $y_2 - y_1$. (Why?) Hence the slope is numerically equal to the change in y for unit change in x. Indicate the change in y corresponding to unit change in x for the linear functions of prob. 3 and Ex. 4.3.2.

11. If we write Δx for the change in x, $x_2 - x_1$, and Δy for the change in y, $y_2 - y_1$, then $\Delta y = m\Delta x$. Find the change in y corresponding to:

(a) $\Delta x = 3$ in $y = 2x + 1$
(b) $\Delta x = 2$ in $y = -0.5x + 3$
(c) $\Delta x = -1$ in $3x - 7y = 2$

(d) $\Delta x = -3.5$ in $y = 3x - 2$
(e) $\Delta x = -4$ in $y = -4x + 10$
(f) $\Delta x = 1/2$ in $x = 2y - 4$

Answers. (a) 6. (c) $-3/7$.

12. Find the slopes of the sides of the triangle determined by $(2, -1)$, $(4, 7)$, $(-3, 6)$.

13. Show that:
$$(y_2 - y_1)/(x_2 - x_1) \equiv (y_1 - y_2)/(x_1 - x_2)$$
and
$$(y_2 - y_1)/(x_2 - x_1) \not\equiv (y_2 - y_1)/(x_1 - x_2)$$

†14. Prove that the three points $(1, -4)$, $(-1, 6)$, $(3, -14)$ lie on the same straight line. (Such points are called **collinear**.)

†15. Prove that the slope remains invariant (unchanged) when the axes are translated.

†16. Our proof of *4.7.6 was based solely on the equation of the line. Give a geometric proof, without reference to the equation, that the ratio $(y_2 - y_1)/(x_2 - x_1)$ is the same for any two points on a straight line.

†17. Find the set of points that satisfy each of the following:

(a) $y \neq 2x + 3$ (d) $y = 0$ and $x = 2y + 4$

(b) $y = 2x$ or $y = x$ (e) $y = x$ or $y = -x$

(c) $y = 2x$ and $y = x$ (f) $y = x$ and $y = -x$

4.8   The equation $Ax + By + C = 0$

The equation $2y + 3x - 6 = 0$ can easily be put in the form $y = mx + b$ by subtracting $3x - 6$ from both sides and dividing both sides by 2. The result is $y = -(3/2)x + 3$, evidently the equation of a line with y-intercept $(0, 3)$ and slope $-3/2$. Since the equations $2y + 3x - 6 = 0$ and $y = -(3/2)x + 3$ are equivalent, we call $2y + 3x - 6 = 0$ another form of the equation of the line. The equation $2y + 3x - 6 = 0$ does not give y explicitly as a function of x, that is, it does not express y as a formula in x. But it does establish y as a function of x, since for every x there is a corresponding y satisfying the equation. We say that it gives y as an **implicit function** of x.

To generalize the results above we consider $Ax + By + C = 0$, where A, B, C are constants. If $B \neq 0$, we can rewrite the equation as

(1) $y = (-A/B)x - C/B$

Since this is in the form $y = mx + b$, it must be the equation of a line with y-intercept $-C/B$ and slope $-A/B$. Evidently $Ax + By + C = 0$ defines y as an implicit function of x if $B \neq 0$. If $B = 0$ and $A \neq 0$, the equation takes the form

(2) $Ax = -C$

(3) $x = -C/A$

This is the equation of a line parallel to the y-axis. The case $B = 0$, $A = 0$ yields a false equation, unless $C = 0$, in which case the equation becomes the identity $0 = 0$. Hence we have

*4.8.1 Every equation of the form $Ax + By + C = 0$, where A and B are not both zero, is the equation of a straight line.

An equation of the form $Ax + By + C = 0$, where A and B are not both zero, is called a **linear equation** in x and y. The left member is a function of the pair of values (x, y), that is, for each pair, the left side has a definite value. Those pairs (x, y) that make the left side zero are the pairs that satisfy the equation and the coordinates of the points that lie on the graph. In fact, the line is just the set of these points. Instead of saying that $Ax + By + C$ is a function of the pair of values (x, y), we usually say that it is a **function of two variables** x and y. Functions of two variables or of number pairs are represented by notations of the form $F(x, y)$, $f(x, y)$, $\theta(x, y)$, etc., which are used in the same way as the corresponding notations for functions of one variable. Thus $f(a, b)$ means the value of $f(x, y)$, when x and y are replaced by a and b. If $f(x, y) = x^2 + 2y$, then $f(1, 3) = 1^2 + 2(3) = 7$ and $f(r, 2t) = r^2 + 4t$.

The foregoing theorem together with those of the previous section are basic. The form $y = mx + b$ is called the **slope-intercept form**. The form $Ax + By + C = 0$ is called the **general form**. We have seen that an equation in the general form can be put in the point-slope form by simple algebra. (Except in what case?) This is the quickest way to find the slope of a line whose equation is given. Suppose now that we wish to find the equation of a line through the point (x_1, y_1) and having the slope m. The equation must be of the form $y = mx + b$, and we have only to evaluate b. Since the line passes through (x_1, y_1), we have:

(4) $$y_1 = mx_1 + b$$

(5) $$b = y_1 - mx_1$$

Hence the equation of the line is:

(6) $$y = mx + y_1 - mx_1$$

or

(7) $$y - y_1 = m(x - x_1)$$

This last equation is called the **point-slope form.** It enables us to write down immediately the equation of any line if we know its slope and one of its points. Moreover, it tells us that given any point (x_1, y_1) on a line, we may find another point (x, y) by going directed distances $x - x_1$ parallel to the x-axis and $y - y_1$ parallel to the y-axis, such that $\dfrac{y - y_1}{x - x_1} = m$.

Perhaps the easiest method of graphing an equation in the form $Ax + By + C = 0$ is to find the intercepts by setting first $y = 0$ and then $x = 0$. If the resulting points are close together, or if one of the constants is zero, other points should be located. Note that we can substitute values for either x or y in order to find pairs of values that satisfy the equation. The equation gives x as an implicit function of y as well as y an implicit function of x.

EXERCISE 4.8

1. Put the following equations in the slope-intercept form and graph:

(a) $2x + 3y - 12 = 0$ (e) $-x - y + 1 = 0$
(b) $-2y + x = 20$ (f) $-13y - 1 = 7x$
(c) $x = 4y - 2$ (g) $2x + 7y = 14$
(d) $15y = 3 - 5x$ (h) $-(2/3)x + (1/2)y = 10$

Answers. (a) $y = (-2/3)x + 4$. (c) $y = (1/4)x + 1/2$.

2. Write the equations of the lines passing through the following points and having the indicated slopes. Rewrite in the form $Ax + By + C = 0$. Graph:

(a) $(2, 4), 5$ (d) $(-2, -2), 1$ (g) $(5, -1), 0$
(b) $(-1, 5), -2$ (e) $(4, 2), -0.2$ (h) $(4, -0.5), 1.3$
(c) $(4, -8), 0.5$ (f) $(1, 0), 10$ (i) $(13, 169), 27$

Answers. (a) $5x - y - 6 = 0$. (c) $x - 2y - 20 = 0$. (e) $x + 5y - 14 = 0$. (g) $y = -1$.

3. Write the equations of the lines passing through each of the following pairs of points. Rewrite in the general form and graph. (*Suggestion.* Find the slope from the two points and use the point-slope form.)

(a) $(4, 3), (-1, 6)$ (d) $(4, -1), (-3, -3)$ (g) $(-3.5, 6.1), (2.8, 1.7)$
(b) $(2.5, 1), (-1.5, 3.2)$ (e) $(0, 5), (8, 8)$ (h) $(-1, -2), (-3, -5)$
(c) $(2, 2), (-2, 0)$ (f) $(0, 1), (1, 0)$ (i) $(a, 0), (0, b)$

Answers. (a) $3x + 5y - 27 = 0$. (c) $x - 2y + 2 = 0$. (e) $3x - 8y + 40 = 0$. (g) $44x + 63y - 230.3 = 0$.

4. Show that the equation of the line through the points (x_1, y_1) and (x_2, y_2) is:

(8) $$\frac{y - y_1}{x - x_1} = \frac{y_2 - y_1}{x_2 - x_1} \qquad \text{(Two-Point Form)}$$

5. In each case find the missing coordinate of the point that lies on the line through the given points and that has the given coordinate.

(a) $(1, 3)$, $(2, 7)$, $x = 1.5$ (e) $(100, 10{,}000)$, $(101, 10{,}201)$,
(b) $(1, 1)$, $(2, 4)$, $x = 1.5$ $x = 101.1$
(c) $(13, 169)$, $(14, 196)$,   (f) $(1, -7)$, $(3, 5)$, $x = 2$
        $x = 13.6$ (g) $(-1, 3)$, $(3, -16)$, $x = 1.5$
(d) $(3, 27)$, $(4, 64)$, $x = 3.5$ (h) $(1, 4)$, $(-9, 3)$, $y = 2$

Answers. (a) 5. (c) 185.2. (e) 10221.1.

6. Why is $B \neq 0$ necessary to equation (1) and why did we not consider the case $A = 0$?

7. Put the slope-intercept form in the general form. What are A, B, and C in the result?

8. Find the slopes and intercepts of the following lines:

(a) $2 - 2x - 3y = 0$ (e) $3y = x$ (h) $(2/3)x - (1/2)y$
(b) $14x - 2 = 9y$ (f) $3x = 5$ $= -17$
(c) $3x + 3y = 3$ (g) $4 = -3y$ (i) $x - 3y = 0$
(d) $-2x - 18y = 1$

Answers. (a) $-2/3$, $(1, 0)$, $(0, 2/3)$. (c) -1, $(1, 0)$, $(0, 1)$. (e) $1/3$, $(0, 0)$. (g) 0, none, $(0, -4/3)$.

9. Show that the equation of the line with intercepts $(a, 0)$ and $(0, b)$ is:

(9) $$\frac{x}{a} + \frac{y}{b} = 1$$ (Intercept Form)

10. Are there any lines that cannot be put in the intercept form?

11. The sum of two positive numbers is 3. Find the locus of possible values.

12. Find the equations of the sides of the triangle whose vertices are $(4, 4)$, $(0, 0)$, and $(0, 6)$. (*Answer.* $x = 0$, $y = x$, and $x + 2y = 12$.)

13. Same as prob. 12 for $(-6, 4)$, $(2, -1)$, $(-3, -4)$.

14. Rewrite the equation as indicated.

(a) $ky - 2bx + c = 0$ in the slope-intercept form.

(b) $\dfrac{x}{r_1} - \dfrac{y}{r_2} = 1$ in the intercept form.

(c) $\dfrac{x + y}{2} = 3a$ in the intercept form.

(d) $\dfrac{x}{a} + \dfrac{y}{b} = 1$ in the slope-intercept form.

(e) $Ax + By + C = 0$ in the intercept form.

15. For what values of k does $y + (2k + 1)x = 3$ have a slope of 2? (*Hint.* What is the slope of this line as a function of k?)

16. For what value of k does the line $3kx - (k + 1)y = 4$ go through the point $(2, -1)$? (*Answer.* $3/7$.)

17. The total cost of production of a certain manufacturing plant is a linear function of the output. When output is zero, cost is $1000. Suppose each additional unit of output costs $5. Letting C be the cost measured in thousands of dollars and u the output measured in hundreds of items, write the relation between C and u. (This is called the *cost function*.) (*Answer.* $C = 0.5u + 1$.)

18. The cost function of a factory is $C = 2u + 8$. Graph it and explain the meaning of the slope and C-intercept. (In the cost function $C = mu + b$, m is called the *marginal cost* and b is called the *overhead*.)

19. Suppose that the amount of a certain product u that the public buys is a linear function of its price P, and that sales are 1000 when the price is 50 and 980 when the price is 52. Find the function. (It is called the *demand function* and its graph is called a *demand curve*.) Graph the demand function and find its slope.

20. If $f(x, y) = 2x - y + 7$, find: (a) $f(0, 0)$. (b) $f(1, -2)$. (c) $f(1, x)$. (d) $f(y, x)$. (e) $f(x, x)$.

21. If $f(x, y) = 1 - 6xy$, find: (a) $f(0, 0)$. (b) $f(s^2, t)$. (c) $f(b, a)$. (d) $f(f(x, y), y)$. (e) $f(a, a)$.

22. If $P(a, b)$ stands for "$a + b = b + a$," find $P(1, 2)$, $P(0, 2)$, $P(-1, 3)$.

†23. If $P(x, y)$ stands for "x is the father of y," can both $P(a, b)$ and $P(b, a)$ be true? What is the relation between x and z if $P(x, y)$ and $P(y, z)$ are true?

†24. If $P(x, y)$ stands for "$x > y$," which of the following are true? $P(2, 1)$, $P(-3, -4)$, $P(-1, 2)$, and $P(0, -1)$.

†25. Graph the following functions:

(a) $f(x) = 0$ when $x \leq 0$
$\quad\quad = x$ when $x > 0$

(b) $f(x) = x$ when $x \leq 1$
$\quad\quad = 2x + 1$ when $1 < x \leq 2$
$\quad\quad = 5$ when $x > 2$

†26. The postage required for a letter is a function of its weight. Graph this function for $0 \leq x \leq 6$, where x is the weight.

4.9 Pairs of lines

A pair of linear equations determines two straight lines, consisting of those points that satisfy at least one of the equations. If the lines meet, the point of intersection lies on both lines and satisfies both equations. For example, the equations

$$(1) \quad\quad\quad\quad\quad\quad y = 2x + 4$$

$$(2) \quad\quad\quad\quad\quad\quad y = -x + 1$$

whose graphs are shown in Fig. 4.9, determine two lines that appear to meet in the second quadrant. We can estimate this point from the graph, but it is easy to find it exactly. Instead of looking at

equations (1) and (2) as determining the lines whose points satisfy
either equation, we look at them as defining the set of points that

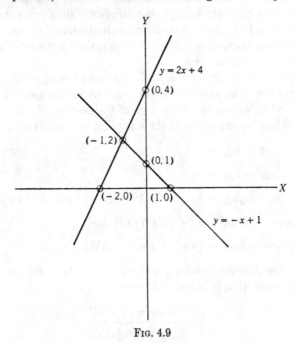

F $_{IG}$. 4.9

satisfy both equations simultaneously. Then we manipulate (1)
and (2) by the rules of Chapter 3 in order to find x and y:

(3) $y - y = 2x - (-x) + 4 - 1$ (Subtracting equation 2
                                      from equation 1)

(4) $3x = -3$ (Why?)

(5) $x = -1$ (Why?)

(6) $y = -1(-1) + 1 = 2$ (Substituting $x = -1$ in
                                    equation 2)

To show that $(-1, 2)$ is a solution we substitute in (1) and (2).
If $x = -1$ and $y = 2$, (1) becomes $2 = 2(-1) + 4$ and (2)
becomes $2 = -(-1) + 1$. Since these are true equations, $(-1, 2)$
is a solution, and (3)–(6) show that it is the only solution.

Two linear equations in x and y, which give two conditions *both*
of which are to be satisfied by a point (x, y), are called a pair of

simultaneous linear equations. If a pair of numbers (x, y) satisfies both equations it is called the solution of the equations. More generally, a pair of values (x, y) that satisfies both of two equations $F(x, y) = 0$ and $G(x, y) = 0$ is called a **solution,** and the equations are called **simultaneous.** The solutions make up the set of points that satisfy both equations.

In order to investigate analytically the solution of simultaneous linear equations, we consider two equations in the general form. To begin with, we assume that none of the coefficients of x and y is zero. Then we can perform the following manipulations:

(7) $a_1 x + b_1 y = c_1$

(8) $a_2 x + b_2 y = c_2$

(9) $a_1 a_2 x + a_2 b_1 y = a_2 c_1$ (Multiplying equation 7 by a_2)

(10) $a_1 a_2 x + a_1 b_2 y = a_1 c_2$ (Why?)

(11) $(a_1 b_2 - a_2 b_1) y = (a_1 c_2 - a_2 c_1)$ (Why?)

We can solve this equation for y *unless* $a_1 b_2 - a_2 b_1 = 0$. Assuming for the moment that it is not, we have:

(12) $y = \dfrac{(a_1 c_2 - a_2 c_1)}{(a_1 b_2 - a_2 b_1)}$

Substitution of this value of y in either (7) or (8) gives:

(13) $x = \dfrac{(c_1 b_2 - c_2 b_1)}{(a_1 b_2 - a_2 b_1)}$

Thus we find just one solution if $a_1 b_2 - a_2 b_1 \neq 0$. If it does equal zero, we can carry the above derivation as far as equation 11, but here the left side is zero. If the right side is not zero, we have a false equation and conclude that there is no solution. On the other hand, if the right side also is zero, there are an infinite number of solutions since any pair of numbers that satisfies one equation satisfies the other. For if both $a_1 b_2 - a_2 b_1$ and $a_1 c_2 - a_2 c_1$ are zero, the coefficients are proportional ($a_1/a_2 = b_1/b_2 = c_1/c_2$), and the equations (7) and (8) are equivalent. In fact, we can obtain (7) by multiplying (8) by a_1/a_2. Since parallel lines are defined as lines that do not meet, another way of stating these results is that the two lines are parallel if $a_1 b_2 - a_2 b_1 = 0$ and

$a_1 c_2 - a_2 c_1 \neq 0$, and coincident if both are zero. Otherwise, the lines meet in just one point.

We have limited ourselves to the case where none of the coefficients of x or y is zero. If just one is zero, its equation can be immediately solved for the other variable. The result can be substituted in the other equation to find the other variable. Evidently $a_1 b_2 - a_2 b_1 \neq 0$ (why?), and the solutions are given by (12) and (13). If $b_1 = b_2 = 0$, the equations take the form

(14)
$$x = c_1/a_1$$
$$x = c_2/a_2$$

It is impossible to find an x satisfying both these unless the right members are equal. Similar remarks apply when $a_1 = a_2 = 0$. In both cases $a_1 b_2 - a_2 b_1 = 0$, and when the lines coincide, $a_1 c_2 - a_2 c_1 = 0$ also. Finally, we may have $a_1 = b_2 = 0$ or $a_2 = b_1 = 0$. Since one equation involves only x and the other only y, the solution is immediate. The equations represent a pair of lines, one parallel to the x-axis and one parallel to the y-axis, and $a_1 b_2 - a_2 b_1 \neq 0$. (Why?) We exclude the possibilities $a_1 = b_1 = 0$ or $a_2 = b_2 = 0$. (Why?)

When simultaneous equations have a solution they are called **consistent**. When they fail to have a solution they are called **inconsistent**. When they have infinitely many solutions they are called **dependent**. The word "dependent" refers to the fact that in this case one equation can be obtained from the other. The quantity $a_1 b_2 - a_2 b_1$, which is the key to determining which case occurs, is called the **determinant** of the pair of equations and is designated by the letter D. With these definitions, we may summarize the foregoing discussion in the following theorem:

*4.9.1 Two linear equations are consistent if $D \neq 0$. If $D = 0$, they are inconsistent, unless the coefficients are proportional, in which case the equations are dependent.

The equation $a_1 b_2 - a_2 b_1 = 0$ has an important geometrical interpretation. If the b's are not zero, it may be rewritten $-a_1/b_1 = -a_2/b_2$. But the two members of this equation are just the slopes of the lines (7) and (8). Hence:

*4.9.2 Two non-vertical lines are parallel or coincident if and only if their slopes are equal.

If either b_1 or b_2 is zero, then $a_1b_2 - a_2b_1 = 0$ implies that the other b is zero also, and that both lines are vertical, parallel to each other, and with undefined slopes. Accordingly:

*4.9.3   A necessary and sufficient condition that two straight lines (7) and (8) be parallel or coincident is that $D = 0$.

EXERCISE 4.9

For all problems prove (by substitution) that your answers are solutions.

1. Graph the simultaneous equations $y = 3x - 5$, and $y = x + 2$. Solve as we did (1) and (2).

2. Graph and solve in the manner of (7) to (13):

(a) $2x + 3y = 7$
$x - y = 1$

(b) $5x - 2y = 6$
$2x + y = 2$

(c) $9x + 7y = -1$
$-2x + y = 10$

(d) $-x - y = 3$
$y + 2x = 14$

3. Solve as above as far as step (11). Graph and explain the result in terms of the discussion of the text. Calculate D and $a_1c_2 - a_2c_1$ in each case.

(a) $x + 2y = 6$
$2x + 4y = 3$

(b) $3x + 2y = -1$
$9x - 6y = 5$

(c) $-9x + 12y = 3$
$3x + 4y = 1$

(d) $2x - 7y = -1$
$21y - 6x = 3$

(e) $2y - 5x = 0$
$16x + 10y = 3$

(f) $x = 14 - 17y$
$34y - 14 + 2x = 0$

4. Solve the following if possible. Graph and calculate D and $a_1c_2 - a_2c_1$ in each case.

(a) $3x - 6 = 0$
$6y + 2x + 4 = 0$

(b) $14x = -7$
$-9y = 18$

(c) $12x = 17$
$-3 - 2x = 0$

(d) $2y - 4x = 5$
$15 - 6y = 0$

(e) $35y - 26 = 0$
$13 - 20y = 0$

(f) $2x - 3y = 6 - 3y$
$15x = 45y$

5. Identify each of the following pairs of equations as consistent, inconsistent, or dependent by calculating D and $a_1c_2 - a_2c_1$. Check your results by graphing.

(a) $2x - 3y = 1$
$4x + 6y = 2$

(b) $y + 3x = 2$
$y - 6x = 3$

(c) $y = 2x - 1$
$y = 4x - 2$

(d) $y - 3 = 0$
$3 + x = 0$

(e) $y - 3x = 1$
$2y - 6x = 5$

(f) $2y + 18x - 3 = 0$
$9x + y = 1$

(g) $x - y = 2$
$y - x = 2$

(h) $45x - 27y = 21$
$15x - 9y = 7$

(i) $0.2x - 1.5y = 4$
$3x + 0.4y = 0.1$

6. In each part of prob. 5 identify the set of points that (a) satisfy both equations, and (b) satisfy at least one equation.

7. Solve each of the following pairs of equations by solving one for one variable and substituting the result in the other:

(a) $128x - 32y = 14$
$2y - 5x = 3$

(b) $2x/3 - 7y = 1$
$3x - y/2 = -2$

(c) $(x - y)/4x = 1$
$8x - 3y + 2 = 0$

(d) $2x - 23 = 4(y - x + 2)$
$4(x - 1) = 2(y - 3)$

(e) $27x - 9y = 36$
$2x + 6y = 14$

(f) $(x - y)/2 = 3x + 1 - 5y$
$x/2 + y = 1 - 3(1 - x)/7$

8. Find two numbers whose sum is 7 and whose difference is 13. (*Hint.* Let one number be x, the other y.)

9. Mr. Chuckle is seven times as old as Debbie. The difference in their ages is 48 years. How old is each? (*Hint.* What are the two unknowns?)

10. An observer at $(0, 0)$ sights an object in the direction of the point $(1, 3)$. An observer at $(100, 0)$ sights the same object in the direction $(98, 1)$. Locate the object graphically and find its exact coordinates by algebra.

11. In a true and false test, one credit is given for each correct answer and one credit is subtracted for each wrong answer. On 100 questions answered, a student answered all questions and received a score of 60. How many questions did he answer correctly?

12. Fill in the following table describing the various cases for a pair of linear equations:

D	$a_1c_2 - a_2c_1$	Equations	Solutions	Lines
$\neq 0$	any value			meet in one point
$= 0$	$\neq 0$	inconsistent		
$= 0$	$= 0$		infinitely many	

13. Substitute (12) and (13) in both (7) and (8) and so verify that they give the solutions.

14. Solve the following and check by substitution:

(a) $2x - ay = c$
$x + by = c$

(b) $x_1x + y_1y = c_1$
$x_2x + y_2y = c_2$

(c) $ax + by = c$
$-bx + ay = c$

(d) $(1 - s)x + sy = 0$
$(s + 3)x + (s + 1)y = s$

15. Solve for R/S and T/S the following equations that occur in genetics:

$$\frac{S}{4} + \frac{R}{16} = r_1\left(\frac{S}{2} + \frac{R}{4} + T\right)$$

$$\frac{S}{4} = r_2\left(\frac{S}{2} + \frac{R}{4} + T\right)$$

16. Prove the statement in the section that $b_1 = 0$ and $a_1b_2 - a_2b_1 = 0$ imply $b_2 = 0$.

17. At a constant price p, the income from the sale of x units of a product is given by $y_I = px$. If the cost of producing and selling x units is given by $y_C = ex + d$, the profit is the vertical distance between the two lines. The point of intersection, where $y_I = y_C$, is called the *break-even point*, and graphs showing these lines are called *break-even charts*. Make such a chart for $p = 1.5$, $e = 1$, $d = 2$. Find the break-even point and indicate on the chart the profit or loss when $x = 3$ and when $x = 6$.

18. Make a break-even chart for $y_I = 4x$ and $y_C = 3x + 2$. At what output does the firm make neither a profit nor a loss? Graph the function $y = y_I - y_C$ on the same axes, and find the profit when $x = 5$.

4.10 Determinants

Formulas (12) and (13) in 4.9 have interesting features that are brought out by comparing them with the coefficients of the equations. We have

$$(1) \qquad\qquad\qquad a_1x + b_1y = c_1$$

$$(2) \qquad\qquad\qquad a_2x + b_2y = c_2$$

$$(3) \qquad\qquad x = \frac{c_1b_2 - c_2b_1}{D}$$

$$\text{where } D = a_1b_2 - a_2b_1$$

$$(4) \qquad\qquad y = \frac{a_1c_2 - a_2c_1}{D}$$

First, we note that D contains only the coefficients of x and y. Moreover, it is obtainable by multiplying these coefficients according to the scheme

and then prefixing a minus sign to the "uphill" product. The numerators $c_1b_2 - c_2b_1$ and $a_1c_2 - a_2c_1$ are given by the same convention applied to

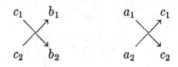

These similarities suggest inventing a new symbol to stand for such expressions:

*4.10.1
$$\begin{vmatrix} a & c \\ b & d \end{vmatrix} = ad - bc \qquad \text{(Definition)}$$

Such an expression is called a **determinant,** and the numbers that appear are called its **elements.** With this new symbol, we rewrite formulas (12) and (13) as follows:

*4.10.2
$$x = \frac{\begin{vmatrix} c_1 & b_1 \\ c_2 & b_2 \end{vmatrix}}{D} \qquad y = \frac{\begin{vmatrix} a_1 & c_1 \\ a_2 & c_2 \end{vmatrix}}{D}$$

where
$$D = \begin{vmatrix} a_1 & b_1 \\ a_2 & b_2 \end{vmatrix}$$

In order to recall these formulas, all we need to remember is the relationship between the determinants and the coefficients in the equations. The denominator D is formed from the coefficients of x and y written down just as they stand in the equations. The numerators are formed by replacing one of the columns in D by the right members of the equations. The column replaced is the one made up of coefficients of the variable being found. Thus, to get x, we replace $\begin{bmatrix} a_1 \\ a_2 \end{bmatrix}$ by $\begin{bmatrix} c_1 \\ c_2 \end{bmatrix}$.

To illustrate the ease with which equations can be solved by *4.10.2, we give the following example:

(5)
$$\begin{aligned} -5x + 2y &= 6 \\ 2x - y &= -3 \end{aligned}$$

(6) $\quad D = \begin{vmatrix} -5 & 2 \\ 2 & -1 \end{vmatrix} = (-5)(-1) - (2)(2) = 5 - 4 = 1$

(7) $\quad x = \dfrac{\begin{vmatrix} 6 & 2 \\ -3 & -1 \end{vmatrix}}{1} = (6)(-1) - (-3)(2) = 0$

(8) $\quad y = \dfrac{\begin{vmatrix} -5 & 6 \\ 2 & -3 \end{vmatrix}}{1} = (-5)(-3) - (2)(6) = 3$

There is no reason why we should not deal with three or more unknowns by the methods of this and the previous section. Suppose, for example, that we wish to find three numbers satisfying the following conditions: Their sum is 20. The first is twice the difference of the other two. The first plus twice the sum of the other two is 30. Letting x, y, z represent these unknowns, we have the following three equations:

$$(9) \qquad\qquad x + y + z = 20$$

$$(10) \qquad\qquad x - 2y + 2z = 0$$

$$(11) \qquad\qquad x + 2y + 2z = 30$$

We can solve such a system of simultaneous equations by methods similar to those used in the last section for two equations. In this case we proceed as follows:

$$(12) \qquad 4y = 30 \qquad \text{[Subtracting (10) from (11)]}$$

$$(13) \qquad y = 15/2 \qquad \text{[Solving (12) for } y\text{]}$$

$$(14) \qquad x + z = 25/2 \qquad \text{[Substituting from (13) in (9)]}$$

$$(15) \qquad x + 2z = 15 \qquad \text{[Substituting from (13) in (10)]}$$

$$(16) \qquad z = 5/2 \qquad \text{[Subtracting (14) from (15)]}$$

$$(17) \qquad x = (25/2) - (5/2) = 10 \qquad \text{[Substituting from (16) in(14)]}$$

That $(10, 15/2, 5/2)$ is the solution can be proved by substitution in each of the equations.

We can always solve simultaneous linear equations in any number of unknowns by methods of this sort if a solution exists. But the process becomes very tedious even for three equations in three unknowns, and the student can easily imagine what it would be like for ten equations in ten unknowns! The solution is considerably simplified if we extend the idea of a determinant so that it can be applied to three or more equations. The general equations in three unknowns appear as follows:

$$a_1x + b_1y + c_1z = d_1$$

$$(18) \qquad a_2x + b_2y + c_2z = d_2$$

$$a_3x + b_3y + c_3z = d_3$$

By analogy with the case of two equations we might think that the solution would perhaps be expressible in terms of some arrays like

$$\begin{vmatrix} a_1 & b_1 & c_1 \\ a_2 & b_2 & c_2 \\ a_3 & b_3 & c_3 \end{vmatrix}$$

This is indeed the case if we define this expression as follows:

$$\textbf{*4.10.3} \quad \begin{vmatrix} a_1 & b_1 & c_1 \\ a_2 & b_2 & c_2 \\ a_3 & b_3 & c_3 \end{vmatrix} = a_1 \begin{vmatrix} b_2 & c_2 \\ b_3 & c_3 \end{vmatrix} - a_2 \begin{vmatrix} b_1 & c_1 \\ b_3 & c_3 \end{vmatrix} + a_3 \begin{vmatrix} b_1 & c_1 \\ b_2 & c_2 \end{vmatrix}$$

(Definition)

Such an expression is called a determinant of the third order. The **order** is equal to the number of rows or columns, so that determinants like those defined in *4.10.1 are said to be of the second order.

It can be shown by direct substitution that the solution of equations (18) is given by:

(19)

$$x = \frac{\begin{vmatrix} d_1 & b_1 & c_1 \\ d_2 & b_2 & c_2 \\ d_3 & b_3 & c_3 \end{vmatrix}}{D}$$

$$y = \frac{\begin{vmatrix} a_1 & d_1 & c_1 \\ a_2 & d_2 & c_2 \\ a_3 & d_3 & c_3 \end{vmatrix}}{D} \qquad \text{where } D = \begin{vmatrix} a_1 & b_1 & c_1 \\ a_2 & b_2 & c_2 \\ a_3 & b_3 & c_3 \end{vmatrix}$$

$$z = \frac{\begin{vmatrix} a_1 & b_1 & d_1 \\ a_2 & b_2 & d_2 \\ a_3 & b_3 & d_3 \end{vmatrix}}{D}$$

These formulas have meaning only if $D \neq 0$. Just as for two equations, $D = 0$ means that the equations have either no solutions (inconsistent) or an infinity of solutions (dependent). Again the formulas are easy to remember, because we simply form D out of the coefficients as they stand, and then form the numerators by replacing appropriate columns in D by the right-hand members in the equations. A geometric interpretation of simultaneous linear equations in three unknowns will be suggested in 14.3.

We illustrate for the following set of equations:

$$
\begin{aligned}
3x + 3y + 4z &= 5 \\
(20) \qquad 2x + y - z &= 3 \\
4x + 2y + 3z &= 8
\end{aligned}
$$

$$
D = \begin{vmatrix} 3 & 3 & 4 \\ 2 & 1 & -1 \\ 4 & 2 & 3 \end{vmatrix} = 3 \begin{vmatrix} 1 & -1 \\ 2 & 3 \end{vmatrix} - 2 \begin{vmatrix} 3 & 4 \\ 2 & 3 \end{vmatrix} + 4 \begin{vmatrix} 3 & 4 \\ 1 & -1 \end{vmatrix}
$$

$$
= 3(3 + 2) - 2(9 - 8) + 4(-3 - 4)
$$

$$
= 15 - 2 - 28 = -15
$$

$$
x = \frac{1}{D} \begin{vmatrix} 5 & 3 & 4 \\ 3 & 1 & -1 \\ 8 & 2 & 3 \end{vmatrix} = \frac{-29}{-15} = \frac{34}{15}
$$

Similarly,

$$
y = \frac{1}{D} \begin{vmatrix} 3 & 5 & 4 \\ 2 & 3 & -1 \\ 4 & 8 & 3 \end{vmatrix} = \frac{-17}{15}
$$

$$
z = \frac{1}{D} \begin{vmatrix} 3 & 3 & 5 \\ 2 & 1 & 3 \\ 4 & 2 & 8 \end{vmatrix} = \frac{2}{5}
$$

EXERCISE 4.10

Verify solutions to simultaneous equations by substitution.

1. Evaluate the following determinants:

(a) $\begin{vmatrix} 2 & 4 \\ 1 & 3 \end{vmatrix}$

(b) $\begin{vmatrix} a & b \\ 4 & 3 \end{vmatrix}$

(c) $\begin{vmatrix} -1 & 2 \\ -3 & -2 \end{vmatrix}$

(d) $\begin{vmatrix} a & c \\ b & a \end{vmatrix}$

(e) $\begin{vmatrix} b & 2c \\ 2a & b \end{vmatrix}$

(f) $\begin{vmatrix} a+b & a \\ a & a-b \end{vmatrix}$

Answers. (a) 2. (c) 8. (f) $-b^2$.

2. Solve the following equations by determinants if possible. If not, indicate whether they are inconsistent or dependent and give the geometrical interpretation.

(a) $4x + 9y = -1$
$x - 3y = 4$

(b) $ax - by = 3$
$2x + y = c$

(c) $(1/2)x - (2/3)y = 5$
$-x + (1/2)y = -1$

(d) $47x + 33y = 18$
$100x + 20y = 1$

(e) $2.5x - y + 7 = 0$
$-5x + 2y = -1$

(f) $3y - 2x + 4 = 0$
$4 - 3x = y$

(g) $\dfrac{x}{2} + \dfrac{y}{3} = 1 - (x - y)$
$(2/3)x + 6 = y$

(h) $2ax + by = 2$
$ax + by = 1 + by/2$

(i) $(a + b)x + (a - b)y = c$
$(a - b)x + (a - b)y = d$

(j) $0.01x - 0.302y = 0.5$
$5.61x + 3.7y = 0.001$

3. Evaluate the following determinants:

(a) $\begin{vmatrix} 2 & 3 & 1 \\ 1 & -2 & 0 \\ -1 & 4 & 1 \end{vmatrix}$

(b) $\begin{vmatrix} 0 & 2 & -2 \\ 0 & 0 & 3 \\ 1 & -1 & -2 \end{vmatrix}$

(c) $\begin{vmatrix} 1 & 6 & 7 \\ 4 & 3 & -8 \\ 5 & 9 & 3 \end{vmatrix}$

(d) $\begin{vmatrix} 1 & 5 & 4 \\ -8 & 6 & 9 \\ -3 & 7 & 2 \end{vmatrix}$

(e) $\begin{vmatrix} a & 2 & 0 \\ b & 1 & 4 \\ a & 3 & 1 \end{vmatrix}$

(f) $\begin{vmatrix} a & a & 1 \\ -b & b & 1 \\ -c & c & 1 \end{vmatrix}$

(g) $\begin{vmatrix} x & y & 1 \\ 2 & 1 & 1 \\ 4 & 3 & 1 \end{vmatrix}$

(h) $\begin{vmatrix} 0 & 0 & 0 \\ a & 0 & 0 \\ b & c & 0 \end{vmatrix}$

(i) $\begin{vmatrix} 1 & 0 & 0 \\ 0 & 1 & 0 \\ 0 & 0 & 1 \end{vmatrix}$

(j) $\begin{vmatrix} 0 & 0 & 1 \\ 0 & 1 & 0 \\ 1 & 0 & 0 \end{vmatrix}$

(k) $\begin{vmatrix} 3 & 2 & 4 \\ 0 & 1 & 5 \\ 0 & 0 & 2 \end{vmatrix}$

(l) $\begin{vmatrix} a & b & c \\ 0 & d & e \\ 0 & 0 & f \end{vmatrix}$

Answers. (a) $2 \begin{vmatrix} -2 & 0 \\ 4 & 1 \end{vmatrix} - 1 \begin{vmatrix} 3 & 1 \\ 4 & 1 \end{vmatrix} + (-1) \begin{vmatrix} 3 & 1 \\ -2 & 0 \end{vmatrix} = -5.$ (c) 84.
(e) $-(3a + 2b)$. (g) $2(-x + y + 1)$. (i) 1. (k) 6.

4. Solve by determinants:

(a) $3x - 4y + z = 2$
$x + y - z = -1$
$2x - y + 4z = 2$

(b) $3x + z = 4$
$x - y = 2$
$3y + 4z = 0$

(c) $x + 2y + 3z = 4$
$2x + 3y + 4z = 1$
$3x + 2y + z = 0$

(d) $3s + 4t + 2u = 0$
$t + 4s - 3 = 0$
$u + t + s + 1 = 0$

(e) $7x - 3y + z = 2$
$2x - y - z = 1$
$x - y - 2z = 0$

(f) $x + y + z = -1$
$4y - 3x + 2z = 5$
$5x - 2y + z = -8$

(g) $x - y - 2z - 1 = 0$
$3x = 3z - y + 8$
$y - 2x + z = -3$

(h) $s + t + u = 1$
$s + t - u = 0$
$s - t - u = 3$

(i) $ax + by + z = d$
$x - y + cz = 2$
$x + ay + bz = d$

(j) $x + y = 3$
$y + z = 5$
$x + z = 6$

(k) $6x + 3y - z = 4$
$2x + y + z = 1$
$3x - y - z = 5$

(l) $0.5x + 2.5y = 0$
$x - y - z + 0.3 = 0$
$z = 15x + 2y - 1$

5. Solve the systems of prob. 4 by manipulation without using determinants.

6. Three numbers are in arithmetic progression. Their sum is 25. The difference of the smallest and the largest is 2 less than twice the other number. Find them.

7. The sum of three numbers is said to be 47. The largest is 500 more than twice the sum of the other two. The middle one is half the sum of the other two. Is this possible?

8. Solve for x and y:

(a) $\dfrac{1}{x} + \dfrac{1}{y} = 2$

$\dfrac{2}{x} + \dfrac{1}{y} = 3$

(b) $x + y = 3xy$
$3x - 4y = xy$

(c) $\dfrac{5y}{x} - 2 = y$

$\dfrac{1}{x} + \dfrac{3}{y} = 1$

(d) $cx - y = 5$
$2x + cy = 1 + s$

(e) $\dfrac{a}{x} + \dfrac{b}{y} = 2$

$\dfrac{b}{x} - \dfrac{a}{y} = 1$

(†f) $2x^2 + 3y^2 = 4$
$x^2 - y^2 = 6$

Suggestions. In part a, let $1/x = u$ and $1/y = v$. In part b, divide both sides by xy to find the non-zero solutions.

9. Solve for x:

(a) $\begin{vmatrix} 1 & 3 & -1 \\ -1 & x & 0 \\ 0 & 2 & 5 \end{vmatrix} = 0$

(b) $\begin{vmatrix} x & 0 & 1 \\ 1 & 3 & 0 \\ 3 & -1 & 1 \end{vmatrix} = 0$

10. Show by substitution that (19) gives the solution of (18).

†11. Suppose that x' and y' are linear functions of x and y. Under what conditions are x and y also linear functions of x' and y'? (*Suggestions.* Start with $x' = a_1x + b_1y + c_1$ and $y' = a_2x + b_2y + c_2$.)

†12. Prove:

*4.10.4 The equation of the straight line through (x_1, y_1) and (x_2, y_2) is

$$\begin{vmatrix} x & y & 1 \\ x_1 & y_1 & 1 \\ x_2 & y_2 & 1 \end{vmatrix} = 0$$

13. If the amount y of a commodity offered by sellers on the market is a function of the price x given by $y = 0.31x - 2$ (the *supply function*) and the amount of the commodity demanded by buyers is given by $y = -0.2x + 15$ (the *demand function*), at what price do supply and demand balance, and how much is sold?

4.11 Properties of determinants

In this section we develop a number of elegant properties that can be used to simplify the handling of determinants. We discuss determinants of the second and third orders, but the theorems apply to those of any order.

Determinants of the second and third orders were defined as follows in *4.10.1 and *4.10.3:

(1)
$$\begin{vmatrix} a_1 & b_1 \\ a_2 & b_2 \end{vmatrix} = a_1b_2 - a_2b_1$$

(2)
$$\begin{vmatrix} a_1 & b_1 & c_1 \\ a_2 & b_2 & c_2 \\ a_3 & b_3 & c_3 \end{vmatrix} = a_1 \begin{vmatrix} b_2 & c_2 \\ b_3 & c_3 \end{vmatrix} - a_2 \begin{vmatrix} b_1 & c_1 \\ b_3 & c_3 \end{vmatrix} + a_3 \begin{vmatrix} b_1 & c_1 \\ b_2 & c_2 \end{vmatrix}$$

$$= a_1(b_2c_3 - b_3c_2) - a_2(b_1c_3 - b_3c_1)$$
$$+ a_3(b_1c_2 - b_2c_1)$$
$$= a_1b_2c_3 + a_2b_3c_1 + a_3b_1c_2 - a_1b_3c_2$$
$$- a_2b_1c_3 - a_3b_2c_1$$

We observe that the notation indicates the location of an element in the square array, the subscript indicating the row and the letter itself indicating the column. Each of the diagonals a_1, b_2 and a_1, b_2, c_3 is called a **principal diagonal.**

If we form a new determinant by rotating the array about its principal diagonal, the elements of the principal diagonal stay in the same position, while others change places, a_2 with b_1, a_3 with c_1, and b_3 with c_2. The determinants (1) and (2) become:

$$(3) \qquad \begin{vmatrix} a_1 & a_2 & a_3 \\ b_1 & b_2 & b_3 \\ c_1 & c_2 & c_3 \end{vmatrix} \qquad \begin{vmatrix} a_1 & a_2 \\ b_1 & b_2 \end{vmatrix}$$

In the new determinants, each row is the same as the corresponding column of the original, and each column is the same as the corresponding row of the original. We may think of the change as interchanging rows and columns or as transposing terms across the diagonal. The interesting fact is that the value of the resulting determinant is identical with the original, as the student may see by expanding (3). Hence:

*4.11.1   The value of a determinant is unaltered by interchanging its rows and columns.

In (1) and (2) every term in the expansion involves just one element from each row and each column. It follows, then, that, if we multiply every element of a row (column) by a number, the result is the same as if we multiplied each term of the expansion by that number. Hence:

*4.11.2   If the elements of a row (column) of a determinant D are multiplied by k, the resulting determinant equals kD.

It follows also that:

*4.11.3   A determinant is zero if it has a row (column) of zeros.

If in (1) we interchange the two columns, we get

$$(4) \qquad \begin{vmatrix} b_1 & a_1 \\ b_2 & a_2 \end{vmatrix} = b_1 a_2 - b_2 a_1 = - \begin{vmatrix} a_1 & b_1 \\ a_2 & b_2 \end{vmatrix}$$

Similarly, the student can verify that if we interchange any two rows (columns) of either (1) or (2), the result is the negative of the original:

*4.11.4   Interchanging two rows (columns) changes the sign of a determinant.

From this theorem, it follows that:

*4.11.5 The value of a determinant is zero if it has two identical rows (columns).

For if we interchange the two identical rows (columns) of D the new determinant equals $-D$. Since the two are identical, $D = -D$, and $D = 0$!

We now come to a theorem that is used repeatedly in evaluating determinants:

*4.11.6 The value of a determinant is unaltered by replacing the elements of any row (column) by the sums of themselves and any multiple of the corresponding elements of another row (column).

For example,

(5)
$$\begin{vmatrix} 5 & 15 & 1 \\ -1 & 5 & 4 \\ 3 & 9 & -2 \end{vmatrix} = \begin{vmatrix} 5 + 2(1) & 15 & 1 \\ -1 + 2(4) & 5 & 4 \\ 3 + 2(-2) & 9 & -2 \end{vmatrix}$$

$$= \begin{vmatrix} 7 & 15 & 1 \\ 7 & 5 & 4 \\ -1 & 9 & -2 \end{vmatrix}$$

Here we replaced the elements of the first column by the sums of themselves and 2 times the corresponding elements of the third column. We describe this usually as "adding 2 times the third column to the first column." The manipulation (5) does not simplify the determinant appreciably, but adding -3 times the first column to the second gives the simpler form:

(6)
$$\begin{vmatrix} 5 & 0 & 1 \\ -1 & 8 & 4 \\ 3 & 0 & -2 \end{vmatrix}$$

Of course the multiple could be just 1 or -1 so that the theorem includes addition or subtraction of rows or columns.

We prove the theorem for determinants of the second order, leaving those of the third order for the exercise. Consider the determinant formed from (1) by adding k times the second column to the first:

(7)
$$\begin{vmatrix} (a_1 + kb_1) & b_1 \\ (a_2 + kb_2) & b_2 \end{vmatrix} \equiv (a_1 + kb_1)b_2 - (a_2 + kb_2)b_1$$

[By definition (1)]

(8)
$$\equiv a_1b_2 - a_2b_1 + k(b_1b_2 - b_2b_1)$$

(9)
$$\equiv \begin{vmatrix} a_1 & b_1 \\ a_2 & b_2 \end{vmatrix}$$

The theorem thus holds in the case considered. Consider now the result of adding a multiple of the first column to the second. We may achieve this by the following three steps: (1) Interchange the first and second columns. (2) Add the multiple of the second column in this determinant to the first column. (3) Interchange the first and second columns in this determinant. These steps cause, respectively, a change of sign (why?), no change (why?), and a second change of sign. Hence the final result is the same as the original. To interchange two rows, we interchange rows and columns, operate on columns, then interchange rows and columns again. By *4.11.1 and the foregoing results there is no change in the value of the determinant. Since we have considered all cases, the theorem is proved.

Before illustrating the use of the above properties, we suggest a convenient way of evaluating a third-order determinant. The expansion (2) is formed by multiplying each element in the first column by the determinant obtained by crossing out the row and column in which the element occurs. Hence, instead of writing out the expansion, you can proceed mentally as indicated in the following diagrams:

$$
\begin{array}{ccc}
(A) & (B) & (C)
\end{array}
$$

(10)
$$
\begin{array}{ccc}
\begin{array}{ccc} 3 & 2 & 7 \\ 1 & 4 & 5 \\ 8 & 6 & 1 \end{array}
&
\begin{array}{ccc} 3 & 2 & 7 \\ 1 & 4 & 5 \\ 8 & 6 & 1 \end{array}
&
\begin{array}{ccc} 3 & 2 & 7 \\ 1 & 4 & 5 \\ 8 & 6 & 1 \end{array}
\end{array}
$$

(11)
$$
\begin{vmatrix} 3 & 2 & 7 \\ 1 & 4 & 5 \\ 8 & 6 & 1 \end{vmatrix}
\overset{(A)\quad(B)\quad(C)}{=} 3(4 - 30) - 1(2 - 42) + 8(10 - 28)
$$

After a little practice, you can evaluate each of the small determinants mentally, and so write simply $3(-26) - 1(-40) + 8(-18)$. Of course, when the numbers are large, it is wiser to write more steps. In any case, all steps that must be written should be included as steps and not as scratchwork somewhere else. This is essential for checking and avoiding careless errors, as well as for guiding anyone who wants to read your work.

We illustrate the discussion by evaluating several determinants:

(12) $\quad \begin{vmatrix} 3 & 0 & 0 \\ 4 & 1 & 6 \\ -7 & 3 & 2 \end{vmatrix} = 3(2 - 18) = -48 \quad$ (The other terms in the expansion are zero because they involve the zeros in the first row.)

(13) $\quad \begin{vmatrix} 2 & -1 & 4 \\ 3 & 91 & 13 \\ -4 & 2 & -8 \end{vmatrix} = \begin{vmatrix} 0 & 0 & 0 \\ 3 & 91 & 13 \\ -4 & 2 & -8 \end{vmatrix} \quad$ (*4.11.6, adding half the last row to the first.)

(14) $\qquad\qquad\qquad = 0 \qquad$ (*4.11.3)

(15) $\quad \begin{vmatrix} 2 & -1 & 4 \\ 3 & 91 & 13 \\ -4 & 2 & -8 \end{vmatrix} = -2 \begin{vmatrix} 2 & -1 & 4 \\ 3 & 91 & 13 \\ 2 & -1 & 4 \end{vmatrix} \quad$ (*4.11.2, with $k = -2$ in the last row.)

(16) $\qquad\qquad\qquad = -2\cdot 0 = 0 \qquad$ (*4.11.5)

(17) $\quad \begin{vmatrix} 4 & 8 & 6 \\ 2 & 3 & -2 \\ 6 & 1 & 7 \end{vmatrix} = 2 \begin{vmatrix} 2 & 4 & 3 \\ 2 & 3 & -2 \\ 6 & 1 & 7 \end{vmatrix} \quad$ (*4.11.2, with $k = 2$, in first row.)

(18) $\qquad\qquad\qquad = 4 \begin{vmatrix} 1 & 4 & 3 \\ 1 & 3 & -2 \\ 3 & 1 & 7 \end{vmatrix} \quad$ (Same with $k = 2$ in first column.)

(19) $\qquad\qquad\qquad = 4 \begin{vmatrix} 1 & 4 & 3 \\ 1 & 3 & -2 \\ 0 & -8 & 13 \end{vmatrix} \quad$ (Subtracting three times second row from last.)

$$(20) \qquad = 4 \begin{vmatrix} 1 & 4 & 3 \\ 0 & -1 & -5 \\ 0 & -8 & 13 \end{vmatrix} \qquad \text{(Subtracting first from second row.)}$$

$$(21) \qquad = 4 \cdot 1 \cdot (-13 - 40) = 4(-53) = -212$$

In the above we have rewritten the determinant anew for each step. The student should be very cautious about consolidating steps, especially those involving the same elements, since the validity of a manipulation may depend on the previous result. Time spent on rewriting the determinant will be made up by time saved on searching for errors!

The **minor** of an element is the determinant formed by striking out the row and column in which the element appears. Thus the minor of b_2 in (2) is $\begin{vmatrix} a_1 & c_1 \\ a_3 & c_3 \end{vmatrix}$. We designate the minor of an element by the corresponding capital letter, for example, B_2 stands for the minor of b_2. With this notation the right member of (2) may be written $a_1 A_1 - a_2 A_2 + a_3 A_3$. We speak of this as the expansion by elements of the first column. The determinant can be expanded by the elements of any row or column by adding the products of the elements and their minors, with signs chosen according to the following scheme:

$$(22) \qquad \begin{vmatrix} + & - & + \\ - & + & - \\ + & - & + \end{vmatrix}$$

Thus (2) equals $a_1 A_1 - b_1 B_1 + c_1 C_1$, $-a_2 A_2 + b_2 B_2 - c_2 C_2$, etc. There are six different expansions of this kind. (See prob. 5 in the Exercise.)

EXERCISE 4.11

1. Verify the following equations by expanding each member according to the definition, and indicate which theorem is illustrated.

(a) $\begin{vmatrix} 3 & 2 \\ -1 & 4 \end{vmatrix} = \begin{vmatrix} 3 & -1 \\ 2 & 4 \end{vmatrix}$ (c) $\begin{vmatrix} 2 & 2 \\ -7 & -7 \end{vmatrix} = 0$

(b) $\begin{vmatrix} 7 & 3 \\ 9 & -2 \end{vmatrix} = - \begin{vmatrix} 3 & 7 \\ -2 & 9 \end{vmatrix}$ (d) $\begin{vmatrix} 2a & 1 \\ 1 & 2a \end{vmatrix} = - \begin{vmatrix} 1 & 2a \\ 2a & 1 \end{vmatrix}$

(e) $\begin{vmatrix} a-3 & 3 \\ b-c & c \end{vmatrix} = \begin{vmatrix} a & 3 \\ b & c \end{vmatrix}$

(h) $\begin{vmatrix} 26 & -2 \\ 7 & 3 \end{vmatrix} = \begin{vmatrix} 0 & -2 \\ 46 & 3 \end{vmatrix}$

(f) $\begin{vmatrix} 25 & 21 \\ 37 & 30 \end{vmatrix} = \begin{vmatrix} 25 & 21 \\ 12 & 9 \end{vmatrix}$

(i) $\begin{vmatrix} 42 & -2 \\ 12 & -1 \end{vmatrix} = -2 \begin{vmatrix} 21 & 2 \\ 6 & 1 \end{vmatrix}$

(g) $\begin{vmatrix} 54 & 9 \\ 12 & 2 \end{vmatrix} = \begin{vmatrix} 0 & 9 \\ 0 & 2 \end{vmatrix}$

(j) $\begin{vmatrix} 27 & 15 \\ 5 & -6 \end{vmatrix} = 3 \begin{vmatrix} 9 & 5 \\ 5 & -6 \end{vmatrix}$

2. In each of the following determinants: (a) Interchange rows and columns. (b) In the result, subtract the first row from the last. (c) In this result, add twice the second row to the first row. (d) In this, interchange the second and third columns and multiply the determinant by −1. (e) In this determinant, interchange the first and second columns and then the second and third rows. Evaluate the first and last determinants to check your work.

(a) $\begin{vmatrix} 2 & 6 & -3 \\ -3 & 7 & 1 \\ 1 & 2 & 5 \end{vmatrix}$

(d) $\begin{vmatrix} 1 & 0 & 1 \\ 0 & 1 & 0 \\ 1 & 0 & 1 \end{vmatrix}$

(b) $\begin{vmatrix} 19 & -8 & -4 \\ 17 & -1 & 5 \\ 1 & 4 & 11 \end{vmatrix}$

(e) $\begin{vmatrix} 1 & 1 & 1 \\ 2 & 2 & 2 \\ 3 & 3 & 3 \end{vmatrix}$

(c) $\begin{vmatrix} 2a & 2a & 1 \\ 3 & s & 2 \\ -a & a & 1 \end{vmatrix}$

(f) $\begin{vmatrix} a & b & c \\ a & b & c \\ 0 & 0 & 0 \end{vmatrix}$

3. Treat the following equations as in prob. 1:

(a) $\begin{vmatrix} 1 & 2 & -1 \\ -1 & 0 & 5 \\ 1 & 3 & 1 \end{vmatrix} = \begin{vmatrix} 1 & -1 & 1 \\ 2 & 0 & 3 \\ -1 & 5 & 1 \end{vmatrix} = \begin{vmatrix} 1 & -1 & 0 \\ 2 & 0 & 1 \\ -1 & 5 & 2 \end{vmatrix}$

(b) $\begin{vmatrix} 5 & -4 & 2 \\ 10 & -8 & 4 \\ 1 & 3 & 5 \end{vmatrix} = \begin{vmatrix} 5 & -4 & 2 \\ 0 & 0 & 0 \\ 1 & 3 & 5 \end{vmatrix} = 0$

(c) $\begin{vmatrix} 1 & 2 & 3 \\ 4 & 4 & 4 \\ 3 & 2 & 1 \end{vmatrix} = \begin{vmatrix} 1 & 1 & 1 \\ 3 & 2 & 1 \\ 3 & 2 & 1 \end{vmatrix} = 0$

(d) $\begin{vmatrix} 1 & 12 & -1 \\ -1 & 6 & 2 \\ 4 & 4 & 8 \end{vmatrix} = 2 \begin{vmatrix} 1 & 6 & -1 \\ -1 & 3 & 2 \\ 4 & 2 & 8 \end{vmatrix} = 4 \begin{vmatrix} 1 & 6 & -1 \\ -1 & 3 & 2 \\ 2 & 1 & 4 \end{vmatrix}$

(†e) $\begin{vmatrix} 1 & 3 & -2 \\ 9 & -1 & 5 \\ 8 & -4 & 7 \end{vmatrix} = \begin{vmatrix} 1 & 3 & -2 \\ 0 & 0 & 0 \\ 8 & -4 & 7 \end{vmatrix} = 0$

4. Evaluate the following determinants by judicious manipulations. Show each step and give reasons. Check results by evaluating in different ways.

(a) $\begin{vmatrix} 142 & 144 \\ 99 & 75 \end{vmatrix}$

(b) $\begin{vmatrix} 284 & 288 \\ 99 & 72 \end{vmatrix}$

(c) $\begin{vmatrix} -8 & 32 \\ -6 & 25 \end{vmatrix}$

(d) $\begin{vmatrix} -116 & 93 \\ 29 & -35 \end{vmatrix}$

(e) $\begin{vmatrix} b & 2c \\ 2a & b \end{vmatrix}$

(f) $\begin{vmatrix} x & y \\ 2x & xy \end{vmatrix}$

(g) $\begin{vmatrix} x_1 & x_2 \\ 1 & 1 \end{vmatrix}$

(h) $\begin{vmatrix} x & y \\ y & x \end{vmatrix}$

(i) $\begin{vmatrix} 4 & 5 & 11 \\ 1 & -1 & 5 \\ 0 & 9 & 3 \end{vmatrix}$

(j) $\begin{vmatrix} x & 3 & 2 \\ 2 & x & 1 \\ 4 & 5 & 2 \end{vmatrix}$

(k) $\begin{vmatrix} 2 & 8 & 1 \\ 4 & -10 & -4 \\ 2 & 5 & 1 \end{vmatrix}$

(l) $\begin{vmatrix} 0 & 4 & 2 \\ 1 & 2 & 3 \\ 3 & -1 & 1 \end{vmatrix}$

(m) $\begin{vmatrix} 2 & 6 & 8 \\ 0 & 1 & 2 \\ 1 & 0 & 1 \end{vmatrix}$

(n) $\begin{vmatrix} a-b & 2a & c \\ b & a & b \\ a & 2a & a \end{vmatrix}$

(o) $\begin{vmatrix} x & y & 1 \\ x^2 & y^2 & 1 \\ x^3 & y^3 & 1 \end{vmatrix}$

(p) $\begin{vmatrix} x-a & 0 & 0 \\ 0 & x-b & 0 \\ 0 & 0 & x-c \end{vmatrix}$

(q) $\begin{vmatrix} x & y & 1 \\ x_1 & y_1 & 1 \\ x_2 & y_2 & 1 \end{vmatrix}$

(r) $\begin{vmatrix} 2A & B & D \\ B & 2C & E \\ D & E & 2F \end{vmatrix}$

(s) $\begin{vmatrix} 1 & 4 & 5 \\ 0 & 2 & 7 \\ 0 & 0 & 3 \end{vmatrix}$

(t) $\begin{vmatrix} -1 & 0 & 0 \\ 8 & 3 & 0 \\ 9 & 4 & -6 \end{vmatrix}$

5. Show that:

$$\begin{vmatrix} 1 & 4 & 7 \\ 2 & 5 & 8 \\ 3 & 6 & 9 \end{vmatrix} = \begin{vmatrix} 11 & 14 & 17 \\ 12 & 15 & 18 \\ 13 & 16 & 19 \end{vmatrix}$$

6. Expand in all six ways the determinants of prob. 2.
Solution for part a:

$$D = 2 \begin{vmatrix} 7 & 1 \\ 2 & 5 \end{vmatrix} - (-3) \begin{vmatrix} 6 & -3 \\ 2 & 5 \end{vmatrix} + 1 \begin{vmatrix} 6 & -3 \\ 7 & 1 \end{vmatrix}$$

$$= -6 \begin{vmatrix} -3 & 1 \\ 1 & 5 \end{vmatrix} + 7 \begin{vmatrix} 2 & -3 \\ 1 & 5 \end{vmatrix} - 2 \begin{vmatrix} 2 & -3 \\ -3 & 1 \end{vmatrix}$$

$$= -3 \begin{vmatrix} -3 & 7 \\ 1 & 2 \end{vmatrix} - 1 \begin{vmatrix} 2 & 6 \\ 1 & 2 \end{vmatrix} + 5 \begin{vmatrix} 2 & 6 \\ -3 & 7 \end{vmatrix}$$

$$= 2 \begin{vmatrix} 7 & 1 \\ 2 & 5 \end{vmatrix} - 6 \begin{vmatrix} -3 & 1 \\ 1 & 5 \end{vmatrix} + (-3) \begin{vmatrix} -3 & 7 \\ 1 & 2 \end{vmatrix}$$

$$= -(-3) \begin{vmatrix} 6 & -3 \\ 2 & 5 \end{vmatrix} + 7 \begin{vmatrix} 2 & -3 \\ 1 & 5 \end{vmatrix} - 1 \begin{vmatrix} 2 & 6 \\ 1 & 2 \end{vmatrix}$$

$$= 1 \begin{vmatrix} 6 & -3 \\ 7 & 1 \end{vmatrix} - 2 \begin{vmatrix} 2 & -3 \\ -3 & 1 \end{vmatrix} + 5 \begin{vmatrix} 2 & 6 \\ -3 & 7 \end{vmatrix}$$

7. Prove *4.11.1 for determinants of the second and third orders by expanding the determinants involved.

8. Verify *4.11.4 for the special case of interchanging the first and third rows in the general determinant of the third order.

9. What would be the effect of multiplying each element of a determinant by a constant?

†**10.** Construct a proof of *4.11.6 for determinants of the third order.

†**11.** Construct a simpler proof of *4.10.4. [*Suggestion.* The equation is linear (why?) and is satisfied by (x_1, y_1) and (x_2, y_2) (why?), and hence $\cdots$.]

†**12.** Three lines are called **concurrent** if they pass through the same point. Show that a necessary and sufficient condition that

$$a_1x + b_1y + c_1 = 0$$

$$a_2x + b_2y + c_2 = 0$$

$$a_3x + b_3y + c_3 = 0$$

be concurrent is that

$$\begin{vmatrix} a_1 & b_1 & c_1 \\ a_2 & b_2 & c_2 \\ a_3 & b_3 & c_3 \end{vmatrix} = 0$$

(*Hint.* Express the solutions of two of the equations and substitute in the third.)

†**13.** Show that:

$$\begin{vmatrix} x & y & 1 \\ x_1 & y_1 & 1 \\ x_2 & y_2 & 1 \end{vmatrix} = - \begin{vmatrix} x - x_1 & y - y_1 \\ x_2 - x_1 & y_2 - y_1 \end{vmatrix}$$

†**14.** Show that:

$$\begin{vmatrix} a_1 + d_1 & b_1 & c_1 \\ a_2 + d_2 & b_2 & c_2 \\ a_3 + d_3 & b_3 & c_3 \end{vmatrix} = \begin{vmatrix} a_1 & b_1 & c_1 \\ a_2 & b_2 & c_2 \\ a_3 & b_3 & c_3 \end{vmatrix} + \begin{vmatrix} d_1 & b_1 & c_1 \\ d_2 & b_2 & c_2 \\ d_3 & b_3 & c_3 \end{vmatrix}$$

15. A definition similar to (2) may be given for determinants of the fourth and higher orders. Suppose that we have a square array of four rows and columns:

(23)
$$A = \begin{vmatrix} a_{11} & a_{12} & a_{13} & a_{14} \\ a_{21} & a_{22} & a_{23} & a_{24} \\ a_{31} & a_{32} & a_{33} & a_{34} \\ a_{41} & a_{42} & a_{43} & a_{44} \end{vmatrix}$$

Here the subscripts indicate the row and column so that a_{ij} is the element in the ith row and the jth column, and i and j are called the **row subscript** and **column subscript,** respectively. By A_{ij} we mean the **minor** of a_{ij} formed by eliminating the ith row and the jth column. Then we define A by

(24) $$A = a_{11}A_{11} - a_{21}A_{21} + a_{31}A_{31} - a_{41}A_{41}$$

All the theorems of this section apply to determinants of the fourth order. Evaluate each of the following in several different ways.

(a) $\begin{vmatrix} 1 & 0 & 1 & 1 \\ 0 & 0 & 1 & 0 \\ 2 & 1 & 0 & 1 \\ 1 & 1 & 0 & 0 \end{vmatrix}$ (c) $\begin{vmatrix} 0 & 2 & 1 & -1 \\ 0 & 0 & 2 & 0 \\ 1 & 3 & 1 & 5 \\ 1 & 0 & 2 & 1 \end{vmatrix}$ (e) $\begin{vmatrix} 5 & 1 & -3 & 4 \\ 0 & 2 & 0 & 1 \\ 10 & 2 & 6 & 4 \\ 0 & 9 & -1 & 8 \end{vmatrix}$

(b) $\begin{vmatrix} 2 & 1 & 2 & -1 \\ 0 & 3 & 3 & 5 \\ 0 & 0 & 1 & -1 \\ 0 & 0 & 0 & 5 \end{vmatrix}$ (d) $\begin{vmatrix} x_1 & y_1 & z_1 & 1 \\ x_2 & y_2 & z_2 & 1 \\ x_2 & y_2 & z_2 & 1 \\ x_3 & y_3 & z_3 & 1 \end{vmatrix}$ (f) $\begin{vmatrix} 1 & 2 & 9 & 3 \\ 0 & 3 & 8 & 7 \\ 0 & 0 & -1 & 4 \\ 0 & 0 & 0 & 1 \end{vmatrix}$

Answers. (a) 0. (c) −22.

†16. Linear equations in four unknowns may be solved by successive elimination of variables as in Ex. 4.10.5 or by determinants of the fourth order according to the natural extension of 4.10(19). Formulate this extension and solve the following by elimination and by determinants.

(a) $x + y - z + w = 0$
$x + 2z - w = 10$
$2x - y = 14$
$3y + z + 2w = 2$

(b) $x + y = 1$
$y + z = 2$
$z + w = 3$
$w - x = 4$

(c) $3x + y - 4z = 2$
$x - y + z + w = -1$
$-x + 5y + 2z - 8w = 0$
$2x - y - z - w = 5$

(d) $r + s + v + p = 100$
$r - s - p - 2v = 10$
$2r + s - 5p + 8v = 1$
$-8r + s + 10v - p = 0$

†17. Show that

$$\begin{vmatrix} x - x_1 & y - y_1 & z - z_1 \\ x_2 - x_1 & y_2 - y_1 & z_2 - z_1 \\ x_3 - x_1 & y_3 - y_1 & z_3 - z_1 \end{vmatrix} = \begin{vmatrix} x & y & z & 1 \\ x_1 & y_1 & z_1 & 1 \\ x_2 & y_2 & z_2 & 1 \\ x_3 & y_3 & z_3 & 1 \end{vmatrix}$$

†18. Suggest a definition for determinants of order 5.

†19. Show that, if a determinant of the third order is written using double subscripts as in prob. 15, then its expansion equals the sum of all products formed in the following way: (a) Choose three elements so that each row subscript appears once and only once and each column subscript appears once and only once. (b) Arrange the elements so that row subscripts are in ascending order and count the number of cases in which a column subscript precedes a smaller one. Call this the number of **inversions**. (c) Prefix a plus or minus sign to the product according as the number of inversions is even or odd. [*Suggestion.* Find all products of three elements according to (a)–(c) and compare with the expansion (2). For example, one possible product is $a_{12}a_{23}a_{31}$. Here the row subscripts are in order, but $2 > 1$ and $3 > 1$ in the column subscripts. Hence there are two inversions and the sign should be plus.] Show that the same thing applies to determinants of the second and fourth orders.

4.12 Linear inequalities

Some inequalities are true for all values of the variables that appear and are called **absolute inequalities.** For example,

(1) $$a + 1 > a$$

for any number a. This is easy to see, because $1 > 0$, and this implies $a + 1 > 0 + a = a$ by *3.4.6.

Other inequalities hold only for some values of the variables and are called **conditional inequalities.** For example,

(2) $$3x + 2 < 0$$

is true for $x = -1$ and false for $x = 0$. To **solve** an inequality means to find the set of values of the variable that satisfy it. We solve (2):

(3) $\qquad\qquad 3x < -2 \qquad$ (*3.4.4)

(4) $\qquad\qquad x < -\tfrac{2}{3} \qquad$ (*3.10.14)

We see that, if $3x + 2 < 0$, then $x < -\tfrac{2}{3}$. But the foregoing steps were reversible (why?), so that $x < -\tfrac{2}{3}$ and $3x + 2 < 0$ are equivalent inequalities. Hence the solution of (2) consists in all numbers less than $-\tfrac{2}{3}$ and no others. We may visualize it as consisting of all the points on the axis to the left of $-\tfrac{2}{3}$ (Fig 4.12.1). This set of points is called the graph of $3x + 2 < 0$

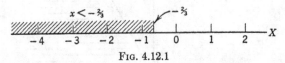

FIG. 4.12.1

More generally, the **graph of an inequality** is the set of points that satisfy it.

The student should compare the foregoing discussion with that in 4.4. To solve an inequality we manipulate much as we did with equations, but we get a solution consisting of an infinite number of points instead of one. Also in multiplying both sides of an inequality by the same number, we must take account of the sign of the multiplier. Thus to solve $-2x < 1$ we may multiply by $-\tfrac{1}{2}$, but we get $x > -\tfrac{1}{2}$, since our multiplier is negative. In dealing with inequalities, multiplication by zero must be avoided since both sides of the inequality will become zero, and 0 is not less than 0! With these cautions, the solution of inequalities is not very different from the solution of equations.

Some care is necessary when letters are involved and we do not know whether they stand for positive or negative numbers. Consider the general linear inequality

(5) $mx + b < 0$

We have $mx < -b$, as for linear equations, but

(6)
$$x < -\frac{b}{m} \quad \text{if } m > 0$$

$$x > -\frac{b}{m} \quad \text{if } m < 0$$

It is essential to consider all possibilities in this way.

Inequalities may involve more than one variable. The inequality

(7) $y < 2x + 1$

is satisfied by some pairs of values (x, y) but not by others. For example, it is a true statement when $x = 1$, $y = 1$, but it is false for $x = 0$, $y = 2$. How can we characterize the pairs of values

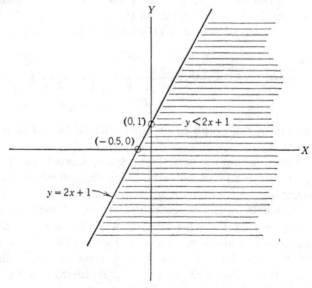

FIG. 4.12.2

that satisfy the inequality?   We know that $y = 2x + 1$ is the
equation of a line with slope 2 and y-intercept $(0, 1)$.   Evidently
any point below this line will have coordinates satisfying the
inequality so that the region shaded in Fig. 4.12.2 is the graph
of (7).

From the example above it appears that the graph of a linear
inequality in two variables is a region.   In general, the line $y = mx
+ b$ divides the plane into two regions, $y < mx + b$ below the
line, and $y > mx + b$ above the line.

We may give a linear inequality in one variable an interpretation
in a coordinate plane.   Just as we interpreted the equations $x = a$
and $y = b$ as lines parallel to the axes, so we can interpret $x > a$,
$x < a, y > b, y < b$ as inequalities whose graphs are regions right,
left, above, and below the corresponding lines parallel to the axes.

Often we are interested in values of variables that satisfy two
or more inequalities simultaneously.   Thus the points that satisfy
$x > 2$ and $x < 5$ make up an interval on the axis as indicated in
Fig. 4.12.3.   We write such a pair of inequalities together as

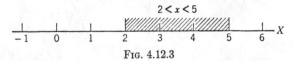

$$2 < x < 5$$

Fig. 4.12.3

$2 < x < 5$.   Note that the graph does not include the end points
of the segment or interval.   If we have $2 \leq x \leq 5$, the end points
are included.   We call the graph of $a < x < b$ an **open interval,**
and the graph of $a \leq x \leq b$ a **closed interval.**   The intervals
$a \leq x < b$ and $a < x \leq b$ are said to be closed on the left and the
right respectively.   If we think of x as the x-coordinate of a point
in a coordinate plane, the graphs of inequalities like $a < x < b$
are strips parallel to the y-axis.   We graph $-1 < x < 3$ in Fig.
4.12.4.   Suppose we wish to write inequalities that define the region
not shaded in Fig. 4.12.4.   A point is in this region if either $x < -1$
or $x > 3$.   Here we cannot write the two inequalities together.   A
statement like $-1 > x > 3$ is nonsense since *no* value of x can
satisfy it.   No number can be less than -1 and greater than 3!

When we have more than one inequality we see that it makes
a great deal of difference whether we want the set of points satisfying *all* of them or the set of points satisfying *at least one*.   If we
mean that all equations or inequalities are to be satisfied we speak

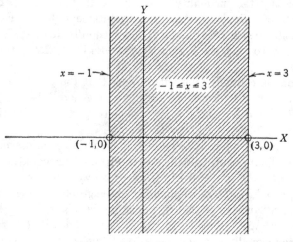

FIG. 4.12.4

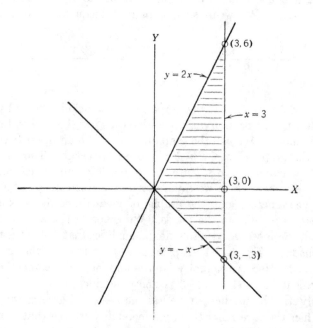

FIG. 4.12.5. $y < 2x$ and $y > -x$ and $x < 3$.

of **simultaneous conditions** or we use the word "and" to connect the conditions. If we mean that at least one condition is to be satisfied we speak of **alternative conditions** and use the word "or." Thus the locus of the simultaneous inequalities $y < 2x$ and $y > -x$ and $x < 3$ is the region shaded in Fig. 4.12.5. Points in this triangle satisfy all three inequalities. The locus of the alternative inequalities $y < 2x$ or $y > -x$ or $x < 3$ is the whole plane, since every point in the plane satisfies at least one of these inequalities.

EXERCISE 4.12

1. Decide whether each of the following is an absolute or conditional inequality. In the first case prove it; in the second case solve it.

(a) $-a + 2 > -a$ (f) $x - a < x$

(b) $2x < 1$ (g) $x < 2(x - 1)$

(c) $2x + 1 < 2x + 2$ (h) $3/(2x + 3) > 2/(1 - x)$

(d) $|x| > 0$ (i) $a - 3 < a - 2$

(e) $14x - 9 < 3 - 2x$ (j) $3 < -2x$

Answers. (a) Absolute since $2 > 0$, and this implies $-a + 2 > -a$ by *3.4.6 with a, b, c replaced by $2, 0, -a$. (b) Conditional. Holds for $x < \frac{1}{2}$.

2. Solve:

(a) $2x + 3 > 5x - 2$ (c) $2x + b > x + c$

(b) $4x - 2 > x + 1$ (d) $a_1 x + b_1 < a_2 x + b_2$

Answers. (a) $x < \frac{5}{3}$. (c) $x > c - b$.

3. Prove that:

(a) $a \geq a$ (d) $x - 3 < x$

(b) $2a > a$ if $a > 0$ (e) $x + 100 > x$

(c) $2a < a$ if $a < 0$ (f) $x - a > x$ if $a < 0$

4. Indicate on the x-axis the set of points determined by

(a) $x = 3$ (g) $x > 2$ or $x < 3$

(b) $x < 3$ (h) $x < 2$ and $x > 1$

(c) $|x| \geq 0$ (i) $-10 < x < 1$

(d) $x < 1.8$ (j) $|x| < 3$

(e) $1.8 < x \leq 1.9$ (k) $|x| > 3$

(f) $x < -3$ (l) $x < 2$ and $x > 3$

Answers. (c) The whole line. (l) Null set.

5. Graph the inequalities of prob. 4 in the x,y-plane.

6. In Ex. 4.4.1, replace $=$ by $\geq$ in each part, solve, and sketch the solution in the x,y-plane.

Answers. (a) $x \geq \frac{8}{3}$. (c) $s \leq -7$. (e) $x \leq 20/9$. (g) $x \leq -(1 + c)$. (i) $x \geq 16/27$. (k) $x \geq 0$.

7. Solve:

(a) $bx > c$	(c) $2x + 1 < ax$	(e) $1/x > -5$
(b) $a^2 x < 1$	(d) $rx \geq sx - 5$	(f) $1/x < 2/x$

Answers. (a) $x > c/b$ if $b > 0$ and $x < c/b$ if $b < 0$.   (c) $x < 1/(a - 2)$ if $a - 2 < 0$ and $x > 1/(a - 2)$ if $a - 2 > 0$.

8. Sketch:

(a) $y < -2x + 1$	(e) $3x - 4y - 2 < 0$	(i) $5 > x > -3$
(b) $2x + y + 1 > 0$	(f) $-3 + x < 1$	(j) $x < y$
(c) $y \leq x$	(g) $0 < y < 3$	(k) $x + y > 3$
(d) $y > 3x - 1$	(h) $-1 < y < 1$	(l) $x - y < 1$

9. Sketch and describe:

(a) $y \neq 3x + 4$	(g) $y > x$ and $y < -x$
(b) $x + a \neq 0$	(h) $y > x$ or $y < -x$
(c) $y \neq -1$	(i) $y > x$ and $x < 3$
(d) $x < -3$ or $x > 3$	(j) $x < 3$ and $y > -x$
(e) $y < 2$ or $y > 4$	(k) $x > 3$, $y > -x$, and $y < x$
(f) $0 < x < 2$ and $0 < y < 2$	(l) $y > -x$ or $x < 3$

10. Solve and sketch:

(a) $\lvert x - 1 \rvert < 1$	(e) $\lvert x + 5 \rvert > 3$	(i) $\lvert x + 3 \rvert < 0$
(b) $\lvert x - 3 \rvert < 2$	(f) $\lvert 2x + 4 \rvert < 2$	(j) $\lvert x \rvert + 3 > 0$
(c) $\lvert x + 2 \rvert < 1$	(g) $\lvert x \rvert < 0$	(k) $\lvert 1 - 4x \rvert < 5$
(d) $\lvert x - 1 \rvert > 2$	(h) $\lvert x - 1 \rvert \leq 0$	(l) $\lvert x - a \rvert < e$

Answers. (a) $0 < x < 2$. (c) $-3 < x < -1$. (e) $x < -8$ or $x > -2$. (g) Null set. (i) Null set. (k) $-1 < x < 3/2$. Note that $\lvert x - 1 \rvert < 1$ means $x - 1 < 1$ if $x - 1 > 0$ and $-(x - 1) < 1$ if $x - 1 < 0$. Both possibilities must be considered.

11. Graph the set of points that satisfy:

(a) $1 < x < 3$ and $1 < y < 2$	(h) $y < 2x$ or $y > 3x$
(b) $1 < x < 3$ or $1 < y < 2$	(i) $\lvert x + y \rvert < 1$
(c) $-1 < x \leq 0$ and $0 < y \leq 3$	(j) $\lvert 2x \rvert < 3$
(d) $x = 3$ and $1 \leq y \leq 2$	(k) $\lvert 2x + 3y \rvert < 6$
(e) $y < 2x + 1$, $y > 0$, and $x < 3$	(l) $y < 3x - 1$, $y < -4x + 6$,
(f) $3x + 2y > 0$, $y < 1$, and	or $y < -2$
$y < 2x + 6$	(m) $y > -1$ or $x > 1$
(g) $2x < y < 3x$	(n) $\lvert x + y \rvert > 2$

12. The sum of two numbers is less than 10. (a) Graph the possibilities. (b) Graph the possibilities if we know also that the numbers are positive, that is, $x > 0$ and $y > 0$. (c) Graph the possibilities if we know that the numbers are positive integers. (Circle the points involved.)

13. Find the inequalities that characterize the points within and on the boundary of a strip two units wide parallel to the y-axis and having its left edge on the y-axis.

14. Find the inequalities that define the triangular region with vertices at $(2, 8)$, $(3, -1)$, $(4, -2)$. (*Suggestion.* Find the three equations, sketch, and decide on the sense of each inequality.)

†**15.** What is the graph of: (a) $|x| < 0$. (b) $|x + y| < 0$. (c) $|x + y| \leq 0$?

†**16.** Show that, if $a > 0$, $b > 0$, $x > 0$, and $a < b$, then

$$\frac{a}{b} < \frac{a + x}{b + x} < 1$$

†**17.** Show that $|a| + |b| \geq \sqrt{a^2 + b^2}$.

5

The Quadratic Function

A **quadratic function** is one of the form $y = ax^2 + bx + c$, where $a \neq 0$. Before dealing with the general quadratic function and equation, we consider the special case $y = x^2$ and its inverse the square root. We also introduce two topics that are very useful in studying functions and that play an important part in mathematics. These are the average rate of change and the exact rate of change.

5.1 The square

By definition, $x^2 = x \cdot x$ (read "x-squared"). The equation $y = x^2$ defines a function that appears so frequently that we investigate its properties in detail. First we note that:

*5.1.1 $\qquad x^2 \geq 0 \qquad$ for any real number x

This follows from the definition of multiplication for positive and negative numbers (3.10). Moreover, from *3.8.4,

*5.1.2 $\qquad\qquad x^2 = 0 \longleftrightarrow x = 0$

From *3.10.3,

*5.1.3 $\qquad\qquad (-x)^2 \equiv x^2$

Finally,

*5.1.4 $\qquad\qquad |x_2| > |x_1| \longleftrightarrow x_2{}^2 > x_1{}^2$

This follows from the monotonic law of multiplication, *3.10.14. (See Ex. 5.1.8.)

These properties enable us to sketch the graph of $y = x^2$ (Fig. 5.1.1). The first indicates that $y < 0$ is an *excluded region*. The second shows that the curve goes through the origin and that this is the only point where $y = 0$. The third shows that for every value of y arising from a value of x, there is an equal value arising from $-x$, for example, $(2, 4)$ and $(-2, 4)$. This property is described by saying that the curve is *symmetric* with respect to

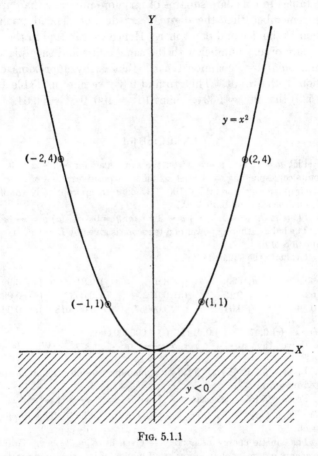

Fig. 5.1.1

the y-axis. (See Ex. 5.1.10.) The last property indicates that the values of y increase as we go to the right or left away from the origin. It appears plausible that a smooth curve drawn through a few plotted points is a good approximation to the graph. The situation is not basically altered if we multiply x^2 by a constant to get $y = ax^2$. If $a > 0$, each y in Fig. 5.1.1 is simply multiplied by a. If $a < 0$, we have a curve that could be obtained from

$y = |a| x^2$ by folding the plane along the x-axis. The graph is called a **parabola.**

The square appears so frequently that tables are available giving the values of $y = x^2$ for various values of x. We may use these tables to calculate squares of far more numbers than appear if we remember that the digits in x^2 do not depend upon the position of the decimal point in x. Hence we can find in the table the square of any number with the same digits and then place the decimal point by "common sense," that is, by approximate calculation of the result. Thus to find 9.8^2, we note in Table I (see page 599) that $98^2 = 9604$. Since $10^2 = 100$, $9.8^2 = 96.04$.

EXERCISE 5.1

1. Plot $y = x^2$ and $y = -x^2$ on the same axes for $-3 \leq x \leq 3$. Plot the points corresponding to at least six different positive values of x.

2. Graph $y = x^2$ for $|x| \leq 100$. (Choose an appropriately small unit length, but the same on both axes.)

3. Sketch: (a) $y = 2x^2$. (b) $y = 3x^2$. (c) $y = 0.5x^2$. (d) $y = -3x^2$.

4. The light-gathering power of a telescope is given by $L = 9a^2$. Graph L as a function of a.

5. Evaluate the squares of:

(a) 69	(d) 690	(g) 8.6	(j) 0.81	(m) 0.99
(b) 6.9	(e) 0.12	(h) 0.07	(k) 0.0037	(n) 0.099
(c) 0.69	(f) 0.045	(i) 0.00067	(l) 0.000018	(o) 0.0011

Answers. (c) 0.4761. (e) 0.0144. (i) 0.0000004489.

6. Prove that $0 < a < x < b \longrightarrow a^2 < x^2 < b^2$. Why is $0 < a$ required?

7. Justify *5.1.1, *5.1.2, *5.1.3 by explaining carefully the application of the cited reasons.

8. Prove *5.1.4. [*Suggestion.* $|x_2| > |x_1| \longrightarrow |x_2| - |x_1| > 0$. (Why?) Then $(|x_2| - |x_1|)(|x_2| + |x_1|) > 0$. (Why?) This implies the conclusion.]

9. The kinetic energy of a body is given by $E = (\tfrac{1}{2})mv^2$. Graph the kinetic energy of a mass of 10 grams as a function of its velocity in centimeters per second.

10. A **figure** is said to be **symmetric with respect to an axis** if, corresponding to every one of its points, there is another point symmetric to it with respect to the axis. (**Two points** are said to be **symmetric with respect to an axis** if the axis is the perpendicular bisector of the segment joining them.) The corresponding points may be considered reflections in the axis. Prove that $y = ax^2$ is symmetric with respect to the axis $x = 0$.

11. Graph: (a) $y > 3x^2$. (b) $y \leq -x^2$. (c) $y > -2x^2$.

12. Graph:

(a) $y - 2 = (x - 1)^2$ (d) $y = (x - 3)^2$ (g) $y + 1 \geq (x - 2)^2$

(b) $y - 1 = (x - 2)^2$ (e) $y - 1 = 2(x + 1)^2$ (h) $y = 1 - (x + 4)^2$

(c) $y - 2 = x^2$ (f) $y - 3 < x^2$ (i) $y - 2(x - 1)^2 = 3$

[*Suggestion for part a.* If we replace $y - 2$ by y' and $x - 1$ by x', the equation becomes $y' = x'^2$, which is easy to graph. But $x' = x - 1$, $y' = y - 2$ means a translation of the origin to the point $(1, 2)$. (See *4.6.1.) Hence we translate the origin to this point and graph $y' = x'^2$ in the new coordinate system as in Fig. 5.1.2.]

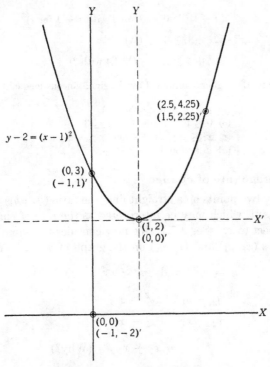

Fig. 5.1.2.   (Prob. 12a.)

13. Find a function of the form $y = ax^2$ whose graph passes through: (a) $(1, 2)$. (b) $(-2, 5)$. (c) $(-2, -3)$. (d) $(2, -1)$. [*Answers.* (a) $y = 2x^2$. (c) $y = -3x^2/4$.]

14. If the force exerted by a river current against a bridge support varies as the square of its velocity and if the force is 20 tons when the water flows at 5 mph and the temperature is 83°F, what is it when the velocity is 8 mph and the temperature is 85°F?

15. Prove that $f(x) = x^2 \longrightarrow f(x^2) = [f(x)]^2$, and show that the same result does *not* hold for $f(x) = 2x^2$.

16. Graph $y = f(x)$ where $f(x) = x^2$ for $x \geq 0$ and $f(x) = x$ for $x < 0$.

†17. Show that the square of an even number is even and the square of an odd number is odd. (*Suggestion.* An even number is always of the form $2n$ and an odd number is always of the form $2n + 1$.)

†18. Prove that for any x:

(a) $x^2 - 1 \geq -1$ \qquad (b) $x^2 + 1 \geq 2x$ \qquad (c) $x^2 + y^2 \geq 2xy$

Solution for part b:

(1) \qquad\qquad $(x - 1)^2 \geq 0$ \qquad (*5.1.1 with $x - 1$ for x)

(2) \qquad\qquad $x^2 - 2x + 1 \geq 0$ \qquad (Why?)

(3) \qquad\qquad $x^2 + 1 \geq 2x$ \qquad (*3.4.4 with ?)

†19. Graph the region determined by the simultaneous inequalities:

(a) $y \geq x^2$ and $y < 3x + 1$
(b) $0 \leq y \leq x^2$ and $0 \leq x \leq 3$
(c) $x^2 \leq y \leq 2x + 1$ and $y \leq -2x + 1$
(d) $y > x^2$ and $y - 3 < -x^2$

5.2   Average rate of change

For any two points on a straight line, the ratio $(y_2 - y_1)/(x_2 - x_1)$ is a constant, which may be interpreted as the rate of change of y with respect to x. (See 4.7.) We now consider the same ratio for two points (x_1, y_1) and (x_2, y_2) on the graph of $y = x^2$. Since

(1) \qquad\qquad $y_1 = x_1{}^2$ \quad and \quad $y_2 = x_2{}^2$ \qquad (Why?)

(2) \qquad\qquad $\dfrac{y_2 - y_1}{x_2 - x_1} = \dfrac{x_2{}^2 - x_1{}^2}{x_2 - x_1}$

(3) \qquad\qquad\qquad\qquad $= x_1 + x_2$ \qquad (Why?)

Evidently the ratio depends upon the location of the points, which is not surprising since it is the slope of the secant line joining the two points, and this slope is by no means the same for all secants (Fig. 5.2.1). If we fix (x_1, y_1) and let the other point move along the curve, the secant rotates about the fixed point. We see both from the figure and from (3) that when the point moves to the right the slope of the secant increases (point A), and when the point moves to the left the slope decreases (points C and

D). When $x_2 = -x_1$, it is zero and the secant is horizontal. When $x_2 < -x_1$, the slope is negative, and as x_2 decreases still further the slope continues to decrease, becoming larger in absolute value (points E and F).

The numerator in $(y_2 - y_1)/(x_2 - x_1)$ represents the change in y when x changes from x_1 to x_2. The denominator is just the change in x. We write Δx for $x_2 - x_1$ and Δy for $y_2 - y_1$. These

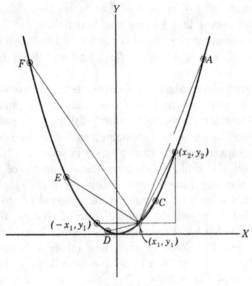

Fig. 5.2.1

symbols are read "delta-x" and "delta-y", where the Greek delta stands for "change in." Then the ratio of the change in y to the change in x may be written $\Delta y/\Delta x$. We call it the **average rate of change** of y in the interval $[x_1$ to $x_2]$. We showed in 4.7 that:

*5.2.1   If $y = mx + b$, then $\dfrac{\Delta y}{\Delta x} = m$

A slight modification of the argument of (1)–(3) shows that:

*5.2.2   If $y = ax^2$, then $\dfrac{\Delta y}{\Delta x} = a(x_1 + x_2)$

The same idea may be applied to any function $y = f(x)$ according to the following definition:

*5.2.3   The average rate of change of $y = f(x)$ in the interval $[x_1$ to $x_2]$ is

$$\frac{\Delta y}{\Delta x} = \frac{f(x_2) - f(x_1)}{x_2 - x_1} \qquad \text{(Definition)}$$

The average rate of change of $f(x)$ in $[x_1$ to $x_2]$ is always the slope of the secant line joining the two points on the graph of $y = f(x)$. Assuming $x_1 < x_2$, it is positive if the function increases in the interval, that is, if $f(x_2) > f(x_1)$, and negative if the function decreases.

The concept of average rate of change is useful whenever we are interested in the way in which one quantity is changing with another. Consider, for example, a body falling from rest according to the physical law $y = 16x^2$, which gives the distance that it falls in feet as a function of the time elasped in seconds. The speed is increasing all the time, so that we cannot speak of a constant velocity during any time interval. But we might ask how fast it is falling on the average in some time interval $[x_1$ to $x_2]$. This average velocity is just the average rate of change, the distance fallen (Δy) divided by the time elapsed (Δx). According to *5.2.2 it is $16(x_1 + x_2)$. Consider the time interval from $x_1 = 1$ to $x_2 = 3$. For $x_1 = 1$, $y_1 = 16$ and for $x_2 = 3$, $y_2 = 144$. Hence $\Delta y = y_2 - y_1 = 144 - 16 = 128$. Since $\Delta x = 2$, the average speed is $\Delta y / \Delta x = 64$ feet per second. Formula *5.2.2 gives the same result immediately, that is, $16(1 + 3) = 64$. Similarly, the average speed in the time interval $[1$ to $2]$ is 48, and in the interval $[1$ to $1.5]$ it is 40.

One of the most frequent applications of average rate of change is to **interpolation,** the process of estimating intermediate values of a function when only certain values are known. Suppose, to take a very simple case, that we wish to find the square of 13.6 from a table of squares of integers. We have $13^2 = 169$ and $14^2 = 196$. The function has changed by $\Delta y = y_2 - y_1 = 196 - 169 = 27$ whereas x has changed by $\Delta x = x_2 - x_1 = 14 - 13 = 1$. Actually, the rate of increase is not constant, but, if it were, we would expect when x changes by 0.6 of 1 that y would change by 0.6 of 27 or 16.2. Hence we estimate 13.6^2 as $169 + 16.2$

$= 185.2 \doteq 185$. The actual value is 184.96. We round off the estimate to the nearest integer because the approximate nature of the calculation makes great "accuracy" illusory.

The student may be familiar with the technique of interpolation, but we are interested now in showing *why* and *how* it "works." The situation is represented in Fig. 5.2.2, which gives an enlarged picture of a portion of the curve $y = x^2$. We are given the points

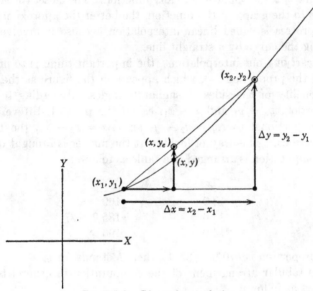

FIG. 5.2.2.   Linear interpolation.

(x_1, y_1) and (x_2, y_2), (13, 169) and (14, 196) in the example above, and we wish to find the y corresponding to some intermediate x. We cannot find the exact value without squaring, but the secant line is very near the curve, and the value of y on this secant line is a good estimate. From 4.8(8) or from similar triangles in the figure, the equation of the secant line is

$$(4) \qquad\qquad y - y_1 = \frac{\Delta y}{\Delta x}(x - x_1)$$

This is the equation of the line passing through (x_1, y_1) and having the slope $\Delta y/\Delta x$. Replacing y by y_e in order to emphasize that this is the y-coordinate of a point on the secant and an estimate of the corresponding y, and solving for y_e, we have:

*5.2.4
$$y_e = y_1 + \left(\frac{\Delta y}{\Delta x}\right)(x - x_1)$$

(Linear Interpolation Formula)

In our numerical example, $y_e = 169 + (27/1)(13.6 - 13)$. The formula *5.2.4 is applicable to any function. The closer the secant line is to the graph of the function, the better the approximation. The process is called linear interpolation because it involves replacing the curve by a straight line.

In carrying out interpolations, the important thing is to understand the proportion (4), which appears in the figure as the proportionality of the sides of similar triangles. According to this proportion, $y_e - y_1$ and $x - x_1$, called the **partial differences,** are proportional to $\Delta y = y_2 - y_1$ and $\Delta x = x_2 - x_1$, the **total differences.** This way of looking at the matter is brought out if the computation is arranged in a table as follows:

$$
\begin{array}{cc}
x & y \\
\left.\begin{array}{l} 13.0 \\ 13.6 \end{array}\right] 0.6 \quad h & \left[\begin{array}{l} 169 \\ 185.2 \end{array}\right] 27 \\
14.0 & 196
\end{array}
$$

(5)

The proportion is $h/0.6 = 27/1$, where h stands for $y_e - y_1$.

The tabular arrangement of the computation in general terms appears as follows:

(6) $\quad \Delta x = x_2 - x_1\left[\begin{array}{l} x_1 \\ x \\ x_2 \end{array}\right] x - x_1 \quad h = y_e - y_1\left[\begin{array}{l} y_1 \\ y_e \\ y_2 \end{array}\right]\Delta y = y_2 - y_1$

Returning to the example of the freely falling body, we interpolate to find the distance it falls in the first 1.8 seconds. Letting $x_1 = 1$ and $x_2 = 3$, we have from *5.2.4:

(7) $y_e = 16 + 64(1.8 - 1) = 67.2 \doteq 67$

The actual distance is $y = 16(1.8)^2 = 51.84$. The approximation is poor because we used the average rate of change for such a long interval. Note that $64(1.8 - 1)$, the second term in the right member of (7), is the estimate of the change in y, the distance fallen, in the interval from $x = 1$ to $x = 1.8$.

EXERCISE 5.2

1. Draw a large graph of $y = x^2$ and label the points corresponding to $x = 1, 3, 0, -1, -2, -3$. Draw and write the equations of the secant lines connecting $(1, 1)$ to each of the other points.

2. Explain why the average rate of change in a given interval is the same regardless of which end point is labeled (x_1, y_1). Illustrate by numerical examples.

3. Find the average rate of change for each of the following functions in the interval indicated, graph the function, show the corresponding secant line, and write its equation:

(a) $y = 2x^2$, $x_1 = 1$, $x_2 = 3$ (e) $y = 3x^2$, $x_1 = -1$, $x_2 = 0$

(b) $y = 4x^2$, $x_1 = 0.5$, $x_2 = 2$ (f) $y = 2$, $x_1 = 1$, $x_2 = 2$

(c) $y = x^2$, $x_1 = -1$, $x_2 = -3$ (g) $y = 7x + 3$, $x_1 = 99$, $x_2 = 178$

(d) $y = x^2/3$, $x_1 = 1.2$, $x_2 = 1.3$ (†h) $y = x^3$, $x_1 = -1$, $x_2 = 2$

Answers. (a) 8. (c) −4. (e) −3. (g) 7.

4. Find the average rate of change of $y = 16x^2$ in two ways, as we did in the section, for the interval $x_1 = 1$ to $x_2 = 2$. Use this average rate of change to estimate the distance traveled in 1.8 seconds, and compare with the estimate in the section and with the true value.

5. A body falls from rest according to $s = 16t^2$, where s is distance and t is time. Find: (a) The distance traveled during the 8th second—do this in two ways. (b) The average speed during the ⅓ second immediately following $t = 4$. (c) Use this result to estimate the distance fallen in the ⅙ second following $t = 4$. (d) Compare this result with the actual distance fallen in this ⅙ second.

6. Complete the following table, which gives $y = x^2$ for the indicated values of x; y_e, the value estimated by linear interpolation without rounding off; y_r, the result of rounding y_e to the nearest integer; and the corresponding errors $y_e - y$ and $y_r - y$. The row for $x = 13.6$ has been filled in. Interpolations are to be done from the average rate of change in the interval [13 to 14].

x	y	y_e	y_r	$y_e - y$	$y_r - y$
13.0	169.00				
13.1					
13.2					
13.3					
13.4					
13.5					
13.6	184.96	185.2	185	0.24	0.04
13.7					
13.8					
13.9					
14.0	196.00				

7. Draw a large graph to illustrate the previous problem. Show the curve $y = x^2$, the secant line, the estimate points, and the points corresponding to the rounded values. (*Suggestion*. Take the origin at $x = 13$, $y = 169$ and choose a large unit for x.)

8. Find the squares of the following numbers by interpolation in Table I:

(a) 25.3	(d) 63.2	(g) 37.4	(j) 37.8	(m) 0.935
(b) 93.7	(e) 56.8	(h) 78.6	(k) 8.3	(n) 0.0187
(c) 41.5	(f) 99.9	(i) 18.7	(l) 0.142	(o) 0.00992

Answers. (a) 640. (c) 1722. (e) 3226. (m) 0.8748.

9. In the indicated parts of prob. 3, interpolate for the y corresponding to the given x: (3a) $x = 1.7$. (3b) $x = 0.6$. (3c) $x = -2$. (3d) $x = 1.24$. (3e) $x = -0.7$. (3f) $x = 1.2$. [*Answers.* (3a) 7.6. (3c) 5. (3e) 2.1.]

10. The following table gives United States national income, I, in billions of dollars (rounded off to two decimals) as a function of time, t. It shows also the population, N, in millions (rounded off to the nearest 100,000).

t	I	N
1850	2.18	23.2
1860	3.60	31.4
1870	6.65	39.8
1880	7.34	50.2
1890	11.97	62.9
1900	17.42	76.0
1910	29.24	92.0

(a) Graph I as a function of t, connecting the plotted points by a smooth curve. (Take the t-axis horizontal and place the origin at the point $t = 1840$, $I = 0$.)

(b) Calculate the average rate of increase (the average annual increase) in each decade.

(c) Estimate national income by linear interpolation in the years 1857, 1885, and 1906. Plot the estimates.

(d) Calculate the average rate of change for the whole 60-year period. In which decades was the average annual increase less than this? greater than this?

(e) Using the values of I in 1850, 1870, 1890, and 1910 interpolate for the values in 1860, 1880, and 1900. Plot and compare your estimates with the actual values.

11. Carry through for N and t the same operations required for I and t in the previous problem.

12. Derive *5.2.2.

13. In each of the following problems we give the values of a function for two values of the independent variable. Assuming that the function has a fairly smooth graph so that the secant line is not far from the curve, estimate the unknown value of the function corresponding to the third value of the independent variable:

(a) $x_1 = 2$, $y_1 = 3$, $x_2 = 3$, $y_2 = 8$, $x = 2.5$

(b) $y = f(x)$, $f(1) = 2$, $f(4) = 17$, $f(3) = ?$

(c) $g_1 = 4$, $x_1 = -1$, $g_2 = -6$, $x_2 = 2$, $x = 0.5$

(d) $g(1.2) = 10.1$, $g(1.3) = 10.8$, $g(1.26) = ?$
(e) $f(5) = 7$, $f(10) = -7$, $f(8) = ?$
(f) $h(167.31) = 59.18$, $h(167.62) = 60.390$, $h(167.56) = ?$
(g) $x_1 = a$, $y_1 = 6$, $x_2 = -a$, $y_2 = -a$, $x = 0$
(h) $f(b) = c$, $f(d) = e$, $f(1) = ?$
(i) $f(x_1) = b$, $f(x_1 + 7c) = 2b$, $f(x_1 + 2c) = ?$
(j) $f(a) = b$, $f(b) = c$, $f(c) = ?$

Answers. (a) 5.5. (c) -1. (e) -1.4. (g) $(6 - a)/2$. (i) $9b/7$.

14. The secant line can be used to estimate values of the dependent variable outside the interval. For example, in prob. 6, the average rate of change in the interval [13 to 14] is 27, and the equation of the secant line is $y = 169 + 27(x - 13)$. The point on this line for which $x = 14.1$ has a y-coordinate of $y_e = 169 + 27(1.1) \doteq 199$. This is a fair estimate of $14.1^2 = 198.81$. This process of estimating values that are not between the known values is called **extrapolation.** Extrapolate for the unknown values in the following:

(a) $14.2^2 = ?$ if $13^2 = 169$ and $14^2 = 196$.
(b) $99^2 = ?$ if 100^2 and 101^2 are known.
(c) $f(4) = 7$, $f(5.5) = 17$, $f(6) = ?$
(d) $f(0.1) = 8.7$, $f(0.2) = 11.2$, $f(0) = ?$
(e) $f(a) = 1$, $f(a + 3) = 5$, $f(a + 4) = ?$
(f) $x_1 = 18.6$, $y_1 = 1000.2$, $x_2 = 19$, $y_2 = 989.3$, $x = 19.1$.
(g) Same as part f with $x = 18$.
(h) $y(x_0) = a$, $y(x_5) = b$, $y(b) = ?$

Answers. (a) 201. (c) 20.3. (e) 6.3. (g) 1016.6.

15. Often we are more interested in the change of a function relative to its original value than in the change itself. The ratio $(y_2 - y_1)/y_1$ is called **relative change.** When expressed in percentage, it is called the **per cent change.** If the price of bread goes up from 15 to 16 cents, the relative change is $\frac{1}{15}$ and the per cent change is 6.7%. (a) In prob. 10, find the per cent increase in population for each decade. (b) Same for national income.

16. The ratio of the relative changes of two variables is called the **average elasticity.** More precisely, the average elasticity of y with respect to x is defined as $[(y_2 - y_1)/y_1] \div [(x_2 - x_1)/x_1]$ or $(\Delta y/y)/(\Delta x/x)$. Find the elasticity in each part of prob. 3.

17. The stretching of a spring is a function of the force applied. If a force of 2 pounds stretches a spring 1 inch, and a force of 6 pounds stretches it 3 inches, find the average elasticity. (*Answer.* 1.)

†18. Show that, if $y = mx$, the elasticity of y with respect to x is 1. (When y is the *stress* and x is the *strain* in an elastic body, m is the *modulus of elasticity*.)

†19. Prove the converse of prob. 18.

†20. If sales (demand) for salt go down 1% when the price rises by 10%, what is the elasticity of the demand for salt? What kind of goods would you expect to have large negative elasticities? Small negative elasticities? Positive elasticities?

21. If the velocity v of a body is a function of time t, its *average acceleration* is defined as the average rate of change of its velocity, that is, $(v_2 - v_1)/(t_2 - t_1)$.

(a) If a body starts from rest ($v = 0$) and accelerates to 100 mph in 8 seconds, what is its average acceleration in mph per second? (b) Suppose a body has the following speeds at the times indicated: $v = v(t)$, $v(0) = 2$, $v(1) = 10$, $v(3) = 15$, $v(4) = 8$, $v(5.2) = 10$, $v(6) = 4$. Find its average acceleration in each interval between successive values of t. (c) If an automobile brakes from 50 mph to a stop in 4 seconds, what is its average acceleration? (*Answer.* -12.5 mph per second.)

22. The average rate of change of $y = f(x)$ is sometimes called the *change in y per unit change in x*, since, if the rate of change were constant, it would be the amount y would change when x changed by one unit. (Why?) (a) What is the change in the area of a triangle per unit change in its altitude when the base is 2 and the altitude changes from 3 to 7? (b) What is the change in area of a circle per unit change in the radius when the radius changes from 1 to 3? (c) Same when the radius changes from 2 to 3? (d) Same when it changes from 3 to 4?

23. Show that, if $g(x) = cf(x)$, then $\dfrac{\Delta g(x)}{\Delta x} = c\,\dfrac{\Delta f(x)}{\Delta x}$.

†**24.** Prove the converses of *5.2.2 and *5.2.1.

5.3 Square root

If $r^2 = b$, we call r a **square root** of b. If $b < 0$, there is no real square root (*5.1.1); if $b = 0$, there is just one (*5.1.2); and, if $b > 0$, there are two real roots, equal in absolute value and opposite in sign (*5.1.3 and Fig. 5.1.1.). We designate the *nonnegative* root by the symbol $\sqrt{b}$ and call it the **principal square root** of b. A formal definition is:

*5.3.1 $y = \sqrt{x} \longleftrightarrow y^2 = x$ and $y \geq 0$ (Definition)

An expression of the form $\sqrt{x}$ is called a **radical** in which x is the **radicand.** We assume $x \geq 0$ for the present in order to avoid complex numbers. We write $-\sqrt{x}$ for the negative solution of $y^2 = x$ and $\pm\sqrt{x}$ when we wish to indicate both solutions. The last expression is read "plus or minus the square root of x."

The equations $y = \sqrt{x}$ and $y = -\sqrt{x}$ define functions of x in which there is just one y corresponding to each x. But $y^2 = x$ defines a function in which two values of y correspond to each value of x. Such functions are called **two-valued.** More generally, if n values of y correspond to each value of x, we call y an **n-valued** function of x. We call the function **single-valued** when $n = 1$ and **multiple-valued** when $n > 1$.

It is instructive to note the analogy between *5.3.1 and the definitions of subtraction and division in 3.7. Subtraction and division were called the inverse operations of addition and multiplication because successive addition and subtraction of the same number or multiplication and division by the same number yielded no change. Similarly, we call taking the square root the inverse operation of squaring because, if we square a positive number and then take the square root or take the square root and then square, the result is the number with which we started, that is,

*5.3.2 $\sqrt{x^2} \equiv (\sqrt{x})^2 \equiv x$ for $x \geq 0$

If $x < 0$ the middle expression is not a real number and $\sqrt{x^2} \equiv -x$. (Why?) Regardless of the sign of x, we can write:

*5.3.3 $\sqrt{x^2} \equiv |x|$

The graph of $y^2 = x$ can be obtained by interchanging x and y for every point on the graph of $y = x^2$ (Fig. 5.1.1). It consists of two branches given by $y = \sqrt{x}$ and $y = -\sqrt{x}$. The branch corresponding to the principal root is drawn with a heavy line in Fig. 5.3.

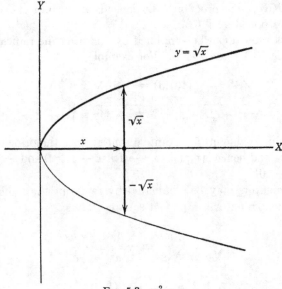

FIG. 5.3. $y^2 = x$.

Square roots appear so frequently that it is worth while to have a certain facility for dealing with them. The following theorems are sufficient for this purpose:

*5.3.4 $$\sqrt{ab} \equiv \sqrt{a}\sqrt{b}$$

*5.3.5 $$(\sqrt{a})^n \equiv \sqrt{a^n}$$

*5.3.6 $$\sqrt{\frac{a}{b}} \equiv \frac{\sqrt{a}}{\sqrt{b}}$$

Theorems *5.3.4 and *5.3.6 follow immediately from the definitions of square root. For example, to prove *5.3.4, it is sufficient to show that the square of the right member is ab, since that is its meaning according to *5.3.1. Now

(1) $$(\sqrt{a}\sqrt{b})^2 \equiv \sqrt{a}\sqrt{b}\sqrt{a}\sqrt{b} \qquad \text{(Why?)}$$

(2) $$\equiv (\sqrt{a})^2(\sqrt{b})^2 \qquad \text{(Why?)}$$

(3) $$\equiv ab \qquad (*5.3.2)$$

We do not write $|ab|$ in (3) because we are assuming radicands to be positive. The proof of *5.3.5 depends on the laws of exponents and is postponed until 6.3.

Often a radical can be simplified by factoring the radicand and then using *5.3.4 and *5.3.3. For example,

(4) $$\sqrt{48} = \sqrt{(16)(3)} = \sqrt{16}\sqrt{3} = 4\sqrt{3}$$

(5) $$\sqrt{x^2 + 2x + 1} \equiv \sqrt{(x+1)^2} \equiv |x+1|$$

The absolute value in (5) is quite essential since the principal root is positive and hence is equal to $x + 1$ if $x + 1 \geq 0$ and $-(x+1)$ if $x + 1 < 0$.

Denominators may be eliminated by rewriting the radicand so that its denominator is a perfect square. Thus

(6) $$\sqrt{\frac{s}{a}} \equiv \sqrt{\frac{as}{a^2}} \equiv \frac{\sqrt{as}}{\sqrt{a^2}} \equiv \frac{\sqrt{as}}{|a|}$$

(7) $$\sqrt{1/2} = \sqrt{2/4} = (1/2)\sqrt{2}$$

Expressions of the form

$$(8) \qquad \frac{A}{\sqrt{a} + \sqrt{b}}$$

may be written without radicals in the denominator by the device of multiplying numerator and denominator by the denominator with the middle sign changed, that is, by $\sqrt{a} - \sqrt{b}$. The result is

$$(9) \qquad \frac{A(\sqrt{a} - \sqrt{b})}{a - b}$$

For example,

$$(10) \qquad \frac{2}{1 - \sqrt{3}} = \frac{2(1 + \sqrt{3})}{(1 - \sqrt{3})(1 + \sqrt{3})}$$

$$(11) \qquad = \frac{2(1 + \sqrt{3})}{1 - 3} = -(1 + \sqrt{3})$$

Rewriting an expression so that no radicals are involved in the denominators is called rationalizing the denominator.

The reader may be familiar with the technique for extracting square roots "by hand." However, it is easier to find them from a table or by approximation. If a table of square roots is available, it can be used to read off square roots and to interpolate for intermediate values as described for squares in 5.1 and 5.2. Square roots can also be found by using a table of squares "backwards." The decimal point must be treated with care when dealing with square roots, since its position *does* make a difference to the digits that appear in the result. Thus $\sqrt{16.0} = 4$ but $\sqrt{160} \doteq 12.6$. More precisely,

$$*5.3.7 \qquad \sqrt{10N} = \sqrt{10}\sqrt{N}$$

$$*5.3.8 \qquad \sqrt{100N} = 10\sqrt{N}$$

Thus, multiplying a number by 100 multiplies its square root by 10, so that moving the decimal point *two* places has no effect on the digits of the square root, whereas moving it one place does have an effect.

We illustrate the technique of interpolating backwards in a table of squares by finding $\sqrt{55}$. In the body of Table I we note that 49 is the square of 7 and 64 is the square of 8. We think of the numbers *in the table* as the values of x and use the interpolation formula *5.2.4, with $x_1 = 49$, $x = 55$, $\Delta x = 15$, $\Delta y = 1$. Then

(12) $$y_e = 7 + (1/15)(55 - 49) = 7.4$$

Actually $(7.4)^2 = 54.76$, and from Table I directly $\sqrt{55} \doteq 7.416$. If the student does not care to use *5.2.4 explicitly, he may visualize the interpolation as follows:

$$x \qquad\qquad y = \sqrt{x}$$

$$15\begin{bmatrix} {\scriptstyle\overset{49}{}} \\ 55 \\ 64 \end{bmatrix}\!{\scriptstyle 6} \qquad d\begin{matrix} 7 \\ \to? \\ 8 \end{matrix}\Big]1$$

Then he can think as follows: y goes 1 while x goes 15. How far does y go while x goes 6? This suggests $d/6 = 1/15$. Of course d is the same as the second term in the right member of (12).

EXERCISE 5.3

1. Justify the following by *5.3.1:

(a) $\sqrt{9} = 3$ (c) $\sqrt{0} = 0$ (e) $\sqrt{12} = 2\sqrt{3}$ (g) $\sqrt{-4} \neq 2$

(b) $\sqrt{9} \neq -3$ (d) $-\sqrt{25} = -5$ (f) $\sqrt{-4} \neq -2$ (h) $\sqrt{(-2)^2} = 2$

2. Graph $y = \sqrt{x}$ by plotting the points for integral values of x from 0 to 10 and joining them with a smooth curve.

3. Sketch a graph of $y = \sqrt{x}$ for $0 \leq x \leq 100$, using the same scale on the axes.

4. Simplify, rationalizing any denominators that appear:

(a) $(\sqrt{a^2})^2$ (f) $2\sqrt{z^2}$ (k) $\sqrt{(a + b)^2 c^4}$ (p) $\sqrt{2x}\sqrt{3x}$

(b) $\sqrt{27}$ (g) $\sqrt{2/3}$ (l) $(\sqrt{2ab})^3$ (q) $\sqrt{1/a}\sqrt{a^2/b}$

(c) $\sqrt{9a}$ (h) $\sqrt{xy/2}$ (m) $\sqrt{50a^3 b^2 x^5}$ (r) $\sqrt{44/3x}$

(d) $\sqrt{12b^2}$ (i) $2\sqrt{1/x}$ (n) $\sqrt{2/(1 - x)}$ (s) $(2\sqrt{3})^3$

(e) $(\sqrt{1 - x^2})^2$ (j) $1/(\sqrt{2} - 1)$ (o) $\sqrt{32a^2 b^5}$ (t) $\dfrac{\sqrt{2/cd}}{c\sqrt{2}}$

Answers. (a) a^2. (c) $3\sqrt{a}$. (e) $|1 - x^2|$. (g) $\frac{1}{3}\sqrt{6}$. (i) $(2/x)\sqrt{x}$. (k) $c^2|a + b|$. (m) $5x^2|ab|\sqrt{2ax}$. (o) $4b^2|a|\sqrt{2b}$. (q) $\sqrt{ab}/b$. (s) $24\sqrt{3}$.

5. Prove: (a) $\sqrt{a^2} \not\equiv a$.   (b) $\sqrt{a^2 + b^2} \not\equiv a + b$.
(c) $\sqrt{x - y} \not\equiv \sqrt{x} - \sqrt{y}$.

6. Rationalize denominators:

(a) $\dfrac{1}{1 + \sqrt{3}}$

(d) $\dfrac{2x}{\sqrt{x} - y}$

(†g) $\dfrac{1}{\sqrt{a} + \sqrt{b} + \sqrt{c}}$

(b) $\dfrac{\sqrt{2}}{\sqrt{3} - \sqrt{2}}$

(e) $\dfrac{\sqrt{x} + \sqrt{y}}{\sqrt{x} - \sqrt{y}}$

(h) $\dfrac{2\sqrt{5} - \sqrt{x}}{3\sqrt{x} + \sqrt{6}}$

(c) $\dfrac{\sqrt{2} + \sqrt{5}}{\sqrt{5} - \sqrt{2}}$

(†f) $\dfrac{1}{\sqrt{2} + \sqrt{3} + \sqrt{5}}$

(i) $\dfrac{1 - \sqrt{a + b}}{\sqrt{a + b} - \sqrt{a - b}}$

Answers. (a) $(\sqrt{3} - 1)/2$. (c) $(10 + 2\sqrt{3})/3$. (e) $(x + y + 2\sqrt{xy})/(x - y)$.

7. Expand and simplify:

(a) $(\sqrt{a} + \sqrt{2})^2$

(j) $a/(\sqrt{b} - \sqrt{c})$

(b) $(\sqrt{2} + \sqrt{3})(\sqrt{2} - \sqrt{3})$

(k) $[1/\sqrt{1 + x}]^2$

(c) $\sqrt{3}(\sqrt{6} - 2\sqrt{2})$

(l) $\sqrt{x^2 + 2x + 1}$

(d) $1/(\sqrt{2} + \sqrt{3})$

(m) $\sqrt{x(x - y)/(x + y)}$

(e) $(2x\sqrt{5} + \sqrt{2})/\sqrt{2}$

(n) $\sqrt{a + b}\sqrt{a - b}$

(f) $1/(\sqrt{2x} - 1)$

(o) $(\sqrt{x + y} - \sqrt{x - y})^2$

(g) $(\sqrt{5} - \sqrt{xy})^2$

(p) $\sqrt{x}/(\sqrt{x} - 1) + \sqrt{x}/(\sqrt{x} + 1)$

(h) $\sqrt{(1/r) + (1/2)}$

(q) $\sqrt{(3/2ax) - (1/x) + x}$

(i) $\sqrt{(b + c)^6 x^4/4}$

(r) $(\sqrt{x} + \sqrt{y})(x - \sqrt{xy} + y)$

Answers.   (a) $a + 2 + 2\sqrt{2a}$.   (c) $3\sqrt{2} - 2\sqrt{6}$.   (e) $x\sqrt{10} + 1$.

(g) $5 + xy - 2\sqrt{5xy}$. (i) $|b + c|^3 x^2 \sqrt{2}$. (k) $1/(1 + x)$. (m) $\dfrac{\sqrt{x(x^2 - y^2)}}{x + y}$.
(o) $2(x - \sqrt{x^2 - y^2})$.

8. Factor into linear factors:

(a) $x^2 - 3$

(e) $\sqrt{3} + \sqrt{27}$

(i) $\sqrt{1/8} - \sqrt{1/2}$

(b) $x^2 - 5y^2$

(f) $\sqrt{2} + \sqrt{8}$

(j) $\sqrt{a^3} - \sqrt{ab}$

(c) $8a^2 - 9s^2$

(g) $\sqrt{12} - \sqrt{18}$

(d) $2x^2 - 2\sqrt{6}x + 3$

(h) $\sqrt{2a} + \sqrt{2b}$

(k) $\sqrt{x^2 - y^2} + 2\sqrt{x - y}$

(l) $\sqrt{1 - (4/x) + (4/x^2)}$

Answers.   (a) $(x + \sqrt{3})(x - \sqrt{3})$.   (c) $(2\sqrt{2}a + 3s)(2\sqrt{2}a - 3s)$.
(e) $\sqrt{3}(1 + \sqrt{3})$.   (g) $\sqrt{6}(\sqrt{2} - \sqrt{3})$.   (i) $(-0.25)\sqrt{2}$.
(k) $\sqrt{x - y}(\sqrt{x + y} + 2)$.

9. Write each of the following as a single radical with coefficient 1:

(a) $2\sqrt{x}$ (c) $(x+y)\sqrt{x-y}$ (e) $(1/2)\sqrt{x+2}$

(b) $a\sqrt{z}$ (d) $2xy\sqrt{x^2}$ (f) $(a/x)\sqrt{x/a}$

Answers. (a) $\sqrt{4x}$. (c) $\sqrt{(x+y)(x^2-y^2)}$. (e) $\sqrt{(x+2)/4}$.

10. Find from Table I by using the column of square roots and *5.3.8 where required:

(a) $\sqrt{45}$ (d) $\sqrt{0.37}$ (g) $\sqrt{0.0051}$ (j) $\sqrt{0.0005}$

(b) $\sqrt{88}$ (e) $\sqrt{0.99}$ (h) $\sqrt{0.67}$

(c) $\sqrt{5400}$ (f) $\sqrt{8300}$ (i) $\sqrt{0.000021}$

Answers. (a) 6.708. (c) 73.48. (e) 0.995. (g) 0.07141.

11. Find by inspection or by using the column of squares in Table I:

(a) $\sqrt{1.44}$ (c) $\sqrt{2500}$ (e) $\sqrt{0.04}$ (g) $\sqrt{0.0121}$ (i) $\sqrt{0.0225}$

(b) $\sqrt{0.36}$ (d) $\sqrt{0.0064}$ (f) $\sqrt{1.69}$ (h) $\sqrt{625}$ (j) $\sqrt{2401}$

Answers. (a) 1.2. (c) 50. (e) 0.2. (g) 0.11.

12. Find the square roots of the following by interpolating in the column of square roots in Table I:

(a) 45.3   (f) 23.5   (k) 0.201   (p) 41.01   (u) 0.998

(b) 88.6   (g) 0.166   (l) 1/20   (q) 0.038   (v) 0.099

(c) 69.2   (h) 5000   (m) 0.235   (r) 75.4   (w) 0.07

(d) 39.1   (i) 0.307   (n) 6.82   (s) 0.0711   (x) 0.00365

(e) 0.513   (j) 0.105   (o) 1152   (t) 0.991   (y) 0.000782

Answers to three digits. (a) 6.73. (c) 8.32. (e) 0.716. (g) 0.407. (i) 0.554. (k) 0.448. (m) 0.485. (o) 33.9. (q) 0.195. (s) 0.267. (u) 0.999. (w) 0.265. (y) 0.0280. The interpolation for (g) may be arranged as follows:

$$\begin{matrix} & x & \sqrt{x} \\ & {-}16.0 & 4.000{-} \\ 1\begin{bmatrix} 0.6\begin{bmatrix} \\ \ \end{bmatrix} \\ \ \end{bmatrix} & {\rightarrow}16.6 & ? \end{bmatrix} d \Big] 0.123 \\ & 17.0 & 4.123 \end{matrix}$$

Then $d/0.123 = 0.6/1$, $d \doteq 0.074$, $\sqrt{16.6} \doteq 4.07$, and $\sqrt{0.166} \doteq 0.407$.

13. Find the square roots of the numbers in prob. 12 by interpolating from the column of squares in Table I, and compare with the previous results. (Note that because of *5.3.8 there is often a choice as to which part of the table to use. Thus in part *a* it is better to interpolate from $67^2 = 4489$ and $68^2 = 4624$ than from $6^2 = 36$ and $7^2 = 49$.)

14. The two methods exhibited in probs. 12 and 13 gave results that were practically the same for the numbers chosen. However, because interpolation is more accurate farther along in the table (why?) and because of *5.3.8, it is sometimes appreciably better to use the square column. Thus interpolation in the column of square roots gives $\sqrt{1.1} \doteq 1.041$ whereas interpolation from the column of squares with $10^2 = 100$ and $11^2 = 121$ gives $\sqrt{110} \doteq 10.5$ and $\sqrt{1.1} \doteq 1.05$. The second result is more accurate, as the student can check by squaring. Find the following square roots by both methods and determine which is more accurate:

(a) $\sqrt{1.2}$ (c) $\sqrt{187}$ (e) $\sqrt{0.0153}$ (g) $\sqrt{0.0451}$ (i) $\sqrt{6.3}$

(b) $\sqrt{215}$ (d) $\sqrt{1.47}$ (f) $\sqrt{3.68}$ (h) $\sqrt{5.5}$ (j) $\sqrt{1.01}$

15. Explain in terms of the graph of $y = \sqrt{x}$ why interpolation for $\sqrt{x}$ near $x = 0$ is inaccurate whereas interpolation when x is large is very accurate.

16. Solve for x:

(a) $x^2 = 16$ (d) $x^2 - 0.011 = 0$ (g) $0.5x^2 - 31 = 0$
(b) $x^2 = 26$ (e) $57 - 3x^2 = 0$ (h) $32x^2 = 1$
(c) $x^2 - 8 = 0$ (f) $x^2 - 1000 = 0$

Answers. (a) ± 4. (c) $\pm 2\sqrt{2}$. (e) $\pm\sqrt{19}$. (g) $\pm\sqrt{62}$.

17. Is it always true that the square root of a number is smaller than the number? Interpret your answer in terms of the graph of $y = \sqrt{x}$.

18. Prove that $0 < a < 1 \longrightarrow a < \sqrt{a}$.

19. Sketch:

(a) $y = \sqrt{2x}$, $y = -\sqrt{2x}$, and $y^2 = 2x$

(b) $y^2 = 4x$ (e) $y = -3\sqrt{x}$ (h) $y = \sqrt{x-1}$

(c) $y = \sqrt{x/2}$ (f) $y^2 = 3.5x$ (i) $y = 2 + \sqrt{x}$

(d) $y^2 = -6x$ (g) $y^2 + 2x = 0$ (j) $y - 1 = 2\sqrt{x+3}$

20. Show that: (a) $\sqrt{2 + \sqrt{3}} = (1/2)(\sqrt{2} + \sqrt{6})$. (b) $\sqrt{18} - \sqrt{16} - \sqrt{9} + \sqrt{8} = \sqrt{50} - \sqrt{49}$.

21. The formula for the torsion pendulum, used by astronomers in computing the mass of the earth, is $T = 2\pi\sqrt{k/f}$. Solve for f. (*Answer.* $f = (2\pi)^2 k/T^2$.)

22. Solve $\sigma_E = \sigma_x\sqrt{1 - r^2}$ for r. (*Suggestion.* Square both sides.)

23. Solve for r and simplify: $V = [(2/9)r^2(d_1 - d_2)g]/n$.

24. From $M = nh/2$ and $M = \sqrt{mrze^2}$ find a formula for r not involving M.

†**25.** Show that $x > 1 \longrightarrow \dfrac{1}{\sqrt{x}} > \dfrac{1}{x}$.

†26. Explain the fallacy in the following:

$$(4/3 - 1)^2 = (4/3 - 5/3)^2$$

Hence $\qquad 4/3 - 1 = 4/3 - 5/3$

and $\qquad\qquad -1 = -5/3 \quad\text{or}\quad -2/3 = 0!$

†27. Show that $p/q = \sqrt{2} \longrightarrow \dfrac{2q - p}{p - q} = \sqrt{2}.$

5.4 Exact rate of change

We return to the body falling according to the formula $y = 16x^2$. Its speed is continually increasing, and therefore it does not have a constant speed during any time interval. However, the very fact that we speak of an increasing speed indicates that we consider it to have a quite definite speed at every instant. What can we mean by this for a body whose speed is changing?

Let us imagine how we might go about estimating the speed of a body "at an instant," say x_1. The natural procedure would be to observe how far it traveled in some short interval of time following x_1, say x_1 to x_2. We observe the distances y_1 and y_2, measured from the point where $x = 0$, and then find the distance traveled, $y_2 - y_1$, in the time interval $x_2 - x_1$. Dividing distance by time elapsed, we have $(y_2 - y_1)/(x_2 - x_1)$. But this is just the average speed considered in the previous section. We have found only an approximation. It would seem, however, that the smaller the interval of time we took, the closer we should come to a value that would correspond to our intuitive idea of the exact speed at an instant. To estimate the exact speed of a moving body we would choose as small an interval of time as possible, within the limits of our instruments, and measure the corresponding distance covered. However, if we know distance as a function of time, we may take our time interval as small as we like.

Why not take $x_2 - x_1 = \Delta x = 0$? The student already knows the answer to this question. If $\Delta x = 0$, then $\Delta y = 0$ also, and we have $\Delta y/\Delta x = 0/0$, a meaningless expression. Hence we can take the interval Δx as small as we like but *not equal to zero*. The smaller we take it, the better the approximation seems.

Suppose we try to find the exact velocity at the time $x = 1$ of the body falling according to $y = 16x^2$. Here $x_1 = 1$, $y_1 = 16$. Letting $x_2 = 2$, we have $y_2 = 64$, and $\Delta y/\Delta x = (64 - 16)/(2 - 1) = 48$. This is the average velocity in the second following $x = 1$.

The corresponding value for the $\frac{1}{2}$ second following $x = 1$ is found by letting $x_2 = 1.5$. It is, from *5.2.2, $16(1 + 1.5) = 40$. Taking x_2 equal to 1.1, 1.01, 1.001, 1.0001, and 1.0000001, we find the corresponding average velocities to be 33.6, 32.16, 32.016, 32.0016, and 32.0000016. These average velocities seem to be getting very near the value 32 as we let x_2 get near x_1. We cannot let $x_2 = x_1$, but it does seem that as we let $\Delta x = x_2 - x_1$ approach arbitrarily near zero, $\Delta y/\Delta x$ approaches arbitrarily near 32. Accordingly, we call 32 the exact speed at $x = 1$.

The student may have noticed that the exact speed was just twice 16. This is not accidental, as we shall show by considering $y = ax^2$. Consider a fixed value of x, say x_1. Then the average rate of change in the interval from x_1 to x_2 is by *5.2.2 equal to $a(x_1 + x_2)$. As we let x_2 approach arbitrarily near x_1, this gets arbitrarily near $a(x_1 + x_1) = 2ax_1$. We say that the average rate of change approaches $2ax_1$ as x_2 approaches x_1, and we call $2ax_1$ the limit of $a(x_1 + x_2)$ as x_2 approaches x_1. This limit is called the exact rate of change of y with respect to x at $x = x_1$. Applying the formula $2ax_1$ to $y = 16x^2$ for $x = 1$, we find the exact rate of change of distance with respect to time, that is, the exact speed, to be $2 \cdot 16 \cdot 1 = 32$, which agrees with the result in the previous paragraph.

These ideas can be generalized to apply to any function. If $y = f(x)$, the average rate of change of y with respect to x is defined (*5.2.3) to be $[f(x_2) - f(x_1)]/(x_2 - x_1) = \Delta y/\Delta x$. Suppose that as we let x_2 approach arbitrarily close to a fixed x_1, the average rate of change approaches arbitrarily close to a definite value. We mean by this that we can make $\Delta y/\Delta x$ as close as we like to this definite value by taking x_2 close enough to x_1. Then we call this value the limit of $\Delta y/\Delta x$ as Δx approaches zero. This limit is called the **exact rate of change of y with respect to x.** It is designated by the symbol $D_x y$, that is,

*5.4.1 $D_x y = $ limit as Δx approaches zero of $\dfrac{\Delta y}{\Delta x}$ (Definition)

We have not given precise definitions of the words "limit" and "approaches a limit." Such questions belong in a course in calculus, where they are a central topic. Here we are satisfied with an intuitive idea and the derivation of simple formulas.

We have shown above that when $y = ax^2$, $D_x y = 2ax$. We omit the subscript 1 on the x, since the formula holds for any value of the independent variable. This result may be written:

*5.4.2 $D_x(ax^2) = 2ax$

Applying this formula to $y = 16x^2$, we find $D_x y = 32x$. If we interpret y and x as distance and time, we have here a formula for the speed of a freely falling body at any time x.

The exact rate of change has an important geometric interpretation. Consider the function $y = f(x)$, a portion of whose graph is shown in Fig. 5.4.1. Let $P_1(x_1, y_1)$ be a fixed point and

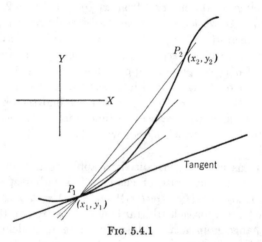

FIG. 5.4.1

$P_2(x_2, y_2)$ another point. Now let P_2 move along the curve toward P_1. The secant line rotates about P_1, and, as x_2 approaches x_1, it approaches coincidence with the tangent line to the graph at P_1. It seems reasonable then that the slope of the tangent line is the limit of the slope of the secant line. Since this limit is the exact rate of change, this means that *the exact rate of change of y with respect to x at $x = x_1$ is the slope of the line tangent to the graph of $y = f(x)$ at the point (x_1, y_1).*

The exact rate of change of y with respect to x, $D_x y$, is called the **derivative** of y with respect to x. Note that it is a function of x. Thus if $y = x^2$, $D_x y = 2x$. If we wish to indicate its value when x has a particular value, we enclose it in parentheses and write the value of x to the right and below. For example, $(D_x y)_{x=2}$ means the value of $D_x y$ when $x = 2$. If $y = 16x^2$, $D_x y = 32x$, and $(D_x y)_{x=2} = 64$. This is the speed of a freely falling body at the end of the second second. It is also the slope of the line tangent to the graph of $y = 16x^2$ at the point $(2, 64)$. Other notations for $D_x y$ are $\dfrac{dy}{dx}$ and y'. $D_x y$ is read "D-x-y," "the derivative of y with respect to x," "d-y, d-x" $\left(\text{for } \dfrac{dy}{dx}\right)$, or "$y$-prime" (for y').

We can now use the point-slope form, 4.8(7), to write the equation of the line tangent to $y = f(x)$ at the point (x_1, y_1):

$*$**5.4.3** The line tangent to the graph of $y = f(x)$ at the point (x_1, y_1) is given by $y - y_1 = (D_x y)_{x=x_1}(x - x_1)$.

For example, the tangent to $y = 16x^2$ at the point $(2, 64)$ is $y - 64 = 64(x - 2)$. The tangent at the point $(1, 16)$ is $y - 16 = 32(x - 1)$. The student should note that $*$5.4.3 is the same as the equation of the secant line, except that $\Delta y/\Delta x$ is replaced by $D_x y$ evaluated at the point. There are, of course, many secants through a given point, but for smooth curves without corners there is only one tangent line at a point.

Evidently the tangent line has positive, negative, or zero slope according to whether the derivative is positive, negative, or zero. Accordingly, if $D_x y > 0$, y is increasing as x increases. If $D_x y < 0$, y is decreasing. If $D_x y = 0$, the tangent is horizontal and y is said to be stationary. We call $D_x y$ the **slope of the curve.**

In summary, the derivative of y with respect to x, $D_x y$, defined by $*$5.4.1 is a function that gives (1) the exact rate of change of y with respect to x, and (2) the slope of the line tangent to $y = f(x)$. It is called the slope of the curve and indicates by its sign whether y is increasing, decreasing, or stationary as x increases.

The following theorem is often useful in finding derivatives:

$*$**5.4.4** $D_x(cy) = cD_x y$

It asserts that the derivative of a constant times a function is the constant times the derivative of the function, and its plausibility is suggested by Ex. 5.2.23.

EXERCISE 5.4

1. In $y = x^2$, let $x_1 = 1$, $y_1 = 1$. Find y_2, $y_2 - y_1$, and $\Delta y/\Delta x$ for $x_2 = 2$, 1.5, 1.1, 1.01, 1.001, 1.0001. What seems to be the limit?

2. Same as prob. 1 for $y = 3x^2$, $x_1 = 0.5$, $x_2 = 1$, 0.7, 0.51, 0.501, 0.5001, 0.50001.

3. Same for $y = 2x^2$, $x_1 = 10$, $x_2 = 12$, 11, 10.1, 10.01, 10.000001.

4. Same for $y = 16x^2$, $x_1 = 1.5$, $x_2 = 2$, 1.7, 1.6, 1.51, 1.501, 1.50001.

5. Graph $y = x^2$ and the tangent lines at the points where $x = -2$ and $x = 2$. Label each tangent with its slope.

6. Same for $y = -x^2/2$ and the tangents where $x = -4$, -1, 3.

7. Same for $y = 2x^2$ and tangents at $x = -3$, 0, 1.

8. Find the exact rate of change of each of the following functions for the indicated value of the independent variable:

(a) $y = x^2$ at $x = 3$ (f) $s = -10t^2$ at $t = 0.01$

(b) $y = 3x^2$ at $x = -1$ (g) $y = ax^2$ at $x = 0$

(c) $y = 1.7x^2$ at $x = -3$ (h) $y = 7Ax^2$ at $x = b$

(d) $y = -4x^2$ at $x = 3$ (i) $y = Bcx^2$ at $x = c$

(e) $y = 16t^2$ at $t = 5$

Answers. (a) 6. (c) -10.2. (e) 160. (g) 0. (i) $2Bc^2$.

9. In prob. 8, parts *a* through *f*, find the equation of the tangent line at the point and sketch. [*Answers.* (a) $6x - y - 9 = 0$. (c) $10.2x + y + 15.3 = 0$.]

10. The number of centimeters covered in t seconds by a body falling from rest is given by $s = 490t^2$. What is the exact velocity of the body after t seconds? At the end of 3 seconds? Compare the last with the average velocities during the third second and fourth second.

11. Estimate how many feet a body falls in the first half of the eighth second of free fall. (*Suggestion.* Use the exact velocity at $t = 7$.) Compare your result with the exact distance.

12. Show that:

*5.4.5 $$D_x(mx + b) = m$$

*5.4.6 $$D_x(b) = 0$$

(The first theorem states that the exact rate of change of a linear function is a constant equal to the slope of its graph. The second simply expresses the common sense idea that the rate of change of a constant is zero.)

13. Find the derivatives of the following: (a) $y = 4x$. (b) $y = 2$. (c) $y = 8x^2$. (d) $s = 3t^2$. (e) $y = -2x$. (f) $y = 4 - 8x$. (g) $y = 0$. (h) $s = 32t$. [*Answers.* (a) 4. (c) $16x$. (e) -2. (g) 0.]

14. Find: (a) $D_x(1 - 3x)$. (b) $D_t(2t^2)$. (c) $D_r(ar + 99)$. (d) $D_t(At + 0.01)$. (e) $D_r(\pi r^2)$. (f) $D_h(bh/2)$. [*Answers.* (a) -3. (c) a. (e) $2\pi r$.]

†15. Show that $D_x(x^2)$ is an increasing function of x. Is this true of $D_x(ax^2)$? Illustrate your answer by considering $D_x(3x^2)$ and $D_x(-2x^2)$.

†16. Show that $y = ax^2$ $(a > 0)$ is increasing for $x > 0$, decreasing for $x < 0$, and stationary for $x = 0$. Make a similar statement for $a < 0$.

17. A body falling from rest has reached a velocity of 100 feet per second. How long has it been falling? How far has it fallen? (*Answer.* 25/8 seconds, 625/4 feet.)

18. The exact rate of change is sometimes described as the change in y per unit change in x, since it equals the change that would occur in y for a unit change in x if the rate of change were to remain constant. (Compare Ex. 5.2.22.) Find the change in: (*a*) The area of a triangle per unit change in the altitude when the base remains equal to 2 and the altitude is 2.5. (*b*) The area of a square per unit change in one side when the side is 4. (*c*) The area of a circle per unit increase in the radius when the radius is 3. (*Answer to part a.* 1.)

19. Show that the exact speed of a freely falling body at the middle of a time interval is equal to the average speed during the interval. (*Suggestion.* Let $y = 16x^2$ and take the time interval to be x_1 to x_2.)

20. The *exact acceleration* of a body is defined as the exact rate of change of its velocity, that is, $D_x(D_xy)$ where x and y are time and distance. Find the acceleration of a body moving according to: (*a*) $y = 16x^2$. (*b*) $y = 490x^2$. (*c*) $y = 3x$. (*d*) $y = 8x^2$. (*e*) $y = -13x^2$. (*f*) $y = 4x - 2$. [*Answers.* (*a*) 32. (*c*) 0. (*e*) -26.]

21. The function $D_x(D_xy)$ is called the **second derivative** of y with respect to x and is written $D_x{}^2(y)$. Find: (*a*) $D_x{}^2(5x^2)$. (*b*) $D_t{}^2(150t^2)$. (*c*) $D_x{}^2(14x - 175)$. (*d*) $D_x{}^2(-232x^2)$. [*Answers.* (*a*) 10. (*c*) 0.]

22. If a body is moving along a curved path and is released from the constraints that hold it in the path, it flies off along the tangent line to the curve at the point. Find and sketch the path of a body moving along the curve $y = 0.25x^2$ from the origin and then released at the point where $x = 4$.

23. The elasticity of y with respect to x is defined to be xD_xy/y. It is the limit of the average elasticity defined in Ex. 5.2.16. (*a*) Show that if y is proportional to x the elasticity is 1. (*b*) Find the elasticity of $y = mx + b$. (*c*) Same for $y = ax^2$. [*Answer to (c).* 2.]

24. In 5.2 we replaced the curve by a secant line in order to estimate values of a function. The student may have observed that the smaller the interval used to calculate the slope of the secant line ($\Delta y/\Delta x$), the more accurate the interpolation. As we let the interval approach zero, the secant line approaches the tangent line, and it would seem reasonable to use the tangent line for estimating values of a function at points near a known point. For example, with reference to Ex. 5.2.6, the tangent line to $y = x^2$ at $x = 13$ is, from *5.4.3, $y - 169 = 26(x - 13)$. Letting $x = 13.6$, we find $y = 184.6$, which is closer to the true value than the estimate by linear interpolation. Letting y_E stand for the estimate by using the tangent line, we have

(1) $$y_E = y_1 + (D_xy)_{x=x_1}(x - x_1)$$

Note that this is the same as *5.2.4 except that the average rate of change is replaced by the derivative. The second term in the right member of (1) is called the **differential** of y. It is equal to the derivative at the point times the change in x, and, when $x - x_1$ is small, it is a good approximation for the change in the function. The relationships are shown in Fig. 5.4.2.

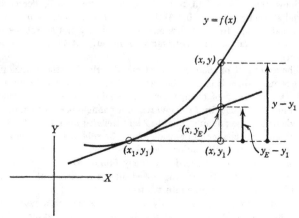

Fig. 5.4.2

Use (1) to estimate the squares in Ex. 5.2.6, and compare the errors with those for linear interpolation.

25. Find the equation of the line tangent to $y = 0.5x^2$ at $x = 2$. Calculate the values of y on the curve and the tangent for $x = 3, 1, 2.5, 2.0, 2.01,$ and compare.

26. Estimate the following by using formula (1): (a) 100.08^2. (b) 8.21^2. (c) 60.3^2. (d) 999^2. (e) 998^2. [*Answers.* (a) 10,016. (c) 3636. (e) 996,008.]

5.5 Graphing the quadratic function

Consider the function

$$(1) \qquad\qquad y = x^2 - 2x - 8$$

If we calculate the y's corresponding to integral values of x from -3 to 5, plot the corresponding points, and join them by a smooth curve, we get the graph shown in Fig. 5.5.1. It looks very much like that of the square considered in 5.1, except that it is in a different position. Let us see what happens if we move the origin to what appears to be the lowest point, that is, $(1, -9)$. Then the new coordinates are related to the old by

$$(2) \qquad\qquad x = x' + 1 \quad \text{and} \quad y = y' - 9$$

and the new equation is

$$(3) \qquad\qquad y' - 9 = (x' + 1)^2 - 2(x' + 1) - 8$$

or

$$(4) \qquad\qquad y' = x'^2$$

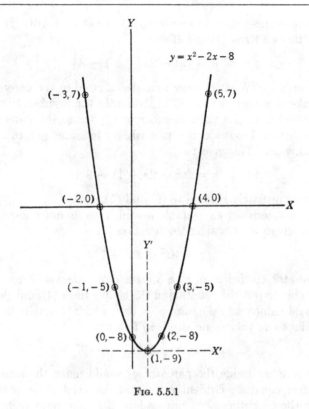

FIG. 5.5.1

We see that the graph *is* just the graph of the square in a different position. Now we are sure that the curve is symmetric about the line $x = 1$ and has its lowest point at $(1, -9)$.

The ease with which we found (4) depended upon our lucky choice of the point $(1, -9)$, which turned out to be the lowest point. But how could we have found this point? In order to see this, we replace x' and y' by their values in terms of the old coordinates. Then (4) becomes:

(5) $$y + 9 = (x - 1)^2$$

or

(6) $$y = (x - 1)^2 - 9$$

The student can easily verify that the right side of (6) is identical with (1). Yet (6) in the form (5) tells us immediately that $y' = y + 9$ and $x' = x - 1$ will put the equation in the form (4). The problem reduces then to rewriting (1) in the form (6). In order to do this we write (1) as follows:

(7) $$y = (x^2 - 2x \qquad) - 8$$

Now we ask: "What number must be inserted in the parentheses to make it a perfect square?" Evidently the number 1 will do. So we add 1 to the right member of (7) inside the parentheses and subtract 1 outside the parentheses in order not to change the function. This gives:

(8) $$y = (x^2 - 2x + 1) - 9$$

which immediately reduces to (6) and (5).

We now consider an example in which we do not know the result in advance. Consider the function

(9) $$y = 3x^2 + 6x - 1$$

Because the coefficient of x^2 is 3 instead of 1 we would not expect to be able to put the equation in the simple form (4), but perhaps we could reduce it to the form $y' = 3x'^2$, which is equally familiar. In order to do this we rewrite it as follows:

(10) $$y = 3(x^2 + 2x \qquad) - 1$$

What number inside the parentheses would make the expression a perfect square? Evidently 1 would do. But if we write a 1 inside the parentheses we have added 3 to the right member of (10) (why?) and so must subtract 3 to leave things as before. Hence we write:

(11) $$y = 3(x + 1)^2 - 4$$

or

(12) $$y + 4 = 3(x + 1)^2$$

Now it can be seen that the translation $x' = x + 1$, $y' = y + 4$ reduces the equation to the form

(13) $$y' = 3x'^2$$

and hence that $(-1, -4)$ is the lowest point, and the curve is symmetric about the line $x = -1$.

The procedure we have carried through in these two examples is called **completing the square**. It always enables us to rewrite a quadratic function $ax^2 + bx + c$ in the form $a(x - h)^2 + k$. To carry through this process in general, we write:

$$(14) \qquad ax^2 + bx + c \equiv a\left(x^2 + \frac{b}{a}x + ?\right) + c$$

$$(15) \qquad\qquad \equiv a\left(x^2 + \frac{b}{a}x + h^2\right) + c - ah^2$$

Now we wish to choose h so that the expression in parentheses is a perfect square, that is, we want:

$$(16) \qquad \begin{aligned} x^2 + (b/a)x + h^2 &\equiv (x - h)^2 \\ &\equiv x^2 - 2hx + h^2 \end{aligned}$$

If this is to be so, we must have $-2h = b/a$ or $h = -b/2a$. Then we have, from (15),

$$(17) \qquad ax^2 + bx + c \equiv a(x - h)^2 + k$$

where $h = -b/2a$ and $k = c - b^2/4a$.

We see that $y = ax^2 + bx + c$ can always be rewritten $y = a(x - h)^2 + k$ or $y - k = a(x - h)^2$. Then a translation to the point (h, k) yields the equation $y' = ax'^2$. We know from 5.1 that this is the equation of a curve that we call a parabola. It is symmetric with respect to the line $x' = 0$, which we call its **axis of symmetry,** and has at $(0, 0)$ its lowest or highest point according to whether $a > 0$ or $a < 0$. In either case we call this point the **vertex** of the curve. In terms of the original coordinates, the axis of symmetry is the line $x = h$ or $x = -b/2a$, and the vertex is at $(-b/2a, c - b^2/4a)$. The discussion may be summarized in the following theorem:

*5.5.1 The graph of the quadratic function $y = ax^2 + bx + c$ is a parabola with axis of symmetry $x = -b/2a$.

We illustrate the use of this result by graphing

$$(18) \qquad y = 2x^2 - 4x + 1$$

From *5.5.1 the axis of symmetry is $x = 1$, and this is also the x-coordinate of the vertex. Substituting $x = 1$ in (18), we find the y-coordinate of the vertex to be -1. Now we draw axes at

$(1, -1)$ and graph $y' = 2x'^2$, which is the equation (18) in terms of the new coordinates (Fig. 5.5.2). We could of course find this result without *5.5.1 by completing the square in (18). Note that in any case it is not worth while to remember the formula for the

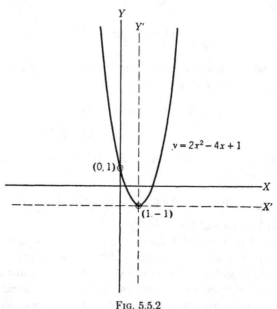

FIG. 5.5.2

y-coordinate of the vertex. It can always easily be found if we remember that the x-coordinate is $-b/2a$.

EXERCISE 5.5

1. Why did we specify $a \neq 0$ in defining quadratic functions as ones of the form $y = ax^2 + bx + c$? Indicate the values of a, b, and c in each of the following: (a) $2x^2 + 3x - 4$. (b) $x^2 - 1$. (c) $-x^2$. (d) $4x - 3 - 2x^2$.

2. Graph $y = x^2 - 2x - 6$ as we did (1). Plot points for all integral values of x such that $|x| \leq 5$.

3. Carry through the algebra to get from (3) to (4). Expand (11) and verify that it is identical with (9). Expand the right member of (17) and verify that it is identical with the left member.

4. Graph each function in prob. 1.

5. Do Ex. 5.1.12 if you have not done it already.

6. In each of the following functions, complete the square, make an appropriate translation, and graph the curve in the new coordinates. Show both old and new axes and plot several points in terms of the old coordinates in order to check your work.

(a) $y = x^2 + 2x + 5$ (e) $y = -x^2 - 2x + 4$ (i) $y = -5x^2 - 10x$
(b) $y = x^2 + 2x + 1$ (f) $y = 2x^2 + 8x + 8$ (j) $y = 4x^2 + 6x - 1$
(c) $y = x^2 + 2x - 1$ (g) $y = -3x^2 + 9x + 8$ (k) $y = -3x^2 + 5x - 4$
(d) $y = 2x^2 - 6x + 7$ (h) $y = x^2 - 2$ (l) $y = 5x^2 + 8x - 1$

7. Graph each of the following by using *5.5.1 to find the axis of symmetry:

(a) $y = 3x^2 - 2x + 14$ (c) $y = 2x^2 - x$ (e) $y = x^2 - 1.5x + 2$
(b) $y = 2x^2 + 4x + 8$ (d) $y = -7x^2 + 6x - 10$ (f) $y = 7x^2 + 4x - 10$

8. Find the quadratic function whose graph passes through:

(a) $(-5, 2)$, $(1, 6)$, $(2, -4)$ (c) $(0, 1)$, $(2, 4)$, $(-3, -2)$
(b) $(0, 3)$, $(4, 0)$, $(6, 5)$ (d) $(1, 1)$, $(-1, 5)$, $(4, -3)$

(*Suggestion.* The coordinates must satisfy $y = ax^2 + bx + c$. This gives three equations to solve for a, b, and c.)

Answers. (a) $y = \dfrac{1}{21}(-32x^2 - 114x + 272)$. (c) $y = \dfrac{1}{10}(x^2 + 13x + 10)$.

9. Complete the square in each of the following expressions:

(a) $3x^2 + ax + 14$ (c) $Au^2 + Bu$ (e) $Ax^2 + Bxy + C$
(b) $Rx^2 - 2Sx + T$ (d) $ax^2 + bx + c$ (f) $-5x^2 + 2x - 15$

Answers. (a) $3\left(x + \dfrac{a}{6}\right)^2 + 14 - a^2/12$. (c) $A\left(u + \dfrac{B}{2A}\right)^2 - B^2/4A$.

10. Find the axis of symmetry and vertex of:

(a) $y = Ax^2 + Bx + C$ (c) $y = 2mx^2 - m^2x + m$
(b) $q = (a + b)u^2 - (a - b)u + c$ (d) $2y = Rx^2 - 3sx$

11. A function $y = f(x)$, considered for a certain range of values of x, is said to have an **absolute maximum (minimum)** at $x = x_0$ if $f(x_0) \geq (\leq) f(x)$ for all values of x in the range. Evidently a quadratic function has an absolute maximum or minimum at its vertex. Locate and evaluate the absolute maximum or minimum of each of the following functions:

(a) $y = 2x^2 - x$ (c) $y = 8x^2 - x + 3$ (e) $u = -4v^2 + 4v + 1$
(b) $y = 3x - 4x^2$ (d) $s = 2t^2 - 2$ (f) $y = -9v^2 + 6v - 1$

Answers. (a) Minimum at $(\frac{1}{4}, -\frac{1}{8})$. (c) Minimum at $(\frac{1}{16}, \frac{95}{32})$. (e) Maximum at $(\frac{1}{2}, 2)$.

†**12.** Find the absolute maximum and minimum of each function in the interval indicated:

(a) $y = 3x^2 - 2x + 1$ in $-1 \leq x \leq 1$ (c) $y = 3x - 4$ in $-2 \leq x \leq 4$
(b) $y = 1 - x^2$ in $0 \leq x \leq 3$ (d) $y = x^2$ in $2 \leq x \leq 5$.

(*Suggestion.* Draw a careful graph and observe that we wish the largest and smallest values *in the restricted interval.*)

13. It is easy to find the derivative of a quadratic function by using the definition *5.4.1 directly or from *5.4.2, *5.4.5, *5.4.6 and the following theorem, which we give without proof:

$$*5.5.2 \qquad D_x[f(x) + g(x)] = D_x f(x) + D_x g(x)$$

This means that the derivative of the sum of two functions is the sum of their derivatives. For example, $D_x(ax^2 + bx + c) = D_x(ax^2) + D_x(bx + c) = 2ax + b$. Find the derivative of each function in prob. 6. [*Answers.* (a) $2x + 2$. (c) $2x + 2$. (e) $-2x - 2$. (g) $-6x + 9$.]

14. Show from *5.5.2 that $D_x(f + g + h) = D_x f + D_x g + D_x h$.

15. Since the derivative gives the slope of the tangent line to the curve, and since this tangent line is horizontal at the vertex of a parabola, we can find the vertex by setting the derivative equal to zero. For example, the derivative of $y = 3x^2 + 2x - 1$ is $D_x y = 6x + 2$. This is zero when $x = -1/3$. Hence the vertex is at $(-1/3, -4/3)$. Find the vertex of each parabola in prob. 7 by using this method.

16. Find the x-coordinate of the vertex of $y = ax^2 + bx + c$ by setting $D_x y$ equal to zero, and compare your result with *5.5.1.

17. Use the derivative to find the maximum or minimum of each of the following:

(a) $y = 5x^2 - 20x + 1$ (f) $u = 3s^2 + 15s - 20$
(b) $s = 1 - 4t - 7t^2$ (g) $v = -1 - 11s - 3s^2$
(c) $U = 3V - V^2 + 8$ (h) $w = 1 + t - 5t^2$
(d) $y = x + x^2/2$ (i) $y = 3(x - 1)^2$
(e) $F = 1 - T^2$

Answers. (a) Minimum is $y = -19$. (c) Maximum is $U = 41/4$. (e) Maximum is $F = 1$. (g) Maximum is $v = 109/12$.

†18. What rectangle with given perimeter has the greatest area? Prove your answer.

19. Find the equations of the tangent lines to $y = 1 + 5x - 10x^2$ at the points where $x = 0$, $x = 1$, and $x = 3$. Sketch.

20. For what value of x is $D_x y = 2$ when $y = 5x^2 - 20x + 1$?

21. Find the equation of the line tangent to each parabola of prob. 6 at the point where $x = 1$.

22. Find the equation of the line tangent to $y = 4x^2 - 2x + 1$ and having a slope of 2.

23. Find the equation of the line tangent to each parabola of prob. 7 at the point one unit to the right of its vertex. Sketch.

24. A body is falling so that its distance from the starting point is given by $s = 10 + 3t + 16t^2$. (a) Find its velocity when $t = 0$ and when $t = 2$. (b) Plot s as a function of t and interpret the two velocities in terms of the graph. (c) Find the time and distance when the velocity is 99. [*Answers.* (a) 3, 67. (c) 3, 163.]

25. The total cost of producing and marketing x units of a certain product is given by the cost function $q = 0.1x^2 + 1.2x + 2$. The exact rate of change of cost with output is called the *marginal cost.* (It is approximately the addi-

tional cost for one additional unit of output. Why?) (a) Find the marginal
cost as a function of x. (b) Graph the cost and marginal cost functions on the
same axes. (c) Find the marginal cost when $x = 2$. (d) Find the added cost
to produce one additional unit when $x = 2$ and compare with the marginal
cost. [Answer. (c) 1.6. (d) 1.7.]

26. If the amount of a product purchased by the public is $x = ap + b$,
where p is the price, the revenue from sales is $R = px = x(x - b)/a$. (Why?)
(a) Graph R as a function of x for $b = 100$ and $a = -0.1$. (b) Find the sales
that give maximum revenue in this case. (c) Find a general formula in terms
of a and b giving the price and sales for maximum revenue.

27. Profit is given by revenue minus cost, that is, $S = R - q$. (a) Graph
R and q as given in probs. 26 and 25 on the same axes. How does S appear on
the graph? (b) Graph $S = R - q$ and find the point where it is maximum.
(c) If $R = x(x - b)/a$ and $q = Ax^2 + Bx + C$, find the x and p for maximum
profit.

28. A number y is the product of two consecutive integers. Graph the
possibilities as a function of the smaller one.

29. Graph the set of points satisfying:

(a) $y + 1 = -x^2$ or $y > 0$ (c) $y = x^2$ or $y = -x^2$
(b) $y + 1 = -x^2$ and $y > 0$ (d) $y > x^2 + 2x - 1$ and $y < 2$

5.6 Quadratic equations

A quadratic equation is one of the form $ax^2 + bx + c = 0$.
Finding its roots is equivalent to finding the values of x for which
the function $y = ax^2 + bx + c$ is zero. We call such a value of x
a zero of the function. More generally, if $f(r) = 0$ we call r a **root**
(or solution) of the equation $f(x) = 0$ and a **zero** of the function
$f(x)$.

Referring to (1) of the previous section, we note that the roots
of $x^2 - 2x - 8 = 0$ are just the x-coordinates of the points where
the curve $y = x^2 - 2x - 8$ crosses the x-axis. Certainly, if the
parabola $y = ax^2 + bx + c$ crosses the x-axis, the intercepts give
the roots. This gives a method of estimating the roots by graphing
the function and observing where it crosses the x-axis. But such
results are only approximate, and the graph may not cross the
x-axis at all! Hence we look for algebraic methods.

We note that $y = x^2 - 2x - 8$ can be rewritten:

(1) $$y = (x - 4)(x + 2)$$

Now we know from *3.8.3 and *3.8.4 that a product is zero if and
only if one of the factors is zero. Hence y is zero if and only if
$x - 4 = 0$ or $x + 2 = 0$. Hence $x = 4$ and $x = -2$ are roots of
$x^2 - 2x - 8 = 0$, and they are the only roots. It is frequently

possible to solve quadratic equations in this way by factoring the left member by trial and error and then using *3.8.4. For example, consider:

(2) $$3x^2 - x - 2 = 0$$

If this factors, its factors look like this in part:

(3) $$(3x \quad)(x \quad) = 0$$

Whatever numbers we put in the blanks must have a product of -2. (Why?) The possibilities are 1, -2; $-1, 2$; 2, -1; and -2, 1. Trying these to see whether they make (3) identical with (2), we find that 2, -1 check. Hence we write:

(4) $$(3x + 2)(x - 1) = 0$$

Then the roots are $x = -2/3$ and $x = 1$.

The method used above may be described in more general terms as follows. We wish to rewrite $ax^2 + bx + c$ as the product of two linear factors, that is, we wish to find r, s, t, u so that

(5) $$ax^2 + bx + c = (rx + s)(tx + u)$$
$$= rtx^2 + (ru + st)x + us$$

Now if this identity is to hold, we must have $rt = a$ and $us = c$. Hence we consider the pairs of integers whose product is a and the pairs whose product is c, and try to find a pair for which $(ru + st) = b$. The method is obviously tedious for values of a and c that give rise to many possible factor combinations, but it always enables us to find rational roots. If the equation has no rational roots the method fails—as it must, since we consider only integral possibilities for r, s, t, u.

When the equation does not have rational roots, its left member will not factor into linear factors with integral coefficients. Nevertheless, any quadratic function will factor into two linear factors! The key to this factorization is the process of completing the square. Consider:

(6) $$5x^2 + 15x - 10 = 0$$

Since we can divide both sides by 5 without changing the roots, we rewrite (6) as

(7) $$x^2 + 3x - 2 = 0$$

The student can easily check to see that the left member does not factor into linear factors with integral coefficients, but we can complete the square as follows:

$$(8) \qquad [x + (3/2)]^2 - (3/2)^2 - 2 = 0$$

or

$$(9) \qquad [x + (3/2)]^2 - 17/4 = 0$$

Now we may look upon the left member of (9) as the difference of two squares if we consider $17/4 = [(1/2)\sqrt{17}]^2$. Hence we can factor it by writing

$$(10) \quad [x + (3/2) + (1/2)\sqrt{17}][x + (3/2) - (1/2)\sqrt{17}] = 0$$

From this it follows that the roots of the equation are $x = (-3 + \sqrt{17})/2$ and $x = (-3 - \sqrt{17})/2$, which are usually written together as $x = (-3 \pm \sqrt{17})/2$.

The student should note that the foregoing solution was still based on the idea of factoring the left member of the quadratic equation. The left member of (10) is identically equal to the left member of (7). We simply used a different method of factoring and discovered that irrational numbers are required to factor the quadratic function in (7). Since we can always complete the square in any quadratic function, it appears that this method should work in all cases. If the roots are rational it leads to the same results as the method of factoring by trial and error.

We observed that if the parabola $y = ax^2 + bx + c$ crosses the axis at two points, their x-coordinates are the roots of the equation $ax^2 + bx + c = 0$. What is indicated if the parabola just touches the x-axis at one point, that is, is tangent to it? The parabola $y = x^2 - 6x + 9$ has its vertex at the point $(3, 0)$, which means that its lowest point is on the x-axis. If we try to solve the equation $x^2 - 6x + 9 = 0$ either by trial and error or by completing the square, we find immediately that the left member is the perfect square $(x - 3)^2$. Hence there is only one root, $x = 3$, or, as we prefer to say, two **coincident roots** or a **double root**. Apparently we may interpret tangency to the x-axis as the geometric counterpart of two equal roots.

What is the algebraic counterpart if the parabola does not touch the x-axis at all? Such is the nature of the parabola $y = x^2$

$- x + 1$, since its vertex is at $(1/2, 3/4)$. The solution of $x^2 - x + 1 = 0$ by completing the square proceeds as follows:

(11) $[x^2 - x + (1/2)^2] - (1/4) + 1 = 0$

(12) $[x - (1/2)]^2 + 3/4 = 0$

This does not appear to be the difference of two squares, but we can put it in the desired form if we write:

(13) $[x - (1/2)]^2 - (-3/4) = 0$

Now this question arises: Is $-3/4$ the square of any number? We know that the square of any real number is positive. Hence if we limit ourselves to the real numbers (and those are all we have so far defined), we would be forced to say that (13) could not be factored and that the equation had no solution. Instead we say that (13) cannot be factored into *real* factors and the equation has no *real* roots. Then we shall define new numbers equal to the square roots of negative numbers. We call these new numbers *imaginary*, although they are no more unreal or imaginary than any other numbers we have invented. These new numbers are defined in such a way that the usual rules of algebra apply to them. We shall carry through this extension of the number system in Chapter 10. Meanwhile we interpret the occurrence of square roots of negatives to mean that the corresponding parabola does not cross the x-axis, and we describe the situation by saying that the roots of the equation are imaginary. With this in mind, we may write:

(14) $[x - (1/2)]^2 - [(1/2)\sqrt{-3}\,]^2 = 0$

(15) $[x - (1/2) - (1/2)\sqrt{-3}\,][x - (1/2) + (1/2)\sqrt{-3}\,] = 0$

Hence the roots are

$$x = (1/2) + (1/2)\sqrt{-3} \text{ and } (1/2) - (1/2)\sqrt{-3}$$

EXERCISE 5.6

1. Find the zeros of:

(a) $y = 3x + 4$ (f) $y = (x + 1)^2$
(b) $y = x^2 - 1$ (g) $v = u(u - b)/a$
(c) $s = t(t - 1)$ (h) $y = 2(x - 1)(2 + x)(2x - 3)$
(d) $y = (2x - 1)(4x + 5)$ (i) $y = x^2$
(e) $s = t^2 - 2$

Answers. (a) $x = -4/3$. (c) $t = 0, 1$. (e) $t = \pm\sqrt{2}$. (g) $u = 0, b$.

2. Prove: (a) If r is a zero of $f(x)$ and $g(x)$, then r is a zero of $af(x) + bg(x)$; (b) If r is a zero of $f(x)$ or $g(x)$, then r is a zero of $f(x)g(x)$.

3. Solve each of the following equations by factoring the left member by trial and error. Sketch the corresponding parabola and label the zeros.

(a) $x^2 - 3x - 4 = 0$ (d) $2x^2 + 5x + 2 = 0$ (g) $x^2 - 14x + 45 = 0$
(b) $x^2 + 8x + 15 = 0$ (e) $3x^2 - 5x - 12 = 0$ (h) $12 - x - 6x^2 = 0$
(c) $x^2 - 7x + 12 = 0$ (f) $2 - 6x + 4x^2 = 0$ (i) $-x^2 + x + 12 = 0$

4. Solve each equation in prob. 3 by completing the square.
5. Solve the following by completing the square, and sketch:

(a) $x^2 - x - 1 = 0$ (c) $2x^2 - 3x + 1 = 0$ (e) $3x^2 + x - 3 = 0$
(b) $x^2 + 4x + 7 = 0$ (d) $2x^2 - 4x - 3 = 0$ (f) $5x^2 + 2x - 1 = 0$

6. Solve by completing the square, interpret the results, and sketch.

(a) $x^2 + x + 1 = 0$ (d) $x^2 + 2x - 1 = 0$ (g) $x^2 + 2x = 0$
(b) $x^2 + 4x + 5 = 0$ (e) $x^2 + 2x + 1 = 0$ (h) $3x^2 - x + 1 = 0$
(c) $x^2 - 8x + 16 = 0$ (f) $x^2 + 2x + 2 = 0$ (i) $5x^2 + 7x + 3 = 0$

7. The legs of a right triangle differ by 2. If the hypotenuse is 10, find the area. (*Suggestion.* Let one of the legs be of length x. Then the other is $\cdots$.)

8. Factor:

(a) $x^2 - 2xy - 8y^2$ (d) $x^2 + (3y - 1)x + y(2y - 1)$
(b) $2x^2 + 3xy + y^2$ (e) $x^2 + (a + b)x + ab$
(c) $18x^2 - 15xy - 3y^2$ (f) $x^2 - (a + b)x + ab$

Answers. (a) $(x + 2y)(x - 4y)$. (c) $3(6x + y)(x - y)$.

9. Simplify:

(a) $\dfrac{x + 3}{x^2 - x - 12}$ (c) $\dfrac{x^3 - 8}{2x^2 - 7x + 6}$ (e) $\dfrac{6x^2 + 5x - 4}{3 - 5x - 2x^2}$

(b) $\dfrac{x^2 + 8x + 15}{x^2 + 10x + 25}$ (d) $\dfrac{a^3 + b^3}{a^3 - a^2b + ab^2}$ (f) $\dfrac{x^3 - y^3}{3x^2 - xy - 2y^2}$

†**10.** Determine k so that the following equations are inconsistent:

$$kx + 2y = 3k$$
$$(1 - k)x + ky = 1$$

Answer: $k = -1 \pm \sqrt{3}$.
†**11.** Solve:

(a) $x^2 + 2x - 15 < 0$ (c) $x^2 + 6x + 9 \geq 0$ (e) $1 - 2x - x^2 > 0$
(b) $x^2 - 1 \leq 0$ (d) $2x^2 - 3x + 1 > 0$ (f) $3x - x^2 < 0$

Suggestion. Draw a graph of the left member. *Answers.* (a) $-5 < x < 3$. (c) $x < -3$ or $x > 3$.

5.7   The quadratic formula

We are going to use the method of completing the square in order to prove the following theorem:

*5.7.1   The quadratic equation $ax^2 + bx + c = 0$ has two and only two roots given by

(1) $$x = \frac{-b \pm \sqrt{b^2 - 4ac}}{2a}$$ (Quadratic Formula)

Proof. The roots of $ax^2 + bx + c = 0$ are the same as the roots of $x^2 + (b/a)x + (c/a) = 0$. (Why?) But

(2) $$x^2 + \frac{b}{a}x + \frac{c}{a} \equiv \left(x + \frac{b}{2a}\right)^2 - \frac{b^2}{4a^2} + \frac{c}{a}$$

(3) $$\equiv \left(x + \frac{b}{2a}\right)^2 - \frac{b^2 - 4ac}{4a^2}$$

(4) $$\equiv \left(x + \frac{b}{2a}\right)^2 - \left[\frac{\sqrt{b^2 - 4ac}}{2a}\right]^2$$

(5) $$\equiv \left[x + \frac{b - \sqrt{b^2 - 4ac}}{2a}\right]\left[x + \frac{b + \sqrt{b^2 - 4ac}}{2a}\right]$$

This last product is zero if and only if one of its factors is zero. Hence the roots of the equation are given by (1).

It is instructive to consider various cases in the above derivation and the resulting formula (1). If $b^2 - 4ac > 0$, then we have in (4) the difference of the squares of two real numbers, and the roots are real numbers given by $-b/2a$ plus and minus the real number $(\sqrt{b^2 - 4ac})/2a$. If $b^2 - 4ac = 0$, the left member of the equation is a perfect square and the two roots coincide. Finally, if $b^2 - 4ac < 0$, the roots are imaginary. The expression $b^2 - 4ac$ plays such an important role in determining the nature of the roots of the quadratic that it is given a special name, the **discriminant,** and is designated by the letter D. We summarize the foregoing discussion in the following table, which also gives the geometric interpretation:

Discriminant	Roots	Parabola
$D > 0$	Real, unequal	Crosses x-axis in two distinct points.
$D = 0$	Real, equal	Tangent to x-axis.
$D < 0$	Imaginary, unequal	Does not touch x-axis.

These results can be used to write down the roots or to find their nature without actually evaluating them. It is a good idea to begin by calculating D. For example, in the equation

$$(6) \qquad 4x^2 + 2x - 1 = 0$$

$$D = 2^2 - 4(4)(-1) = 20 \quad \text{and} \quad \sqrt{D} = 2\sqrt{5}$$

Hence the roots are real, different, and irrational. Their values are

$$(7) \qquad x = \frac{-2 \pm 2\sqrt{5}}{8} = \frac{-1 \pm \sqrt{5}}{4}$$

Usually, in elementary work, the coefficients of quadratic equations are integers, but there is no reason why they should not be fractions or irrational numbers—or even imaginary numbers. The derivation of *5.7.1 does not depend on any special assumptions about the nature of the coefficients, except of course that they are subject to the usual laws of the algebra of numbers. Consider, for example, the quadratic equation

$$(8) \qquad 2\pi x^2 - (\sqrt{31}\,)x + 1.832 = 0$$

The solutions are

$$(9) \qquad x = \frac{\sqrt{31} \pm \sqrt{31 - 8\pi(1.832)}}{4\pi}$$

which can be evaluated as accurately as desired.

The quadratic formula suggests interesting relationships between the roots and coefficients of a quadratic equation. Let the roots be r_1 and r_2. Then (5) becomes:

$$(10) \qquad x^2 + (b/a)x + (c/a) \equiv (x - r_1)(x - r_2)$$

$$(11) \qquad \equiv x^2 - (r_1 + r_2)x + r_1 r_2$$

From (11) it follows that

$$r_1 + r_2 = -\frac{b}{a}$$

(12)   (Why?)

$$r_1 r_2 = \frac{c}{a}$$

We have proved:

*5.7.2 The sum of the roots of $ax^2 + bx + c = 0$ is $-b/a$ and the product is c/a.

Another way of stating *5.7.2 is that when the quadratic equation is written so that the coefficient of x^2 is one, then the product of the roots is the constant term and the sum of the roots is the negative of the coefficient of x.

EXERCISE 5.7

1. Carry through the derivation of the quadratic formula with a, b, c replaced by: (a) a_1, a_2, a_3. (b) 1, p, q.

2. For each of the following equations, calculate D and the roots:

(a) $x^2 + x - 1 = 0$
(b) $4x^2 - 5x - 1 = 0$
(c) $4x^2 + 5x + 1 = 0$
(d) $-3x^2 - 4x + 4 = 0$
(e) $8x^2 + 3x = 1$

(f) $\sqrt{5}x^2 + 2x - 3 = 0$
(g) $1.2x^2 - 3.5x - 0.01 = 0$
(h) $3x^2 - 28 = 0$
(i) $x^2 + 5 = 0$

Answers. (a) $D = 5$, $x = (-1 \pm \sqrt{5})/2$. (c) $D = 9$, $x = -\frac{1}{4}$, -4. (e) $D = 41$, $x = (-3 \pm \sqrt{41})/16$.

3. Use the quadratic formula to find the zeros of the functions you plotted in Ex. 5.5, and compare your algebraic and graphical results.

4. Use the quadratic formula to solve the equations that appear in Ex. 5.6, and compare with previous results.

5. Without solving the following equations, find whether their roots are real or imaginary, rational or irrational, equal or unequal:

(a) $10x^2 - x + 1 = 0$
(b) $2x^2 - 17x - 1 = 0$
(c) $9x^2 + 100x + 1 = 0$
(d) $3x^2 - 9x - 5 = 0$
(e) $3x^2 - 9x + 5 = 0$

(f) $15x^2 - 11x + 2 = 0$
(g) $\sqrt{6}x^2 - \sqrt{10}x + 2/6 = 0$
(h) $2x^2 - 2\sqrt{2}x + 1 = 0$
(i) $\sqrt{2}x^2 + 2\sqrt{6}x + \sqrt{3} = 0$

6. Factor the left member in each part of prob. 5.

7. Factor the following quadratic functions: (a) $x^2 + 2x + 2$. (b) $3x^2 + x + 2$. (c) $x^2 + 1$. (d) $x^2 + 3$. [*Answers.* (a) $(x + 1 + \sqrt{-1})(x + 1 - \sqrt{-1})$. (c) $(x + \sqrt{-1})(x - \sqrt{-1})$.]

8. Evaluate the roots of (7) to three significant digits.

9. Write formulas for the solutions of: (a) $x^2 + rx + s = 0$. (b) $m^2x^2 + mx + 1 = 0$. (c) $rsx^2 - sx + x^2 = 0$.

10. Solve for s: $m = Es/(1 + s)(1 - 2s)$.

11. In graphing $y = ax^2 + bx + c$, we substituted values of x to find the corresponding y's. Now we can find the two values of x that correspond to any y. For $y = 2x^2 - 3x - 4$, find the x's for which: (a) $y = 0$. (b) $y = 1$. (c) $y = 3$. (d) $y = -1$. (e) $y = -4$. (f) $y = -5$. (g) $y = -6$. Plot the points.

12. Find the intersections of the line $y = 3$ with the parabola $y = 2x^2 - x - 1$.

13. Show that the line $y = -9/8$ is tangent to the parabola of prob. 12.

14. Find the intersections of the following pairs of curves and check by graphing. (*Suggestion.* Solve the linear equation for one variable and substitute in the other.)

(a) $y = 2x + 1$, $y = x^2 + 3x - 4$ (d) $2x - y + 1 = 0$,
(b) $y = x - 1$, $y = x^2 + x + 5$ $6x - 2x^2 - 1 + y = 0$
(c) $x - y = 1$, $y = -3x^2 + x$

15. Solve for y: $y\theta + x\phi + x'\phi' + \dfrac{f^2\theta}{y}\left(1 - \dfrac{x}{b} - \dfrac{x'}{b'}\right) = \dfrac{fk}{y}$. (This equation is involved in air photography.)

16. The velocity of a wave on the surface of a liquid is given by $V = \sqrt{\dfrac{g\lambda}{2\pi} + \dfrac{2\pi T}{\lambda d}}$, where g is the acceleration of gravity, T the surface tension, d the density of the liquid, and λ the wavelength. Solve for λ. (*Answer.* $\dfrac{\pi}{dg}[dV^2 \pm \sqrt{d(dV^2 - 4gt)}\,]$.)

17. Solve for m the following equation, which appears in the theory of optics:

$$\frac{r_1}{r_2} = \frac{m + 4 - 2m^2}{m + 2m^2}$$

18. Solve for n: $V = R\left(0.25 - \dfrac{1}{n^2}\right)$ (Physics)

19. Solve for N: $r = N/\sqrt{N + P}\sqrt{N + R}$ (Psychological Testing)

20. Factor the left member of $x^3 - 1 = 0$ and so find the three cube roots of unity. [*Answer.* $1, (-1 \pm \sqrt{-3})/2$.]

21. Find the general formulas giving the solutions of the simultaneous equations $x^2 + y^2 = a^2$ and $xy = b$.

22. A pilot wishes to fly 500 miles and return in 4 hours. If the wind is blowing at 50 mph from the direction of his destination, at what constant air speed should he fly? (*Answer.* 2.60×10^2 mph.)

†23. Is it correct to say that the roots of a quadratic are rational if $D = 0$? Illustrate your answer by finding the roots of $2x^2 - \sqrt{8}x + 1 = 0$. If the coefficients are rational, what is the condition on D for the roots to be rational?

†24. Solve $y = ax^2 + bx + c$ for x in terms of y. Show that, if $a > 0$, the smallest value of y for which x is real is $c - b^2/4a$ and that for this value $x = -b/2a$. Show also that when $a < 0$, these values correspond to the largest value of y for which x is real. What is the connection between these results and our discussion in 5.5.

25. For what values of k does each of the following equations have the kind of roots indicated?

(a) $x^2 + kx + 3 = 0$, equal
(b) $kx^2 - 2x + 1 = 0$, real
(c) $2x^2 - 4x + k = 0$, imaginary
(d) $x^2 - kx + 2 = 0$, rational
(e) $(k + 1)x^2 + kx + 1 = 0$, real
(f) $k^2x^2 + kx - 4 = 0$, imaginary

Answers. (a) $k = \pm2\sqrt{3}$. (c) $k > 2$. (e) $k \le 2 - 2\sqrt{2}$ or $k \ge 2 + 2\sqrt{2}$.

†26. What must be the value of k in part e of Prob. 25 so that one root is the negative of the other?

27. The sum of two numbers is 14, and their product is 6. Find them.

28. Write quadratic equations with the following roots: (a) 6, -2. (b) $\sqrt{3}$, 2. (c) $2 + \sqrt{5}$, $2 - \sqrt{5}$. (d) 0, $\sqrt{7}$. (e) 1, 1.

29. Find the values of n for which $nr - \dfrac{r(n-1)}{2} \le \dfrac{n(n-1)}{2}$. (This inequality arises in psychology.)

30. Solve the following inequalities and indicate the solutions graphically:

(a) $x(x - 1) < 0$
(b) $x(x + 2) > 0$
(c) $(x + 1)(x - 3) \le 0$
(d) $x^2 - 2x + 1 \ge 0$
(e) $2x^2 + 4x - 3 \le 0$
(f) $x + 3x^2 \ge -4$
(g) $1 + 3x - x^2 < 0$
(h) $1 - x^2 \ge 0$
(i) $x^2 + x + 1 > 0$
(j) $x^2(x^2 - 3x - 4) \le 0$
(k) $(x + 1)/(x - 2) < 4$
(l) $(x + 2)/x^2 \le 1$

Answers. (a) $0 < x < 1$. (c) $-1 \le x \le 3$. (e) $(-2 - \sqrt{10})/2 \le x \le (-2 + \sqrt{10})/2$.

†31. Show that there are no real values of a and b for which:

$$\frac{1}{a} + \frac{1}{b} = \frac{1}{a + b}$$

6

The Power Functions

Power functions of a single variable are those of the form $y = ax^n$. In this chapter we begin by considering only positive integral values for n, and then extend its range to include all rational numbers. We introduce and make use of the principle of finite induction and consider various topics associated with power functions.

6.1 The cube and higher powers

The **cube,** or **third power,** is defined by $y = x^3$. Like the square, it appears so frequently that it is available in tables. Table I gives the cubes of integers from 1 to 100. Cubes of numbers formed from these by shifting the decimal point can be found from the table by ignoring the decimal point and then placing it in the answer by approximate computation. (Compare 5.1 and 3.14.) Cubes of still other numbers can be found by linear interpolation. In applying *5.2.4 and Fig. 5.2.2, Δy is now the difference of two entries in the cube column, Δx equals 1 as before, and the secant line is drawn between two points on the graph of $y = x^3$.

The graph of $y = x^3$ near the origin is shown in Fig. 6.1. Its appearance may be verified by plotting numerous points or by the following considerations. Since $x^3 \equiv x \cdot x^2$ and $x^2 \geq 0$, it is evident that x^3 is positive, zero, or negative according as x is positive, zero, or negative. Also,

$$(1) \qquad (-x)^3 \equiv -(x^3) \qquad \text{(Why?)}$$

It follows that if (x, y) is on the graph, so is $(-x, -y)$. This property is called symmetry with respect to the origin. As in the case of the square, the absolute value of x^3 increases with the absolute value of x since $|x_2| > |x_1| \longrightarrow |x_2|^3 > |x_1|^3$. (Why?) In order to get a still better picture of the way in which y changes with x, we find the average and exact rates of change.

225

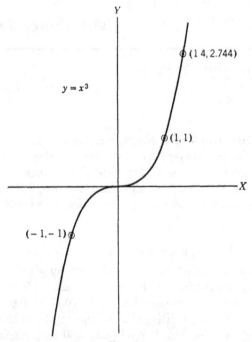

$$y = x^3$$

$(1\,4, 2.744)$

$(1, 1)$

$(-1, -1)$

FIG. 6.1

We have:

(2)
$$\frac{\Delta y}{\Delta x} = \frac{x_2{}^3 - x_1{}^3}{x_2 - x_1}$$

In order to simplify this we make use of an identity that the student can easily verify, namely,

*6.1.1 $a^3 - b^3 \equiv (a - b)(a^2 + ab + b^2)$

By this identity, (2) becomes:

(3)
$$\frac{\Delta y}{\Delta x} = x_2{}^2 + x_2 x_1 + x_1{}^2$$

From this we see that as x_2 approaches x_1, the average rate of change approaches $x_1{}^2 + x_1 x_1 + x_1{}^2$ or $3x_1{}^2$. Hence it appears plausible that $D_x x^3 = 3x^2$. Using *5.4.4, we have:

*6.1.2 $D_x(ax^3) = 3ax^2$

It follows that the slope of $y = x^3$ is $3x^2$, which is everywhere positive, except at the origin. There it is zero, and the curve is tangent to the x-axis. Hence x^3 always increases except that it "levels off" at $x = 0$. The student should note that without the derivative we could not be quite sure of the shape of the curve near the origin. As in the case of the square, the constant in the graph of ax^3 has the effect of multiplying each y-coordinate of $y = x^3$ by a. Also the graphs of $y = ax^3$ and $y = -ax^3$ are reflections, each being obtainable from the other by rotating the plane about the x-axis.

The **fourth power**, $y = x^4$, is similar in its properties to the square. It is never negative, increases when x increases in absolute value, and has a graph that is symmetric with respect to the y-axis and is tangent to the x-axis at the origin. The fifth power resembles the cube again. In general, $y = x^n$ resembles the square or the cube according as n is even or odd. The even powers have graphs above the x-axis, tangent to the x-axis at the origin, and symmetric to the y-axis. The odd powers have graphs above or below the x-axis according to the sign of x, tangent to the x-axis at the origin (except for $n = 1$), and symmetric with respect to the origin. Also for any n,

*6.1.3 $$D_x(ax^n) = nax^{n-1}$$

Proofs are postponed until 6.3.

EXERCISE 6.1

1. Find the cubes of the following numbers. Round off interpolated values to three digits.

(a) 37	(g) 0.37	(m) −86.2	(s) 1362
(b) 6.5	(h) 0.26	(n) 91.7	(t) −1.345
(c) −8.3	(i) −0.8	(o) 0.458	(u) 0.00382
(d) 0.94	(j) 0.013	(p) 0.771	(v) −0.00121
(e) 0.88	(k) 1.45	(q) −392	(w) 8.001
(f) −0.59	(l) 29.3	(r) 0.0934	(x) 0.000587

Answers. (c) −571.787. (e) 0.681487. (i) −0.512. (k) 2.10. (m) −6.40 × 10⁵. (o) 0.0961.

2. Graph:

(a) $y = x^3$	(f) $y = x^4, y = x^2$	(k) $y - 2 = (x - 3)^3$
(b) $y = 2x^3$	(g) $y = x^5, y = x^3$	(l) $y = (x + 2)^4$
(c) $y = x^3/3$	(h) $y = 0.1x^2, y = 0.1x^3$	(m) $y - 2 = 2(x - 1)^5$
(d) $y = -x^3, y = -3x^3$	(i) $y = x^3$	
(e) $y = x^3, y = 4x^3$	(j) $y = (x - 1)^3$	

3. Graph $y = f(x)$, where $f(x) = 1$ when $x < -3$
$= x^3$ when $-3 \leq x \leq 0$
$= x$ when $x > 0$.

4. Make a large-scale graph of $y = x^2$ and $y = x^3$ on the same axis for $0 \leq x \leq 1.5$.

5. A figure is said to be **symmetric with respect to a point,** called the **center,** if to every one of its points there corresponds another such that the segment joining them is bisected by the center. Justify calling $y = x^3$ symmetric with respect to the origin. [*Suggestion.* Consider (a, b) on the curve, and show that the line joining it to $(-a, -b)$ passes through the origin and is bisected by it.]

6. Prove that $x > 0 \longrightarrow x^3 > 0$ and $x < 0 \longrightarrow x^3 < 0$.

7. Prove (1).

8. Show that $(10x)^3 \equiv 1000x^3$.

9. Show that the absolute value of x^3 increases with the absolute value of x.

10. If the wind resistance R to an automobile is proportional to the cube of the speed v, write the relation between R and v. Compare the resistances at speeds of 45 and 60 mph.

11. On a graph of $y = 0.5x^3$ draw tangent lines at the points where $x = 1$, $x = -1$, $x = 2$, and $x = -3$. Find the equation of each. (*Partial solution:* $D_x y = 1.5x^2$. When $x = 1$, $D_x y = 1.5$. Hence the tangent line at this point is $y - 0.5 = (1.5)(x - 1)$. Similarly for the other points.)

12. Find the exact rates of change of the following functions at the value indicated:

(a) $y = x^3$ at $x = 12.3$ (d) $y = 3x$ at $x = 0$
(b) $y = -2x^3$ at $x = 2$ (e) $y = 14x^3$ at $x = 1$
(c) $y = x^2$ at $x = 3.2$ (f) $y = -bcx^3$ at $x = a$

Answers. (a) 453.87. (c) 6.4. (e) 42.

13. A body is moving according to $s = 2t^3$, where s is distance and t is time. Find its speed when $t = 3$. (*Answer.* 54.)

14. Show that:

(a) $x^4 = (x^2)^2$ (d) $x^3/x^1 = x^2$ (g) $x^3/x^6 = 1/x^3$
(b) $x^6 = (x^2)^3$ (e) $x^7 = x \cdot x^6$ (h) $(x^3)^3 = x^9$
(c) $x^3 \cdot x^5 = x^8$ (f) $x^6/x^2 = x^4$ (i) $(x^2)^5 = (x^5)^2$

Suggestion. Refer to 3.5(1).

15. Prove *6.1.1.

16. Prove:

*6.1.4 $a^3 + b^3 \equiv (a + b)(a^2 - ab + b^2)$

*6.1.5 $a^4 - b^4 \equiv (a - b)(a^3 + a^2b + ab^2 + b^3)$

*6.1.6 $a^5 - b^5 \equiv (a - b)(a^4 + a^3b + a^2b^2 + ab^3 + b^4)$

*6.1.7 $a^5 + b^5 \equiv (a + b)(a^4 - a^3b + a^2b^2 - ab^3 + b^4)$

*6.1.8 $(a + b)^3 \equiv a^3 + 3a^2b + 3ab^2 + b^3$

*6.1.9 $(a + b)^4 \equiv a^4 + 4a^3b + 6a^2b^2 + 4ab^3 + b^4$

17. Use the identities of this and previous sections to expand:

(a) $(2x + y)^3$

(b) $(x - y)^3$

(c) $(x + y)^4$

(d) $(x - 2y)^4$

(e) $(a + 2b)^4$

(f) $(3x + b)^3$

(g) $(ax + b)(a^2x^2 - abx + b^2)$

(h) $(2x - 3t)^3$

(i) $(s - t)^2(s + t)^3$

(j) $(s^2 + t^2)(s^2 - t^2)$

(k) $(s^2 - 2x)^3$

(l) $(a + b + c)^3$

Answers. (a) $8x^3 + 12x^2y + 6xy^2 + y^3$. (g) $a^3x^3 + b^3$.

18. Rewrite without denominators:

(a) $\dfrac{8x^3 - 1}{2x - 1}$

(b) $\dfrac{8x^3 + s^3}{2x + s}$

(c) $\dfrac{32x^4 - 2y^4}{2x - y}$

(d) $\dfrac{x^4 - y^4}{x + y}$

(e) $\dfrac{s^5 - 32t^5}{s - 2t}$

(f) $\dfrac{x^6 - s^6}{x^3 - s^3}$

(g) $\dfrac{8x^3 - 27}{2x - 3}$

(h) $\dfrac{a^3b^3 + 1}{ab(ab - 1) + 1}$

(i) $\dfrac{x^6 + b^6}{x^2 + b^2}$

Answers. (a) $4x^2 + 2x + 1$. (c) $2(8x^3 + 4x^2y + 2xy^2 + y^3)$.

19. If the power required to drive an ocean liner varies as the cube of the speed, and the cost of power increases with the square of the amount, compare the cost of a 4-day Atlantic crossing with that of a 9-day crossing.

20. Fermat's theorem states that if p is a prime number and a does not have p as a factor, then $a^{p-1} - 1$ is divisible by p. Test this for: $a = 2$, $p = 5$; $a = 2$, $p = 7$; $a = 4$, $p = 3$; and $a = 6$, $p = 7$.

21. Stefan's law states that the intensity of radiation I from a "black body" is given by $I = kT^4$, where T is the absolute temperature and k is a constant. Graph I as a function of T for $k = 1.36$.

22. A textbook on zoology states that a starfish possesses "radial symmetry" and a frog, man, or fly has "bilateral symmetry." Define these words in terms of the types of symmetry considered in this book. Is the statement strictly true about man?

23. Find the slope of each of the following curves at the point indicated and write the equation of the tangent line there:

(a) $y = 3x^3$ at $x = -1$

(b) $y = x^4$ at $x = 2$

(c) $y = -2x^4$ at $x = 3$

(d) $2y = 7x^5$ at $x = 1$

(e) $y = 7x^7$ at $x = 0$

(f) $y = -4x^6$ at $x = -2$

(g) $y = (1.93)x^4$ at $x = -1.5$

(h) $y = x^{10}$ at $x = 1$

Answers. (a) 9, $y = 9x + 6$. (c) -216, $y + 216x = 486$. (e) 0, $y = 0$.

24. The side of a cube is $a = 4t$, where t is the time. How fast is its volume increasing when $t = 2$?

25. How fast is the volume of a sphere increasing with respect to its radius, when the radius is 2? 4?

†**26.** Show that for all $x > 1$:

$$(a)\ \frac{x^2 + 1}{x^3 + 1} > \frac{1}{x} \qquad\qquad (b)\ \frac{x + 1}{x^3} < \frac{2}{x^2}$$

†**27.** Show that $(-x)^4 \equiv x^4$; $x^4 \geq 0$; and $|x_1| > |x_2| \longrightarrow |x_1|^4 > |x_2|^4$.

28. Find $f(x + h) - f(x)$ for:

(a) $f(x) = x^3 - 2$ (d) $f(x) = 2x^4$
(b) $f(x) = x^2 + 1$ (e) $f(x) = 3x^5 + 1$
(c) $f(x) = x^3 - x^2$ (f) $f(x) = 4x - 3$

†**29.** Give a plausibility proof that $D_x(ax^4) = 4ax^3$ and $D_x(ax^5) = 5ax^4$ without using *6.1.3.

6.2   The principle of finite induction

We have met a number of theorems that are easy to verify for special cases but for which we have not given a general proof. For example, we found the formula for the nth term of an arithmetic progression (*4.5.2) by considering the form of the first few terms and then guessing at the general formula. But such a conjecture is not a proof, for it is easy to cite formulas that are true for several positive integral values of a variable but not for all positive integral values. For example, $2n + (n - 1)(n - 2)(n - 3)$ takes the values 2, 4, 6 for $n = 1, 2, 3$, but it does not have the value 8 for $n = 4$. What we need is a principle that enables us to pass from specific cases to a proof for *all* cases.

Suppose that we could prove that a certain theorem involving a variable n is true for $n = 1$ and also that it is always possible to prove it for the *next* integral value of n if it is *assumed* for any particular value of n. It would certainly seem reasonable to state that the theorem is true for all positive integral values of n. Given any value of n, we could prove the theorem for $n = 1$, then from this show it for $n = 2$, then for $n = 3$, etc., until we eventually arrived at the given value of n. We state this idea as follows:

***6.2.1**   **Principle of finite induction.** Let $T(n)$ be a theorem involving a positive integer n. If

(I) $T(1)$ is true, and

(II) $T(k) \longrightarrow T(k+1)$ for any positive integral k,

then $T(n)$ is true for all positive integral values of n.

In this theorem, the values of n are the positive integers, and the values of $T(n)$ are statements. Thus if $T(n)$ stands for "$n + n = 2n$," $T(1)$ is "$1 + 1 = 2 \cdot 1$" and $T(8)$ is "$8 + 8 = 2 \cdot 8$."

We illustrate the use of the principle by proving *4.5.2. We are given an arithmetic progression, that is, a sequence in which $y_{n+1} = y_n + d$ for all n. Since $y_1 = y_1 + (1-1)d$, $T(1)$ is true. [Note that $T(n)$ stands for $y_n = y_1 + (n-1)d$.] Hence we have proved part I of the hypothesis in *6.2.1. Now we show that $T(k) \longrightarrow T(k+1)$.

(1) $\qquad\qquad$ If $y_k = y_1 + (k-1)d$

then

(2) $\qquad y_{k+1} = y_k + d \qquad$ (Why?)

(3) $\qquad\qquad = y_1 + (k-1)d + d \qquad$ [From (1)]

(4) $\qquad\qquad = y_1 + (k+1-1)d \qquad$ (Why?)

But (1) is just $T(k)$ and (4) is $T(k+1)$. Hence we have proved hypothesis II, and we can assert the formula for all positive integral n.

Many interesting identities can be proved by finite induction. For example,

*6.2.2 $\quad 1^2 + 2^2 + 3^2 + \cdots + n^2 \equiv \dfrac{n(n+1)(2n+1)}{6}$

It is easy to verify that the theorem holds when $n = 1, 2, 3$. (Of course, it is sufficient to show that it holds for $n = 1$.) Now, if

(5) $\qquad 1^2 + 2^2 + 3^2 + \cdots + k^2 \equiv \dfrac{k(k+1)(2k+1)}{6}$

(6) $\quad 1^2 + 2^2 + 3^2 + \cdots + k^2 + (k+1)^2$

$$\equiv \frac{k(k+1)(2k+1)}{6} + (k+1)^2$$

(7) $$\equiv \frac{(k+1)(k+2)(2k+3)}{6}$$

But (7) is just the theorem for $n = k + 1$. Hence the theorem is proved for all positive integral n by the principle of finite induction.

The principle is by no means limited to the proof of algebraic identities. For example, it is involved in the proof of the following theorem in geometry: The sum of the interior angles of a polygon of $n + 2$ sides is $n(180°)$. When $n = 1$, this becomes the familiar theorem about a triangle. Suppose now, for the purpose of proving hypothesis II of *6.2.1, that the theorem holds for $n = k$, that is, for a polygon of $k + 2$ sides, and consider a polygon of $k + 3$ sides (Fig. 6.2). We draw a diagonal connecting two

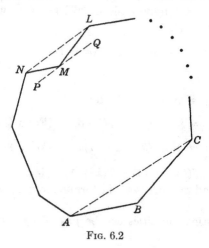

FIG. 6.2

alternate vertices and dividing the polygon into a triangle and a polygon of $k + 2$ sides. The sum of the angles of the triangle is 180°, and by the induction assumption the sum of the angles of the polygon is $k(180°)$. Hence the sum of the interior angles of the polygon of $k + 3$ sides is $k(180°) + 180° = (k + 1)(180°)$. But this is just the theorem for $n = k + 1$. Hence hypotheses I and II hold, and *6.2.1 justifies asserting the theorem for all positive integral n.

Finally we mention an application to logic. Suppose that in a sequence of statements $A_0, A_1, A_2, A_3, \cdots A_n$, each one implies the following, that is, $A_i \longrightarrow A_{i+1}$ for $i = 0, 1, \cdots n - 1$. Then we assert that for any n, $A_0 \longrightarrow A_n$. Hypothesis I holds since, when $n = 1$, the theorem becomes $A_0 \longrightarrow A_1$. If the theorem holds for $n = k$, then $A_0 \longrightarrow A_1$, $A_1 \longrightarrow A_2$, $A_2 \longrightarrow A_3$, $\cdots A_{k-1} \longrightarrow A_k$ imply that $A_0 \longrightarrow A_k$. For $n = k + 1$ the theorem states that $A_0 \longrightarrow A_1$, $A_1 \longrightarrow A_2$, $\cdots A_{k-1} \longrightarrow A_k$, $A_k \longrightarrow A_{k+1}$ imply $A_0 \longrightarrow A_{k+1}$. By the induction assumption, the hypothesis of this last theorem implies that $A_0 \longrightarrow A_k$. This, together with $A_k \longrightarrow A_{k+1}$ implies $A_0 \longrightarrow A_{k+1}$ by *2.3.1. Hence hypothesis II holds and the theorem is proved for any n.

In devising proofs by finite induction it is well to begin by writing out the theorem for $n = 1$, $n = k$, and $n = k + 1$. Then there are two tasks to be done: (a) Show that the theorem for $n = 1$ is true. (b) Derive the theorem for $n = k + 1$ from the theorem for $n = k$. Task b is usually the key part of the proof. Where $T(n)$ is an identity, $T(k + 1)$ may sometimes be derived from $T(k)$ by adding the same expression to both sides of $T(k)$. The expression is chosen so as to make one side of the resulting equation obviously identical with one side of $T(k + 1)$. Then it is a question of showing that the other side reduces to the other side of $T(k + 1)$. This is the method we used in proving *6.2.2. However, as illustrated in the other examples above, other methods are appropriate in other cases. The student should approach each proof as an opportunity to develop his heuristic ability.

The use of the word "induction" in the name of *6.2.1 is perhaps unfortunate, since *the principle of finite induction is as much a part of the deductive method as any other theorem of logic or mathematics.* It permits us to assert a theorem for all positive integral values of n if we can prove hypotheses I and II. The principle is based on a fundamental property of the set of positive integers.[1] It should not be confused with the heuristic or "inductive" methods by which we guess or conjecture a general theorem from specific instances. The student should compare the way in which we proved *4.5.2, using strict deduction, with the way in which we found these formulas in 4.5 by using heuristic ("inductive") arguments. The "inductive method" in science, by which general laws are conjectured from observation and experiment, is similar to the way in which we found *4.5.2 and is quite distinct from the method of proof based on the principle of finite induction.

[1] For a proof, see *A Survey of Modern Algebra*, by Garrett Birkhoff and Saunders MacLane, Macmillan, New York, 1946, pp. 9–13.

EXERCISE 6.2

1. Rewrite the proofs of *4.5.2 and *6.2.2, with all steps and algebraic manipulations shown.

2. Prove *4.5.4, *4.5.5, and *4.5.6 directly by finite induction without using *4.5.3.

3. Fill in the reasons in the following proof of *4.5.3:

(a) The formula holds for $n = 1$.

(b) $T(k)$ is $s_k = \dfrac{k}{2}(y_1 + y_k)$ and $T(k + 1)$ is $s_{k+1} = \dfrac{(k + 1)}{2}(y_1 + y_{k+1})$.

(c) $s_{k+1} = s_k + y_{k+1}$.

(d) Hence, if $T(k)$ holds, then

$$s_{k+1} = \frac{k}{2}(y_1 + y_k) + y_{k+1}$$

(e) $$= \frac{k}{2}(y_1 + y_{k+1} - d) + y_{k+1}$$

(f) $$= \frac{k}{2}(y_1 + y_{k+1}) - \frac{kd}{2} + y_{k+1}$$

(g) $$= \frac{k}{2}(y_1 + y_{k+1}) - \frac{(y_{k+1} - y_1)}{2} + y_{k+1} \qquad (*4.5.2)$$

(h) $$= \frac{(k + 1)}{2}(y_1 + y_{k+1})$$

But (h) is just the theorem for $n = k + 1$, i.e., $T(k + 1)$, and hence we have proved hypothesis II of the principle of finite induction.

4. In a certain sequence $\{y_n\}$, $y_{n+1} = (-1)y_n$. Show that $y_n = (-1)^{n+1}y_1$.

5. Letting $T(n)$ stand for each formula in prob. 6, write $T(1)$, $T(2)$, $T(3)$, $T(6)$, $T(k)$, $T(k + 1)$, $T(s - 1)$, and $T(x^2)$.

Answer for part a. $T(1)$ is $1 = 1(2)/2$, $T(2)$ is $1 + 4 = 2(5)/2$, $T(6)$ is $1 + 4 + 7 + 10 + 13 + 16 = 6(17)/2$, $T(k)$ is $1 + 4 + 7 + 10 + \cdots + (3k - 2) = k(3k - 1)/2$, $T(k + 1)$ is $1 + 4 + 7 + 10 + \cdots + (3k - 2) + (3k + 1) = (k + 1)(3k + 2)/2$, $T(s - 1)$ is $1 + 4 + 7 + 10 + \cdots + (3s - 5) = (s - 1)(3s - 4)/2$, and $T(x^2)$ is $1 + 4 + 7 + 10 + \cdots + (3x^2 - 2) = x^2(3x^2 - 1)/2$.

6. Prove each of the following for all positive integral n:

(a) $$1 + 4 + 7 + 10 + \cdots + (3n - 2) \equiv \frac{n(3n - 1)}{2}$$

(b) $$1 + 5 + 9 + 13 + \cdots + (4n - 3) \equiv n(2n - 1)$$

(c) $$1^3 + 2^3 + 3^3 + \cdots + n^3 \equiv \frac{n^2(n + 1)^2}{4}$$

(d) $$2^2 + 4^2 + 6^2 + \cdots + (2n)^2 \equiv \frac{2n(n + 1)(2n + 1)}{3}$$

7. Use induction to give rigorous proofs of: (a) *3.10.7.   (b) *3.10.13.
(c) *3.14.3.   (d) *3.11.10.

8. If in the sequence $\{y_n\}$, $y_0 = 1$, and $y_n = ny_{n-1}$, show that $y_n = n(n-1)(n-2)(n-3)\cdots 5\cdot 4\cdot 3\cdot 2\cdot 1$.

9. Prove for all positive integral n:

(a) $1\cdot 3 + 2\cdot 4 + 3\cdot 5 + \cdots + n(n+2) \equiv \dfrac{n(n+1)(2n+7)}{6}$

(b) $\dfrac{1}{1\cdot 2} + \dfrac{1}{2\cdot 3} + \dfrac{1}{3\cdot 4} + \cdots + \dfrac{1}{n(n+1)} \equiv \dfrac{n}{n+1}$

(c) $\dfrac{1}{1\cdot 3} + \dfrac{1}{3\cdot 5} + \dfrac{1}{5\cdot 7} + \cdots + \dfrac{1}{(2n-1)(2n+1)} \equiv \dfrac{n}{2n+1}$

10. In a sequence $\{y_n\}$, $y_n = y_{n+1}$, show that $y_n = y_0$.

11. Prove that

$$1\cdot 2 + 2\cdot 2^2 + 3\cdot 2^3 + 4\cdot 2^4 + \cdots + n\cdot 2^n \equiv 2 + (n-1)2^{n+1}.$$

†12. Prove *3.6.3. [*Suggestion.* Restate it in the following form:

$$a(b_1 + b_2 + b_3 + \cdots + b_{n-1} + b_n) \equiv ab_1 + ab_2 + ab_3 + \cdots + ab_{n-1} + ab_n.]$$

†13. Prove that the product of n numbers is zero if and only if at least one of them is zero.

14. Letting $T(n)$ stand for each of the following, show that $T(k) \longrightarrow T(k+1)$, but that $T(n)$ is *false* for all positive integral n.

(a) $2(n+1) = 2n(?)$ (c) $n > n(?)$ (e) $n - n = 1(?)$

(b) $n+1 = n-1(?)$ (d) $n \neq n(?)$

(f) $\dfrac{1}{n} = \dfrac{1}{n+1}(?)$

(g) The sum of the interior angles of a polygon of n sides is $180n$ degrees.(?)

15. It is possible to cite numerous cases in which a theorem involving n is true for several values of n but not for all values. For example, $1 + 3 + 5 + \cdots + (2n+1) + (n-1)(n-2)(n-3)(n-4)$ identically equals n^2 for $n = 1, 2, 3, 4$ but not for any other values of n. Several examples are given by C. S. Carlson in the *National Mathematics Magazine* of October, 1944, p. 36. Verify the statements made in this reference, and construct other examples.

16. Extend the result of Ex. 3.11.18 to n ratios and prove it.

17. Prove that if $a_0 > a_1$, $a_1 > a_2$, $a_2 > a_3$, $\cdots$, $a_{n-1} > a_n$, then $a_0 > a_n$.

†18. Prove by induction that

$$(n+1)^3 \equiv 1 + 3(1^2 + 2^2 + 3^2 + \cdots + n^2) + 3(1 + 2 + 3 + \cdots + n) + n$$

and use this result to prove *6.2.2.

†19. If A_1, A_2, $\cdots A_n$ are n points on a straight line, show that the sum of the directed distances

$$A_1A_2 + A_2A_3 + A_3A_4 + \cdots A_{n-1}A_n = A_1A_n$$

20. In each of the following show that the recurrence relation (see Ex. 4.5.23) implies the formula for the nth term:

(a) $y_{n+1} = y_n - 3 \longrightarrow y_n = y_1 - (n-1)3$

(b) $y_{n+1} = ny_n/(n+1) \longrightarrow y_n = y_1/n$

(c) $y_{n+1} = (n+1)y_n \longrightarrow y_n = n(n-1)(n-2) \cdots 5 \cdot 4 \cdot 3 \cdot 2y_1$

21. Prove that, if in two sequences $\{a_n\}$ and $\{b_n\}$, $a_i < b_i$ for all i, then $a_1 + a_2 + a_3 + \cdots + a_n < b_1 + b_2 + b_3 + \cdots + b_n$ for all n.

22. Under the hypothesis of prob. 21 and the additional assumption that $a_i > 0$ for all i, show that

$$a_1 a_2 a_3 \cdots a_n < b_1 b_2 b_3 \cdots b_n$$

†**23.** Prove that:

(a) $|a_1 + a_2 + a_3 + \cdots + a_n| \le |a_1| + |a_2| + |a_3| + \cdots + |a_n|$

(b) $|a_1 a_2 a_3 \cdots a_n| = |a_1| \, |a_2| \, |a_3| \cdots |a_n|$

†**24.** Prove that a polygon of n sides has $(n^2 - 3n)/2$ diagonals.

25. Prove that the recurrence relation implies the formula for the nth term.

(a) $y_{n+1} \le y_n \longrightarrow y_n \le y_1$

(b) $y_{n+1} = y_n\sqrt{\dfrac{n}{n+1}} \longrightarrow y_n = y_1/\sqrt{n}$

(c) $y_{n+1} = (\sqrt{y_n} + 1)^2$ and $y_1 = 1 \longrightarrow y_n = n^2$

(d) $y_{n+1} < 2y_n \longrightarrow y_n < 2^{n-1}y_1$

†**26.** Prove that the sum of the digits of any multiple of 9 is divisible by 9.

†**27.** Suppose that a determinant of the nth order is defined by

$$A = \begin{vmatrix} a_{11} & a_{12} & a_{13} & \cdots & a_{1n} \\ a_{21} & a_{22} & a_{23} & \cdots & a_{2n} \\ a_{31} & a_{32} & a_{33} & \cdots & a_{3n} \\ a_{41} & \cdot & \cdot & \cdots & \cdot \\ \cdot & \cdot & \cdot & \cdots & \cdot \\ \cdot & \cdot & \cdot & \cdots & \cdot \\ a_{n1} & a_{n2} & a_{n3} & \cdots & a_{nn} \end{vmatrix} = a_{11}A_{11} - a_{21}A_{21} + \cdots + (-1)^{n-1}a_{n1}A_{n1}$$

where A_{ij} is the minor of a_{ij}. Show that the theorems of 4.11 apply to determinants of the nth order. (*Suggestion.* Begin by proving *4.11.2 and *4.11.3, which are easy. For the others you may find it convenient to extend the results of Ex. 4.11.19 to determinants of any order.)

6.3 Positive integral exponents

We now give a precise definition of positive integral exponents:

*6.3.1 $b^{n+1} = b \cdot b^n$ and $b^1 = b$ (Definition)

From this definition we have immediately (by setting $n = 1$) b^2 $= b \cdot b^1 = b \cdot b$. Similarly, $b^3 = b^{2+1} = b \cdot b^2 = b \cdot b \cdot b$. In general,

*6.3.2 $\qquad\qquad b^n = b \cdot b \cdot b \cdot b \cdots b \qquad (n \text{ factors})$

The expression b^n is called the **nth power** of b or b to the nth, n is called the **exponent,** and b the **base.**

The following general laws of exponents can be proved for positive integral exponents from *6.3.1:

*6.3.3 $\qquad\qquad b^m \cdot b^n \equiv b^{m+n}$

*6.3.4 $\qquad\qquad (b^m)^n \equiv b^{mn}$

*6.3.5 $\qquad\qquad (ab)^n \equiv a^n b^n$

*6.3.6 $\qquad\qquad b^m \div b^n \equiv b^{m-n} \text{ if } m > n$

$\qquad\qquad\qquad\qquad \equiv \dfrac{1}{b^{n-m}} \text{ if } m < n$

*6.3.7 $\qquad\qquad \left(\dfrac{1}{b}\right)^n \equiv \dfrac{1}{b^n}$

*6.3.8 $\qquad\qquad \left(\dfrac{a}{b}\right)^n \equiv \dfrac{a^n}{b^n}$

We leave it to the student to illustrate the laws for particular values of m and n. For the present, m and n are assumed to be positive integers, but the laws hold when the range of m and n is extended to include all real and complex numbers.

*Proof of *6.3.3 by Finite Induction.* We wish to prove this for any positive integral values of m and n. It is certainly true for any m and for $n = 1$, since this is just the definition *6.3.1. Now if it is true for any m and $n = k$, that is,

(1) $\qquad a^m \cdot a^k \equiv a^{m+k} \qquad$ (Induction Assumption)

then

(2) $\qquad a^m \cdot a^{k+1} \equiv a^m \cdot a^k \cdot a \qquad$ (*6.3.1)

(3) $\qquad\qquad \equiv a^{m+k} \cdot a \qquad$ (Step 1)

(4) $\qquad\qquad \equiv a^{m+k+1} \qquad$ (Why?)

But (4) is just the theorem for $n = k + 1$. Hence we have established both hypotheses of the principle of finite induction, and the theorem holds for all positive integral m and n.

Proof of *6.3.4. It is obvious for $n = 1$.

(5)   If $(a^m)^k \equiv a^{mk}$ (Induction Assumption)

(6) $(a^m)^{k+1} \equiv (a^m)^k \cdot (a^m)$ (*6.3.1)

(7) $\equiv a^{mk} \cdot a^m$ (Why?)

(8) $\equiv a^{mk+m}$ (*6.3.3)

(9) $\equiv a^{m(k+1)}$ (Why?)

But this is just the theorem for $n = k + 1$. Hence the proof is complete.

Proof of *6.3.5. We leave it to the student to complete the proof, of which only a part is given below:

(10) $(ab)^{k+1} \equiv (ab)^k \cdot (ab)$

(11) $\equiv a^k \cdot b^k \cdot a \cdot b$

(12) $\equiv (a^k \cdot a)(b^k \cdot b)$

(13) $\equiv a^{k+1} \cdot b^{k+1}$

These laws of exponents together with previous identities are sufficient for manipulating expressions involving exponents. Sometimes, in dealing with fractions it is easier to use $ca/cb \equiv a/b$ directly, instead of *6.3.6, which is derived from it in any case. We give numerous examples in the exercise. Here we shall use the laws to generalize formulas for the sums and differences of like powers, which we found in 5.4.

*6.3.9 $a^n - b^n \equiv (a - b)(a^{n-1} + a^{n-2}b$

$$+ a^{n-3}b^2 + \cdots + a^2b^{n-3} + ab^{n-2} + b^{n-1})$$

*6.3.10 $a^{2n+1} + b^{2n+1} \equiv (a + b)(a^{2n} - a^{2n-1}b$

$$+ a^{2n-2}b^2 - \cdots - ab^{2n-1} + b^{2n})$$

Special cases of the first of these identities are *6.1.1, *6.1.5, and *6.1.6. It means that $a^n - b^n$ is always divisible by $(a - b)$ in the sense that $(a^n - b^n)$ always equals $(a - b)$ times another polynomial given by *6.3.9. The only special cases of *6.3.10 that we have encountered are *6.1.4 and *6.1.7. This theorem states that the sum of two like powers of a and b is divisible by $(a + b)$ *if the powers are odd.* No such result exists for sums of even powers. Hence we have given the exponents as $2n + 1$, which assures that they be odd.

*Proof of *6.3.9.* The theorem is obviously true for $n = 1$. Now assume the theorem for $n = k$. Then

(14) $\quad a^{k+1} - b^{k+1} \equiv a^{k+1} - ab^k + ab^k - b^{k+1}$ (Why?)

(15) $\qquad\qquad \equiv a(a^k - b^k) + b^k(a - b)$ (Why?)

(16) $\qquad\qquad \equiv a(a - b)(a^{k-1} + a^{k-2}b + \cdots + ab^{k-2}$

$\qquad\qquad\qquad + b^{k-1}) + b^k(a - b)$ (Why?)

(17) $\qquad\qquad \equiv (a - b)(a^k + a^{k-1}b + \cdots + a^2 b^{k-2} + ab^{k-1}$

$\qquad\qquad\qquad + b^k)$ (Why?)

But this is just the theorem for $n = k + 1$. Hence it holds for all n by the principle of finite induction.

We can use *6.3.9 to give a plausibility proof of *6.1.3. The average rate of change of ax^n is given by

(18) $\quad \dfrac{\Delta y}{\Delta x} = a\left(\dfrac{x_2^{\,n} - x_1^{\,n}}{x_2 - x_1}\right)$

(19) $\qquad = a(x_2^{\,n-1} + x_2^{\,n-2}x_1 + x_2^{\,n-3}x_1^{\,2} + \cdots + x_2^{\,2}x_1^{\,n-3}$

$\qquad\qquad + x_2 x_1^{\,n-2} + x_1^{\,n-1})$

We see that there are n terms in this expression, each one of which approaches $x_1^{\,n-1}$ as x_2 approaches x_1. Hence $\Delta y/\Delta x$ appears to approach $nax_1^{\,n-1}$, which is what we wanted to show.

EXERCISE 6.3

1. Apply the laws of exponents to the following in order to simplify or carry out indicated operations:

(a) $2^2 \cdot 2^4$
(b) $(2^2)^4$
(c) $a^3 \cdot a^3$
(d) $(a^3)^3$
(e) $(2t)^2$
(f) $(2x)(5x^2)$
(g) $2c^4/c$

(h) $(1/2)^5$
(i) $(cd)^3/(cd)^7$
(j) $(1/st)^4(st)^2$
(k) $6^6/6^7$
(l) $(2xy)^3 \cdot (xy)^3$
(m) $(y^c)^2$
(n) $(-3x^2)^2$

(o) $(1/3c)^3(c^2)(12c)$
(p) $(ab/c)^2$
(q) $(x/b)^3(x/b)^4$
(r) $(s^2 a^3/ts)(s^3 t/s)$
(s) $a^b a^c - (a^b)^c$
(t) $a^4 - a^2 - a^3$
(u) $a^{2a} - a^a - a^a$

Answers. (a) 2^6. (c) a^6. (e) $4t^2$. (g) $2c^3$. (i) $1/(cd)^4$. (k) $1/6$. (m) y^{2c}. (o) $4/9$. (q) $(x/b)^7$. (s) $a^{b+c} - a^{bc}$.

2. Perform indicated operations and simplify:

(a) $(2xy)^3$ (d) $(xy^2)^2(x^3)^2$ (g) $(3x^2y^a)^3$
(b) $(stuv)^3$ (e) $(ax)^a(xy)^b$ (h) $(a/b)^{3a}$
(c) $(2a^2b^3)^3$ (f) $(2xy)^c(y^2)^c$ (i) $ab^3cd^4/b^2c^3d^2$

Comment. It is in problems like this that the identity $ca/cb \equiv a/b$ is useful directly. We have only to remember that the powers represent products. Then this identity enables us to cancel *factors* common to numerator and denominator. In fraction *i*, b^2, c, and d^2 are common. Hence the fraction reduces to abd^2/c^2.

(j) $(x^4y)(2x^2y)(-4xy^3)$ (r) $(-6mn)^2(1/n)^3(1/3m)^4$
(k) $2a^3s^4t/4s^2ta^4$ (s) $(xy/a)/(x^2a/y^3)$
(l) $(2a/bc)^3$ (t) $8x^2yxs^3/3xys^2t$
(m) $(x/2)^2(y/2)^3$ (u) $[2(x+y)^2]^3$
(n) $(a/2b)^2/(a/2b)^3$ (v) $c^3(a-b)^4/c(a-b)^2$
(o) $[(xy)^2]^3$ (w) $(a^3x^2-ax)/a^2x$
(p) $a/b^3 + 2/b^5$ (x) $(14xy^3 - 7x^2y^2)/7xy$
(q) $x^{20} + x^{30}$ (y) $(36a^6b^6 - 12a^3b^3 + a^3b^2)/12a^2b^3$

Answers. (a) $8x^3y^3$. (c) $8a^6b^9$. (e) $a^ax^{a+b}y^b$. (g) $27x^6y^{3a}$. (k) $s^2/2a$. (m) $x^2y^3/32$. (o) x^6y^6. (q) $x^{20}(x^{10}+1)$. (s) y^4/a^2x.

3. Write without denominators, indicating the identities used. Check by multiplication.

(a) $\dfrac{x^4 - y^4}{x - y}$ (e) $\dfrac{R^5 + T^5}{R + T}$ (i) $\dfrac{1 - s^{n+1}}{1 - s}$

(b) $\dfrac{8a^3 - s^3}{2a - s}$ (f) $\dfrac{x^6 - y^6}{x^2 - y^2}$ (j) $\dfrac{1 + x^{2n+1}}{1 + x}$

(c) $\dfrac{a^6 - b^6x^3}{a^2 - b^2x}$ (g) $\dfrac{1 - x^3}{1 - x}$ (k) $\dfrac{1 + x^9}{1 + x^3}$

(d) $\dfrac{a^6 + b^6}{a^2 + b^2}$ (h) $\dfrac{1 - x^6}{1 - x}$ (l) $\dfrac{s^{12} + t^{12}}{s^4 + t^4}$

Answer to part a. $x^3 + x^2y + xy^2 + y^3$ by *6.1.5 with x, y for a, b.

4. Expand, indicating identities and substitutions used:

(a) $(a^2 + 2c^3)^2$ (i) $(x^b - y^a)^2$
(b) $(a^3 + b^3)(a^3 - b^3)$ (j) $(5a^3x + 3a^2x^2 - 5x^5)^2$
(c) $a^7(ax - 7)$ (k) $(0.3ab^2 - c/b)^2$
(d) $(a^2 + b^2 + c^2)^2$ (l) $(2x^2 + x + 1)(x^2 - 4x + 3)$
(e) $(x^2 - y^2)^2$ (m) $(s^2q^2 - b^3)^2$
(f) $(x^7 + 2y^5)^2$ (n) $(a^2 + v^2 - c^3)^2$
(g) $a^x(a^x + 1)$ (o) $(5c - w^3)^3$
(h) $(x^a + y^b)^2$

Answers. (a) $a^4 + 4a^2c^3 + 4c^6$. (c) $a^8x - 7a^7$. (e) $x^4 - 2x^2y^2 + y^4$. (g) $a^{2x} + a^x$. (i) $x^{2b} - 2x^by^a + y^{2a}$. (k) $0.09a^2b^4 - 0.6abc + c^2/b^2$.

5. Factor:

(a) $a^6 + 2a^3 + 1$

(b) $a^{50}x^{50} + 10a^{25}x^{25} + 25$

(c) $x^{20} - y^{10}$

(d) $x^{2n} - y^{2n}$

(e) $a^4 + b^4 + c^4 + 2a^2b^2 + 2a^2c^2 + 2b^2c^2$

(f) $x^{16} - 1$

Answers. (a) $(a^3 + 1)^2$. (c) $(x^2 + y)(x^2 - y)(x^8 - x^6y + x^4y^2 - x^2y^3 + y^4)(x^8 + x^6y + x^4y^2 + x^2y^3 + y^4)$.

6. Use *6.3.5 to show that $(abc)^n \equiv a^n b^n c^n$. [*Hint. abc* $\equiv$ (*ab*)*c*.]

7. Write *6.3.10 for $n = 2, 3, 4$. Prove it by finite induction.

8. Find formulas for $(a + b)^3$, $(a + b)^4$, $(a + b)^5$, and $(a + b)^6$ by expanding.

9. From the results of prob. 8 write formulas for the 3rd, 4th, 5th, and 6th powers of $(a - b)$.

10. Prove *6.3.6, 7, 8.

†**11.** Prove by finite induction that *6.3.5 holds for any number of factors, i.e., that $(a_1a_2 \cdots a_n)^m \equiv a_1^m a_2^m \cdots a_n^m$.

†**12.** Show that, if $(a/b)^3 - 3(a/b) - 1 = 0$, then $a^2 - 3b^2 = b^2(b/a)$ and $3a + b = a(a/b)^2$.

13. If $y = x(a + bx)^m$ and $z = a + bx$, find y as a function of z.

14. Prove *5.3.5.

†**15.** Show that the following recurrence relations imply the formulas for the nth terms:

$$(a) \quad y_{n+1} = y_n^2 \quad \longrightarrow \quad y_n = y_1^{2^{n-1}}$$

$$(b) \quad y_{n+1} = y_n^3 \quad \longrightarrow \quad y_n = y_1^{3^{n-1}}$$

16. Graph the region in which the points satisfy $x^3 \leq y \leq x$ or $x \leq y \leq x^3$.

†**17.** If n is a positive integer, show that the last digit in (a) 10^n is 0; (b) 5^n is 5; (c) 2^n is 2, 4, 6, or 8; (d) 3^n is 1, 3, 7, or 9; (e) 4^n is 4 or 6.

†**18.** Show that when n is even $x^n \geq 0$ and $(-x)^n = x^n$.

†**19.** Show that when n is odd $(-x)^n = -x^n$, and $x \geq 0$ $\longleftrightarrow$ $x^n \geq 0$.

†**20.** Show that $|x_1| > |x_2| \longrightarrow |x_1|^n > |x_2|^n$.

6.4 Permutations and combinations

By a **permutation** we mean simply an ordered set, that is, a set of objects arranged in a definite order in a row. Two permutations are called the same if they consist of the same elements arranged in the same order. Thus a, b, c and a, c, b are different permutations because the order is different, whereas a, b, c and a, b, d are different because they do not contain the same elements.

By a **combination** we mean a set, without regard to the order of its elements. Two combinations are called the same if they contain the same elements. Thus, a, b, c and a, c, b are the same combination, whereas a, b, c and a, b, d are different.

It is often of interest in mathematics and in applications (especially in statistics) to be able to find out how many different permutations (ordered sets) or how many different combinations (sets) can be formed by choosing r objects from a set of different objects. The number of different permutations of r objects chosen from n objects is designated by P_r^n. The number of different combinations of r objects chosen from n objects is designated by C_r^n or $\binom{n}{r}$. Sometimes P_r^n is called "the permutations of n things r at a time" and C_r^n is called the "combination of n things r at a time." Suppose we have the three objects a, b, c. How many permutations of two objects can we form? The possibilities are ab, ac, bc, ba, ca, cb. Evidently $P_2^3 = 6$. Also we observe that there are only three different combinations. Hence $C_2^3 = 3$.

In order to write general formulas for P_r^n and C_r^n, we find it convenient to make use of the symbol $n!$, called **n-factorial**. It may be defined by the recurrence relation

*6.4.1 $n! = n \cdot (n-1)!$ and $0! = 1$ (Definition)

From this definition we see immediately that $1! = 1 \cdot 0! = 1 \cdot 1 = 1$, $2! = 2 \cdot 1! = 2$, $3! = 3 \cdot 2! = 6$, etc. By induction (see Ex. 6.2.8), it is easy to show that

*6.4.2 $n! = n(n-1)(n-2) \cdots 3 \cdot 2 \cdot 1$

Evidently, $n!$ is just the product of the first n positive integers, except when $n = 0$, in which case it is by definition equal to 1.

In order to find P_r^n and C_r^n in any particular case, we might write out all the possibilities, but this would soon become tedious. Moreover, it is quite unnecessary if we make use of the following principle, which follows from the definition of multiplication:

*6.4.3   If a first choice can be made in c_1 ways and then a second choice in c_2 ways, then the pair of choices in order can be made in $c_1 c_2$ ways.

Consider again the trivial problem of finding P_2^3. In order to form a permutation we make two choices, first for the letter to occupy the first place, then for the letter to occupy the second.

We can make the first choice in three ways, then the second choice in two ways (since one letter has already been chosen). Hence $P_2^3 = 3 \cdot 2 = 3!$ Consider now the problem of finding the number of permutations of three things chosen from four. Here we must make three choices in order to form a permutation. These can be made in four, three, and two ways. If we assume a principle like *6.4.3 to hold for more than two choices, we can write $P_3^4 = 4 \cdot 3 \cdot 2 = 4!$ Actually *6.4.3 can easily be extended by finite induction to any number of choices.

*6.4.4 If in n successive choices, the first can be made in c_1 ways, the second in c_2 ways, and in general the ith in c_i ways, then the set of n choices in order can be made in $c_1 c_2 c_3 \cdots c_i \cdots c_n$ ways.

We can now work out a formula for P_r^n. In order to form a permutation of r objects from n objects, we must make r successive choices of the elements to fill the r positions. These can be made in $n,\ n-1,\ n-2,\ \cdots$ ways. When $r-1$ positions have been filled, there are $n - (r-1)$ objects left to choose from, and the last choice may be made in this many ways. Hence

*6.4.5 $P_r^n = n(n-1)(n-2) \cdots (n-r+1) = \dfrac{n!}{(n-r)!}$

A special case of this formula is the number of permutations of n things taken n at a time:

*6.4.6 $P_n^n = n!$

Having found the general formula for P_r^n, we can derive a formula for C_r^n by observing another way in which a permutation can be formed. We may form a permutation of r things chosen from n things by first choosing a combination of r things and then forming a permutation of these r things. The first choice can be made in C_r^n ways, the second in $P_r^r = r!$ ways. Hence $P_r^n = C_r^n \cdot r!$ But we already have a formula for P_r^n from *6.4.5. Hence $n!/(n-r)! = C_r^n \cdot r!$ and

*6.4.7 $C_r^n = \dfrac{n!}{r!(n-r)!}$

To illustrate these formulas, consider four objects. In how many ways can we arrange them? According to *6.4.6, $P_4^4 = 4! = 4\cdot3\cdot2\cdot1 = 24$. How many permutations could we form of two objects chosen from the four? According to *6.4.5, $P_2^4 = 4!/2! = 12$. How many different combinations of three objects can we choose from the four objects? From *6.4.7, $C_3^4 = 4!/3!1! = 4\cdot3\cdot2/3\cdot2 = 4$.

Permutations and combinations are elementary topics in a branch of mathematics called *combinatorial analysis*. The typical question of this subject is of the form: "In how many ways can a certain event take place?" The fundamental principle of combinatorial analysis is *6.4.4. The student should rely on it rather than try to make all problems fit the permutation or combination formulas. For example, suppose it is asked how many different ways we can arrange the letters a, a, b, c, d, e. It would not do to say that there are six objects and so the answer is P_6^6, because the formula for permutations is based on the assumption that all the objects involved are different. But it is easy to give an answer by noting that any arrangement may be formed by making two choices. First we choose where to put the a's. That is, we choose two positions out of the six possible positions. This can be done in C_2^6 ways. Having done this, we can put the remaining four letters in the remaining four places in P_4^4 ways. Hence the desired number is $C_2^6 P_4^4$.

EXERCISE 6.4

1. Find P_3^3 by listing all the arrangements of the letters a, b, c.

2. Find P_4^4, P_2^4, and C_3^4 by listing and counting the appropriate permutations and combinations of the numbers 1, 2, 3, 4. Compare with the results in the text.

3. Review or do now Ex. 3.5.14c and Ex. 6.2.8.

4. There are three roads from Omsk to Tomsk and five roads from Tomsk to Stomsk. In how many ways can a person travel from Omsk to Stomsk? How many routes are there from Omsk to Stomsk and return without traveling the same road twice? (Use *6.4.3 and *6.4.4.)

5. Out of seven subjects a person wishes to become expert in three. How many different triples could he choose? (*Answer.* 35.)

6. Verify the second form of the formula for P_r^n by writing out the factorials which appear. Show how both forms reduce to $n!$ when $r = n$.

7. Obviously the number of combinations of n things that can be chosen from n things is just 1. Show that *6.4.7 gives this result for $r = n$.

8. Evaluate:

(a) P_3^5 (d) $C_2^6 P_4^4$ (g) $\dbinom{6}{4}$ (j) $\dbinom{n}{3}$ (m) $\dbinom{n}{n-2}$

(b) C_2^5 (e) C_3^6 (h) $\dbinom{n}{1}$ (k) $\dbinom{n}{r-1}$ (n) $\dbinom{n}{4}$

(c) $\dbinom{5}{3}$ (f) P_6^8 (i) $\dbinom{n}{2}$ (l) $\dbinom{n}{n-1}$

Answers. (a) 60. (c) 10. (e) 20. (g) 15. (i) $n(n-1)/2$.

9. Show that

*6.4.8 $$\dbinom{n}{n-r} \equiv \dbinom{n}{r}$$

*6.4.9 $$\dbinom{n}{r-1} + \dbinom{n}{r} \equiv \dbinom{n+1}{r}$$

10. Four colors are used to color the same number of countries on a map, no color being used twice. How many different ways can this be done? (*Answer.* 24.)

11. In how many points do six straight lines meet if no three are concurrent?

12. In how many ways can $a(bc)$ be written by the commutative and associative laws, i.e., in how many ways can the multiplication of three numbers be performed? (*Answer.* 12.)

13. A battalion commander is going to place one company on the right, one on the left, and one in reserve. How many ways can he do this if he has three companies at his disposal?

14. In a book on archeology it is stated that in order to find which of n similar objects are alike, it is necessary to compare each one with every other and hence make $[(n-1)^2 + (n-1)]/2$ comparisons. Explain this formula, and also decide whether it is really necessary to compare every pair of objects.

15. In how many ways can a line of eight people be arranged?

16. If four couples at a dance plan to trade partners so that each girl will dance with each boy, how many dances will be required?

17. A committee of 5 is to be chosen from a club of 100 members. In how many different ways can it be done?

18. A club of 10 members decides to rotate the three offices among all the members. If officers change every year, how many years will it require to assure that each member has occupied some office? Every office? How many years would it take to assign offices in every possible way? (*Answers.* 4, 10, 720.)

19. In how many ways can two dice come up? (The two dice are different objects and hence the result 3, 2 may come up in two distinct ways.)

20. Show that the number of permutations of r objects chosen from n objects, where the same object may be chosen more than once, is n^r. Explain how prob. 19 is a special case of this. (This is called the permutations of n things taken r at a time **with repetitions.**)

21. In how many different ways can three dice come up? (*Answer*. 216.)

22. In how many different ways may the six letters a, b, c, c, d, c be arranged? (*Answer*. 120.)

23. In how many orders can seven people be seated around a circular table? (We consider as the same any two arrangements that can be obtained from each other by rotating the entire circle.)

24. Prove that the number of different ways n objects can be arranged in a circle is $(n - 1)$!

25. How many three-digit numbers are there that contain only the digits 3, 4, 6, or 7 but no one of them twice?

26. Answer the question of prob. 25 if the numbers can contain any digit more than once.

27. How many different permutations can be formed from the letters of the word "philomathian"?

28. How many symbols can be constructed with dots and dashes where each symbol contains four or less? (*Answer*. 30.)

29. If a signal man can use a flag in each hand and put each one in four positions, how many different signals can he send by one position of his two flags? By two successive positions?

30. Information may be recorded on "punched cards" by punching holes in appropriate places. If there are 80 columns in a card, 12 positions in a column, and not more than one hole may be made in any one column, how many different messages may be recorded in this way? (*Answer*. $13^{80} \doteq 130{,}380{,}$-000,000,000,000,000,000,000,000,000,000,000,000,000,000,000,000,000,-000,000,000,000,000,000,000,000,000.)

31. The following triangular arrangement of the values of $\binom{n}{r}$ is known as **Pascal's triangle:**

$$\binom{0}{0}$$

$$\binom{1}{0} \quad \binom{1}{1}$$

$$\binom{2}{0} \quad \binom{2}{1} \quad \binom{2}{2}$$

$$\binom{3}{0} \quad \binom{3}{1} \quad \binom{3}{2} \quad \binom{3}{3}$$

$$\binom{4}{0} \quad \binom{4}{1} \quad \binom{4}{2} \quad \binom{4}{3} \quad \binom{4}{4}$$

Replace each term in the triangle by its numerical value. Fill in the next two rows, i.e., $\binom{5}{0}$, $\binom{5}{1}$ etc., and $\binom{6}{0}$, $\binom{6}{1}$, etc. Note how the numbers reflect *6.4.8 and *6.4.9. Use those properties to write the lines for $n = 7, 8, 9, 10$. Compare the value of $\binom{10}{5}$ found from the triangle with its value found from *6.4.7.

†**32.** Prove *6.4.5 by finite induction, using *6.4.3 but not *6.4.4.

†33. There are n points in space of which p are in one plane, and there is no other plane that contains more than three of these points. How many different planes are there, each containing at least three of the points?

†34. Show that the number of permutations of n things of which n_1 are alike is $n!/n_1!$. (*Hint.* Compare with the permutations of the same objects, but with the n_1 like objects numbered.)

†35. Extend the foregoing proof by finite induction to show that the number of permutations of n things of which $n_1, n_2, \cdots n_s$ are alike is $n!/n_1!n_2! \cdots n_s!$

†36. Justify *6.4.3 by reference to Ex. 3.5.18–19.

6.5 The binomial theorem

We have already found the following formulas for the first few powers of a binomial:

$$(a + b)^2 \equiv a^2 + 2ab + b^2$$
$$(a + b)^3 \equiv a^3 + 3a^2b + 3ab^2 + b^3$$
$$(a + b)^4 \equiv a^4 + 4a^3b + 6a^2b^2 + 4ab^3 + b^4$$
$$(a + b)^5 \equiv a^5 + 5a^4b + 10a^3b^2 + 10a^2b^3 + 5ab^4 + b^5$$

If the student compares the coefficients in these expansions with the numbers in Pascal's triangle (Ex. 6.4.31) he may reasonably conjecture that there is a connection between combinations and the coefficients in the expansion of $(a + b)^n$ for positive integral n. He might even guess that the coefficient of $a^{n-r}b^r$ in the expansion is $\binom{n}{r}$, that is,

*6.5.1 $$(a + b)^n \equiv a^n + \binom{n}{1} a^{n-1}b + \binom{n}{2} a^{n-2}b^2 + \cdots$$

$$+ \binom{n}{r} a^{n-r}b^r + \cdots + b^n \qquad \text{(Binomial Theorem)}$$

$$\text{where } \binom{n}{r} = \frac{n!}{r!(n - r)!}$$

To prove the theorem for positive integral values of n, we write:

(1) $(a + b)^n \equiv (a + b)(a + b)(a + b)\cdots(a + b)$ (n factors)

In order to get a term in the expansion before collecting similar terms we must multiply together n factors found by choosing either a or b from each factor. (The student is asked to prove this by finite induction in Ex. 6.5.15.) For example, if we choose a from each factor we get a^n, and if we choose a from the first $n - 1$

factors and b from the last, we get $a^{n-1}b$. The terms involving b^r
arise from the choice of b from r factors and a from $n-r$ factors.
But we can choose these r factors in $\binom{n}{r}$ ways. Hence, when
terms are collected, the coefficient of $a^{n-r}b^r$ is $\binom{n}{r}$. This argument holds for any r from 0 to n. Hence the theorem is proved.

A knowledge of the formula for the general term $\binom{n}{r}a^{n-r}b^r$ is
all that is necessary to expand any binomial. However, it is
convenient to know *6.5.1 in a form that gives the first few terms
without the combination notation. This is:

$$(2)\quad (a+b)^n \equiv a^n + na^{n-1}b + \frac{n(n-1)}{2!}a^{n-2}b^2$$

$$+ \frac{n(n-1)(n-2)}{3!}a^{n-3}b^3$$

$$+ \frac{n(n-1)(n-2)(n-3)}{4!}a^{n-4}b^4 + \cdots$$

Each term can be formed from the previous one by reducing by
one the exponent of a, increasing by one the exponent of b, multiplying by one less than the last factor in the numerator of the
coefficient, and dividing by one more than the last factor in the
denominator. When n is not integral, the combinatorial notation
is undefined, but (2) still holds. Binomials of the form $(a-b)^n$
can be handled by noting $a-b = a+(-b)$, and multinomials
can be expanded by grouping, for example, $[a+b+c]^n = [a+(b+c)]^n$.

We may also prove the binomial theorem as follows. It certainly holds when
$n = 1$, since we have in this case $(a+b)^1 \equiv a^1 + b^1$. Suppose that it holds
for $n = k$. Then

$$(3)\qquad (a+b)^{k+1} \equiv (a+b)(a+b)^k \qquad \text{(Why?)}$$

$$(4)\qquad \equiv (a+b)\left[a^k + \binom{n}{1}a^{k-1}b + \cdots + \binom{k}{r-1}a^{k-r+1}b^{r-1}\right.$$

$$\left. + \binom{k}{r}a^{k-r}b^r + \cdots + b^k\right]$$

Now, when we carry through this multiplication, the only terms in the expansion that involve the rth power of b arise from multiplying $\binom{k}{r} a^{k-r}b^r$ by a and $\binom{k}{r-1} a^{k-r+1}b^{r-1}$ by b. (Why?) Hence, in $(a+b)^{k+1}$ the term involving b^r is

(5) $$\left[\binom{k}{r-1} + \binom{k}{r} \right] a^{k-r+1}b^r = \binom{k+1}{r} a^{(k+1)-r}b^r \qquad (*6.4.9)$$

But this is just the term involving b^r in the theorem with $n = k + 1$. This argument holds for $0 < r < k + 1$. For $r = 0$ and $r = k + 1$, it is evident that the second member of (4) gives the terms a^{k+1} and b^{k+1} as desired. Hence we have proved hypotheses I and II of the principle of finite induction, and so proved the theorem for all positive integral n.

Sums involving numerous terms, such as the right member of *6.5.1, are tiresome to write. Moreover, all we really need to remember in many sums is the typical term. Thus in the binomial theorem, the typical term is $\binom{n}{r} a^{n-r}b^r$, and we get all the terms of the expansion by letting r take the values 0, 1, 2, 3, 4, $\cdots$ n. We have met other sums where each term was given by a formula involving the number of the term. (See 4.5, 6.2.) In order to be able to express such sums concisely, we adopt the **summation sign, Σ** ("capital sigma," a Greek letter), defined as follows:

*6.5.2 $$\sum_{x=m}^{x=n} f(x) \equiv f(m) + f(m+1) + f(m+2)$$

$$+ f(m+3) + \cdots + f(n-1) + f(n) \qquad \text{(Definition)}$$

The symbol means the sum of the terms found by substituting successively for x each integer from m to n inclusive. The function $f(x)$ is the typical term. Often the letters i or j are used in place of x and the typical term is written in the form y_i. The numbers m and n are called the **limits of summation.**

Using this notation, we can write the sum of the first n terms of a sequence (compare 4.5):

(6) $$s_n = \sum_{i=1}^{i=n} y_i$$

The binomial theorem may be written:

(7) $$(a + b)^n = \sum_{r=0}^{r=n} \binom{n}{r} a^{n-r} b^r$$

The definition of the mean given in Ex. 4.5.8 becomes:

(8) $$\bar{x} = \frac{1}{n} \sum_{i=1}^{i=n} x_i$$

Summations have the following attractive properties that follow from the commutative, associative, and distributive laws:

*6.5.3 $$\sum_{x=a}^{x=b} cf(x) = c \sum_{x=a}^{x=b} f(x)$$

*6.5.4 $$\sum_{x=a}^{x=b} [f(x) + g(x)] \equiv \sum_{x=a}^{x=b} f(x) + \sum_{x=a}^{x=b} g(x)$$

*6.5.5 $$\sum_{x=a}^{x=c} f(x) + \sum_{x=c+1}^{x=b} f(x) \equiv \sum_{x=a}^{x=b} f(x) \text{ where } a < c < b$$

EXERCISE 6.5

1. Verify that the coefficients in the expansions of the first five powers of $(a + b)$ are the same as those given by the binomial theorem.

2. Find $(a + b)^6$ by multiplying $(a + b)^5$ by $(a + b)$. Verify that its coefficients conform to the numbers in the appropriate row in Pascal's triangle.

3. Repeat the first proof of the binomial theorem for $n = 4$. Find each term in the expansion by the combinatorial argument.

4. Carry through the manipulations of the second proof of the binomial theorem for the special case of $k = 4$.

5. Evaluate:

(a) $\binom{5}{1}$ (e) $\binom{5}{5}$ (i) $\binom{n}{3}$ (m) $\binom{10}{5}$ (q) $\binom{15}{8}$

(b) $\binom{5}{2}$ (f) $\binom{n}{0}$ (j) $\binom{n}{4}$ (n) $\binom{100}{3}$ (r) $\binom{n-2}{3}$

(c) $\binom{5}{3}$ (g) $\binom{n}{1}$ (k) $\binom{n}{5}$ (o) $\binom{7}{r}$ (s) $\binom{10}{x-1}$

(d) $\binom{5}{4}$ (h) $\binom{n}{2}$ (l) $\binom{50}{48}$ (p) $\binom{7}{n-r}$ (t) $\binom{n}{n-2}$

Answers. (a) 5. (c) 10. (e) 1. (g) n. (i) $n(n-1)(n-2)/3!$ (k) $n(n-1)(n-2)(n-3)(n-4)/5!$ (m) 252. (o) $7!/r!(7-r)!$

6. Use the binomial theorem to expand:

(a) $(a + b)^7$

(b) $(a + 2c)^4$

(c) $\left(1 + \dfrac{x}{y}\right)^6$

(d) $(102)^5$

(e) $(a - b)^5$

(f) $(1.01)^8$

(g) $(2 + x)^9$

(h) $(1 - x)^8$

(i) $\left(x + \dfrac{1}{x}\right)^4$

(j) $(a^2 + 1)^4$

(k) $(a^b + c)^3$

(l) $(ab - c^2)^6$

(m) $(1 + x^2)^9$

(n) $(1 + x^3)^{10}$

(o) $(1 - 3x)^5$

(p) $(a + b + c)^3$

(q) $(x + y - 1)^4$

(r) $(a^2 + a + 1)^5$

(s) $(a^2 - b^2)^7$

(t) $\left(1 - \dfrac{b}{a}\right)^5$

(u) $(x^n + 1)^5$

Answers. (e) $a^5 - 5a^4b + 10a^3b^2 - 10a^2b^3 + 5ab^4 - b^5$. (i) $x^4 + 4x^2 + 6 + 4/x^2 + 1/x^4$. (k) $a^{3b} + 3a^{2b}c + 3a^bc^2 + c^3$.

7. Write out the first four terms and the tenth term of:

(a) $(a + b)^{13}$

(b) $(1 - 2x)^{12}$

(c) $(m - s)^{17}$

(d) $(p + q)^{20}$

(e) $(0.99)^{15}$

(f) $(0.5 + 0.5)^{30}$

(g) $(ax + b)^{50}$

(h) $(2x - 3t)^{100}$

(i) $(x^2 + y^2)^{41}$

(j) $(1 - y^3)^{31}$

(k) $(x^3 - 2)^{20}$

(l) $(mn - 3)^b$

(m) $(0.5 - s)^{21}$

(n) $(1.05)^{32}$

(o) $\left(y + \dfrac{1}{y}\right)^{23}$

Tenth terms. (a) $715a^4b^9$. (c) $-24310m^8s^9$. (e) -0.000000000000005005.
(g) $\dbinom{50}{9}(ax)^{41}b^9$. (i) $\dbinom{41}{9}x^{64}y^{18}$. (k) $-\dbinom{20}{9}x^{33}2^9$.

8. Write the formula for the general term (involving the rth power of the second term in the binomial) of each expansion in probs. 6 and 7.

9. The binomial theorem can often be used to approximate large powers of numbers. For example,

(9) $(1.02)^8 = (1 + 0.02)^8$

(10) $= 1 + 8(0.02) + \dfrac{8 \cdot 7}{2}(0.02)^2 + \dfrac{8 \cdot 7 \cdot 6}{2 \cdot 3}(0.02)^3 + \cdots$

(11) $= 1 + 0.16 + 0.0112 + 0.000448 + \cdots$

(12) $\doteq 1.17$

Since the second term in the binomial is small, its powers get smaller, and the corresponding terms may be neglected when they are so small as not to affect the accuracy desired. The foregoing result gives us also $(102)^8 \doteq 1.17 \times 10^{16}$. (Why?) Estimate by using four terms of the binomial expansion:

(a) $(1.02)^7$

(b) $(1.01)^{11}$

(c) $(1.005)^6$

(d) $(1.03)^8$

(e) $(1.001)^{12}$

(f) $(1012)^{10}$

(g) $(10.01)^8$

(h) $(1.004)^{32}$

10. Find the results in prob. 9 exactly and determine the error in your approximation.

11. The following might be terms in the expansion of a binomial to some power. Indicate in each case two possible binomials in whose expansion the term would appear.

(a) $3a^2b$ (c) $12p^{11}q$ (e) $40a^3b^2$ (g) $-792p^5q^7$

(b) $3a^4b^2$ (d) $10a^6b^4$ (f) $495p^8q^4$ (h) $15x^4y^4c^2$

12. Write a formula for the expansion of $(a - b)^n$.

†13. Show that the total number of different subsets of a set of n members is 2^n.

†14. Show that the total number of different combinations of n things taken $1, 2, 3, \cdots n$ at a time is $2^n - 1$.

†15. Show by finite induction that $(a + b)^n = (a + b)(a + b) \cdots (a + b)$ equals the sum of all products formed by choosing either a or b from each of the n factors.

†16. Show that there are 2^n terms in the expansion of $(a + b)^n$ before collecting similar terms.

17. Expand the following:

(a) $\sum_{i=1}^{i=5} i$ (d) $\sum_{j=2}^{j=7} a_j$ (g) $\sum_{x=1}^{x=5} (-1)^x$ (j) $\sum_{i=1}^{i=3} \binom{6}{i}$

(b) $\sum_{i=1}^{i=5} (2i + 1)$ (e) $\sum_{n=1}^{n=4} (a_n + y_n)$ (h) $\sum_{s=1}^{s=6} 2^s$ (k) $\sum_{n=2}^{n=7} n!$

(c) $\sum_{i=1}^{i=n} i^2$ (f) $\sum_{x=1}^{x=4} cf(x)$ (i) $\sum_{t=2}^{t=5} ta^t$ (l) $\sum_{x=0}^{x=4} f(x^2)$

Answers. (a) $1 + 2 + 3 + 4 + 5$. (c) $1 + 4 + 9 + \cdots + n^2$. (g) $-1 + 1 - 1 + 1 - 1$. (i) $2a^2 + 3a^3 + 4a^4 + 5a^5$. (k) $2! + 3! + 4! + 5! + 6! + 7!$

18. Write with the summation notation: (a) The sum of the first n odd numbers. (b) The sum of the first n even numbers. (c) The sum of the first n terms of an arithmetic progression with $y_1 = 1$ and $d = 3$. (d) $(x_1 + x_2)/2$. (e) *3.6.3. (f) *3.6.5. (g) *3.6.7. (h) *3.10.13. (i) *3.11.10. (j) *6.3.9. (k) *6.3.10. (l) The sum of the first n terms of the sequences in Ex. 4.5.22.

†19. Prove by induction that:

(a) $\sum_{i=1}^{i=n} i^4 = \frac{1}{5}n^5 + \frac{1}{2}n^4 + \frac{1}{3}n^3 - \frac{1}{30}n$

(b) $\sum_{i=1}^{i=n} i^{10} = \frac{1}{11}n^{11} + \frac{1}{2}n^{10} + \frac{5}{6}n^9 - n^7 + n^5 - \frac{1}{2}n^3 + \frac{5}{66}n$

20. Illustrate each of *6.5.3, 4, 5 with several examples of your own construction.

†21. Prove *6.5.3, *6.5.4, *6.5.5.

†22. Prove that:

$$\sum_{r=0}^{r=n} \binom{n}{r} = 2^n$$

23. Prove that:

$$\sum_{i=1}^{i=k} n_i(x_i + 1)^2 \equiv \sum_{i=1}^{i=k} n_i x_i{}^2 + 2 \sum_{i=1}^{i=k} n_i x_i + \sum_{i=1}^{i=k} n_i$$

24. Prove that if c is a constant independent of i, then

$$\sum_{i=1}^{i=n} c = nc$$

†25. A function $f(x_1, x_2, x_3 \cdots x_n)$ of n variables is called a mean or **averaging function** if $f(m, m, m, \cdots m) \equiv m$. Show that the following are averaging functions:

(a) $\sqrt{\dfrac{\sum\limits_{i=1}^{i=n} x_i{}^2}{n}}$ (b) $\dfrac{\sum\limits_{i=1}^{i=n} x_i}{n}$ (c) $n \Big/ \sum\limits_{i=1}^{i=n} \dfrac{1}{x_i}$ (d) $\sqrt[n]{x_1 x_2 \cdots x_n}$

†26. Prove *Abel's identity:*

$$\sum_{i=1}^{i=n} a_i u_i = \sum_{i=1}^{i=n-1} s_i(a_i - a_{i+1}) + s_n a_n, \text{ where } s_n = \sum_{i=1}^{i=n} u_i$$

†27. Look up *Abel's inequality* in the *Mathematics Dictionary* (James and James, editors), and prove it.

†28. Show that:

$$\frac{\sum\limits_{i=1}^{i=n} p_i}{\sum\limits_{i=1}^{i=n} q_i} \not\equiv \sum_{i=1}^{i=n} \frac{p_i}{q_i}$$

6.6 Geometric progressions

Geometric progressions are sequences in which the *quotient* of any two successive terms is a constant. Each term arises from the preceding one by multiplication by this constant. For example, in the sequence 1, 2, 4, 8, 16, 32, $\cdots$, each term is obtained from the previous one by multiplying by two. More generally,

*6.6.1 A geometric progression is a sequence of numbers $\{y_n\}$ such that $y_{n+1} = y_n \cdot r$.   (Definition)

The constant quotient is called the **common ratio**. Geometric progressions may be contrasted with arithmetic progressions in which the difference of successive terms is constant and each term is obtained by adding this difference to the previous term.

In a geometric progression with **first term** y_1 and common ratio r, *6.6.1 gives $y_2 = y_1 r$, $y_3 = y_2 r = y_1 r^2$, etc. The first few terms of the progression are:

$$y_1, \; y_1 r, \; y_1 r^2, \; y_1 r^3, \; y_1 r^4, \; \cdots$$

We note that the power of r is one less than the number of the term (compare 4.5), and we conjecture:

*6.6.2   The nth term of a geometric progression with common ratio r is given by $y_n = y_1 r^{n-1}$.

The proof by induction is left to the reader.

The formula *6.6.2 shows that the nth term of a geometric progression is a power function of the common ratio, but the variable in which we are interested is n. In the next chapter we shall call functions of the form $y = a r^x$ exponential functions of x. With this definition we may say that the nth term of a geometric progression is an exponential function of n. Thus we are here studying exponential functions in which the variable takes only positive integral values, just as in 4.5 we studied linear functions in which the variables took only positive integral values.

It is often of interest to calculate the sum of the first n terms of a geometric progression. Letting s_n stand for **the sum of the first n terms** as in 4.5, we have:

(1) $s_n = y_1 + y_1 r + y_1 r^2 + y_1 r^3 + \cdots + y_1 r^{n-2} + y_1 r^{n-1}$

(2) $\quad = y_1 (1 + r + r^2 + r^3 + \cdots + r^{n-2} + r^{n-1})$ (Why?)

(3) $\quad = y_1 \left(\dfrac{1 - r^n}{1 - r} \right)$ (*6.3.9 with $a = 1$ and $b = r$)

Hence

*6.6.3   The sum of the first n terms of a geometric progression is given by

$$s_n = y_1 \left(\frac{1 - r^n}{1 - r} \right)$$

The best way to remember this formula is to understand steps (2) and (3) in the derivation.

It is helpful to interpret the geometric progression and its sum graphically. In Fig. 6.6 we have sketched the progression $y_n = 2(1.5)^{n-1}$. Each y_n is represented by the ordinate of a circled

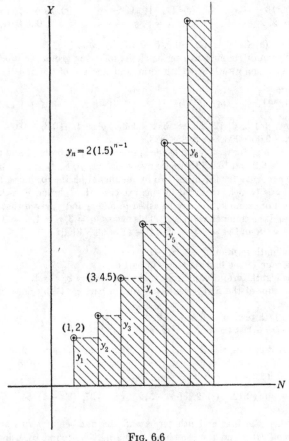

$y_n = 2(1.5)^{n-1}$

FIG. 6.6

point and also by the area of a rectangle of base 1 and height equal to this ordinate. The sum of the first six terms, s_6, is represented by the shaded area. Note that the points do not lie on a straight line. (Compare Fig. 4.5.)

EXERCISE 6.6

1. Find the common ratio and the formula for the nth term in each of the following geometric progressions:

(a) 1, 4, 16, $\cdots$ (c) 1, $-\frac{1}{2}$, $\frac{1}{4}$, $\cdots$ (e) $y_2 = 5$, $y_3 = 45$
(b) 2, -2, 2, $\cdots$ (d) $y_1 = 8$, $y_2 = -4$ (f) 1, 0.2, 0.04, $\cdots$

Answers. (a) 4, 4^{n-1}. (c) -0.5, $(-0.5)^{n-1}$. (e) 9, $5 \cdot 9^{n-2}$.

2. Write out the first five terms of the following geometric progressions. Find the common ratio, the eighth term, and the sum of the first ten terms.

(a) $\{3 \cdot 2^{n-1}\}$ (c) $\{(-1)2^{n-1}\}$ (e) $\{(0.5)^{n-1}\}$ (g) $\{-8(0.1)^{n-1}\}$
(b) $\{2 \cdot 2^{n-1}\}$ (d) $\{(-1)^{n-1}\}$ (f) $\{(-0.25)^{n-1}\}$ (h) $\{(1.5)^{n-1}\}$

Answers. (a) 3, 6, 12, 24, 48; 384; 3069. (g) -8, -0.8, -0.08, -0.008, -0.0008; -0.0000008; -8.888888888.

3. Graph y_n for $n \leq 5$ and shade s_4 in each part of probs. 1 and 2.

4. In *6.6.2 and *6.6.3 there are five variables, y_1, y_n, n, r, s_n. Given any three, we can solve for the other two by manipulating the formulas, although in some cases (which we avoid in this exercise) the solution is awkward to compute. For example, we may be asked to find s_3 and y_3 given that the first term is 3 and the common ratio is 4. This means $y_1 = 3$, $r = 4$, $n = 3$. Hence $y_3 = 3 \cdot 4^2 = 48$ and $s_3 = 3(1 - 4^3)/(1 - 4) = 63$. Find:

(a) The ninth term of 2, -4, 8, -16, $\cdots$.
(b) The sum of the first six terms in part a.
(c) The eighth term of a progression in which $y_1 = 2$, $r = 1$.
(d) The sum of the first seventeen terms in part c. (Why does *6.6.3 not apply here?)
(e) The 11th term of -0.5, 1, -2, 4, $\cdots$.
(f) The $(n + 1)$st term of a, $a^2/2$, $a^3/4$, $\cdots$.

(g) y_1, s_6 if $y_6 = 9$, $r = 3$ (j) n and s_n if $y_1 = -2$,
(h) s_5, r if $y_1 = 3$, $y_5 = 3$ $y_n = -\frac{1}{8}$, $r = -0.5$
(i) y_1, y_3 if $s_3 = 14$, $r = 3$

Answers. (a) 512. (c) 2. (e) -512. (g) $1/27$, $(3^6 - 1)/54$. (i) $14/13$, $126/13$.

5. When the first and nth terms and the number of terms are given, the common ratio and the remaining terms may be found by using *6.6.2. This is called **inserting geometric means** between y_1 and y_n. The first and last terms are called the **extremes,** and the intermediate terms are called the **means.** Insert the indicated number of geometric means between the given numbers:

(a) Three between 1 and 81.   (d) Two between 1 and 64.
(b) Four between 8 and 0.25.   (e) Two between 1 and -1.
(c) Six between -2 and 256.   (f) One between 2 and 2.

Solution for part a. $y_1 = 1$, $y_5 = 81$, and $n = 5$. Then $81 = r^4$, $r = \pm 3$, $y_2 = \pm 3$, $y_3 = 9$, $y_4 = \pm 27$.

6. When just one geometric mean is inserted between two numbers, it is called **the geometric mean.** (a) Show that the geometric mean of two numbers a and b is given by $x_g^2 = ab$. (b) Show that the geometric mean of two numbers is the mean proportional between them.

7. Find the geometric means of:

(a) 4 and 9   (c) 1 and 2   (e) a^3b and ab^3
(b) 4 and 4   (d) 3 and 12   (f) $4x$ and $16x^3$

Answers. (a) 6. (c) $\sqrt{2}$. (e) a^2b^2.

8. Bacteria multiply by dividing (!?) in two. Suppose one bacterium is allowed to divide and redivide unchecked, and assume that division takes place daily. How many bacteria would there be at the end of a month?

9. Suppose that each time a test tube is washed we eliminate 20% of the impurities. How many times must it be washed to reduce the impurities to less than one-third of their original amount? (*Answer.* 5.)

10. When money is deposited at compound interest, interest is earned on the total amount on deposit, including previously earned interest. If A_n represents the amount on deposit at the end of the nth year, the amount at the end of the next year is $A_{n+1} = A_n + A_n i = A_n(1 + i)$, where i is the *rate of interest*. Evidently the amounts form a geometric progression whose first term is the original amount P, called the *principal*, whose common ratio is $(1 + i)$, and whose $(n + 1)$st term is the *amount after n years.* Hence by *6.6.2:

*6.6.4 $A_n = P(1 + i)^n$ (Compound Interest Law)

(a) If $i = 50\%$, find the value of \$100 after 3 years.

(b) If $i = 10\%$, find the principal which deposited now will accumulate to \$100 in 3 years. (In such problems, P is called the *present value* of A_n.)

(Further problems are postponed until the next chapter.)

11. Find a formula for the value of an annuity of n payments of R each at the moment of making the last payment if interest is compounded at the rate i per period. [See Ex. 4.5.13. *Answer.* $S_n = R[(1 + i)^{n+1} - 1]/i$.]

12. If rabbits double in number every month, how many months would it take a pair to become 1,000,000? (Assume no deaths.) Answer the same question if they double every 2 months and every s months.

13. Malthus argued in the early nineteenth century that population would outrun food supplies in the world because food supply increases only in arithmetic progression whereas population increases in geometric progression. Make a graph to illustrate this argument. How do you explain that his predictions have so far not been realized? (*Suggestion.* Examine his assumptions.)

14. In a book on flour milling, it is stated that the female grain weevil lives 4 or 5 months and lays 100 or 200 eggs, so that four generations can produce a million or more weevils. If a million are produced in three generations, what conclusion can you draw about the number of surviving offspring per female?

†15. A rabbit farmer starts with two pairs of rabbits and fills his farm to capacity in x years. Assuming that rabbits double their numbers every 3 months, how long would it have taken him if he had started with twice as many rabbits? With half as many rabbits? With one quarter as many rabbits?

†16. Find a formula for r in terms of y_1, y_n, and s_n only and not involving any roots.

†17. If a, b, $c > 0$ and if $a^{2/b}$, $b^{1/ac}$, and $c^{2/b}$ are in geometric progression, show that: (a) a^{2/b^n}, $b^{1/a^n c^n}$, c^{2/b^n} are also in geometric progression. (b) $ac = b$.

†18. Show that in a geometric progression, the difference between successive terms is always increasing or decreasing in absolute value according as $|r| > 1$ or $|r| < 1$.

19. Graph s_n as a function of n for: (a) $y_1 = 3$, $r = 2$. (b) $y_1 = 3$, $r = 0.5$. (c) $y_1 = 3$, $r = -1$.

20. Find the first six terms in the following sequences:

(a) $\{1 - 1/n\}$
(b) $\{n^3/(n^3 + 1)\}$
(c) $\{n/(n + 1)\}$
(d) $\{n(n + 1)/n^2\}$
(e) $\left\{\dfrac{1 \cdot 3 \cdot 5 \cdots (2n - 1)}{2 \cdot 4 \cdot 6 \cdots 2n}\right\}$

21. For each of the sequences in prob. 20 find: (a) y_3/y_2. (b) y_{10}/y_9. (c) y_{20}/y_{19}. (d) y_{n+1}/y_n.

6.7 Geometric series

As we mentioned in Ex. 3.12.17, any rational number is equal to a repeating decimal. For example,

(1)
$$\tfrac{1}{3} = 0.3333\cdots$$
$$= 3(1/10) + 3(1/100) + 3(1/1000) + 3(1/10{,}000) + \cdots$$

The dots indicate that the sum continues indefinitely or, as is sometimes said, "to an infinite number of terms." But since we cannot actually perform more than a finite number of additions, what can this mean?

Evidently the numbers in (1) form a geometric progression with first term $3/10$ and common ratio $1/10$. We could easily find the sum of any finite number of terms from *6.6.3 or by cutting off the decimal at any point. For example,

$$s_2 = 33/100 = 0.33$$
$$s_7 = 3333333/10{,}000{,}000 = 0.3333333$$

For any n, s_n is a number smaller than $\tfrac{1}{3}$ (why?), yet s_n increases with n and appears to get closer to $\tfrac{1}{3}$. In order to see precisely what happens, we use *6.6.3 to write:

(2) $$s_n = \frac{3}{10}\left[\frac{1 - (1/10)^n}{1 - 1/10}\right]$$

(3) $= 1/3 - 1/3(10)^n$ (Why?)

We see that s_n equals $\frac{1}{3}$ minus a quantity that is divided by 10 each time n increases by one unit. Rewriting (3) as

(4) $1/3 - s_n = 1/3(10^n)$

we see that the difference between $\frac{1}{3}$ and s can be made as small as we like by taking n big enough. Hence we say that the limit of s_n as n approaches infinity is $\frac{1}{3}$. We cannot actually add up an infinite number of terms, but we can find the limit of s_n, and this limit is what we mean by $0.333\cdots$.

More generally, suppose we have the indicated sum of the terms of a geometric progression of an infinite number of terms:

(5) $a + ar + ar^2 + ar^3 + ar^4 + \cdots + ar^{n-1} + \cdots$

Such an expression is called a **geometric series**. The sum of the the first n terms, called the **nth partial sum**, is given by

(6) $$s_n = a\left(\frac{1 - r^n}{1 - r}\right) = \frac{a}{1 - r} - \frac{ar^n}{1 - r}$$

If, as n increases without limit ("approaches infinity"), s_n approaches a finite limit s, then s is called the **sum of the geometric series**.

Under what conditions will (3) approach a limit, and what will that limit be? Suppose first that $|r| < 1$. Then as n increases r^n gets smaller in absolute value (why?), just as it did when $r = 1/10$ in the example opening this section. In fact by taking n big enough we can in this case make r^n as small as we like. Evidently then, s_n approaches $a/(1 - r)$ as n approaches infinity, and we have:

*6.7.1 When $|r| < 1$, the sum of a geometric series with common ratio r and first term a is given by

(7) $$s = \frac{a}{1 - r}$$

What happens if $|r| \geq 1$? If $r = 1$, the series is $a + a + a + a + a + \cdots$. Since $s_n = na$, as n increases its absolute value becomes larger than any given number. Hence s_n cannot approach a limit except for the trival case $a = 0$. If $r = -1$, we have $a - a + a - a + a - a + \cdots$, and $s_n = a$ or 0, according to whether n is odd or even. Obviously s_n oscillates between these two values and cannot approach a limit (except in the trivial case where $a = 0$). If $|r| > 1$, then as n increases r^n gets larger without limit in absolute value; hence s_n evidently cannot approach a definite limit. Thus the phrase "when $|r| < 1$" is important in *6.7.1, for only in this case does the sum, that is, the limit, exist. When $|r| < 1$ the series is said to be **convergent**, and when $|r| \geq 1$ it is said to be **divergent**.

Now we can find the fraction (rational number) represented by any repeating decimal, for it follows from the discussion above that any such decimal is just an infinite geometric series whose common ratio is some power of 0.1. For example, to evaluate $d = 2.127272727\cdots$, we separate the repeating part by writing:

$$d = 2.1 + 0.0272727\cdots$$

The repeating part means:

$$0.027 + 0.00027 + 0.0000027 + \cdots$$

or

$$0.027 + 0.027/100 + 0.027/10{,}000 + \cdots$$

Evidently this is a geometric series with first term 0.027 and common ratio $1/100$, and its sum is

$$(8) \qquad s = \frac{0.027}{1 - (1/100)} = \frac{27}{990} = \frac{3}{110}$$

Finally,

$$(9) \qquad d = \frac{21}{10} + \frac{3}{110} = \frac{117}{55}$$

The result can be checked by long division.

EXERCISE 6.7

1. Consider the geometric series $1 + 1/2 + 1/4 + 1/8 + \cdots$.

(a) Find s_1, s_2, s_3, s_4, s_5, and s.

(b) Plot these values on an axis of s_n.

(c) Letting $y = s_n$ and $x = n$, plot y as a function of x and compare with the line $y = s$.

2. Do the same for $2 + 4/3 + 8/9 + 16/27 + \cdots$.

3. Consider the decimal $0.6666\cdots = 0.\dot{6}$ (compare Ex. 3.12.17). Write the geometric series to which it is equal. In this series, what is the first term? What is the common ratio? Evaluate by *6.7.1.

4. Express each of the following repeating decimals in rational form as the ratio of two integers:

(a) $0.\dot{2}$	(d) $1.\dot{5}\dot{1}$	(g) $0.\dot{0}\dot{9}$	(j) $3.\dot{5}1\dot{1}$	(m) $-2.\dot{8}\dot{5}$
(b) $2.\dot{4}$	(e) $8.23\dot{8}$	(h) $0.\dot{9}$	(k) $0.7\dot{0}7\dot{1}$	(n) $16.7\dot{3}\dot{1}$
(c) $0.2626\cdots$	(f) $0.4\dot{3}\dot{2}$	(i) $397.\dot{3}9\dot{7}$	(l) $8.\dot{9}$	(o) $1.76\dot{2}9\dot{0}$

Answers. (a) 2/9. (c) 26/99. (e) 7415/900. (g) 1/11. (i) 397,000/999.

5. Find the exact error when the following approximations are used:

(a) $1/6 \doteq 0.1667$
(b) $2/3 \doteq 0.667$
(c) $1/9 \doteq 0.11111$
(d) $1/7 \doteq 0.1429$
(e) $25/11 \doteq 2.27$

(f) $0.00\dot{1}\dot{8} \doteq 0.001818$
(g) $-0.00\dot{1}3\dot{1} \doteq -0.00131$
(h) $-35.5\dot{3} \doteq -35$
(i) $1\dot{4}285\dot{7} \doteq 0.142857$

Answers. (a) 1/30,000. (c) $-1/900,000$.

6. A *perpetuity* is an unending sequence of equal payments at equal intervals of time. (a) How much should we be willing to pay for a promise to pay $10 a year in perpetuity, if the compound interest rate is 5% and the first payment is to be made in one year? [*Hint.* The present value of the first payment is $10/(1 + i)$, of the second is $10/(1 + i)^2$, etc. *Answer.* $200.] (b) Write a general formula for the present value of a perpetuity of R per period at compound interest rate of i per period.

7. Find:

(a) $2 + 1/3 + 1/9 + 1/27 + \cdots$
(b) $3 - 0.3 + 0.09 - 0.027 + \cdots$
(c) $0.5 + 0.1 + 0.02 + 0.004 + 0.0008 + \cdots$
(d) $1.5 - 0.6 + 0.24 - 0.96 + \cdots$

Answers. (a) 5/2. (c) 5/8.

†8. Find the values of x for which the following geometric series are convergent:

(a) $1 + x + x^2 + x^3 + \cdots$
(b) $3 + 3x + 3x^2 + 3x^3 + \cdots$
(c) $1 + (x - 1) + (x - 1)^2 + (x - 1)^3 + \cdots$
(d) $1 + x/2 + x^2/4 + x^3/8 + \cdots$
(e) $a + a(x - 3) + a(x - 3)^2 + a(x - 3)^3 + \cdots$
(f) $2 + 2(x + 2) + 2(x + 2)^2 + \cdots$
(g) $a + a(bx + c) + a(bx + c)^2 + \cdots$

Answers. (a) $-1 < x < 1$. (c) $0 < x < 2$. (e) $2 < x < 4$.

6.8 Inverse variation

It frequently happens that two variables are related so that their product is constant. For example, according to Boyle's law, if a certain quantity of gas is kept at a constant temperature, the product of the volume occupied by the gas and the pressure exerted by it is constant. If x and y are the variables, the relationship may be written in the form

(1) $xy = c$ or $y = c/x$

An equation of this form defines y as a function of x, for all values of x except $x = 0$. (Why is $x = 0$ excluded?) It is said that y **varies inversely** as x. The student should note that "y varies inversely as x" means the same as "y varies directly as $1/x$." (Compare 4.1.) It is also said that y is **inversely proportional** to x and c is the **constant of proportionality.** We leave it to the student to show that for two pairs of corresponding values (x_1, y_1) and (x_2, y_2):

(2) $$\frac{y_1}{y_2} = \frac{x_2}{x_1}$$

Consider the special case:

(3) $xy = 1$ or $y = 1/x$

We observe as before that y is undefined when $x = 0$. Evidently y always decreases as x increases. (Why?) When $x > 0$, $y > 0$, and when $x < 0$, $y < 0$. Using the functional notation $y(x) = 1/x$, we may write $y(-x) = -y(x)$; hence the graph is symmetric with respect to the origin. (Compare 6.1 and Ex. 6.1.4.) By plotting points corresponding to a few positive values of x and using symmetry, we find the graph sketched in Fig. 6.8.

What happens to the curve as $|x|$ gets very large? Clearly, when $|x|$ is large, $|y| = 1/|x|$ is small, and, indeed, $|y|$ can be made as small as desired by taking $|x|$ large enough. On the other hand, when x is near zero, $|y| = 1/|x|$ is large and can be made as large as we like by taking $|x|$ small enough. The geometrical intrepretation is that the curve approaches the y-axis when x approaches 0 and the x-axis when $|x|$ "approaches infinity." We call the axes **asymptotes** of the curve. The fact that $1/|x|$ gets large when x gets small is sometimes described by writing $1/0 = \infty$. However, the symbol ∞ does not stand for a number, and this is merely shorthand for "$1/|x|$ may be made larger than any given number by taking $|x|$ small enough." Similarly $1/\infty = 0$ is shorthand for "$1/|x|$ may be made arbitrarily small by taking $|x|$ large enough."

The graph of any function of the form $y = c/x$ is similar to Fig. 6.8.   If c is positive, the curve lies in the first and third quadrants and may be obtained from Fig. 6.8 by multiplying each y by c.

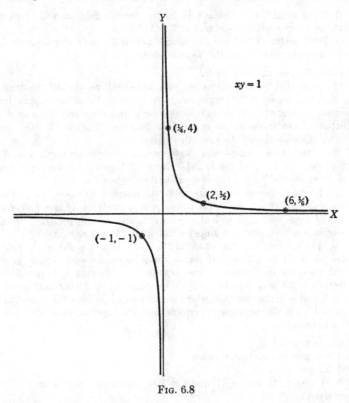

Fɪɢ. 6.8

If c is negative, the curve lies in the second and fourth quadrants and may be obtained by rotating $y = |c|/x$ around the origin through 90°.   Curves of this type are called **hyperbolas**; they will be studied further in Chapter 11.

EXERCISE 6.8

1. Find in decimal form the values of $y = 1/x$, corresponding to $x = 1$, -1, 2, 5, 10, -10, 100, 10^6, 0.5, 0.3, 1/10, $1/10^3$, $1/10^6$.

2. Graph:

(a) $xy = 1$ (c) $xy = 1/3$ (e) $xy = -2$ (g) $xy = 0.1$

(b) $xy = 5$ (d) $xy = 2$ (f) $xy = 10$ (h) $xy = 0.01$

3. Derive (2).

4. If a *lever arm* is balanced on a *fulcrum* by two weights, $WD = wd$, where W and w are the weights and D and d are the distances of these weights from the fulcrum. Hence for fixed w and d, W varies inversely with D.

(*a*) What weight at a distance of 10 feet from a fulcrum will balance a weight of 150 pounds at a distance of 3 feet?

(*b*) Two weights of 100 and 345 pounds are on the ends of a beam 20 feet long. At what point would the beam balance?

Answer for part a. 45 pounds.

5. Show that for constant weight, volume varies inversely with the density.

6. The *intensity of light* varies inversely as the square of the distance from the source. Suppose that an individual using a sunlamp wants twice as much radiation and so moves to half the original distance. What will be the result? To what distance should the sunbather move?

7. If the piston of a gasoline engine on the compression stroke goes 0.88 of the way to the top of the cylinder, by what factor is the pressure in the cylinder multiplied? (The result is called the *compression ratio*.)

8. *Electrical current* varies directly as the *potential difference* and inversely as the *resistance* (*Ohm's law*). The constant of proportionality is one when the units are amperes, volts, and ohms, respectively. What voltage difference corresponds to a current of 5 amperes through a 25-ohm resistance?

9. The *measure of a magnitude* varies inversely as the size of the *unit of measurement*. Thus a foot is 12 inches and a length measured in feet is one-twelfth times the same length measured in inches. We may write this (length in feet) $= (1/12)$(length in inches). More generally, if M is the measure in one unit and m in another unit, where the first unit is k times the second, then $M = m/k$. Use this principle to perform the following changes of units. Indicate k in each case.

(*a*) 5 inches to feet.

(*b*) 8 square feet to square inches.

(*c*) 5 pints to quarts.

(*d*) 11 meters to inches (1 meter $= 100$ centimeters and 1 inch $= 2.54$ centimeters).

(*e*) 18 cubic inches to cubic yards.

(*f*) 1 acre to square feet (1 acre $= 160$ square rods and 1 rod $= 5.5$ yards).

(*g*) 1 foot per second to miles per hour.

(*h*) 8 liters to gallons (1 gallon $= 3.785$ liters).

(*i*) 5 square millimeters to square centimeters (1 centimeter $= 10$ millimeters).

(*j*) 12 ounces to grams (1 gram $= 0.035$ ounce).

(*k*) 60 miles per hour to feet per second.

Answers. (*a*) 5/12 feet. (*c*) 5/2 quarts. (*e*) $18/36^3$ cubic yards. (*g*) 15/22 mph. (*i*) 0.05 square centimeter. (*k*) 88 feet per second.

10. How many miles per hour is a man running when he does the 100-yard dash in 10 seconds flat?

11. Mont Blanc (the tallest peak in Europe) is 4809 meters high. What is its height in feet? in miles? (*Answer:* 15,781 feet.)

12. A man breathes in about 1/2 liter in a normal breath. How many pints is this?

13. A human egg is about 1/175 inch in diameter. (a) If 1 micron = 0.001 millimeter, show that the diameter of the human egg is about 140 microns. (b) If it weighs about 0.0015 milligram (1 gram = 1000 milligrams), find its weight in pounds. (1 pound = 453.6 grams.)

14. An angstrom unit (used to measure wavelengths) equals $1/10^8$ centimeters. How many angstrom units are there in one inch?

15. Suppose that the total manufacturing cost of a plant is given by $C = mu + b$, where u is the output, b the overhead, and m is the marginal cost. (Compare Ex. 4.8.17, 18, 19.) Then the cost per unit output, or *unit cost*, is $C/u = m + b/u$. This unit cost is made up of a constant part (*constant unit cost*) and a variable part (*variable cost*). Evidently the variable unit cost varies inversely as the output in this case. Graph the variable unit cost for Ex. 4.8.17, 18.

16. Write a formula relating intensity (I) and distance (d) from a light source for which $I = 1$ when $d = 1$. (a) Find I when $d = 2.7$. (b) Find the distance required to reduce intensity to $1/2.512$. (*Answer.* 1.585.) (c) Show that for any light source, if the distance is multiplied by 1.585, the intensity is multiplied by $1/2.512$.

17. Acoustical engineers have found that the *reverberation time* of a room (the time for sound to die out to one millionth of its original intensity) varies inversely as the coefficient of absorption of the surface materials. If wood has an absorption coefficient of 0.2 and masonry of 0.03, how much longer would it take for sound to die out in a masonry room as compared with a wood room? (*Answer.* $6\frac{2}{3}$ times as long.)

18. The ratio of ships lost to the total number of ships in a convoy in wartime was found to be given by $r = k/NC$, where N is the number of ships, C the number of escort vessels, and k a constant. (a) Write a formula for the number of ships lost. (b) Compare the losses resulting from sending 100 ships and 10 escort vessels in a single convoy with those resulting from (1) dividing ships and escort vessels into two equal convoys, (2) sending two convoys of 50 ships, *each* with the same number of convoy vessels as the original 100, and (3) sending two convoys of 50 ships, *each* with twice as many escorts as the original 100.

†19. According to a handbook of bricklaying: "To find the capacity of a cistern in barrels, multiply the square of the diameter by the depth (all in feet) and this product by 0.1865." Write a formula expressing this rule. Deduce the number of cubic feet in a barrel.

†20. Prove that the hyperbola $xy = c$ is symmetric with respect to the lines $y = x$ and $y = -x$.

21. Graph:

(a) $y - 2 = 1/x$

(b) $y + 1 = 2/x$

(c) $y = 1/(x - 3)$

(d) $(y - 1)(x + 1) = 3$

(e) $x(y + 3) = 6$

(f) $x - 1 = 2/y$

(g) $x + 1 = 2(y - 1)$

(h) $xy < 1$

(i) $-1/x < y < 1/x$

(j) $(x - 4)(y + 4) = 1$

(k) $xy \geq -1$

(l) $|xy| < 1$

Suggestion. Translate.

22. Graph the unit cost functions of prob. 15.

23. For $f(x) = 1/x$, show that $f(1/x) \equiv 1/f(x)$. Show that this does not hold for $f(x) = 1/(x + 1)$.

24. Calculate $y = 1/x$ for several terms of the following sequences of values of x. $(a)\, x_n = 10^n$. $(b)\, x_n = -10^n$. $(c)\, x_n = 1/10^n$. $(d)\, x_n = -1/10^n$. How large would x have to be to assure that $|y| < 1/10^6$? How small would it have to be so that $|y| > 5000$?

†25. When electrical conductors with resistance r_1, r_2, $\cdots r_n$ are placed in parallel, the resistance of the entire circuit R is given by $1/R = 1/r_1 + 1/r_2 + \cdots + 1/r_n$. Show that R is less than the smallest of the r's.

26. Estimate $1/1.5$, $1/2.7$, $1/3.3$ by linear interpolation from the values of $1/1$, $1/2$, $1/3$, $1/4$. Work to three significant digits. Find the error in each interpolation.

27. The derivative of $y = c/x$ is given by

$$*6.8.1 \qquad D_x\left(\frac{c}{x}\right) = -\frac{c}{x^2}$$

(a) Find the equations of the lines tangent to $xy = 1$ at $x = 0.5$, 1, and 2.

(b) At what points does the tangent line to $xy = c > 0$ have a slope of -1?

(c) Is there a point on the graph of $xy = c$ where the tangent line is horizontal?

(d) What happens to the tangent line as $|x|$ gets large? as x approaches zero?

28. Find the derivatives of the following:

(a) $y = 1 + 1/x$ (c) $y = -5/x$ (e) $y = 2c - d/x$

(b) $y = 3/x - 2$ (d) $y = 2 - 4/x$ (f) $y = x^2 + 1/x$

6.9 Negative exponents

Although an expression like 2^{-3} is meaningless in terms of the definition of exponents given in 6.3, it is not necessary to limit exponents to positive integral values. We now define negative and zero exponents in such a way that the laws previously developed for positive exponents still hold good.

$$*6.9.1 \qquad a^{-n} \equiv \frac{1}{a^n} \qquad n > 0 \qquad \text{(Definition)}$$

$$*6.9.2 \qquad a^0 \equiv 1 \quad \text{when } a \neq 0 \qquad \text{(Definition)}$$

Note that 0^0 is not defined. According to the first definition, 2^{-3} is simply another way of writing $1/2^3$ or $1/8$. Obviously, negative exponents enable us to write fractions with greater simplicity. Thus

$$\frac{1}{x^2 + 3xy - ab} = (x^2 + 3xy - ab)^{-1}$$

The second form takes only one line. Similarly,

$$\frac{1}{a} + \frac{1}{b^2} = a^{-1} + b^{-2}$$

The most important thing about these definitions is that, with them, the laws of exponents stated in 6.3 hold good where m and n are any integers. We consider a few cases, leaving complete proof to the student. First we note that *6.9.1 holds for $n < 0$. (Why?) Next we see that it is no longer necessary to consider the two cases in *6.3.6. For by definition:

(1) $$a^{m-n} \equiv 1/a^{-(m-n)}$$

(2) $$\equiv 1/a^{n-m}$$

Thus, in division, we may always subtract the exponent of the denominator from that of the numerator. The definition of negative exponents guarantees that the two forms of the right member of *6.3.6 have the same meaning.

*Proof of *6.3.3 for All Integral m and n.* When $m > 0$ and $n < 0$, *6.3.3 becomes:

(3) $$a^m \cdot a^{-r} \equiv a^{m+(-r)}, \text{ where } n = -r \text{ and } r > 0$$

To prove this, we write:

(4) $$a^m \cdot a^{-r} \equiv a^m \cdot \frac{1}{a^r} \qquad (\text{*}6.9.1)$$

(5) $$\equiv \frac{a^m}{a^r} \qquad (\text{Why?})$$

(6) $$\equiv a^{m-r} \qquad (\text{*}6.3.6)$$

(7) $$\equiv a^{m+(-r)} \qquad (\text{Why?})$$

Hence *6.3.3 holds in this case. Since $a^m a^n = a^n a^m$, the case $m < 0$, $n > 0$ follows immediately. If $n < 0$, $m < 0$,

(8) $$a^m \cdot a^n \equiv a^{-r} \cdot a^{-s}, \text{ where } r = -m > 0 \text{ and } s = -n > 0$$

(9) $\qquad \equiv \dfrac{1}{a^r} \cdot \dfrac{1}{a^s}$ (Why?)

(10) $\qquad \equiv \dfrac{1}{a^{r+s}}$ (Why?)

(11) $\qquad = a^{-(r+s)}$ (Why?)

(12) $\qquad = a^{m+n}$ (Why?)

We leave to the exercise the case when m or n is zero.

The student should observe how we use the definition of negative exponents to restate the expression in terms of positive exponents only. Then the laws for positive exponents are applied to give the desired result.

The definition $a^0 = 1$ may seem a little mysterious, but the student can easily see that it is necessary. For example, if we are to have $a^m \div a^n = a^{m-n}$ for all m and n, it must hold for $m = n$. But this yields $a^m \div a^m \equiv a^{m-m} \equiv a^0$. On the other hand, we know that $a^m \div a^m \equiv 1$. (Why?) The definition *6.9.2 permits treating the exponent zero just like any other exponent.

In dealing with negative exponents, it is necessary to apply the definitions carefully. For example,

(13) $\qquad\qquad a^{-1} + b^{-1} \not\equiv \dfrac{1}{a+b}$

as can easily be proved by letting $a = b = 1$.

(14) $\qquad\qquad a^{-1} + b^{-1} \equiv \dfrac{1}{a} + \dfrac{1}{b}$

(15) $\qquad\qquad\qquad\quad \equiv \dfrac{a+b}{ab}$

EXERCISE 6.9

1. Write without negative exponents and simplify:

(a) $(a+b)^{-1}$	(e) $(2^0)^{-2}$	(i) $(3x^{-1})^{-2}$	(m) $(x^{-1})^0$
(b) 10^{-2}	(f) $(-1)^{-1}$	(j) $a^{-3} \cdot a^{-4}$	(n) $(-2s)^{-3}$
(c) $(ab)^{-3}$	(g) $(3^{-2})^2$	(k) $(1/2)^{-2}$	(o) $(a^{-1}+b^{-1})^{-3}$
(d) $1/2^{-3}$	(h) $1/3x^{-2}$	(l) $(2/3)^{-3}$	(p) $a^{-1} + x^{-1}$

Answers. (a) $1/(a+b)$. (c) $1/a^3b^3$. (e) 1. (g) $1/81$. (i) $x^2/9$. (k) 4. (m) 1. (o) $[ab/(a+b)]^3$.

2. Write without positive exponents:

(a) a^2 (c) $(a-b)^3$ (e) $1/y^{16}$ (g) $(x/2a)^2$

(b) $a^2 - b^2$ (d) x^3/y (f) $a^2c^3/(x+Y)^2$ (h) $x + 1/y$

Answers. (a) $1/a^{-2}$. (c) $1/(a-b)^{-3}$. (e) y^{-16}. (g) $(2a/x)^{-2}$.

3. Expand and simplify:

(a) $(2a^{-2}b^2)(0.3ab^{-3})$ (j) $(1/a)^{-1}(b/a^{-1})^{-2}$

(b) $(x^{-1}y + xy)y^{-1}$ (k) $1/(x^{-1} - y^{-1})$

(c) $(a^2b^{-3})/(a^{-1}b^2)$ (l) $(a^{-2} - b^{-2})/(a^{-1} - b^{-1})$

(d) $(a^{-1} + b^{-1})^2$ (m) $(a^{-1} + b^{-1})(a^{-1} - b^{-1})$

(e) $(x^{-2} - y^{-2})^{-1}$ (n) $(a^{-2} + b^{-2})(a^{-2} - b^{-2})$

(f) $(x^{-3} + y^{-3})^{-1}$ (o) $2x^{-2}(x^3 + 3x^4 - x^5)$

(g) $(a+b)^{-1}(a^{-1} + b^{-1})$ (p) $(x^{-1} - x^{-2})(2x^{-1} + 3x^{-2} + 1)$

(h) $(a + a^{-1})^3$ (q) $a^4(a^{-1} + a^{-2} + a^{-3} + a^{-4})^0$

(i) $(s^{-1} + t^{-1})^{-1}$ (r) $(a + b)^{-3}(a^{-1} + b^{-1})^3$

Answers. (a) $0.6(ab)^{-1}$. (c) a^3b^{-5}. (e) $x^2y^2(y^2 - x^2)^{-1}$. (g) $(ab)^{-1}$. (i) $st(s+t)^{-1}$. (k) $xy(y-x)^{-1}$. (m) $a^{-2} - b^{-2}$.

4. Factor:

(a) $x^{-2} - y^{-2}$ (e) $x^{-3} - y^{-3}$

(b) $x^{-2} + 4x^{-1} + 4$ (f) $a^{-1}b^{-2} - a^{-2}b^{-1}$

(c) $x^{-2} + 2x^{-1}y^{-1} + y^{-2}$ (g) $16x^{-4} - y^{-4}$

(d) $x^{-3} + y^{-3}$ (h) $x^{-6} - y^{-6}$

Answers. (a) $(x^{-1} + y^{-1})(x^{-1} - y^{-1})$. (c) $(x^{-1} + y^{-1})^2$.

5. Prove that x^{-1} is the reciprocal of x.

6. Newton's law of universal gravitation states that the *force of attraction* of two bodies varies directly as the product of their masses and inversely as the square of their distance apart. To show how the attraction falls off with distance, graph $F = x^{-2}$, which is the force attracting two masses at a distance x, where units are chosen to make the constant of proportionality one.

7. Graph:

(a) $y = 3x^{-2}$ (c) $y = x^{-4}$ (e) $y = -2x^{-3}$

(b) $y = x^{-3}$ (d) $y = 3x^{-5}$ (f) $y - 3 = x^{-2}$

8. Prove that:

$$(a) \quad (a^{-1} + b^{-1})^{-1} \neq a + b$$

$$(b) \quad cx^{-n} \neq (cx)^{-n}$$

$$(c) \quad 1 + x^{-1} \neq \frac{1}{1+x}$$

9. Prove that the laws of exponents hold for zero as an exponent.

10. Prove that *6.9.1 holds for $n \leq 0$.

11. Prove for all integral m and n: (a) *6.3.4. (b) *6.3.5. (c) *6.3.6. (d) *6.3.8. (e) *6.3.7 as a special case.

12. If an amount P is deposited at compound interest rate i for n periods and amounts to A_n at the end of this time, show that $P = A_n(1 + i)^{-n}$. (The coefficient of A_n is called the *discount factor*.)

13. Show by induction that for any number of factors the reciprocal of the product is the product of the reciprocals.

14. Show that $a > b > 0 \longrightarrow a^{-n} < b^{-n}$ for any positive integral n.

15. The formula for the derivative of a power (*6.1.3) holds when n is a negative integer. Use it to find:

(a) $D_x(x^{-4})$

(b) $D_x(x^{-2})$

(c) $D_x(x^{-3})$

(d) $D_x(2/x^2)$

(e) $D_x(1 + 1/x)$

(f) $D_x(1 + x^2 + x^{-1})$

(g) $D_x(3x^{-4})$

(h) $D_x(-1/2x^2)$

(i) $D_x(-1/x^3)$

(j) $D_x(x^{-1} + x^{-2}/2 + x^{-3}/3 + 10)$

Answers. (a) $-x^{-2}$. (c) $-3x^{-4}$. (e) $-x^{-2}$. (g) $-12x^{-5}$.

16. Show that *6.8.1 is a special case of *6.1.3.

17. Find the equation of the line tangent to $x^2y = 4$ at the point $(2, 1)$ and sketch. Same for the point $(-2, 1)$.

18. Show that the slope of the tangent line to $y = 4x^{-3}$ increases as x increases.

19. At what point on the curve $x^2y = 1$ is the tangent line perpendicular to the line from the origin?

20. Find the rate of change of the attraction between two bodies as a function of the distance between them. Compare the rate of change at a distance d with that at a distance $2d$.

6.10 Roots and inverse functions

If $r^3 = b$, we call r a **cube root** of b. We shall show in Chapter 10 that there are three cube roots of any number. It is evident from the discussion in 6.1 that when b is real there is just one real cube root, which we call the **principal cube root** and designate by the symbol $\sqrt[3]{b}$.

*6.10.1 $y = \sqrt[3]{x} \longleftrightarrow y^3 = x$ and y is real   (Definition)

More generally, if $r^n = b$, we call r an **nth root** of b. As we shall show in Chapter 10, any number has n nth roots. Thus $y^n = x$ determines y as an n-valued function of x. The discussion in 6.1 suggests that when x is real and positive and n is even, there are two real nth roots, equal in absolute value and opposite in sign. For example, 2 and -2 are fourth roots of 16, since $2^4 = (-2)^4 = 16$. Also when x is real and n is odd, there is just one real root. Thus -2 is a cube root of -8, since $(-2)^3 = -8$. The positive real root, if there is one, or the negative real root, otherwise, is called the **principal nth root** and is represented by the symbol $\sqrt[n]{x}$.

*6.10.2 $y = \sqrt[n]{x} \longleftrightarrow y^n = x, y$ is real, and

$$y \geq 0 \text{ if } n \text{ is even} \qquad \text{(Definition)}$$

The symbol $\sqrt[n]{}$ is to be read "nth root of." The bar (vinculum) over the expression following the root sign indicates the extent of the operation. Thus $\sqrt[3]{3} + 2$ is quite different from $\sqrt[3]{3 + 2}$. An expression of the form $\sqrt[n]{x}$ is called a **radical.** The expression x is said to be under the radical and is called the **radicand.** The number n is called the **index.** When it is 2, it is usually omitted as in 5.3. To avoid complex numbers for the present, we exclude negative radicands when n is even.

Simple manipulations with radicals can be handled by using the following theorems, which follow from *6.10.2 and the laws of exponents. (See Ex. 6.10.32.)

*6.10.3 $\sqrt[n]{ab} \equiv \sqrt[n]{a}\sqrt[n]{b}$

*6.10.4 $(\sqrt[n]{a})^m \equiv \sqrt[n]{a^m}$

*6.10.5 $\sqrt[n]{\dfrac{a}{b}} \equiv \dfrac{\sqrt[n]{a}}{\sqrt[n]{b}}$

*6.10.6 $\sqrt[kn]{a^{km}} \equiv \sqrt[n]{a^m} \qquad (a \geq 0)$

*6.10.7 $\sqrt[k]{\sqrt[n]{a}} \equiv \sqrt[kn]{a}$

However, operations with radicals are greatly simplified by using fractional exponents, which we define in the next section. Cube and higher roots can be evaluated numerically by using tables as we did for square roots or by methods that we introduce in Ex. 6.11.21, 7.4, and 12.4.

In the nth root we have another example of the concept of inverse, which we introduced in connection with the definitions of subtraction, division, and square root extraction. Since this concept plays an important role in mathematics, we are going to define it precisely and discuss its properties. To introduce the definition, we consider more carefully the relations between addition and subtraction and between multiplication and division.

The function $y = x + a$ determines a correspondence between two sets S and T in which the number y in T corresponding to a number x in S is given by $y = x + a$. (See Fig. 6.10.1.) Now we may look at this correspondence the other way round and ask for the function that gives the number y in S corresponding to a given number x in T. This requires that $x = y + a$ or $y = x - a$. Evidently the functions $F(x) = x + a$ and $G(x) = x - a$ represent the same correspondence between S and T, but with the independent variable x assigned to S in the first case and T in the second.

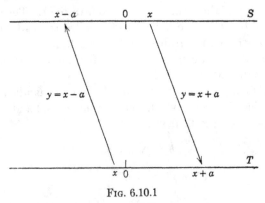

FIG. 6.10.1

Similarly, $y = ax$ determines a correspondence in which the x corresponding to any y is given by $x = y/a$. Interchanging x and y, we find $y = x/a$. The functions $F(x) = ax$ and $G(x) = x/a$ represent the same correspondence, but with independent and dependent variable interchanged. We use this idea to formulate a general definition.

*6.10.8 The **inverse** of a given function is the function that results from interchanging dependent and independent variables.

Thus x/a is the inverse of ax, since interchanging x and y in $y = ax$ yields $x = ay$ or $y = x/a$. Again $y^2 = x$ or $y = \pm\sqrt{x}$ is the inverse of $y = x^2$, and $y^n = x$ determines the inverse of $y = x^n$. We see from these examples that the inverse function of a single-valued function may not be single-valued. Thus $y = \sqrt[n]{x}$ is only part of the inverse of $y = x^n$, since it gives only one of n values. Also we see that the range of the independent variable in the inverse may not be the same as in the given function.   In $y = x^2$, x may have any real value, but in $y^2 = x$ it is restricted to non-negative values.   As another example, consider $y = F(x) =$ (the army serial number of x).   Interchanging x and y, we have $x =$ (the army serial number of y), or $y =$ (the person whose army serial number is x).   Hence $G(x) =$ (the person whose army serial

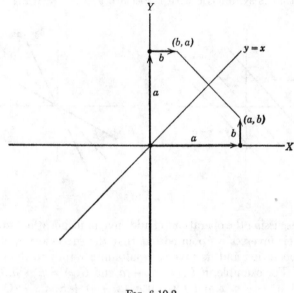

FIG. 6.10.2

number is x) is the inverse function.   Note that in $F(x)$ the values of x are soldiers, whereas in $G(x)$ the values of x are serial numbers.

It is evident from the definition that if $G(x)$ is the inverse of $F(x)$, then $F(x)$ is the inverse of $G(x)$.   Hence we may speak of them as inverses.   However, sometimes we speak of one of them as the **direct function** and the other as the **inverse function**. Since it frequently happens that one of the two functions is easier to deal with than the other, the simpler one is often called the direct function.   Thus $y = x^2$ is single-valued, and its inverse is two-valued, so that we usually think of the former as the direct function.

The relation between a function and its inverse has a simple geometric interpretation.   Since the inverse $y = G(x)$ results from interchanging x and y in $y = F(x)$, a point (a, b) lies on the graph of one function if and only if the point (b, a) lies on the graph of the other. But (a, b) and (b, a) are symmetric with respect to the line $y = x$, that is, this line is the perpendicular bisector of the segment joining them. (See Fig. 6.10.2.) Hence the graph of the inverse may be obtained by reflecting the graph of the direct function in $y = x$, or by rotating it about this line. The two graphs form a figure that is symmetric with respect to $y = x$. (See Fig. 6.10.3.)

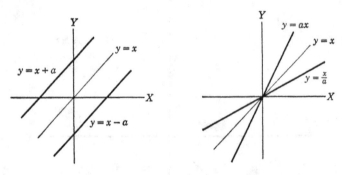

FIG. 6.10.3

In discussing the operations of addition, multiplication, squaring, and their inverses we pointed out that the successive application of an operation and its inverse results in a return to the starting point. For example, if $F(x) = x + a$ and $G(x) = x - a$, $G[F(x)] \equiv (x + a) - a \equiv x$ and $F[G(x)] \equiv (x - a) + a \equiv x$. (Compare *3.7.3.) Again, if $F(x) = ax$ and $G(x) = x/a$, $G[F(x)] \equiv (ax)/a \equiv x$ and $F[G(x)] \equiv a(x/a) \equiv x$. (Compare *3.7.5.) We now state:

*6.10.9 If $F(x)$ and its inverse $G(x)$ are single-valued,

(1) $G[F(x)] \equiv x$ and $F[G(x)] \equiv x$

The identities hold of course only for the values of x that give meaning to the left members. To prove the theorem we refer to Fig. 6.10.4. The direct function $F(x)$ gives the element in T corresponding to the element x in S, and $G(x)$ gives the element in S corresponding to the element x in T. Accordingly $G[F(x)]$ means that we go from x in S to the corresponding element in T and then to the element in S that corresponds to this element in T. Since F and G are single-valued, we return to our starting point. Similarly, $F[G(x)]$ means that we start with an element in T, go to the corresponding element in S, and back to the original element in T. The range of x in the two identities is not necessarily the same. In the first identity, x is an element in S; in the second it is an element in T.

To illustrate the theorem, we return to $F(x)$ = (the army serial number of x), $G(x)$ = (the soldier whose serial number is x). Then $G[F(x)]$ ≡ (the soldier whose serial number is the army serial number of x) ≡ x. Also $F[G(x)]$ ≡ (the serial number of the

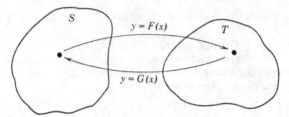

FIG. 6.10.4.   Inverse functions.

soldier whose serial number is x) ≡ x. As a final example, we consider $F(x) = x^2$ for $x \geq 0$. The inverse is $G(x) = \sqrt{x}$. Both are single-valued, and (1) holds by *5.3.2. On the other hand, $F(x) = x^2$ for $x < 0$ has the inverse $G(x) = -\sqrt{x}$. Both are single-valued, and we have $G[F(x)] \equiv -\sqrt{x^2} \equiv -|x| \equiv x$ since $x < 0$. The student can easily verify that $F[G(x)] = x$.

EXERCISE 6.10

1. Justify by reference to *6.10.1 and *6.10.2:

(a) $\sqrt[3]{8} = 2$

(b) $\sqrt[4]{81} = 3$

(c) $\sqrt[4]{81} \neq -3$

(d) $\sqrt[3]{a^6} = a^2$

(e) $\sqrt[4]{a^4} = |a|$

(f) $\sqrt[5]{\frac{1}{32}} = \frac{1}{2}$

(g) $\sqrt[3]{-27} = -3$

(h) $\sqrt[4]{16x^8} = 2x^2$

(i) $\sqrt[n]{1} = 1$

(j) $\sqrt[n]{1} \neq -1$

(k) $\sqrt[19]{-1} = -1$

(l) $\sqrt[6]{a^4} = \sqrt[3]{a^2}$

(m) $\sqrt[3]{a^9} = a^3$

(†n) $\sqrt[4]{a^{12}} \not\equiv a^3$

(o) $\sqrt[2a]{x^a} = \sqrt[2]{x}$

2. In each expression in prob. 1 indicate the radicand and the index.

3. Justify by reference to *6.10.3 through *6.10.7, or by previous identities:

(a) $\sqrt[3]{16} = \sqrt[3]{8}\sqrt[3]{2}$

(b) $\sqrt[4]{16x} = \sqrt[4]{16}\sqrt[4]{x}$

(c) $\sqrt[3]{54} = 3\sqrt[3]{2}$

(d) $\sqrt[5]{32y^2} = 2\sqrt[5]{y^2}$

(e) $(\sqrt[3]{x^2})^2 = \sqrt[3]{x^4}$

(f) $(\sqrt{2})^3 = \sqrt{8}$

$(g)\ \sqrt[3]{8} = 2\sqrt{2}$ $(j)\ \sqrt[3]{\tfrac{1}{2}} = (\tfrac{1}{2})\sqrt[3]{4}$ $(m)\ \sqrt[4]{a^2} = \sqrt{a}$

$(h)\ \sqrt[3]{\tfrac{1}{2}} = \sqrt[3]{\tfrac{4}{8}}$ $(k)\ \sqrt[3]{a^6} = a^2$ $(n)\ \sqrt[6]{a^{10}} \neq \sqrt[3]{a^5}$

$(i)\ \sqrt[3]{\tfrac{4}{8}} = \sqrt[3]{4}/\sqrt[3]{8}$ $(l)\ \sqrt[3]{\sqrt{a}} = \sqrt[6]{a}$ $(o)\ \sqrt[3]{9} = \sqrt{3}$

4. Make a large graph of $y = \sqrt[3]{x}$ for $-10 \le x \le 10$. (Use Table I.)

5. Graph $y = \sqrt[3]{x}$ for $-100 \le x \le 100$, using the same scale on the two axes.

6. Find, from Table I, interpolating where required:

(a) $\sqrt[3]{29}$ (c) $\sqrt[3]{-21.7}$ (e) $\sqrt[3]{85.9}$ (g) $\sqrt[3]{-69.2}$

(b) $\sqrt[3]{8.3}$ (d) $\sqrt[3]{41.3}$ (f) $\sqrt[3]{-50.3}$ (h) $\sqrt[3]{78.6}$

Answers rounded to four digits. (a) 3.072. (c) −2.789. (e) 4.412. (g) −4.106.

7. Show that $\sqrt[3]{1000N} \equiv 10\sqrt[3]{N}$.

8. Problem 7 shows that the digits in a cube root are not changed by moving the decimal point *three* places. Use this idea to find the cube roots of the following from the cube root column in Table I:

(a) 0.083   (e) −0.0183   (i) 58,300   (m) −0.0058

(b) 57,600   (f) 7010   (j) 9110   (n) 0.09356

(c) 92,600   (g) −8562   (k) −3,500,000   (o) 0.0000772

(d) −0.003   (h) 10,314   (l) 65,000,000   (p) −0.03571

Answers. (a) 0.4362. (c) 45.24. (e) −0.2635. (g) −20.46. (i) 38.78. (k) −151.4. (m) −0.1796.

9. If a number is between 100 and 1000, its cube root cannot be found from the cube root column in Table I, and it is necessary to use the column of cubes. Even when a cube root can be obtained by interpolation in the column of cube roots, it may be better to use the column of cubes. (See Ex. 5.3.14.) For example, interpolation in the column of cube roots yields $\sqrt[3]{1.5} \doteq 1.13$, whereas interpolation in the column of cubes, using 1500, yields $\sqrt[3]{1.5} \doteq 1.14$. Since $1.13^3 = 1.442897$ and $1.14^3 = 1.481544$, the second value is better. Find to three significant digits the cube roots of the following:

(a) 285   (d) 2.14   (g) 943.28   (j) −0.0010051

(b) 462   (e) −28.95   (h) 1.704

(c) 1.61   (f) 561.9   (i) −0.01357

Answers. (a) 6.58. (c) 1.17. (e) −3.07. (g) 9.81. (i) −0.239.

10. Find the fourth roots of the following numbers by double use of Table I:

(a) 50   (c) 1030   (e) 289   (g) 110   (i) 2,000,000

(b) 29.3   (d) 5   (f) 0.0658   (h) 0.00913   (j) 1000

Answers. (a) 2.66. (c) 5.66. (e) 4.12. (g) 3.24. (i) 37.6.

11. Show that:

(a) $\sqrt[3]{\sqrt[3]{a}} \neq \sqrt[5]{a}$ (c) $(\sqrt[n]{x})^n \neq x$ (e) $\sqrt[2]{a}\sqrt[3]{a} \neq \sqrt[5]{a}$

(b) $\sqrt[3]{a^3 + b^3} \neq a + b$ (d) $\sqrt[n]{x^n} \neq x$ (f) $\sqrt{a}\sqrt[3]{a} \neq \sqrt[6]{a}$

12. Show that:

$$(a) \quad \sqrt[3]{-x} \equiv -\sqrt[3]{x}$$

$$(b) \quad \sqrt[n]{-x} \equiv -\sqrt[n]{x} \quad \text{when } n \text{ is odd}$$

13. Factor:

(a) $2x^3 - y^3$

(b) $8x^3 + 3y^3$

(c) $1 + 3\sqrt[3]{2} + 3\sqrt[3]{4} + 2$

(d) $5x^3 + 3\sqrt[3]{25}\,x^2 + 3\sqrt[3]{5}\,x + 1$

(e) $x^2 + 2xy + y^2 - 3$

(f) $x^4 - 5$

Answers. (a) $(\sqrt[3]{2}\,x - y)(\sqrt[3]{4}\,x^2 + \sqrt[3]{2}\,xy + y^2)$. (c) $(1 + \sqrt[3]{2}\,)^3$. (e) $(x + y + \sqrt{3}\,)(x + y - \sqrt{3}\,)$.

14. Simplify:

(a) $\sqrt[3]{\sqrt[2]{8}}$

(b) $\sqrt[4]{4}$

(c) $\sqrt{\sqrt{a}}$

(d) $(\sqrt[3]{2}\,)(\sqrt[3]{4}\,)$

(e) $(\sqrt{4}\,)^3$

(f) $(\sqrt[4]{x^2}\,)(\sqrt[4]{3x^2}\,)$

(g) $\sqrt[3]{8x}$

(h) $\sqrt[4]{2z^8}$

(i) $\sqrt[5]{32(x^5 + y^5)}$

(j) $\sqrt[3]{x/27}$

(k) $\sqrt[4]{2x^{-4}}$

(l) $\sqrt[5]{-5x^{-10}}$

(m) $\sqrt[3]{x^4 y^7}$

(n) $\sqrt[4]{x^5 y^{-8}}$

(o) $\sqrt[3]{(x + y)^{-3}}$

(p) $\sqrt[5]{-x^{-1}}$

Answers. (a) $\sqrt{2}$. (c) $\sqrt[4]{a}$. (e) 8. (g) $2\sqrt[3]{x}$. (i) $2\sqrt[5]{x^5 + y^5}$. (k) $(1/x)\sqrt[4]{2}$. (m) $xy^2\sqrt[3]{xy}$. (o) $(x + y)^{-1}$.

15. Rewrite without radicals in any denominator. (*Suggestion.* The best technique is to use *3.7.10 so that the resulting denominator is a perfect power, and then to use *6.10.5 to "take it out.")

(a) $\sqrt[3]{a/b}$

(b) $\sqrt[4]{ab^{-1}}$

(c) $\sqrt[4]{a^2b^{-1}}$

(d) $\sqrt[4]{\frac{1}{2}}$

(e) $\sqrt[3]{(x - y)/(x + y)^2}$

(f) $\sqrt[5]{x^{-3}}$

(g) $\sqrt[3]{a^4c^{-5}}$

(†h) $1/(\sqrt[3]{a} - \sqrt[3]{b})$

(i) $\sqrt[3]{54a^4x^6/(a - b)}$

Answers. (a) $(1/b)\sqrt[3]{ab^2}$. (c) $(1/b)\sqrt[4]{a^2b^3}$. (e) $(x + y)^{-1}\sqrt[3]{x^2 - y^2}$. (g) $ac^{-2}\sqrt[3]{ac}$. (i) $3ax^2(a - b)^{-1}\sqrt[3]{2a(a - b)^2}$.

16. Expand:

(a) $(\sqrt[3]{a} + 1)^3$

(b) $(\sqrt[3]{x} - 1)(\sqrt[3]{x} + 1)$

(c) $(\sqrt[3]{2} + \sqrt[3]{3}\,)(\sqrt[3]{4} - \sqrt[3]{6} + \sqrt[3]{9}\,)$

(d) $(\sqrt[4]{a} + \sqrt[4]{b}\,)^4$

(e) $(\sqrt[4]{a} + \sqrt{a}\,)^4$

(f) $(\sqrt[4]{2} + x)^2$

Answers. (a) $a + 3\sqrt[3]{a^2} + 3\sqrt[3]{a} + 1$. (c) 5.

17. Kepler's third law of planetary motion states that the cubes of the distances d of the planets from the sun are proportional to the squares of the times t of one revolution around the sun. (a) Write the relation between t and d. (b) If planet A is twice as far from the sun as planet B, compare their times of rotation. (*Answer.* $t_A = t_B 2\sqrt{2}$.)

18. If in a sequence $\{y_n\}$, $y_1 = 1$ and $y_{n+1} = (\sqrt[b]{y_n} + 1)^b$, show that $y_n = n^b$.

278 THE POWER FUNCTIONS

19. Graph:

(a) $y = \sqrt[3]{2x}$ (d) $y = \sqrt[3]{-x}$ (g) $y = \sqrt[4]{x-3}$

(b) $y = \sqrt[4]{x}$ (e) $y = \sqrt[3]{x+2}$ (h) $y = \sqrt[3]{1-x}+1$

(c) $y = \sqrt[4]{-x}$ (f) $y-1 = \sqrt[3]{x-2}$

20. Compare the graphs of: (a) $y = \sqrt[4]{x}$ and $y^4 = x$. (b) $y = -\sqrt[4]{x}$ and $y^4 = x$.

21. Graph: (a) $y = x^2$ for $x \geq 0$ and $y = \sqrt{x}$. (b) $y = x^2$ for $x < 0$ and $y = -\sqrt{x}$.

22. In each part below we give a direct function $F(x)$. Find the inverse $G(x)$ by solving $x = F(y)$ for y. Ignore complex solutions.

(a) $y = 5x$ (g) $F(x) = 3x - 5$ (m) $y = x + x^2$

(b) $y = x + 1$ (h) $F(x) = -4x + 2$ (n) $y = x^3 - 1$

(c) $y = 1 - 2x$ (i) $F(x) = -x$ (o) $y = x^4$

(d) $y = 2x/3$ (j) $F(x) = -x^2$ (p) $y = 1 + x - x^2$

(e) $F(x) = 2x + 1$ (k) $F(x) = x^3$

(f) $F(x) = (x - 1)/2$ (l) $F(x) = 4x^2$

Answers. (a) $y = x/5$. (c) $y = (1 - x)/2$. (e) $G(x) = (x - 1)/2$. (g) $G(x) = (x + 5)/3$. (i) $G(x) = -x$. (k) $G(x) = \sqrt[3]{x}$. (m) $G(x) = \dfrac{-1 \pm \sqrt{1 + 4x}}{2}$. (o) $G(x) = \pm \sqrt[4]{x}$.

23. Graph each pair of inverses in prob. 22. Draw the line $y = x$ on each graph and verify the symmetric character of the figure.

†**24.** Show that the points (a, b) and (b, a) are symmetric with respect to the line $y = x$.

25. Graph the following by graphing their inverses and using the relation between a function and its inverse: (a) $y = 2\sqrt[3]{x}$. (b) $y - 1 = \sqrt[4]{x}$. (c) $y = (-3 \pm \sqrt{9 + 4x})/2$. (d) $y = 3 + \sqrt[3]{x + 1}$.

26. Find the inverse of $F(x) = ax^2 + bx + c$.

27. In which parts of prob. 22 are both the direct and inverse function single-valued? Verify *6.10.9 for these cases.

28. Why does $-\sqrt{x^2} \equiv x$ when $x < 0$?

†**29.** Find the inverses of the following functions, show that both are single-valued, and verify *6.10.9.

(a) $y = -x^2$ for $x \geq 0$ (d) $y = 1 - 1/x$ (g) $y = \sqrt{x}$

(b) $y = -x^2$ for $x < 0$ (e) $y = 1/2x$ (h) $y = 1 - \sqrt[3]{x}$

(c) $y = 1 - x^2$ for $x \geq 0$ (f) $y = x/(1 - x)$

Answers. (a) $y = \sqrt{-x}$. (c) $y = \sqrt{1 - x}$. (e) $y = 1/2x$. (g) $y = x^2$ with $x \geq 0$.

†**30.** Are $F(x) = x$ and $F(x) = -x$ the only functions that are their own inverses? Justify your answer.

†31. Find the inverse of each function and verify that *6.10.9 holds in the appropriate cases:

(a) y = the day before x (c) y = the birthday of x

(b) y = the husband of x (d) y = the senator from the state of x

†32. In order to prove that $y = \sqrt[n]{x}$, it is necessary and sufficient to prove $y^n = x$ and *in addition* that y is real and $y \geq 0$ if n is even. Thus to prove *6.10.3 we must show that $(\sqrt[n]{a}\sqrt[n]{b})^n = ab$ and that $\sqrt[n]{a}\sqrt[n]{b}$ is real and also non-negative if n is even. For the first we have:

(2) $$(\sqrt[n]{a}\sqrt[n]{b})^n \equiv (\sqrt[n]{a})^n(\sqrt[n]{b})^n \qquad \text{(Why?)}$$

(3) $$\equiv ab \qquad \text{(Why?)}$$

Certainly $\sqrt[n]{a}\sqrt[n]{b} = y$ is real since we exclude negative radicands when n is even. When n is even, each factor of y is by definition non-negative. Hence their product y is non-negative, and the proof is complete. Use similar methods to prove *6.10.4 through *6.10.7.

†33. Give an example for $a < 0$ in which *6.10.6 does not hold even though both sides are real.

6.11 Rational exponents

In 6.3 and 6.9 we defined $y = b^x$ for all integral values of x. We now extend the definition of exponents to the case where x is any rational number, that is, a number of the form m/n where m and n are integers and $n > 0$. We shall do this in such a way that the laws of exponents still hold. When $n = 1$, m/n reduces to the integer m. Accordingly we define:

*6.11.1 $b^{m/1} \equiv b^m$ (Definition)

For $n \geq 2$ we define:

*6.11.2 $b^{m/n} \equiv \sqrt[n]{b^m}$ (Definition)

Thus $b^{m/n}$ is defined as the principal nth root of b^m. By *6.10.4 this is the same as the mth power of the principal nth root of b, provided both expressions exist. It follows from the definition that $b^{1/n} = \sqrt[n]{b}$ and $b^{m/n} = (b^m)^{1/n} = (b^{1/n})^m$. Also

*6.11.3 $(b^{m/n})^n \equiv b^m$

We exclude negative values of b when n is even. When n is odd, $(-b)^{1/n} = -b^{1/n}$. Hence we can always write exponential expressions in terms of positive bases. Accordingly, for simplicity, we usually assume $b \geq 0$. In this case $b^{m/n} \geq 0$.

Fractional exponents are usually neater than the corresponding radical notation. Thus

(1) $$\sqrt[6]{(x-2y)^5} \equiv (x-2y)^{5/6}$$

(2) $$\sqrt[3]{x^2} + 2\sqrt[3]{x} + 1 \equiv x^{2/3} + 2x^{1/3} + 1$$

The forms on the right are simpler. However, the important thing about fractional exponents is that they obey the laws of exponents given in 6.3. Suppose that we wish to simplify $(\sqrt[3]{x^2})(\sqrt[5]{x^3})$. None of the identities of 6.10 applies directly, but we may write:

(3) $$(\sqrt[3]{x^2})(\sqrt[5]{x^3}) \equiv (x^{2/3})(x^{3/5})$$

(4) $$\equiv x^{2/3+3/5} \quad \text{(Assuming *6.3.3 holds)}$$

(5) $$\equiv x^{19/15}$$

The last expression means $\sqrt[15]{x^{19}}$, but the exponential form is preferable.

To prove that the laws of exponents hold for rationals and positive bases, we refer to *6.11.2, the identities *6.10.3 through *6.10.7, and the laws of exponents in the integral case. Thus to prove *6.3.4, we write:

(6) $$(a^{p/q})^{r/s} \equiv \sqrt[s]{(\sqrt[q]{a^p})^r} \quad \text{(Why?)}$$

(7) $$\equiv \sqrt[s]{\sqrt[q]{a^{pr}}} \quad \text{(*6.10.4 and *6.3.4)}$$

(8) $$\equiv \sqrt[qs]{a^{pr}} \quad \text{(Why?)}$$

(9) $$\equiv a^{pr/qs} \quad \text{(Why?)}$$

The last equation is just *6.3.4 with the integral exponents replaced by rationals. The proofs of the other laws are similar. The theorems hold also for negative bases, provided all expressions are real and unambiguous. The difficulties that may arise are suggested by the fact that $(-3)^{5/3} < 0$, whereas $(-3)^{10/6} > 0$ according to *6.11.2 and is undefined if we interpret it as $[(-3)^{1/6}]^{10}$.

The following examples illustrate the technique of manipulating expressions involving rational exponents:

(10) $\qquad (\sqrt[7]{ax^2})^4 \equiv [(ax^2)^{1/7}]^4 \qquad$ (*6.11.2)

(11) $\qquad\qquad\quad \equiv (ax^2)^{4/7} \qquad$ (*6.3.4)

(12) $\qquad\qquad\quad \equiv a^{4/7}x^{8/7} \qquad$ (*6.3.5)

(13) $\quad (a^{1/2} - 2^{1/3})^2 \equiv (a^{1/2})^2 - 2a^{1/2}2^{1/3} + (2^{1/3})^2 \qquad$ (Why?)

(14) $\qquad\qquad\quad \equiv a - 2^{4/3}a^{1/2} + 2^{2/3} \qquad$ (Why?)

(15) $\quad \sqrt[3]{\sqrt[4]{(\sqrt{x^2})^5}} \equiv \{[(x^{2/2})^5]^{1/4}\}^{1/3} \qquad$ (Why?)

(16) $\qquad\qquad\quad \equiv x^{5/12} \qquad$ (Why?)

(17) $\quad \sqrt[5]{128c^7} \equiv (2^7c^7)^{1/5} \qquad$ (Why?)

(18) $\qquad\qquad\quad \equiv (2c)^{7/5} \qquad$ (Why?)

EXERCISE 6.11

1. Answer the following questions, giving reasons:

(a) $2 = 8^{1/3}$?

(b) $4 = 6^{2/3}$?

(c) $3 = 27^{1/6}$?

(d) $-8 = 16^{3/4}$?

(e) $9 = 27^{2/3}$?

(f) $a^2 = (a^{10})^{1/2}$?

(g) $-4 = 16^{1/2}$?

(h) $1/2 = (1/16)^8$?

(i) $(0.25)^{1/2} = 0.5$?

(j) $3 = (-27)^{1/3}$?

(k) $-a^2 = (-a^6)^{1/3}$?

(l) $3 = (1/9)^{-1/2}$?

(m) $-1/a = (-a^5)^{-1/5}$?

(n) $a + b = [(a + b)^2]^{1/2}$?

(o) $27 = (81)^{3/4}$?

Answers. (a) Yes, since $2^3 = 8$. (c) No, since $3^9 \not\equiv 27$. (e) Yes, since $(\sqrt[3]{27})^2 = 9$. (g) No, since $-4 \not> 0$. (k) Yes, since $(-a^6)^{1/3} = -(a^6)^{1/3} = -a^2$. (m) Yes. (o) Yes.

2. Rewrite each expression in prob. 1 in terms of radicals without fractional exponents, and describe each in words in terms of powers and roots. For example, in part b, $6^{2/3} = \sqrt[3]{6^2}$, which is the cube root of the square of 6.

3. Rewrite in terms of fractional exponents, and simplify where possible:

(a) $\sqrt[3]{a^6/b^3}$

(b) $\sqrt[4]{2^6}$

(c) $\sqrt{\sqrt{\sqrt{2}}}$

(d) $\sqrt[3]{\sqrt{x} - \sqrt{y}}$

(e) $\sqrt[8]{16a^6b^2}$

(f) $\sqrt[3]{x - y^3}$

(g) $\sqrt[4]{xz^3}$

(h) $\sqrt[3]{\sqrt{\sqrt[3]{x^4}}}$

(i) $\sqrt[6]{27x^4}$

(j) $\sqrt[4]{125x/y^2}$

(k) $\sqrt[3]{1/\sqrt{x}}$

(l) $\sqrt{2\sqrt{2\sqrt{2}}}$

Answers. (a) a^2b^{-1}. (c) $2^{1/8}$. (e) $(4a^3b)^{1/4}$. (g) $x^{1/4}z^{3/4}$. (i) $3^{1/2}x^{2/3}$. (k) $x^{-1/6}$.

4. Rewrite without using fractional exponents:

(a) $(x/y)^{-\frac{1}{2}}$ (c) $x^{\frac{1}{3}}(ab)^{-\frac{2}{3}}$ (e) $b^{-\frac{3}{4}}$

(b) $2x^{\frac{1}{2}}b^{\frac{3}{2}}$ (d) $\left(\dfrac{x-y}{x+y}\right)^{-\frac{1}{3}}$ (f) $(a^{\frac{1}{2}}+b)^{c/2}$

Answers. (a) $\sqrt{(y/x)}$.   (c) $\sqrt[3]{x/a^2b^2}$.   (e) $1/\sqrt[4]{b^3}$.

5. Graph:

(a) $y = x^{\frac{2}{3}}$ (c) $y = x^{\frac{3}{4}}$ (e) $y = x^{-\frac{1}{2}}$

(b) $y = x^{-\frac{2}{3}}$ (d) $y = x^{\frac{3}{2}}$ (f) $y = x^{-\frac{1}{3}}$

6. Carry out indicated operations and simplify:

(a) $(1/2)^0$

(b) $x^{\frac{1}{3}}x^{\frac{2}{3}}$

(c) $x^{\frac{1}{2}}x^{-\frac{1}{2}}$

(d) $a^{\frac{1}{2}}b^{\frac{1}{2}}$

(e) $a^{c/d}a^{d/c}$

(f) $(b^x)^{x/2}$

(g) $(c^x)^{1/x}$

(h) $[(xy)^{\frac{2}{3}}]^{\frac{1}{2}}$

(i) $[\sqrt[3]{(x+y)^2}]^6$

(j) $x^2/x^{\frac{1}{3}}$

(k) $x^{-\frac{1}{3}}(x^2+x)$

(l) $\sqrt{a+b}\,(a+b)^{-\frac{1}{3}}$

(m) $(a^{\frac{1}{3}}b^{-2})(ca^{-3}b^{\frac{1}{2}})$

(n) $(a^2n^{\frac{1}{4}})(a^{\frac{1}{2}}b)$

(o) $(a/b^2)^{\frac{1}{2}}$

(p) $\sqrt{(a+b)^{\frac{2}{3}}}$

(q) $(5a^{\frac{1}{2}})^{-1}(ab)^{\frac{2}{3}}$

(r) $(a+b)^{\frac{1}{2}}(a-b)^{\frac{1}{2}}$

(s) $(32x^4y^{10}z^{-15})^{\frac{1}{5}}$

(t) $(x^{10}x^{\frac{2}{3}}t^{-\frac{1}{3}})^{-\frac{9}{5}}$

(u) $[(1+t)^{1/t}]^n$

Answers. (a) 1.   (c) 1.   (e) $a^{(c^2+d^2)/cd}$.   (g) c.   (i) $(x+y)^4$.   (k) $x^{\frac{5}{3}}+x^{\frac{2}{3}}$.
(m) $a^{-\frac{5}{3}}b^{-\frac{3}{2}}c$.   (o) $a^{\frac{1}{2}}b^{-1}$.   (q) $(a^{\frac{1}{6}}b^{\frac{2}{3}})/5$.

7. Simplify, showing your work and giving reasons:

(a) $\sqrt[3]{6a^3}$

(b) $\sqrt[4]{10a^4}$

(c) $\sqrt[3]{54}$

(d) $\sqrt[3]{16}$

(e) $\sqrt[5]{64}$

(f) $\sqrt[3]{-5c^3}$

(g) $\sqrt[3]{-32}$

(h) $\sqrt[5]{a^5c^{10}}$

(i) $\sqrt[4]{a^2c^4}$

(j) $\sqrt[4]{81x^2}$

(k) $\sqrt{(\frac{2}{3})}\sqrt[3]{\frac{3}{16}}$

(l) $x^{\frac{2}{3}}x^{\frac{3}{4}}$

(m) $a^{\frac{1}{2}}(ab)^{\frac{1}{5}}$

(n) $x^{-\frac{3}{5}}/x^{-\frac{2}{3}}$

(o) $\sqrt[3]{(\sqrt[5]{a^2})^{\frac{1}{2}}}$

(p) $(\sqrt[3]{x}+3)(\sqrt[3]{x}-3)$

(q) $[\sqrt[9]{3x/(1-x)}\,]^{-3}$

(r) $\sqrt[3]{4x}\,\sqrt[2]{2x}$

(s) $(x\sqrt{2a}\,)(y\sqrt{6a}\,)$

(t) $\sqrt[3]{(2x-2y)/x^3}\,[(x-y)/x]^{-\frac{1}{3}}$

(u) $\sqrt{x+2\sqrt{xy}+y}$

(v) $\sqrt{3\sqrt{3\sqrt{3}}}$

(w) $32^{-\frac{1}{5}}+27^{\frac{2}{3}}+61^0$

Answers. (a) $\sqrt[3]{6a^3} = 6^{\frac{1}{3}}a^{\frac{3}{3}} = a\sqrt[3]{6}$.   (c) $3\sqrt[3]{2}$.   (e) $2\sqrt[5]{2}$.   (g) $-2\sqrt[3]{4}$.
(i) $|c|\sqrt{a}$.   (k) $2^{-\frac{1}{3}}3^{-\frac{1}{3}}$.   (m) $a^{\frac{13}{10}}b^{\frac{1}{5}}$.   (o) $a^{\frac{1}{15}}$.   (q) $\sqrt[3]{(1-x)/3x}$.   (s)
$2axy\sqrt{3}$.   (u) $\sqrt{x}+\sqrt{y}$.

8. Expand or simplify:

(a) $x^{\frac{1}{2}}(x^{\frac{1}{2}} + x^{-\frac{1}{2}})$

(b) $xy^{\frac{1}{4}}(x^{\frac{1}{4}} - y^{\frac{1}{4}})$

(c) $(a^2b^{\frac{1}{3}}c^{-\frac{1}{2}})(bac^{\frac{1}{4}})$

(d) $(Rs^2u^{\frac{2}{3}})^{\frac{1}{6}}$

(e) $(x^{\frac{1}{2}} - y + z^{\frac{1}{2}})(x^{\frac{1}{2}} + y - z^{\frac{1}{2}})$

(f) $(a^{\frac{1}{2}} + b^{\frac{1}{2}})^2$

(g) $(x^{\frac{1}{3}} + 1)^3$

(h) $(x^{\frac{1}{2}} + 1)(x^{\frac{1}{2}} - 1)$

(i) $(x^{\frac{1}{4}} + y^{\frac{1}{4}})(x^{\frac{1}{4}} - y^{\frac{1}{4}})$

(j) $x(x - 1)^{-\frac{1}{2}}(x - 1)^{\frac{1}{2}}$

(k) $(ab^2c^{\frac{2}{3}}e^{-1}f^{-\frac{1}{2}}b^{\frac{1}{3}})^{-6}$

(l) $(x + y)^{\frac{1}{8}}(x - y)^{\frac{1}{6}}$

(m) $(x - 1)^{\frac{1}{2}} - x^2(x - 1)^{-\frac{3}{2}}$

(n) $x(x - 1)^{-\frac{1}{2}} - (x - 1)^{\frac{1}{2}}$

(o) $(\sqrt{a} + \sqrt{b} + \sqrt{c})$
$\times (\sqrt{a} + \sqrt{b} - \sqrt{c})$

(p) $(x^{\frac{1}{2}} + y^{\frac{1}{2}} + 1)^2$

Answers. (a) $x + 1$. (c) $a^3b^{\frac{4}{3}}c^{-\frac{1}{4}}$. (e) $x - y^2 + 2yz^{\frac{1}{2}} - z$. (g) $x + 3x^{\frac{2}{3}}$
$+ 3x^{\frac{1}{3}} + 1$. (i) $\sqrt{x} - \sqrt{y}$. (m) $(1 - 2x)(x - 1)^{-\frac{3}{2}}$.

9. Factor:

(a) $x + 2x^{\frac{1}{2}}$

(b) $4x - 4x^{\frac{1}{2}} + 1$

(c) $x^{\frac{2}{3}} - a^{\frac{2}{3}}$

(d) $x^{\frac{3}{2}} - y^{\frac{3}{2}}$

(e) $z^{\frac{2}{3}} + 6z^{\frac{1}{3}} + 9$

(f) $a^2x - a^{\frac{1}{2}}xy + bx^{\frac{2}{3}}a^{\frac{3}{4}}$

(g) $x^{-1} + 2x^{-\frac{1}{2}}y^{-\frac{1}{2}} + y^{-1}$

(h) $x^3 + x^{\frac{1}{2}}$

(i) $x^{\frac{2}{3}} - 1$

(j) $x^{3i} + 4x^{\frac{1}{3}i} + 4$

Answers. (a) $x^{\frac{1}{2}}(x^{\frac{1}{2}} + 2)$. (c) $(x^{\frac{1}{3}} + a^{\frac{1}{3}})(x^{\frac{1}{3}} - a^{\frac{1}{3}})$. (e) $(z^{\frac{1}{3}} + 3)^2$. (g)
$(x^{-\frac{1}{2}} + y^{-\frac{1}{2}})^2$. (i) $(x^{\frac{1}{3}} + 1)(x^{\frac{1}{3}} - 1)$.

10. Solve the following equations:

(a) $x - 2x^{\frac{1}{2}} + 1 = 0$

(b) $x^{\frac{2}{3}} + 3x^{\frac{1}{3}} - 4 = 0$

(c) $x^{\frac{2}{5}} - 5x^{\frac{1}{5}} - 14 = 0$

(d) $x^{\frac{2}{3}} + 6x^{\frac{1}{3}} - 1 = 0$

Suggestions. In part a, let $t = x^{\frac{1}{2}}$ and solve for t. Then $t = 1$ and $x = t^2$
$= 1$. Answer to part c is $x = 7^5$ or $x = -32$.

11. Show that:

(a) $(a^{\frac{1}{2}} + b^{\frac{1}{3}})^3 \not\equiv a + b$

(b) $a^{\frac{1}{2}} \not\equiv a/2$

(c) $a^m b^n \not\equiv (ab)^{m+n}$

(d) $(x/y)^{\frac{1}{3}} \equiv (1/y)(xy^2)^{\frac{1}{3}}$

(e) $\sqrt[3]{-a/b} \equiv (-1/b)\sqrt[3]{ab^2}$

(f) $\sqrt[3]{(a - b)c^3} \equiv c(a - b)^{\frac{1}{3}}$

(g) $ab^m \not\equiv (ab)^m$

(h) $(x^{\frac{2}{3}} - y^{\frac{2}{3}})(x^{\frac{1}{3}} - y^{\frac{1}{3}})^{-1} \equiv x^{\frac{1}{3}} + y^{\frac{1}{3}}$

(i) $[(x + y)/(x - y)]^{\frac{2}{3}} \equiv (x - y)^{-1}\sqrt[3]{(x^2 - y^2)(x + y)}$

(j) $x^{-\frac{3}{8}} \equiv (1/x)\sqrt[8]{x^3}$

12. Write each of the following sets of radicals as radicals with the same index:

(a) $\sqrt[3]{6}, \sqrt{2}, \sqrt[4]{9}$

(b) $\sqrt[6]{8}, \sqrt[3]{xy}, \sqrt{y}$

(c) $\sqrt[4]{8x^5}, \sqrt[3]{ac}, \sqrt{2}, \sqrt[6]{36x}$

(d) $\sqrt{2a}, (x + y)^{\frac{1}{3}}, a^{\frac{2}{3}}, \sqrt[5]{x - y}$

Suggestion. Write with fractional exponents and rewrite so that exponents have a common denominator. Answer to part *a* is $\sqrt[12]{6^4}$, $\sqrt[12]{2^6}$, $\sqrt[12]{9^3}$.

13. If $T = (1.135)p^{1/2}P^{-5/6}E^{1/3}$, find formulas for p, P, and E. [*Partial answer.* $P = (1.135)^{1.2}p^{0.6}E^{0.4}T^{-1.2}$.]

14. If $R = F\theta^a e^{-b\theta}$, $\theta = r - s$, and $r = s + (a/b)$, show that $R = Fa^a e^{-a}b^{-a}$.

15. Show that $b^{km/kn} \not\equiv b^{m/n}$ for negative values of b.

†**16.** Give examples for negative b to show that $(b^n)^{1/m}$ may be real although $b^{1/m}$ is not.

17. Prove *6.3.3 and *6.3.5 for fractional exponents. Why is it not necessary to prove *6.3.6, 7, 8 directly for fractional exponents? (*Answer.* They follow from the previous laws.)

18. Show from the definition *6.11.2 and the definition of negative integral exponents that $b^{-m/n} = 1/b^{m/n}$.

†**19.** Prove that:

$$a > b > 0 \longrightarrow a^r > b^r \text{ for any positive rational } r$$

†**20.** State a similar theorem for negative r. Prove it.

21. The binomial theorem applies when the power is negative, fractional, or indeed any number. In this case 6.5(2) must be used. The expansion does not terminate when the power is not a positive integer (why?), and the partial sums give good approximations only under certain conditions, which are fully considered in calculus. Here we simply state that, if the binomial theorem is applied to expand $(1 + x)^n$ where $|x| < 1$, the result gives approximations that become as accurate as desired by taking enough terms. Write the first four terms of:

(a) $(1 + x)^n$	(d) $(1 + x)^{2/3}$	(g) $(1 - x)^{-1}$	(j) $(1 + 0.5)^{1/2}$
(b) $(1 + x)^{p/q}$	(e) $(1 - x)^n$	(h) $(1 + x)^{-1/2}$	(k) $(1 - 0.1)^{-2}$
(c) $(1 + x)^{1/2}$	(f) $(1 + x)^{-1}$	(i) $(1 + x)^{-2}$	(l) $(1 - 0.01)^{1/3}$

Answers. (a) $1 + nx + \dfrac{n(n-1)}{2}x^2 + \dfrac{n(n-1)(n-2)}{2 \cdot 3}x^3 + \cdots$. (c) $1 + \frac{1}{2}x - \frac{1}{8}x^2 + \frac{1}{16}x^3 \cdots$. (i) $1 - 2x + 3x^2 - 4x^3 \cdots$.

22. The limitation of the binomial theorem to $(1 + x)^n$ with $|x| < 1$ is not serious because $(a + b)^r \equiv a^r\left(1 + \dfrac{b}{a}\right)^r$. Thus to find $\sqrt{2}$, we write:

(19)
$$\sqrt{2} = (4 - 2)^{0.5} = 2(1 - 0.5)^{0.5}$$

(20)
$$= 2\left[1 + (0.5)(-0.5) + \frac{(0.5)(-0.5)(-0.5)^2}{2}\right.$$
$$\left. + \frac{(0.5)(-0.5)(-1.5)(-0.5)^3}{3!} \cdots\right]$$

(21)
$$\doteq 2(0.71) = 1.42$$

Estimate the following by the binomial theorem, using four terms and rounding off to three digits:

(a) $\sqrt{3}$	(d) $(0.99)^{0.5}$	(g) $(0.7)^{1.5}$	(j) $(53)^{0.5}$	(m) $(0.99)^{0.1}$
(b) $\sqrt{5}$	(e) $(1.04)^{\frac{2}{3}}$	(h) $(0.99)^{-1}$	(k) $25^{0.2}$	(n) $(1.01)^{0.1}$
(c) $\sqrt[3]{3}$	(f) $2^{0.3}$	(i) $(1.1)^{0.4}$	(l) $(1.08)^{\frac{1}{6}}$	(o) $(1.01)^{0.01}$

Answers. (a) 1.73. (e) 1.03. (i) 1.04.

†23. Expand $1/(1-x)$ by long division and by the binomial theorem and compare.

†24. Under what conditions is the approximation $(1+x)^n \doteq 1 + nx$ a close one? Why?

25. Find the inverses of the following functions, where the independent variable is limited to non-negative values. Show that all functions are single-valued. Graph each function and its inverse and verify *6.10.9.

(a) $y = x^{0.5}$	(c) $y = x^{-0.5}$	(e) $F(x) = 1 + x^{0.4}$
(b) $y = x^{1.5}$	(d) $y = x^4$	(f) $F(x) = (1+x)^{0.4}$

7

The Exponential and Logarithmic Functions

Exponential functions are those of the form $y = b^x$, where b is a constant. They are similar to the power functions except that the exponent rather than the base is the variable. The inverse of the exponential function is called the logarithm. In this chapter we develop the properties of these functions, and indicate the practical importance of logarithms in numerical computations.

7.1 The exponential function

We are now able to attach a definite meaning to an expression of the form b^x, where x is any rational number. For instance, $2^{1.321}$ means $2^{1321/1000}$, which is the thousandth root of the 1321st power of 2. The numerical value would be tedious to find with methods so far considered, but it is clear that it has a definite value. In order to avoid complications with even roots of negative numbers (what complications?), we limit b to real positive values. Then for every rational value of x (positive or negative), b^x has a definite numerical value.

But what about irrational values of x? What would we mean by $3^{\sqrt{2}}$ or 3^{π}? Since neither $\sqrt{2}$ nor π are exactly equal to the ratio of two integers, we cannot interpret these exponents directly in terms of powers and roots. However, we can approximate $\sqrt{2}$ as closely as desired by rational numbers. If S is such an approximation, 3^S has a definite decimal value. By taking an approximation of $\sqrt{2}$ that is too large we get an approximation of $3^{\sqrt{2}}$ that is too large, and by taking an approximation of $\sqrt{2}$ that is too small we get an approximation of $3^{\sqrt{2}}$ that is too small. Thus two approximations that "bracket" $\sqrt{2}$ give two approximations that "bracket" $3^{\sqrt{2}}$. By taking the bracketing approximations of

$\sqrt{2}$ close enough together we can get the bracketing approximations of $3^{\sqrt{2}}$ as close together as we like. It seems reasonable then that $3^{\sqrt{2}}$ represents a definite number that can be approximated as closely as we like. Similar ideas can be used to define b^x, where x is any irrational number, and it can be proved that the

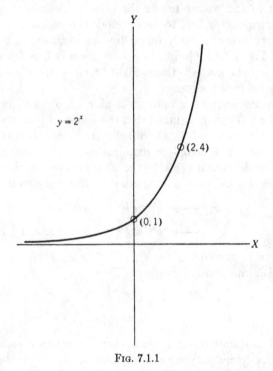

FIG. 7.1.1

laws of exponents hold when the exponents are irrational. The precise statement and justification of these ideas are not attempted here. With this understanding, we may think of b^x as having a definite numerical value for every real value of x, rational or irrational, and we may operate with real exponents according to the laws that were stated first in 6.3.

We now examine and graph the function defined by $y = b^x$. First we note that when $x = 0$, $y = 1$. Accordingly we plot the point $(0, 1)$ on the graph (Fig. 7.1.1). For convenience we take $b = 2$. Then $(1, 2)$, $(2, 4)$, $(3, 8)$, $(4, 16)$ are on the graph. It seems clear that y increases as x increases. What happens when

x is negative? If $x = -1$, $y = 2^{-1} = 1/2$. Similarly $(-2, 1/4)$, $(-3, 1/8)$, $(-4, 1/16)$ lie on the graph. It seems that, as x gets less, y gets smaller. By taking x large enough in absolute value and negative we can make y as near zero as we like (why?), so that we call the x-axis an asymptote of the curve. However, since b is positive and we are taking the principal values of any roots, b^x is never negative, and we can exclude the region $y < 0$. If we connect the points already found by a smooth curve we get the graph of Fig. 7.1.1. Graphs of $y = b^x$ for any $b > 1$ are similar. When $b = 1$ the graph is the straight line $y = 1$. The case $b < 1$ is considered in the exercise.

In constructing the graph we evaluated y only for integral values of x. We than assumed that the values of y at intermediate rational and irrational values of x lie on a smooth curve joining these points. The complete justification of this assumption requires more advanced ideas, but it is easy to see, in the case considered, that y increases as x increases. More generally,

*7.1.1
$$x_2 > x_1 \longrightarrow b^{x_2} > b^{x_1} \qquad \text{if } b > 1$$
$$\longrightarrow b^{x_2} < b^{x_1} \qquad \text{if } 0 < b < 1$$

To show this, consider $b^{x_2} - b^{x_1}$. If $x_2 > x_1$, then $x_2 = x_1 + h$, where h is some positive number. Then

(1) $$b^{x_2} - b^{x_1} = b^{x_1 + h} - b^{x_1}$$

(2) $$= b^{x_1}(b^h - 1)$$

Now b^{x_1} is certainly positive, since b is positive and we always take the positive (principal) value of any power. If b is greater than 1, then any positive power of b is greater than 1 (Ex. 6.11.19). Hence $b^h - 1$ is positive. It follows that $b^{x_2} - b^{x_1} > 0$ and hence $b^{x_2} > b^{x_1}$. On the other hand, if b is less than 1, then any positive power of it is less than 1, and we have $b^h - 1 < 0$, $b^{x_2} - b^{x_1} < 0$, and $b^{x_2} < b^{x_1}$. When $b = 1$, of course $b^{x_2} = b^{x_1}$.

The base that appears most frequently in exponential functions in both pure and applied mathematics is a certain irrational number called e. To five decimal places it is given by $e = 2.71828$. It is defined as the limit as t approaches infinity of $\left(1 + \dfrac{1}{t}\right)^t$. Since $\left(1 + \dfrac{1}{t}\right)$ evidently approaches 1 as t approaches infinity, it might appear that any power would approach 1. This, however, is not the case, as the student can verify by expanding the binomial for increasing values of t. The function $y = \left(1 + \dfrac{1}{x}\right)^x$ is graphed in Fig. 7.1.2 for $x > 0$. Various applications and properties of e

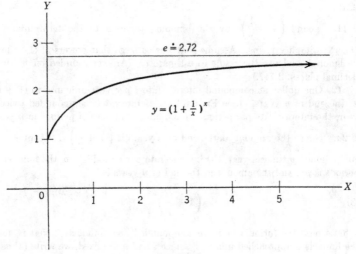

Fig. 7.1.2

are indicated in the exercise. The following table of values of e^x is useful for doing simple problems.

x	e^x	x	e^x
1	2.72	-1	0.368
2	7.39	-2	0.135
3	20.1	-3	0.0498
4	54.6	-4	0.0183

EXERCISE 7.1

1. Make a large graph of $y = 2^x$. Find the values of y for $x = 1/2, 1/3, -1/2, -1/3$, and check to see that the corresponding points appear to lie on the graph.

2. Graph: (a) $y = 3^x$. (b) $y = 3 \cdot 2^x$. (c) $y = -2^x$.

3. What point is common to the graphs of all functions of the form $y = b^x$?

4. Graph: (a) $y = 2^{-x}$. (b) $y = 2^{x/2}$. (c) $y = 2^{-x^2}$. (d) $y = 2^{3x}$.

5. Graph $y = (1/2)^x$ and compare with the graph of $y = 2^x$.

6. Show that the graph of $y = b^x$ is the reflection in the y-axis of the graph of $y = b^{-x}$.

7. Graph: (a) $y = (0.25)^x$. (b) $y = (1/3)^x$. (c) $y = (0.1)^x$.

8. Describe the graph of $y = b^x$ for $b < 1$.

9. From the graph of $y = 2^x$ estimate $2^{\sqrt{2}}$, $2^{\sqrt{3}}$, $2^{\sqrt{17}}$, 2^π.

10. Use the binomial theorem to estimate $(1 + \frac{1}{2})^2$, $(1 + \frac{1}{3})^3$, $(1 + \frac{1}{4})^4$, $(1 + \frac{1}{5})^5$, $(1 + 0.1)^{10}$, $(1 + 0.01)^{100}$. (*Answers.* 2.25, 2.37, 2.44, 2.48, 2.59, 2.70.)

11. Expand $\left(1 + \dfrac{1}{x}\right)^x$ by the binomial theorem to the term involving $(1/x)^5$. Simplify terms. Assume that x is so large that powers of $(1/x)$ may be ignored, and use the result to estimate e. (Answer rounded off to three decimal places: 2.717.)

12. One dollar at compound interest rate i per year amounts to $(1 + i)^n$ at the end of n years. (See Ex. 6.6.10.) If interest is compounded twice a year, the interest rate per period is $i/2$, and the number of periods in n years is $2n$. Hence the amount at the end of n years is $\left(1 + \dfrac{i}{2}\right)^{2n}$. If interest is compounded p times a year, the interest rate per period is i/p, the number of periods is pn, and the amount at the end of n years is:

$$(3) \qquad \left(1 + \frac{i}{p}\right)^{pn}$$

What would the formula be if we compounded "continuously," that is, took the limit as p approached infinity? Since i and n are fixed, we write (3) as

$$(4) \qquad \left[\left(1 + \frac{i}{p}\right)^{\frac{p}{i}}\right]^{in}$$

The student should verify that (3) and (4) are identical. Now let $p/i = t$. Then (4) becomes $\left[\left(1 + \dfrac{1}{t}\right)^t\right]^{ni}$. But as p approaches infinity, t approaches infinity, and the quantity in the square bracket approaches e. Hence, when interest is compounded continuously at interest rate i, the amount at the end of n years is $A = Pe^{ni}$. The exponential e^i takes the place of $(1 + i)$ in *6.6.4, and is called the *force of interest*. (a) How much does \$10 amount to in 3 years at continuous compound interest of $33\frac{1}{3}\%$? (b) How much does \$500 amount to in 20 years at continuous compound interest of 10%? (c) How much money would have to have been deposited 50 years ago to amount to \$1000 if interest is compounded continuously at the rate of 6% per year. [*Answers to three significant figures.* (a) \$27.20. (b) \$3700. (c) \$49.80.]

13. Graph: (a) $y = e^x$. (b) $y = e^{-x}$. (c) $y = -2e^x$.

14. A certain human population increases at the rate of 10% per decade, that is, its size at the end of each decade is 1.10 times its size at the beginning. (a) If the original population is 1.00×10^9, find the population at the end of 10, 20 and 30 years. (b) Write a formula for the population as a function of n, the number of decades elapsed. (c) Graph the function found in part b.

Do you think it would give good estimates of the population in intermediate years? [*Answers.* (*a*) 1.10×10^9, 1.21×10^9, 1.33×10^9.

(*b*) $P_n = (1 \times 10^9) (1.1)^n$.]

15. In the previous problem, the formula was derived by considering the population only at the ends of each decade. Actually the population increases almost continuously at a fairly steady rate. By an argument similar to that of prob. 12, it follows that the population at the end of n periods is given by $P_n = Pe^{nr}$, where P is the original population and r is the rate of increase per period. Find the population in prob. 14 at the end of 10, 20, and 30 years if the continuous rate of increase is 10%. (Note that the answers are not the same as in prob. 14, illustrating the fact that an equal continuous rate of increase gives a larger growth.)

16. Biologists use the formula $W = W_0 e^{rt}$ to approximate the growth in weight of an organism. Here W is the weight at time t, W_0 is the initial weight, r is the instantaneous (continuous) rate of growth, and t is the time. If the initial weight of a plant is 5 and it grows at the rate of 0.1 per day, what is its weight in: (*a*) 10 days? (*b*) 20 days? (*c*) 30 days? (*d*) 5 days? (Answer for part *a*. 13.6.)

17. Use Table I to estimate $e^{1/2}$ and $e^{1/3}$ and check against your graph of $y = e^x$.

18. The function $y = \dfrac{e^{-m}m^x}{x!}$ plays an important role in statistics, where it is called the *Poisson distribution*. The range of x is the positive integers. Graph this function.

19. The function $y = \dfrac{1}{\sqrt{2\pi}} e^{-x^2/2}$ is called the *normal distribution* in statistics. Graph it.

20. The Heinis curve, used in mental testing, is given by $y = b(1 - e^{x/d})$. Graph it for $b = 429$, $d = 6.675$.

21. Graph:

(*a*) $y = 2e^{3x}$

(*b*) $y = e^{-4x}$

(*c*) $y = -e^x$

(*d*) $y = \dfrac{e^x - e^{-x}}{2}$

(*e*) $y = \dfrac{e^x + e^{-x}}{2}$

(*f*) $y = 1 - e^x$

(*g*) $y = x + e^x$

(*h*) $y = e^x - x$

(*i*) $y = xe^{-x^2}$

22. The number e has the remarkable property that

*7.1.2 $$D_x(ae^x) = ae^x$$

(*a*) Find $D_x(e^x)$, $D_x(3e^x)$, $D_x(-re^x)$.

(*b*) What is the slope of $y = e^x$ when $x = 0$?

(*c*) Write the equations of the tangent lines to $y = e^x$ at the points where $x = 0$, -1, 2, 4, and sketch.

†**23.** Show that $(b^{x_1}b^{x_2}b^{x_3}b^{x_4} \cdots b^{x_n})^{1/n} \equiv b^{\bar{x}}$ where $\bar{x}$ is the average of the x's.

†**24.** Explain the fallacy in the following. Let x be such that $e^x = -1$. Then $e^{2x} = (e^x)^2 = (-1)^2 = 1 = e^0$. Hence $2x = 0$ and $x = 0$. It follows that $e^x = e^0 = 1$. Finally $-1 = 1$!?

7.2 The logarithm

We have seen in the previous section that $y = b^x$ has a unique value for every real x when $b > 0$. If we draw a vertical line through any point on the x-axis, it meets the graph of $y = b^x$ in just one point. A glance at Fig. 7.1.1 suggests that it is equally true that, if we draw a horizontal line through any point on the positive y-axis, it meets the curve in just one point, unless $b = 1$. This means that $y = b^x$ determines x as a single-valued function of y. This function is called the logarithm of y to the base b.

We are now going to give a precise definition of the logarithm along lines suggested by the foregoing discussion, but we are going to interchange x and y so that x is the independent variable.

*7.2.1 $y = \log_b x \longleftrightarrow x = b^y$ (Definition)

The symbol "$\log_b x$" is read **"the logarithm of x to the base b"** or "log x to the base b." Sometimes x is called the "antilogarithm of y to the base b" and is written "antilog$_b$ y." The logarithmic function $\log_b x$ and the exponential function b^x are evidently inverse functions in the sense of 6.10.

According to this definition, $\log_2 8$ is 3 because $2^3 = 8$. Similarly, $2 = \log_3 9$ since $3^2 = 9$, $1 = \log_2 2$ since $2^1 = 2$, $0 = \log_a 1$ since $a^0 = 1$, $-2 = \log_4 (1/16)$ since $4^{-2} = 1/16$, and $0.5 = \log_{81} 9$ since $\sqrt{81} = 9$.

In order to become acquainted with the most important properties of logarithms, we consider the special case of $b = 2$. We make a table of values of $y = \log_2 x$, or what is the same thing, of $x = 2^y$. It is as follows:

$x = 2^y = $ antilog$_2$ y	$y = \log_2 x$
64	6
32	5
16	4
8	3
4	2
2	1
1	0
1/2	−1
1/4	−2
1/8	−3
1/16	−4

The curve is shown in Fig. 7.2. It is the reflection in the line $y = x$ of $y = 2^x$. We observe that $\log_2 1 = 0$, and this is true regardless of the base, since $b^0 = 1$ for any $b \neq 0$. Hence

*7.2.2 $\log_b 1 = 0$ for any $b \neq 0$

Suppose we multiply two numbers in the x column in the table, say 2 and 8. The product is 16. The logarithms of these numbers

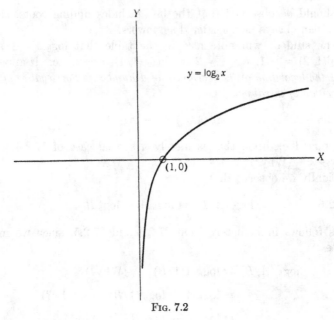

Fig. 7.2

are 1, 3, and 4, and $1 + 3 = 4$. Similarly, $(1/16)(8) = 1/2$, and $-4 + 3 = -1$. It appears that *the logarithm of the product is the sum of the logarithms*, and this is in fact a general property of logarithms, that is,

*7.2.3 $\log_b (AB) \equiv \log_b A + \log_b B$

It is this property that makes logarithms so useful in doing computations. In order to multiply numbers we have only to add their logarithms and then see what number has the sum as its logarithm. For example, in the above table we may multiply 64 and 1/4 by adding $6 - 2 = 4$ and noting that the number whose logarithm is 4 is just 16.

By experimenting with the table the student will discover that
*the logarithm of a power of a number equals the exponent times the
logarithm of the number.* Thus $(1/4)^2 = 1/16$ and $2(-2) = -4$,
or $\log (1/4)^2 = 2 \log_2 (1/4)$. Similarly, $8^{\frac{1}{3}} = 2$ and $(1/3)3 = 1$,
i.e., $\log_2 8^{\frac{1}{3}} = (1/3) \log_2 8$. This law of logarithms is

*7.2.4 $\log_b A^k \equiv k \log_b A$

It should be observed that the law includes finding roots, since
roots can always be considered as powers.

The student will note also in the table that $\log_2 2 = 1$ and
$\log_2 (1/2) = -1, \log_2 4 = 2$ and $\log_2 (1/4) = -2$, etc. It appears
that *the logarithm of the reciprocal of a number is the negative of the
logarithm of the number,* i.e.,

*7.2.5 $\log_b (1/A) \equiv - \log_b A$

We may look upon this as merely a special case of *7.2.4 with
$k = -1$. (Why?)
    Finally we observe that

*7.2.6 $\log_b (A/B) \equiv \log_b A - \log_b B$

This follows immediately from *7.2.3 and *7.2.5, since we may
write

(1) $\log_b (A/B) \equiv \log_b A(1/B)$ (Why?)

(2) $\equiv \log_b A + \log_b (1/B)$ (Why?)

(3) $\equiv \log_b A - \log_b B$ (Why?)

We have not yet proved the fundamental properties *7.2.3 and
*7.2.4. Why should it happen that the logarithm of a product is
the sum of the logarithms and the logarithm of a power is that
power times the logarithm? Really, it is not surprising. A logarithm is defined as an exponent, the exponent we must attach to
the base b in order to get the given number. But we know that we
add exponents in multiplication, that is, $b^{y_1} b^{y_2} \equiv b^{y_1 + y_2}$. We also
know that in taking a power, we multiply the power times the
exponent, that is, $(b^y)^k \equiv b^{ky}$. We shall base our proofs on these
ideas, but first we state an "obvious" property of the logarithm:

*7.2.7 $b^{\log_b x} \equiv x$

This is obvious because, according to *7.2.1, $\log_b x$ is a number y such that $b^y = x$.

Proof of *7.2.3:

(4) $$AB \equiv (b^{\log_b A})(b^{\log_b B}) \qquad \text{(Why?)}$$

(5) $$\equiv b^{\log_b A + \log_b B} \qquad \text{(Why?)}$$

But this last equation is equivalent, by *7.2.1, to:

(6) $$\log_b AB \equiv \log_b A + \log_b B$$

Heuristic Discussion. The student should note how we used the definition of the logarithm, the property *7.2.7, and the laws of exponents. Since (5) means *7.2.3 by definition, we have only to demonstrate (5) in order to prove the theorem.

Proof of *7.2.4:

(7) $$A^k \equiv (b^{\log_b A})^k \qquad (\ast 7.2.7)$$

(8) $$\equiv b^{k \log_b A} \qquad (\ast 6.3.4)$$

But (8) means *7.2.4 by definition *7.2.1.

EXERCISE 7.2

1. Test the following statements by *7.2.1:

(a) $\log_2 4 = 2$?   (f) $\log_2 2 = -1$?   (k) $\log_{10} 10 = 1$?
(b) $3 = \log_3 3$?   (g) $-1 = \log_2 (0.5)$?   (l) $2 = \log_{10} 100$?
(c) $2 = \log_3 6$?   (h) $\log_{(0.5)} 8 = -3$?   (m) $-3 = \log_{10} (10^{-3})$?
(d) $\log_3 27 = 3$?   (i) $\log_2 1 = 0$?   (n) $0.25 = \log_{16} 2$?
(e) $0.5 = \log_4 2$?   (j) $1 = \log_5 5$?   (o) $1/3 = \log_{1000} 10$?

Answers. (a) True, since $2^2 = 4$. (c) False, since $3^2 \neq 6$.

2. Evaluate:

(a) $2^{\log_2 4}$ (c) $5^{\log_5 25}$ (e) $2^{\log_2 5}$ (g) $8^{\log_8 2}$
(b) $3^{\log_3 3}$ (d) $3^{\log_7 49}$ (f) $2^{\log_3 9}$ (h) $16^{\log_{16} 4}$

Answers. (a) 4. (c) 25. (e) 5. (g) 2.

3. Find:

(a) antilog$_2$ 5   (d) antilog$_4$ (0.5)   (g) antilog$_2$ x
(b) antilog$_2$ 3   (e) antilog$_3$ 3   (h) antilog$_{(0.5)}$ 3
(c) antilog$_3$ 2   (f) antilog$_a$ 2   (i) antilog$_e$ 1

Answers. (a) $2^5 = 32$. (c) 9. (e) 27. (g) 2^x. (i) e.

4. Find:

(a) $\operatorname{antilog}_2 (\log_2 4)$

(b) $\operatorname{antilog}_b (\log_b 2)$

(c) $\log_3 (\operatorname{antilog}_3 2)$

(d) $\log_c (\operatorname{antilog}_c 7)$

(e) $\log_2 (\log_3 9)$

(f) $\log_3 (\operatorname{antilog}_9 1)$

(g) $\log_e (\operatorname{antilog}_e e^2)$

(h) $\operatorname{antilog}_2 (\log_3 9)$

Answers. (a) 4. (c) 2. (e) 1. (g) e^2.

5. Explain why $\log_b (\operatorname{antilog}_b x) \equiv \operatorname{antilog}_b (\log_b x) \equiv x$.

6. Explain why $\operatorname{antilog}_b x \equiv b^x$.

†**7.** Why would it be awkward to use negative bases? Why would a base less than or equal to 1 be inconvenient?

8. Fill in the reaons in the proof of *7.2.6.

9. Explain the justification of *7.2.5.

†**10.** Does $\log_b x$ exist for negative x? (*Hint.* What can you say about $x = b^y$ for x negative?)

11. Make up a table for $b = 3$ of $y = \log_b x$ as we did in the section for $b = 2$. Test the various properties and theorems.

12. Justify the following:

(a) $\log_3 8 = 3 \log_3 2$

(b) $\log_e 14 = \log_e 7 + \log_e 2$

(c) $\log_{10} \sqrt{2} = (1/2) \log_{10} 2$

(d) $\log_3 49 = 2 \log_3 7$

(e) $\log_4 3 = (1/3) \log_4 27$

(f) $\log_6 5 = \log_6 30 - \log_6 6$

(g) $\log_b (x + y) \not\equiv \log_b x + \log_b y$

(h) $\log_b (Ax) \not\equiv A \log_b x$

13. Find x:

(a) $x = \log_2 64$

(b) $\log_2 x = 3$

(c) $3x - 1 = \log_4 2$

(d) $\log_3 x = 1$

(e) $\log_4 x = 0$

(f) $\log_{(0.5)} x = -1$

(g) $\log_2 x = -3$

(h) $\log_x 4 = 2$

(i) $\log_x 5 = 3$

(j) $\log_{(1-x)} 100 = 2$

(k) $\log_x x = 2$

(l) $\log_x x = x$

Answers. (a) 6. (c) 0.5. (e) 1. (g) 1/8. (i) $\sqrt[3]{5}$. (k) 1 or 0.

14. Since $y = \log_b x$ is the inverse of $y = b^x$ it can be graphed by reflecting the latter in the line $y = x$. (See 6.10.) Use this idea to plot (a) $y = \log_3 x$. (b) $\log_{10} x$. (c) $y = \log_e x$.

15. Graph:

(a) $y = \log_2 (x - 3)$

(b) $y = 2 \log_2 (x + 1)$

(c) $y + 2 = \log_2 (x - 1)$

(d) $y = \log_3 (-x)$

(e) $y = x + \log_2 x$

(f) $y = -x^2 + \log_2 x$

(*Suggestion.* In parts e and f, a convenient method is to graph each term in the right members and then find points by adding the ordinates graphically.)

16. Since $x = \log_b c$ is the solution of $b^x = c$, we can write solutions for equations of this type, which are called **exponential equations.** (Numerical evaluation will be considered in the next two sections.) Solve:

(a) $5^x = 3$

(b) $10^x = 4$

(c) $e^x = 2$

(d) $4 \cdot 6^x = 5$

(e) $cb^x = a + b$

(f) $4e^{2x} = 21$

(g) $e^{-x} = 4$ (i) $10^{4x+5} = 13$ (k) $e^{x^2-x} = 3$

(h) $e^{1-x} = 15$ (j) $4e^{x^2} = 7$ (l) $10^x + 10^{-x} = 8$

Answers. (a) $\log_5 3$. (c) $\log_e 2$. (e) $\log_b [(a + b)/c]$. (g) $-\log_e 4$. (i) $(\log_{10} 13 - 5)/4$. (k) $(1 \pm \sqrt{1 + 4\log_e 3})/2$.

17. Find and plot the inverse of each function in parts *a* through *f* in Ex. 7.1.21. (*Suggestion.* In part *d* write $e^y - e^{-y} = 2x$, multiply both sides by e^y, and solve the resulting quadratic for e^y. Then solve for *y*.)

18. Solve for *x*:

(a) $\log_{10}(1 - x) = 0.13$ (c) $\log_u x^2 = D$

(b) $\log_c 5x = d$ (d) $\log_3 (x^2 + 1) = 3$

19. Simplify:

(a) $\log_b A + \log_b C - \log_b D$ (c) $\log_c A^2 + \log_c \sqrt{A}$

(b) $2\log_b x - \log_b x^2 + \log_b 1$ (d) $\log_a A - 0.5\log_a A + \log_a 2A$

Answers. (a) $\log_b (AC/D)$. (c) $\log_c A^{5/2}$.

20. Expand:

(a) $\log_b AB^2$ (c) $\log_b \sqrt{AB^3}$ (e) $\log_a (Ab^{-2}a^3)$

(b) $\log_b (A^2 - B^2)$ (d) $\log_c (xy^2 z^{-2})$ (f) $\log_b (x + y)^{1.3}c$

Answers. (a) $\log_b A + 2\log_b B$. (c) $(1/2)(\log_b A + 3\log_b B)$. (e) $\log_a A - 2\log_a b + 3$.

21. Rewrite as a single logarithm in simplest form:

(a) $\log_b (x^2 - 1) - \log_b (x + 1)$

(b) $\log_b (x^2 + 2x + 1)$

(c) $\log_b (x^2 + 1) + \log_b (x^2 - 1) - \log_b x^2$

(d) $\log_b \sqrt{2x} + \log_b (2x) + \log_b (4x^2)$

Answers. (a) $\log_b (x - 1)$. (c) $\log_b (x^2 - x^{-2})$.

22. Solve for *x*: $\log_e (x + a) - \log_e (x + b) = \log_e c$.

23. Solve for *t*: $p = Pe^{-(t-t_0)/\theta}$.

24. Solve for *E*: $M = \dfrac{1}{1.8} \log_b \left(\dfrac{E}{E_0}\right)$.

25. Solve for *P*: $sRT \log_e (p/P) = \dfrac{2a}{r} - \dfrac{e^2}{8\pi r^4}$.

26. Show that if $x = Bt$ and $xt = A$, then

$$(1/A) \log_b \left(1 + \frac{x}{B}\right) = (1/x) \log_b (1 + t)^{1/t}$$

†**27.** Show that, if *n* numbers are elements in a geometric progression, their logarithms form an arithmetic progression.

28. The following formula is used by some plant physiologists to approximate the relation between time and the growth of a plant:

$$\log_e \left(\frac{x}{A - x} \right) = K(t - t_0)$$

Here A is the maximum weight, x the growth in time t, t_0 the time for half the maximum growth, and K a constant indicating the rate of growth. Solve for x.

29. Show that the logarithm of the geometric mean of two numbers is the arithmetic mean of their logarithms.

†**30.** If the **geometric mean** of n numbers, $x_1, x_2, \cdots x_n$, is defined by

$$x_g = \text{antilog}_b \left[(\log_b x_1 + \log_b x_2 + \cdots + \log_b x_n)/n \right]$$

show that

$$x_g = \sqrt[n]{x_1 x_2 \cdots x_n}$$

†**31.** Discover the fallacy in the following: $(-1)^2 = 1$. Hence $\log_e (-1)^2$ $= \log_e 1 = 0$. Hence $2 \log_e (-1) = 0$, $\log_e (-1) = 0$, and $-1 = e^0$. Hence $-1 = 1$!?

†**32.** Show that $y_1 > y_2 > 0$ and $b > 1 \longrightarrow \log_b y_1 > \log_b y_2$.

†**33.** Starting from *7.2.3, prove by induction:

(a) *7.2.4 for positive integral k.

(b) $\log_b (A_1 A_2 \cdots A_n) \equiv \log_b A_1 + \log_b A_2 + \cdots + \log_b A_n$.

7.3 Common logarithms

Logarithms to the base 10 are called **common logarithms.** Suppose that we wish to find the common logarithm of some number that is not an integral power of 10, $\log_{10} 2.5$, for example. We wish to find a y, such that $10^y = 2.5$. Since $10^0 = 1$ and $10^1 = 10$, y evidently lies between 0 and 1. From Table I, $10^{0.5} = 3.16$. Hence 0.5 is too large. Also $10^{1/3} = 2.15$. Hence $1/3$ is too small. Already we see that $0.33 < y < 0.5$, but it would be tedious to continue in this way. Much more efficient methods of finding logarithms have been developed and have been used to construct tables that give the logarithms of enough numbers for all practical purposes.

In Table II are listed to four-digit accuracy the logarithms of all three-digit numbers between 1 and 10. From this table we can approximate the logarithm of any number, but in this section we consider only numbers from 1 to 10. We observe that all the logarithms listed are decimals between 0 and 1. (The decimal point is omitted in both numbers and logarithms in the table.) Also it is easy to see that the figures given in the table are merely approximations to the logarithms, except for $\log_{10} 1$ and $\log_{10} 10$.

Suppose, for instance, that $\log_{10} 2$ were exactly equal to a finite or repeating decimal, that is, a rational number. This would mean that $\log_{10} 2 = m/n$ or $10^{m/n} = 2$, where m and n are positive integers. From this follows $10^m = 2^n$. (Why?) But the left member of this equation must end in a zero, and the right member must end in 2, 4, 6, or 8. (Why?) Hence our assumption that $\log_{10} 2$ is rational must be false. A similar proof could be given for almost all numbers. Obviously, the logarithms in Table II must be merely approximations to irrational numbers.

In order to illustrate the way in which Table II can be used to carry out computations, we find $2\cdot4$, $6/2$, and 3^2. First we observe from *7.2.3, *7.2.6, and *7.2.4 that

(1) $$\log (2\cdot4) = \log 2 + \log 4$$

(2) $$\log (6/2) = \log 6 - \log 2$$

(3) $$\log 3^2 = 2 \log 3$$

Note that we have omitted the base, as we shall do from now on when it is 10. We look up the logarithms indicated on the right sides of these equations and arrange the computations as follows:

$$\log 2 = 0.3010 \qquad\qquad \log 6 = 0.7782$$
$$+ \log 4 = 0.6021 \qquad\qquad - \log 2 = 0.3010$$
$$\text{antilog } 0.9031 = 8 \qquad\qquad \text{antilog } 0.4772 = 3$$

$$\log 3 = 0.4771$$
$$\times 2$$
$$\text{antilog } 0.9542 = 9$$

The result 9 is found by looking in Table II for the number whose logarithm is 0.9542, and similarly for the others. Of course using logarithms to do the foregoing problems is like using a sledge hammer to crack a peanut, but it is no harder to use them to do very much more complicated computations.

By interpolating in Table II we can find the logarithms of numbers involving four digits. (For greater accuracy than this, tables to five or more places are required.) For example, to find $\log 3.122$, we note that $\log 3.12 = 0.4942$ and $\log 3.13 = 0.4955$. The tabular difference is 0.0013, but we ignore the decimal point and think just of 13. We may think of the situation as follows.

While the number goes a distance of 10 in the fourth place, the logarithm goes a distance of 13. Hence, while the number goes a distance of 2 in the fourth place, the logarithm will go a distance of approximately $(2/10)13 = 2.6 \doteq 3$. Hence we add 3 in the last place to the logarithm of 3.12 to get log $3.122 = 0.4945$. This is exactly equivalent to using *5.2.4 with $x_1 = 3.12$, $y_1 = 0.4942$, $\Delta x = 0.01$, $\Delta y = 0.0013$, and $x = 3.122$.

Interpolation is often involved also in finding antilogarithms. We simply interpolate "backwards" in the table, considering the values in the body of the table as the independent variable. The interpolation to find antilog 0.7107 may be arranged as follows:

$$\text{antilog } x \qquad\qquad x$$

$$10\begin{bmatrix} {-}5.130{-} \\ ? \quad \hookleftarrow \end{bmatrix}h \qquad 6\begin{bmatrix} {-}0.7101{-} \\ {\hookrightarrow}0.7107 \end{bmatrix}9$$
$$\qquad {\hookrightarrow}5.140 \qquad\qquad 0.7110{\hookleftarrow}$$

We wish to find h, so that $h/10 = 6/9$. The closest integral value is 7. Hence we write log $5.137 = 0.7107$ or antilog $0.7107 = 5.137$. The interpolation formula *5.2.4 gives the same result with $x_1 = 0.7101$, $y_1 = 5.13$, $\Delta x = 0.0009$, $\Delta y = 0.01$, $x = 0.7107$.

EXERCISE 7.3

1. In Table II opposite 54 and under 2 we find 7340. Just what does this mean?

2. Show from Table I that $0.33 < \log_{15} 3 < 0.5$.

3. Show that $\log_{10} 3$ is not a rational number.

4. Is it strictly correct to write $\log_{10} 4 = 0.6021$? (*Answer.* No, the equality is only approximate. However, the equality sign is always used with the understanding that the logarithm is accurate to the number of decimal places shown.)

5. From Table II find the common logarithms of the following numbers:

(a) 1.01   (c) 6.13   (e) 7.95   (g) 4.115   (i) 3.888
(b) 2.04   (d) 4.99   (f) 3.010   (h) 5.893   (j) 9.931

6. Find the antilogarithms of the following:

(a) 0.8388 (c) 0.9133 (e) 0.6990 (g) 0.8717 (i) 0.6154
(b) 0.0934 (d) 0.4425 (f) 0.6994 (h) 0.2840 (j) 0.9601

7. Solve for x:

(a) $10^x = 2$ (c) $10^x = 4.861$ (e) $10^x = 1.007$ (g) $10^x = 1.346$
(b) $10^x = 3$ (d) $10^x = 7.113$ (f) $10^x = 8.009$ (h) $10^{2x} = 5.124$

Answers. (a) 0.3010. (c) 0.6867. (e) 0.0030. (g) 0.1290.

8. Find:

(a) $10^{0.7212}$ (c) $10^{0.1}$ (e) $10^{0.3}$ (g) $\sqrt[10]{10^9}$ (i) $10^{\sqrt{2}-1}$

(b) $10^{0.5314}$ (d) $\sqrt[5]{10}$ (f) $10^{0.7}$ (h) $10^{0.5555}$ (j) $10^{0.3987}$

Answers. (a) 5.262. (c) 1.259. (e) 1.995. (g) 7.943. (i) 2.594.

9. Compute by logarithms:

(a) $(2.5)(2)$ (d) $(1.9)(4.519)$ (g) $(1.99)^3$

(b) $(8.6)/2$ (e) $(1.002)^3$ (h) $(2\pi)^2$

(c) $(1.5)^2$ (f) $(8.887)/(3.125)$ (i) $\sqrt[3]{4}$

(Check that your answers are accurate to four digits by finding the results differently.)

10. Compute:

(a) $(3.894)(2.111)(1.003)$ (e) $(5.674)^{1/3}(1.275)^{-2}$

(b) $(5.117)(4.662)/(9.916)$ (f) $(1.999)^{0.1}$

(c) $(2.003)^3/(1.05)^2$ (g) $(1.734)^{-1}(4.617)^{-1}(9.344)$

(d) $\sqrt{(4.518)(1.139)^3}$ (h) $(1.017)^{15}$

Answers. (a) 8.245. (c) 7.290. (e) 1.097.

11. How much will \$100 amount to in 100 years at 2% compound interest?

12. When negative numbers appear in a computation, we rewrite the computation so that only the logarithms of positive numbers are involved. (Why?) For example, to find $(-8)^{1/3}$ by logarithms, we write $(-8)^{1/3} = -(8)^{1/3}$, compute $8^{1/3}$, and prefix a minus sign to the result. Find:

(a) $(-9)^{1/3}$

(b) $(-3.375)^{1/3}$

(c) $(-9)^{1/5}$

(d) $(-3.182)(-5.999)(2.467)^{-1}$

(e) $(2.881)(-1.983)(-7.201)/(4.911)(-5)$

(f) $(3.004)(2.988)/(-1.165)(-2.556)(-1.099)$

Answers. (a) Check from Table I. (c) -1.552. (e) -1.676.

13. Use logarithms to do the computations for Ex. 7.1.10. Find also

$$\left(1 + \frac{1}{500}\right)^{5.0} \text{ and } \left(1 + \frac{1}{1000}\right)^{1000}.$$

14. The computations in the last part of the previous problem were not very accurate. (Why?) Using the fact that $\log(1.001) = 0.00043408$, $\log(1.0001) = 0.00004343$, and $\log(1.00001) = 0.00000434$, compute $(1.001)^{1000}$, $(1.0001)^{10,000}$, and $(1.00001)^{100,000}$. Note that the last result is accurate to only three digits (why?), whereas the second is accurate to four.

†**15.** Prove that the following numbers are irrational: (a) $\log_{10} 4$. (b) $\log_{10} 5$. (c) $\log_{10} 6$. (d) $\log_{10} 27$.

16. When interest is compounded several times a year, it is customary to give an annual interest rate (called the *nominal rate*) and the number of conversions per year. If the nominal rate is j converted p times per year, one dollar will amount to $\left(1 + \dfrac{j}{p}\right)^{p}$ in one year. This is not the same as $(1 + j)$, which is what one dollar would amount to at the rate j. If $(1 + i)$ $= \left(1 + \dfrac{j}{p}\right)^{p}$ we call i the *effective rate*, because it is the rate such that one dollar at that rate would amount to the same as one dollar at the nominal rate converted p times. Find: (a) The effective rate corresponding to 12% compounded monthly. (b) The effective rate corresponding to 5% compounded quarterly. (c) The nominal rate compounded quarterly corresponding to an effective rate of 10%.

7.4 Computations with logarithms

The ideas applied in the previous section to numbers between 1 and 10 require no essential modification for dealing with other numbers. To find the common logarithm of a number, we make use of the fact that any number can be written in scientific notation, that is, in the form $N \times 10^{n}$, where n is an integer and $1 \leq N < 10$. From $x = N \times 10^{n}$ it follows that

$$(1) \qquad \log x = \log 10^{n} + \log N = n + \log N \qquad \text{(Why?)}$$

Evidently the common logarithm of any number equals an integer plus the logarithm of a number between 1 and 10. The integer is called the **characteristic** of the logarithm and may be found quickly by the following rule:

> *Rule for Finding the Characteristic.* Starting at the right of the first significant digit, count the number of spaces to the decimal point, positive to the right and negative to the left. The result is the characteristic.

The decimal part of the logarithm is called the **mantissa.** It is the same for all numbers that differ only in the position of the decimal point and is found from Table II. For example, to find log 0.0138 we ignore the decimal point and find in Table II the mantissa corresponding to 138. It is 0.1399. Then we note that we must count two spaces to the left in order to get to the decimal point from the space at the right of the first significant digit in 0.0138. Hence

$$(2) \qquad \log 0.0138 = -2 + 0.1399$$

We cannot write this as -2.1399, for this means $-2 - 0.1399$. We could carry out the subtraction and write -1.8601, but then

the mantissa would be negative also.   Since only positive mantissas are tabulated, it is more convenient to write:

(3)   log 0.0138 = 8.1399 − 10

We have added and subtracted 10.   It is standard procedure to
write negative characteristics in this way as some positive integer
minus 10 or some multiple of 10.   Then all decimals are positive,
and all negative numbers are integers.

In order to find antilogarithms, we reverse the above procedure.
The mantissa serves to determine the digits in the number.   Then
the characteristic indicates the position of the decimal point according to the following rule:

> *Rule for Placing the Decimal Point.*   Starting at the space to
> the right of the first signficant digit in the number, count a
> number of spaces equal to the characteristic, positive to the
> right and negative to the left.   Place the decimal point in the
> resulting position.

To find the antilogarithm of 7.4430 − 10, we first enter Table
II with the mantissa 4430 and find 2773.   Since the characteristic
is 7 − 10 = −3, we count to the left three spaces from the position
at the right of 2.   Hence

(4)   antilog 7.4430 − 10 = 0.002773

Success in logarithmic computations depends on systematic
methods.   To illustrate these we compute:

$$\text{(5)} \qquad\qquad A = \sqrt[3]{\frac{(399.2)(2.991)}{(0.0811)}}$$

It is advisable to begin by making a rough estimate of the result.
This is most easily done by writing all numbers in scientific notation, rounding off to two or even one digit, and calculating approximately.   We have:

$$\text{(6)} \qquad A \doteq \sqrt[3]{(4 \times 10^2)(3)/8 \times 10^{-2}}$$

$$\text{(7)} \qquad\qquad \doteq \sqrt[3]{\frac{12}{8} \times 10^4} \doteq \sqrt[3]{10 \times 10^3} \doteq 2 \times 10 = 20$$

Such an estimate guards against gross errors.   Since we wish only a rough estimate, there is considerable room for judgment in rounding off.

The next step is to write out a formula for the computation by using the laws of logarithms. In this case we have:

(8) $\log A = (1/3) [\log 399.2 + \log 2.991 - \log 0.0811]$

From the formula we prepare a form that is both a plan of action and a neat arrangement of the work. This should be made out completely *before* looking up any logarithms. A satisfactory form for this problem is:

$$\log 399.2 = \cdots\cdots$$
$$+ \log 2.991 = \cdots\cdots$$

$$\cdots\cdots$$

(9) $$- \log 0.0811 = \cdots\cdots$$

$$\cdots\cdots$$

$$\div 3$$

$$\log A = \cdots\cdots \qquad A =$$

Now we look up and record all logarithms. Then we carry through the computations and look up the antilogarithm. When completed, the work appears as follows:

$$\log 399.2 = 2.6012$$
$$+ \log 2.991 = 0.4758$$

$$13.0770 - 10$$
(10) $$- \log 0.0811 = 8.9090 - 10$$

$$4.1680$$
$$\div 3$$

$$\log A = 1.3893 \qquad A = 24.51$$

The result agrees with the estimate. The student should note that we wrote the characteristic of the sum of the first two logarithms as $13 - 10$ instead of 3. This enabled us to subtract the next logarithm without getting a negative mantissa. In subtraction, the characteristic should always be adjusted so that the subtrahend (ignoring the -10's) is smaller. Also, when a log is to be divided in order to find a root, if the characteristic is negative it should be adjusted so that after division the negative part is still 10. Thus, if we had $4.1680 - 10$ instead of 4.1680 above, we would have rewritten it $24.1680 - 30$. Then division by 3 would yield $8.0560 - 10$.

In computations involving many divisions it is convenient to use **cologarithms,** which are defined as follows:

*7.4.1 $\operatorname{colog}_b x = \log_b (1/x)$ (Definition)

It is immediate that

*7.4.2 $\operatorname{colog}_b x = - \log_b x$

If we wish to perform a division, we can add the cologarithm of the divisor instead of subtracting its logarithm. The cologarithm can be found by subtracting the logarithm from $10.0000 - 10$ as in the following examples:

(11)

$$
\begin{array}{cc}
10.0000 - 10 & 10.0000 - 10 \\
1.4522 & 8.2211 - 10 \\
\hline
8.5478 - 10 & 1.7789
\end{array}
$$

Or the mantissa may be read directly from the table by replacing each digit except the last in the mantissa of the logarithm by its difference from 9 and replacing the last by its difference from 10. The characteristic is then obtained by subtracting from -1 the characteristic of the logarithm. The use of cologarithms simplifies the form of some computations. For example, the form for $A = BC/DE$ is:

(12)

$$
\begin{aligned}
\log B &= \cdots\cdots \\
\log C &= \cdots\cdots \\
\operatorname{colog} D &= \cdots\cdots \\
\operatorname{colog} E &= \cdots\cdots \\
\hline
\log A &= \cdots\cdots
\end{aligned}
$$

EXERCISE 7.4

1. Do now, or review, Ex. 3.14.3.

2. Find the common logarithm of 3281 by rewriting it in scientific notation and writing an equation like (1). Check to see that the result is the same as that found by using the rule for finding the characteristic.

3. Why does the mantissa always lie between zero and one?

4. Find the characteristic of the logarithm of each of the following numbers:

(a) 3829	(d) 99.11	(g) 2032	(j) 24.14×10^7
(b) 2.994	(e) 0.00059	(h) 0.100002	(k) 0.0010012
(c) 0.0132	(f) 1,000,001	(i) 0.0018	(l) 852×10^{-3}

Answers. (a) 3. (c) -2. (e) -4. (g) 3. (i) -3. (k) -3.

5. Find the logarithm of each number in prob. 4. [*Answers.* (a) 3.5831. (c) $8.1206 - 10$. (e) $6.7709 - 10$. (g) 3.3079. (i) $7.2553 - 10$. (k) $7.0004 - 10$.]

6. Find the antilogarithms of:

(a) 0.3356	(d) 6.1000	(g) $8.2358 - 10$	(j) $3.9782 - 10$
(b) 1.8291	(e) $6.5483 - 10$	(h) 3.6909	(k) $5.4485 - 10$
(c) $9.4407 - 10$	(f) 2.9114	(i) 0.0018	(l) 5.7364

Answers. (a) 2.166. (c) 0.2759. (e) 0.0003534. (g) 0.01721.

7. Find the cologarithm of each number in prob. 4. [*Answers.* (a) $6.4169 - 10$. (c) 1.8794. (e) $3.2291 - 10$. (g) $6.6921 - 10$. (i) 2.7447. (k) 2.9996.]

8. Find the reciprocals of: (a) 48.18. (b) 0.0818. (c) 4287. (d) 8.991×10^{-8}. (e) 0.9473. [*Answers.* (a) 0.02075. (c) 0.0002332.]

9. Find:

(a) $10^{2.4478}$	(c) $10^{-0.1137}$	(e) $10^{-4.8643}$
(b) $10^{3.8992}$	(d) $10^{-0.4935}$	(f) $10^{1.9786}$

(*Suggestion for part c:* $-0.1137 = 9.8863 - 10$.)

10. Compute by logarithms:

(a) $(2894)(0.01391)$ (b) $(84.29)(23.72)^{-1}$

(c) $(3.018)(1,000,500)(99.11)(0.000961)$

(d) $(2.835)^2(-0.003)$ (e) $(9.377)(7.139)/(8.212)$

(f) $(1,000,000)/(354.3)(-0.2194)(1000.4)$

(g) $\sqrt{(54.83)(0.3182)}$ (h) $\sqrt{(513.29)^2 - (482.4)^2}$

(i) $[(3.183)(0.00181)]^{1/6}$ (j) $2^{1/2}3^{1/3}4^{1/4}5^{1/5}$

(k) $(28.99)^{1/5}(18.103)^2/(0.018)^{1/6}(301.01)^3$

(l) $(0.9999)^{50}$ (m) $(0.5)^{50}$ (n) $(1.1)^{50}$

Answers. (c) 2.875×10^5. (e) 8.150. (g) 4.177. (m) 8.912×10^{-16}. (*Suggestion for part h.* Look before you leap!*)*

11. Find: (a) log (log 15). (b) log (log 5). (c) log (| log 0.5 |). (*Discussion.*
log 0.5 = 9.6990 − 10 = −0.3010. Hence log (| log 0.5 |) = log 0.3010.)

12. Compute:

(a) $15^{0.1398}$ (c) e^π (e) 10^π (g) $(0.01754)^{0.9832}$

(b) $(0.5)^{2.348}$ (d) π^e (f) 3^e (h) $(236.7)^{1.3895}$

[*Suggestions.* log x^m = m log x, but we may calculate the right member by
logs, that is, log (log x^m) = log m + log (log x).]

13. Taking e = 2.718, find $\log_{10} e$, $\log_{10} \sqrt{e}$, $\log_{10} (1/e)$.

14. Do again parts b, e, f, k of prob. 10, using cologarithms if you did not
use them the first time, and not using them if you did.

15. How long will it take money to triple itself at 2% compound interest?
At 4%? (*Answers.* 56− and 28+.)

16. At what interest rate would money double itself in half the time required at the compound interest rate of 2%?

17. The following formulas are from various fields of science and engineering. Evaluate for the given values.

(a) $s = 2\pi\sqrt{(a^2 + b^2)/2}$, when a = 136.5, b = 48.22.

(b) $y = M(b - a)^5(2n)^{-4}/180$, when M = 3, n = 26, a = 1.827, b = 8.359.

(c) $A = \dfrac{R}{i}[1 + i - (1 + i)^{-n+1}]$, when R = 1247, i = 0.03, n = 15.

(d) $(r/r_0)^3 = W/W_0$. Find r/r_0 when W/W_0 = 1.7834.

(e) $I = k'w^{1/3}q^{-5/2}$, when k' = 6.64, w = 88,592, q = 9.813.

(f) $u = u_0[1 + (r/h)]^{-1.07}$. Find u/u_0 when r/h = 3.

(g) $P = A_n(1 + r)^{-n} + s(1 + r)^{-n}$, when A_n = 200,000, s = 50,000,
r = 0.02, n = 20.

Answers. (a) 643.3. (c) 1.530×10^4. (e) 0.9812.

18. Find the geometric mean of 357.2; 5211; 2.894; 0.9916; and 43.5.

19. Assuming that the earth travels in a circle with a radius of 9.29×10^7
miles, compute its speed in miles per second. (*Answer.* 18.5.)

20. A *light year* is the distance traveled by light in one year. If light
travels 186,284 miles per second, find the length in miles of one light year.
(*Answer.* 5.876×10^{12}.)

21. The parallax p of a star is related to its apparent magnitude m and
its absolute magnitude M by the formula 5 log p = M − m − 5. The star
"Wolf 359" has m = 13.5 and M = 16.6. Find its parallax. (*Answer.* 0.4171.)

22. The brightness of a star is measured by its *magnitude* defined by
M = −2.5 log L, where L is the intensity of light from the star. (a) Does a
fainter star have a larger or smaller magnitude? (b) Solve the definition for L.
(c) Show that a star of the first magnitude is 100 times brighter than one of
the sixth magnitude. (d) Magnitudes vary from −5 to +15. Show that
the ratio of the intensity of the brightest to dullest star is of the order of 10^8
to 1. (e) Show that when the magnitude increases by 1 the intensity is multiplied by $(2.512)^{-1}$.

23. In chemistry the pK of an acid is defined as the logarithm of the reciprocal of its ionization constant k. Graph pK as a function of k.

24. If a human egg is $1/175$ of an inch in diameter, calculate the total volume of the eggs from which the present world population (about 2×10^9) has grown. If one gallon equals 231 cubic inches, how many quarts would suffice to hold all these eggs?

25. Ownership of 51% of the stock of a corporation permits certain control. By owning 51% of a "holding company" that owns 51% of another company, it is possible to control the second company with capital equal to $(0.51)^2$ of its capital stock. Suppose a chain of six holding companies is organized, each owning 51% of the next. What percentage of the value of the stock of the sixth company is required to control it by controlling 51% of the first?

26. In a geometric progression, $y_1 = 5.32$ and $y_7 = 73.1$. Find r and s_7. (*Answers.* $r = 1.55$, $s_7 = 197$.)

27. A certain bacteria culture grows according to the law $y = 5e^{rt}$. If $y = 8.34$ when $t = 5$, find r.

28. The following are from statistics:

(a) P.E. $= 0.6745\sigma$. Find P.E. if $\sigma = 53.88$.

(b) $S = \sqrt{1 - r^2}$. Find S if $r = 0.91$.

(c) $y = \dfrac{1}{\sigma\sqrt{2\pi}} e^{-x^2/2\sigma^2}$. Find y for $\sigma = 15.23$, $x = 2\sigma$.

†**29.** From the fact that $2^{10} = 1024$, show that $\log_{10} 2 = 0.3$ correct to one decimal place.

†**30.** From the fact that there are 3010299956 digits in $2^{(10^{10})}$, find $\log_{10} 2$ correct to 10 decimal places.

31. Make a table of antilogarithms of 4 digit mantissas from 0.4810 to 0.4830 by interpolating in Table II.

32. Find the discount factor (see Ex. 6.9.12) when (a) $r = 0.02$, $n = 20$. (b) $r = 3.5\%$, $n = 6$. (c) $r = 2.1\%$, $n = 8.5$.

33. The slide rule is simply a device for adding logarithms mechanically. Write a brief essay explaining this statement. (*A Course in the Slide Rule and Logarithms*, by E. J. Hill, Ginn and Co., 1943, is one of many possible sources of information.)

†**34.** A book on physiology states that the following formula, based on empirical studies, gives the surface area of a man in terms of his height and weight:

$$A = 71.84(\text{height})^{0.725}(\text{weight})^{0.425}$$

No indication is given of the units. On the following page the book states that a man weighing 175 pounds and standing 6 feet tall has a surface area of 2 square meters. What are the appropriate units for this formula?

35. When n is large, $n!$ is hard to compute. However, it is given approximately by $n! \doteq n^n e^{-n} \sqrt{2n\pi}$. Use this formula (called *Stirling's formula*) to estimate $7!$ and compare with the true value. Do the same for $10!$ and $25!$ Find the error and per cent error in each case. (See Ex. 3.5.14.)

7.5 Natural logarithms and change of base

Logarithms to the base e are called **natural logarithms.**
Their importance derives from the special properties of e and its
frequent appearance in pure and applied mathematics. The use
of natural logarithms is, of course, based on the same principles
as common logarithms. However, the use of tables to find logarithms and antilogarithms is slightly less simple because powers of
ten no longer have integral logarithms.

Table III gives the natural logarithms to five-digit accuracy of
three-digit numbers between 1 and 10. Logarithms of four-digit
numbers and antilogarithms can be found by interpolation. For
numbers outside the range of the table, we can write, as in 7.4,
$x = N \times 10^n$ and hence $\log x = \log_e N + \log_e 10^n$. But $\log_e 10$
is not an integer, so the best we can do is

$$(1) \qquad\qquad \log x = \log_e N + n \log_e 10$$

The procedure is the same as for common logarithms except that
now we add $n \log_e 10$ to the logarithm found from the table. Thus
to find $\log_e 15.73$ we find $\log_e 1.573 = 0.45255$. To this we add
$\log_e 10 = 2.30259$. The result is $\log_e 15.73 = 2.75514$. To find
antilogarithms of logarithms that do not appear in Table III we
follow the above procedure in reverse. First we add some integral
multiple (positive or negative) of $\log_e 10$ to the logarithm so that
the result is within the table. We find the antilogarithm of this
and place the decimal point by reference to the multiple we added.
Thus to find antilog$_e$ 5.61147, we subtract 2(2.30259) to get
1.00629. Since antilog$_e$ 1.00629 = 2.738, antilog 5.61147 = 273.8.
A rule could be made up for placing the decimal point, but the
easiest way is to place it by common sense. Obviously if the
logarithm is too big (small) for the table, the antilogarithm will
be also.

It is often convenient to use logarithms to some base for which
tables do not exist or to express logarithms to one base in terms of
another. This is made possible by the following identity:

$$\star 7.5.1 \qquad\qquad \log_a x \equiv \frac{\log_b x}{\log_b a}$$

We shall prove the equivalent identity $\log_b x \equiv \log_a x \log_b a$.

(2) $b^{\log_a x \log_b a} \equiv (b^{\log_b a})^{\log_a x}$ (Why?)

(3) $\equiv a^{\log_a x}$ (Why?)

(4) $\equiv x$ (Why?)

But the last equation is by definition what we want to prove. Letting $x = b$, we get the special case

*7.5.2 $\log_a b \equiv \dfrac{1}{\log_b a}$

By using *7.5.1 we could compute Table III from Table II or Table II from Table III. Actually it is easier to calculate natural logarithms directly, so that Table III is more fundamental than Table II.

In order to check the foregoing statements and see how *7.5.1 is used, we find $\log_e 2$ from Table II. We have:

(5) $\log_e 2 = (\log_{10} 2)/(\log_{10} e)$

(6) $= 0.3010/.4343$

This may be computed by arithmetic or on a machine. It may also be done by logarithms as follows:

$$\log 0.3010 = 9.4786 - 10$$
$$- \log 0.4343 = 9.6378 - 10$$
(7) $\overline{\qquad\qquad\qquad\qquad}$
$$\log_{10} (\log_e 2) = 9.8408 - 10$$
$$\log_e 2 = 0.6931$$

The last result agrees with the value in Table III.

EXERCISE 7.5

1. From Table III find the natural logarithms of:

(a) 1.9	(c) 6.03	(e) 9.51	(g) 8.293	(i) 1.582
(b) 2.68	(d) 4.79	(f) 9.515	(h) 3.714	(j) 5.238

Answers. (a) 0.64185. (c) 1.79675. (e) 2.25234. (g) 2.11541. (i) 0.45868.

2. From Table III find the antilogarithms of:

(a) 2.00283	(c) 1.60980	(e) 0.25	(g) 0.19745
(b) 1.31104	(d) 2.23536	(f) 1.5	(h) 0.99541

3. Find: (a) $e^{1.378}$. (b) $e^{2.13572}$. (c) $e^{0.67}$.

4. Use Table III to find e correct to four digits.

5. Find the natural logarithms of:

(a) 15	(c) 100	(e) 1158	(g) 0.00146
(b) 0.5964	(d) 52.71	(f) 0.08937	(h) 82.95

Answers. (a) 2.70806. (c) 4.60518. (e) 7.05446. (g) -6.52932.

6. Verify the values in the table at the end of 7.1.

7. Find: (a) $e^{4.75}$. (b) $e^{-1.48}$. (c) $e^{\sqrt{15}}$. (d) e^{π}. (e) e^{e^2}.

8. Find from Table II: (a) $\log_e 3$. (b) $\log_e 5$. (c) $\log_e 10$. (d) $\log_e 47.21$.

9. From Table III find: (a) $\log_{10} 5$. (b) $\log_{10} 172$. (c) $\log_{10} 0.4693$.

10. Use natural logarithms to do Ex. 7.3.9.

11. Solve the following equations for x:

(a) $e^x = 4$	(d) $3e^{x-1} = 5$
(b) $e^{2x} = 3.57$	(e) $e^{x^2-3x-1} = 14$
(c) $e^{x^2} = 8.237$	(f) $\log_e (2x + 1) = 2.8146$

Answers. (a) 1.38629. (c) ± 1.452. (e) -0.927 and 3.927.

12. One of the most important special properties of natural logarithms is:

*7.5.3 $$D_x (\log_e x) = \frac{1}{x}$$

This formula holds *only* when the base is e. It means that the slope of the line tangent to $y = \log_e x$ at any point is equal to the reciprocal of the x-coordinate of the point.

(a) Graph $y = 1/x$ and $y = \log_e x$ on the same coordinate plane.

(b) Find to the nearest tenth the root of $\log_e x = 1/x$.

(c) Find the slope of the curve $y = \log_e x$ at $x = 0.5, 1, 2, 3$.

(d) Draw the tangent lines to the curve at each point in part c.

(e) Write the equations of the tangent lines in part d.

(f) Since $D_x y$ is the limit of $\Delta y/\Delta x$ as Δx approaches zero, $\Delta y/\Delta x$ for very small Δx should approximate $D_x y$. Find $\Delta y/\Delta x$ for $y = \log_e x$ for $\Delta x = 0.01$ and $x_1 = 3.5$. Do the same for $\Delta x = 0.001$ and $x_1 = 7.2$. In each case compare with $1/x_1$.

†13. Show from *7.5.3 that $D_x (\log_b x) = \dfrac{1}{x} \log_b e$.

14. Find:

(a) $\log_e e$	(c) $\log_5 4$	(e) $\log_4 17$	(g) $\log_8 2.01$	(i) $\log_{25} 4$
(b) $\log_2 7$	(d) $\log_4 6$	(f) $\log_3 6$	(h) $\log_2 41$	(j) $\log_{0.1} 2$

Answers. (c) 0.8614. (e) 2.044. (g) 0.3357. (i) 0.4306.

15. Solve:

(a) $x \log_{10} 7 = \log_{10} 36$	(d) $10^{2x} + 10^x = 1$	(g) $(1.2)^x = 3^{x^2}$
(b) $\log_{10} 5^x = 315.8$	(e) $\log_x 9 = 5.886$	(h) $2^x - 2^{-x} = 1$
(c) $5^x = 10$	(f) $47^{2x} = 100$	(i) $3^x + 3^{-x} = 10$

Answers. (c) 1.431. (e) 1.452. (g) 0 and 0.1660.

16. Solve for x:

(a) $1/c = a + bc^x$

(b) $p = ab^{x/(s+x)}$

(c) $z = (1/2) \log_e \left(\dfrac{1 + x}{1 - x} \right)$

(d) $(-L/R) \log_e (E - Rx) = t + C$

(e) $x^{5.34} = 0.1$

(f) $x^{8.2} = 50$

17. Graph:

(a) $y = e^{\sqrt{x}}$

(b) $y = \log_e |x|$

(c) $y = |\log_e x|$

(d) $y = x^{4.7}$

(e) $y = \log_x e$

(f) $\log_{10} x < y < \log_e x$

(g) $\log_{10} x < y < 1$

(h) $\log_e x < y < x^{1/3}$

(i) $y = 10 + 5e^x$

(j) $y = x + \log_e (-x)$

(k) $y = \log_e x^3$

(l) $y = \log_e (x^2 - 1)$

18. Planck's equation:

$$E = c_1^{-5}\lambda(e^{c_2/\lambda T} - 1)$$

gives the rate of emission of radiation of a body in terms of wavelength, λ, and absolute temperature T. Plot E as a function of λ for $c_1 = 3.71 \times 10^{-5}$, $c_2 = 1.435$, and (a) $T = 100$, (b) $T = 10^8$.

19. Show that antilog $(A + B) \equiv$ (antilog A)(antilog B) and antilog (kA) $\equiv$ (antilog $A)^k$.

8

Circular Functions

The circular functions, often called the trigonometric functions, are among the most important in mathematics. Trigonometry, the study of the properties and the application of these functions, is the main topic of this chapter. We consider first some topics in analytical geometry that are necessary to a precise and general definition of the circular functions.

8.1 Distance

In this section we derive an analytic formula for the distance between two points with coordinates (x_1, y_1) and (x_2, y_2). (See Fig. 8.1.) We draw lines through the points and parallel to the

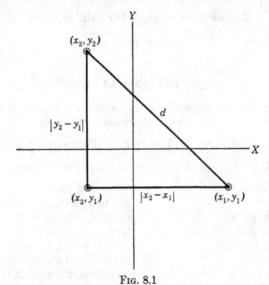

FIG. 8.1

axes, thereby completing a right triangle whose hypotenuse is the desired distance, and whose sides are $(x_2 - x_1)$ and $(y_2 - y_1)$.

313

(Recall 4.6.)   Hence

(1) $d^2 = (x_2 - x_1)^2 + (y_2 - y_1)^2$ (Why?)

We have proved:

*8.1.1   The distance d between the points (x_1, y_1) and (x_2, y_2) is

$$d = \sqrt{(x_2 - x_1)^2 + (y_2 - y_1)^2}$$

The student should note that this distance is always positive (since we take the principal square root) unless the two points are the same, in which case it is zero.   We are here talking about distance, and not directed distance.   (Compare 4.6.)   In order to have directed distance on a line not parallel to one of the axes we have to choose a positive direction on the line.   Then the directed distance from (x_1, y_1) to (x_2, y_2) is $+d$ or $-d$ according as the direction from the first to the second is the same or contrary to the positive direction.

If one of the points is the origin, we have the special case

*8.1.2   The distance of (x, y) from the origin is

$$d = \sqrt{x^2 + y^2}$$

EXERCISE 8.1

1. Draw diagrams similar to Fig. 8.1 for the given pairs of points.   Illustrate the proof of *8.1.1 by calculating the quantities involved.

(a) $(2, 3), (5, 6)$ (b) $(2, 4), (-1, 2)$ (c) $(2, -3), (-1, 5)$

2. Find the distances between the following pairs of points and sketch:

(a) $(2, 1), (3, 5)$
(b) $(2, 1), (1, 6)$
(c) $(-3, 4), (0, 7)$
(d) $(3, -2), (-1, -5)$
(e) $(1.9, 4), (-2.1, 1)$
(f) $(a, 1), (2, 3)$
(g) $(a - b, c + d), (a, d)$
(h) $(2, 5), (2, 8)$

(i) $(-1, 6), (3, 6)$
(j) $(\frac{1}{3}, 2), (-0.2, -0.25)$
(k) $(\sqrt{2}, \sqrt{3}), (0, 0)$
(l) $(1 - \sqrt{2}, 1 + \sqrt{2}),$
    $\quad (1 - \sqrt{2}, 1 - \sqrt{2})$
(m) $(1, \sqrt{5}), (\sqrt{2}, 1)$
(n) $(2 + a, \sqrt{2} + b), (2, b)$
(o) $(a + b, c), (a, c)$

Answers.   (a) $\sqrt{17}$.   (c) $3\sqrt{2}$.   (e) 5.   (g) $\sqrt{b^2 + c^2}$.   (i) 4.   (k) $\sqrt{5}$.
(m) $\sqrt{9 - 2\sqrt{2} - 2\sqrt{5}}$.

3. Find the distance from the origin to each point in prob. 2. [*Answers.*
(a) $\sqrt{5}$, $\sqrt{34}$. (c) 5, 7. (e) $\sqrt{19.61}$, $\sqrt{5.41}$.]

4. Consider two points with the same y-coordinate. Show that *8.1.1 gives the same result as that indicated in 4.6. Show a similar result for two points with the same x-coordinate.

5. Prove, using the notion of distance, that $(2, -3)$, $(-4, 2)$, and $(-1, 1/3)$ are collinear.

6. Prove that $(-2, 3)$ lies on a circle with center at $(0, 0)$ and passing through the point $(\sqrt{5}, \sqrt{8})$.

7. Prove that the triangle determined by $(-3, -1)$, $(-2, \sqrt{6})$, and $(2, -\sqrt{6})$ is a right triangle.

8. Prove that $([a + b]/2, 0)$ is equidistant from (a, c) and $(b, -c)$.

9. Prove that the quadrilateral $(1, -10)$, $(8, -7)$, $(5, 4)$, $(-2, 1)$ is a parallelogram.

10. The point $(a, 2a)$ is twice as far from (c, d) as from (d, c). Find a.

†**11.** Find the circumference of the circle circumscribed about the triangle $(-1, 1)$, $(3, 5)$, $(-8, 8)$.

†**12.** Prove that the distance $d(P_1P_2)$ between two points P_1 and P_2 given by *8.1.1 has the following properties:

(a) $d(P_1P_2) = d(P_2P_1)$.

(b) $d(P_1P_2) \geq 0$.

(c) $d(P_1P_2) = 0 \longleftrightarrow P_1$ coincides with P_2.

(d) $d(P_1P_2) \leq d(P_1P_3) + d(P_3P_2)$, where P_3 is any third point.

(The last of these is the **triangular inequality** and says that any side of a triangle is less than the sum of the other two sides. It can be proved simply by using *8.1.1.)

†**13.** Suppose that we agree to choose on every line the positive direction so that the directed distance from (x_1, y_1) to (x_2, y_2) is positive if $y_2 > y_1$, or if $y_1 = y_2$ and $x_2 > x_1$. With this definition find the directed distance from the first to the second of each pair of points in prob. 2. Explain the meaning of this convention about the choice of positive direction on the line. (*Answer.* The positive direction is always up unless the line is horizontal, in which case it is toward the right.)

8.2 The circle and arc length

The student will recall from plane geometry that a **circle** is the locus of points equidistant from a given point, where the distance is called the **radius** and the given point is called the **center.** In order to treat a circle from the point of view of analytic geometry, we must find its equation. (See 4.6.) Consider a circle of center (h, k) and radius r. (Fig. 8.2.1.) We wish to find an equation that will be satisfied by the coordinates of every point at a distance r

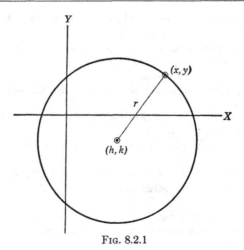

Fɪɢ. 8.2.1

from (h, k) and only by such points. But the distance of a point from (h, k) is given by

(1) $\sqrt{(x - h)^2 + (y - k)^2}$

This equals r if and only if the point lies on the circle. (Why?) Hence the desired equation is:

(2) $\sqrt{(x - h)^2 + (y - k)^2} = r$

This gives the theorem:

*8.2.1   The equation of the circle with radius r and center (h, k) is

(3) $(x - h)^2 + (y - k)^2 = r^2$

If the center of the circle is at the origin, the equation becomes:

(4) $x^2 + y^2 = r^2$

Later in the book, we shall study various properties of the circle, but at present we are interested in the idea of the length of portions of the circumference. These are called **circular arcs**. The entire circumference is one such circular arc. In plane geometry the student learned that the length of the circumference is $C = 2\pi r$, where π is a certain irrational number. We wish to look into this matter a little more carefully here in order to understand what is meant by the length of a circular arc, so that we can discuss intelligently the idea of an angle. Obviously we cannot measure the length of an arc directly, since the axes upon which we have set up a unit of length are straight lines and so are the rulers we would use to measure. What can we mean by the length of a curved path? This is important because until we have defined such lengths it is futile to talk about finding them!

Consider a circular arc AB. (See Fig. 8.2.2.) We mark off on the arc several points (C, D, and E in the figure). Then the length

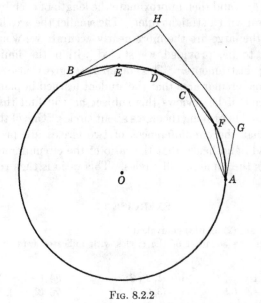

FIG. 8.2.2

of the broken line joining these points is simply the sum of the chord lengths, each one measurable and given by *8.1.1. If we insert more points along the arc (for example, F in the figure), the new broken line is longer than the old one. (For example, $AC < AF + FC$.) We can continue to insert points indefinitely, always getting broken lines whose lengths are well defined and larger than the preceding ones. Yet obviously the length of the broken line cannot get indefinitely large. In fact, the length of any broken line inscribed in this way is certainly less than the length of any broken line outside the circle and joining the two end points of the arc. Thus we have a sequence of broken lines whose lengths are increasing and yet all less than a fixed number. (In the figure we might take this fixed number to be the length of $AGHB$.) It can be proved that, if the number of chords is increased indefinitely so that the maximum chord length approaches zero, the length of the broken line approaches a definite limit, which is independent of the choice of points. We define the **arc length** to be this limit.

This definition of arc length can be used for any curve, whether circular or not. It agrees with common sense and experience, for if we wished to measure the length of an arc we would divide it into small arcs and then approximate the length of each by measuring its chord with a straight ruler. The smaller the arcs into which we broke the large arc the more nearly accurate we would expect the result to be, provided we stayed within the limits of our measuring instruments. This definition agrees also with the theorems on circular arcs that the student learned in plane geometry. In fact, if he reviews this subject he will find that similar ideas are used in proving theorems about circles. One of these is the theorem that the circumferences of two circles are proportional to their radii and hence that the ratio of the circumference to the diameter is the same for all circles. This ratio is then called π.

EXERCISE 8.2

1. Why are (2) and (3) equivalent?

2. Write the equation of the circles with indicated centers and radii. Graph each one.

(a) (0, 0), 1   (d) (1, 2), 3   (g) $(-4, -8)$, $\sqrt[3]{5}$
(b) $(-2, -3)$, 0.5   (e) (3, 0), 3   (h) $(2, -1)$, $\sqrt{3}$
(c) $(-1, 4)$, 5   (f) (0, 1), 1

Answers. (a) $x^2 + y^2 = 1$. (c) $(x + 1)^2 + (y - 4)^2 = 25$.

3. Test the following points to see whether they lie on the circle found in prob. 2d: (1, 2), (4, 2), (1, 5), (0, 0), (3/5, 2).

4. The circle with center at the origin and radius one is called the **unit circle**. (a) Write its equation and graph it. (b) What points with integral coordinates lie on this circle? (c) What is its circumference?

†*5.* Prove that there are only four points with integral coordinates that lie on the unit circle.

6. Show that the following points lie on the unit circle, and plot each one: $(0.5, 0.5\sqrt{3})$, $(-0.5, 0.5\sqrt{3})$, $(0.5\sqrt{2}, 0.5\sqrt{2})$, $(0.5\sqrt{3}, -0.5)$.

7. Show that if (a, b) is a point on the unit circle, $(-a, b)$, $(a, -b)$, and $(-a, -b)$ are also on the circle. Illustrate by plotting four such points. Explain why this fact enables us to say that the circle is symmetric with respect to both axes and to the origin.

8. Show that the statements in prob. 7 are true of any circle with center at the origin.

9. Graph:

(a) $x^2 + y^2 = 9$
(b) $x^2 + y^2 = 2$
(c) $(x - 2)^2 + (y + 1)^2 = 1$

(d) $x^2 + y^2 = 0$
(e) $(x - 2)^2 + (y - 3)^2 = 5$
(f) $(x + 1)^2 + (y - 5)^2 = 7$

10. In one or more of the equations in probs. 2 and 9 translate the origin to the center and write the equation in terms of the new coordinates. Graph, showing both old and new axes.

11. Graph:

(a) $x^2 + y^2 \geq 4$
(b) $(x + 1)^2 + (y + 2)^2 = 0$
(c) $(x + 1)^2 + (y - 5)^2 < 5$

(d) $x^2 + y^2 > 0$
(e) $x^2 + y^2 \geq 0$
(f) $x^2 + y^2 = -1$

12. Graph: (a) $y = \sqrt{1 - x^2}$. (b) $x = \sqrt{2 - y^2}$. (c) $y = 2 + \sqrt{3 - x^2}$.

13. A circle has a radius of 7.23. Find its circumference and area accurate to two decimal places.

14. Graph the annulus defined by $r \leq \sqrt{x^2 + y^2} \leq R$. Find its area.

15. An equation of the form $x^2 + y^2 + Dx + Ey + F = 0$ is a circle since it can be put in the form (3) by completing the square on the terms in x and y. Thus $x^2 + 2x + y^2 - 6y - 2 = 0$ becomes $(x^2 + 2x + 1) + (y^2 - 6y + 9) = 2 + 1 + 9$ or $(x + 1)^2 + (y - 3)^2 = 12$. Graph the following circles, indicating the center and radius of each:

(a) $x^2 + y^2 + 2x - 2y + 1 = 0$
(b) $x^2 + y^2 + 6x - 4y = 0$
(c) $x^2 + y^2 - 5x + 8y + 4 = 0$

(d) $x^2 + y^2 - 7x - 6y = 8$
(e) $x^2 + y^2 + 2x = 4$
(f) $x^2 + y^2 - 3y = 5$

Answers. (a) $(-1, 1)$, $r = 1$. (c) $(5/2, -4)$, $r = 0.5\sqrt{73}$.

16. Show that our definition of arc length gives the ordinary length if we apply it to a straight line or a broken line.

17. Graph the set of points that satisfy: $(x - 1)^2 + (y - 1)^2 \leq 1$; or $(x + 1)^2 + (y - 1)^2 \leq 1$;   or $(x + 1)^2 + (y + 1)^2 \leq 1$;   or $(x - 1)^2 + (y + 1)^2 \leq 1$; or $-1 \leq x \leq 1$ and $-1 \leq y \leq 1$. (The figure is called a **quatrefoil.**)

8.3 Angles

It is the purpose of this section to give a precise definition of "angle." The student probably has a fairly clear idea of angles and how to measure them in degrees. However, if he tries to explain what an angle is and why, for example, a right angle is 90°, he may have difficulty. Such "definitions" as "an angle is the opening between two straight lines" are not satisfactory. In Fig. 8.3.1, which is the "opening" and how are we to measure it? Again, it is not satisfactory to say a "degree of angle is 1/360 of the whole angular space" about a point. What is the "whole angular space" and how are we to find 1/360 of it?

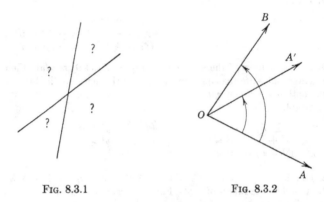

FIG. 8.3.1   FIG. 8.3.2

The word "angle" suggests a picture like AOA' in Fig. 8.3.2. The point O is called the **vertex** and OA and OA' are called the **sides.** The sides are called **half-lines** since they extend only in one direction from O. In order to compare two angles we can place their vertices and one side together. Then if the other sides coincide we call the angles congruent. If not, the one whose other side lies "outside" we call larger. Thus in the figure, $AOB >$ AOA'. But what do we mean by "outside"? An easy way to explain the idea precisely is to say that if we rotate OA about O in a counterclockwise direction it comes to OA' before it comes to OB. We make use of these ideas in order to give a precise definition of an angle.

Consider a vector OA with its initial point at the origin of a coordinate system and its terminal point on the positive x-axis (Fig. 8.3.3). If this vector rotates about the origin, its terminal point traces out a portion of the circumference of a circle of radius $r = OA$. We call the vector OA the **radius vector.** It may trace out part or all of the circumference of the circle, or it may trace out the circumference more than once. Since it may rotate in either direction, we agree to measure arc lengths from A positively in a counterclockwise direction and negatively in a clockwise direction. After performing the rotation, the radius vector lies along some line through O, say OA'. We say that the radius vector has generated the **angle** AOA', which consists in the half-line OA, called the **initial side;** the half-line OA', called the **terminal** side; the point O, called the **vertex;** and the circular arc traced out by the terminal point of the radius vector in rotating from the initial to the terminal side. The direction of rotation is indicated by putting an arrow on the arc at its terminal end. If the rotation covers the circumference more than once, this can be indicated by separating the arcs where they overlap. When the angle is placed with its vertex at the origin and its initial side along the positive

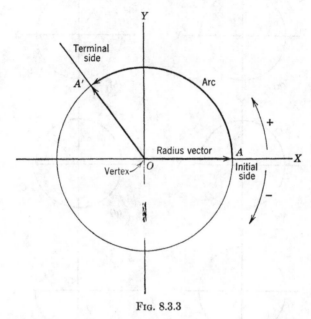

FIG. 8.3.3

x-axis, as in Fig. 8.3.3, it is said to be in **standard position,** and we say that it lies in the quadrant in which its terminal side is located. In Fig. 8.3.4 are sketched the angles corresponding to arcs that are various multiples of the circumference C.

Since our discussion indicates that an angle is just the geometric picture of a rotation, it would seem reasonable to define the magnitude of an angle by the number of revolutions in the corresponding rotation. Since the number of revolutions is given by the ratio of arc to circumference, we have:

*8.3.1   Angle in **revolutions** $= \dfrac{\text{arc}}{\text{circumference}} = \dfrac{\text{arc}}{2\pi r}$

(Definition)

Here and later we write "angle" in place of "magnitude of angle." According to the definition, if the radius vector traces out an arc equal to twice the circumference, it has swept out an angle of two revolutions. Similarly an arc of one-third of the circumference

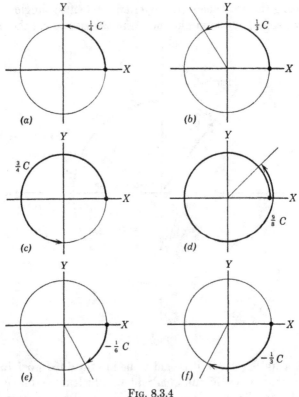

Fig. 8.3.4

corresponds to an angle of one-third revolution. The angle is positive or negative according to the direction of the arc. Two angles are called **equal** if they are generated by the same rotation, that is, have the same magnitude. Two angles are called **congruent** if their terminal sides coincide when they are in standard position. Equality implies congruence, but not conversely. (See Ex. 8.3.10.)

The student should note that the ratio of arc to circumference is not changed if we take a radius vector of different length, for two half-lines through the center of concentric circles cut off arcs that

are proportional to the circumferences (and also to the radii).
Thus, in Fig. 8.3.5,

(1)
$$\frac{\text{arc } AA'}{2\pi(OA)} = \frac{\text{arc } BB'}{2\pi(OB)}$$

It follows that magnitude is independent of the length of the radius
vector.

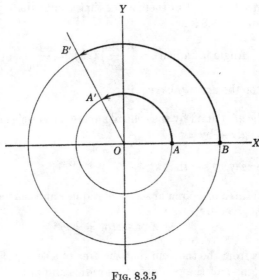

FIG. 8.3.5

Natural as is the definition *8.3.1, angles are not usually measured in revolutions. The most common system in elementary work
is the degree system, in which 360° represents one revolution.
Hence:

*8.3.2 Angle in **degrees** = (revolutions)(360°)   (Definition)

For example, one revolution is 360°, one-eighth revolution is 45°,
etc.

The degree system is often convenient for numerical computations, but in theoretical work the radian system is superior. It
results from letting 2π (the circumference of the unit circle) represent one revolution. Then:

*8.3.3 Angle in **radians** = (revolutions)(2π) (Definition)

One revolution is 2π radians, one-eighth revolution is $\pi/4$ radians, etc. Radian measure is sometimes indicated by "rad" or by a raised r in parentheses as in $2^{(r)}$; but *when an angle is given without any units, the units are understood to be radians.* Thus an angle of 1 means 1 radian, not $1°$, and an angle θ (theta) means θ radians. There are many advantages in using radian measure, most of which will become apparent only when the student continues with more advanced work. But there is one that is immediately evident. We may rewrite *8.3.3 as

(2)   Angle in radians $= \left(\dfrac{\text{arc}}{2\pi r}\right)(2\pi) = \dfrac{\text{arc}}{\text{radius}}$

Solving for the arc, we have:

*8.3.4   The arc cut off by the central angle θ on a circle of radius r is given by
$$\text{Arc} = \theta r$$
It is also easy to see that

*8.3.5   The area of a circular sector of radius r and central angle θ is given by
$$\text{Area of sector} = \tfrac{1}{2}\theta r^2$$

This follows from the theorem that the area of a sector (Fig. 8.3.6) is given by one-half the product of its radius and arc. Thus if we know an angle in radians we can get the corresponding arc or area by very simple formulas.

It is often necessary to change from one system of measuring angles to another. The easiest way to do this in simple cases is to visualize the angle as a number of revolutions and then use *8.3.2 or *8.3.3 to express it in the other system. Thus an angle of $3\pi/2$ radians means $(3\pi/2)/2\pi =$ three-fourths of a revolution. Hence it is $(3/4)(360°) = 270°$. Or, in the other direction, an angle of $135°$ is $135/360 =$ three-eighths of a revolution and hence is $(3/8)(2\pi)$ or $3\pi/4$ radians. The relationship between radians and degrees is given by the following:

*8.3.6 $\dfrac{\text{Angle in radians}}{2\pi} = \dfrac{\text{angle in degrees}}{360°}$

$= \text{angle in revolutions}$

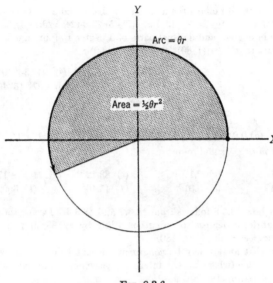

Fɪɢ. 8.3.6

This follows from *8.3.2 and *8.3.3. By solving for "angle in radians" and "angle in degrees" we find formulas for changing from either system to the other.

EXERCISE 8.3

1. Sketch the following angles in standard position, being sure to include the axes, initial and terminal sides, arc, and the size of the angle:

(a) 0.5 rev.   (c) 2.5 rev.   (e) ⅙ rev.   (g) 0.75 rev.
(b) −⅔ rev.   (d) ⅓ rev.   (f) 0.2 rev.   (h) 1.1 rev.

2. Same as prob. 1 for:

(a) 180°   (c) 450°   (e) 60°   (g) 45°   (i) 120°   (k) −181°
(b) −90°   (d) 135°   (f) 30°   (h) −85°   (j) 400°   (l) 236°

3. Same as prob. 1 for:

(a) π (c) $-\pi/2$ (e) $-3\pi/4$ (g) 3π (i) $7\pi/8$ (k) $-\pi$
(b) $\pi/4$ (d) $3\pi/4$ (f) $\pi/8$ (h) $5\pi/4$ (j) -7π (l) $\pi/6$

4. Indicate how many revolutions are represented by each angle in probs. 2 and 3. [*Answers.* (2*a*) 0.5. (2*c*) 5/4. (2*e*) 1/6. (3*a*) 0.5. (3*c*) −0.25. (3*e*) −3/8. (3*g*) 1.5.]

5. Transform to degrees the angles in probs. 1 and 3. [*Answers.* (1*a*) 180°. (1*c*) 900°. (1*e*) 60°. (1*g*) 270°. (3*a*) 180°. (3*c*) −90°. (3*e*) −135°. (3*g*) 540°.]

6. Transform to radians the angles in probs. 1 and 2. [*Answers.* (1*a*) π. (1*c*) 5π. (1*e*) π/3. (1*g*) 3π/2. (2*a*) π. (2*c*) 5π/2. (2*e*) π/3. (2*g*) π/4.]

7. Find in degrees and decimal parts of a degree to four significant digits, using the fact that $1' = (1/60)°$ and $1'' = (1/60)'$:

(*a*) 1'	(*e*) 50° 10'	(*i*) 94° 36' 19''
(*b*) 1''	(*f*) 41° 17'	(*j*) 127° 15' 40''
(*c*) 6'	(*g*) 10° 5' 25''	
(*d*) 36''	(*h*) −83° 42'	

Answers. (*a*) 0.01667°. (*b*) 0.0002778°. (*e*) 50.17°. (*g*) 10.09°.

8. Find in degrees and minutes:

(*a*) 31.27°	(*c*) 0.1°	(*e*) 83.12°	(*g*) −179.23°
(*b*) 18.94°	(*d*) 0.01°	(*f*) 45.46°	(*h*) 359.11°

9. Show how *8.3.6 follows from *8.3.2 and *8.3.3. Derive the following rule: To transform degrees to radians multiply by π/180, and to transform radians to degrees multiply by 180/π.

10. Show that angles may be congruent without being equal by showing that θ and $\theta + 2k\pi$ (where k is any integer) have coincident initial and terminal sides when placed in standard position.

11. Find in degrees and hundredths of a degree, and in degrees, minutes, and seconds:

(*a*) 3.1416	(*c*) −1	(*e*) 1.2	(*g*) π/5	(*i*) 4π/9
(*b*) 2	(*d*) π/10	(*f*) 1.57	(*h*) 5π/6	(*j*) 7π/10

Answers. (*a*) 180.00°, 180° 0'. (*c*) −57.30°, −57° 18'.

12. Find in radians, using logs where needed:

(*a*) 27.6°	(*e*) 10°	(*i*) 57.3°
(*b*) 45° 34'	(*f*) 36.71°	(*j*) 1°
(*c*) 100°	(*g*) 59° 20'	
(*d*) 50°	(*h*) 88° 21' 43''	

Answers. (*a*) 0.4817. (*c*) 1.745. (*e*) 0.1745. (*g*) 1.035. (*i*) 1.000.

13. In a circle of radius 6 there is a central angle of 3π/4. Find the arc and the area of the sector that it subtends.

14. Find to four significant figures the number of degrees in one radian, and record for reference. Express also in degrees and minutes, in minutes, and in seconds. (*Answer.* 57.30°.)

15. Find to four significant digits the number of radians in one degree, and record for reference.

16. Recall or look up the geometric theorem that states that the ratio of arc to circumference is the same no matter what circle we draw with center at the vertex of an angle. Write out a formal proof.

17. Find the arc and area of a sector whose central angle is 36° in a circle of radius 2. (*Answer.* Both are 1.257.)

18. Derive a formula for the arc and the area of a sector determined by a central angle A in degrees. (*Suggestion.* Replace θ by an equal expression involving A.)

19. In a circle of radius 3 a central angle cuts off an arc of length 2. Find the angle in radians and degrees.

20. Show that, if we draw a circle of unit radius, central angles in radians are equal to their arcs, and thus we may think of angles in radians or arcs on a unit circle.

21. If a body is rotating about a center its *angular velocity* is the angle passed through per unit time, and its *linear velocity* is the arc length passed over per unit time, that is, these velocities are the rates of change of angle and arc with time. If the velocity is constant, we have what is called *uniform angular motion.* Then linear velocity v is given by $v = s/t$, where s is the arc covered in a time t, and angular velocity w is given by $w = v/r$, where r is the radius. A wheel is rotating 100 times a minute. What is its angular velocity in rotations per minute? radians per minute? degrees per minute? If the radius is 6, what is the linear velocity of a point on the circumference?

22. At a time $t = 1$ a body has rotated through $20°$. At $t = 2.3$ it has rotated $26° \, 30'$. Find the average angular velocity in radians per unit time and the average linear velocity of a point at a distance of 10 inches from the center of rotation.

23. A car is traveling 60 mph. Its wheels have a radius of 1 foot. How many times per second are the wheels revolving? (*Suggestion.* Assume that the distance traveled on the ground is the same as the arc length covered by a point on the circumference of a wheel. What is the angular velocity in radians per second of each wheel?)

24. The angular velocity of the earth about its axis is $\pi/12$ per hour. (Why?) If the radius of the earth is 3964 miles, how fast is a point on the equator moving? Ignore the motion of the earth about the sun. (*Answer.* 1038 mph.)

25. A body moves in a circle so that the angle through which it has rotated is $180°$ when time is 2 seconds and $720°$ when time is 4 seconds. Find its average angular velocity in radians per second. Suppose $\theta = \pi t^2/4$. (*a*) Show that the given values satisfy this relation. (*b*) Find the exact velocity at $t = 2, 3, 4$.

26. Angles are sometimes measured in a unit called the **mil.** One mil equals 1/6400 of a revolution. Find in mils the angles in probs. 1, 2, and 3.

27. How many degrees are there in one mil? How many mils in one degree?

28. Show that an angle of one mil subtends an arc approximately equal to 10^{-3} times the radius.

29. When an angle is small its arc and chord are almost equal. (Why?) (*a*) Suppose the sun is at a distance of 93,000,000 miles from the earth and that it subtends an angle of $31' \, 50''$ at the earth. Find the sun's diameter. (*b*) If the planet Mars subtends an angle of $24''$ at the earth and is 36,354,000 miles away, what is its diameter?

30. The angle through which a flywheel has rotated is given by $\theta = 0.5at^2$, where t is the time and a is a constant. (*a*) Find the angular velocity w. (*b*) Show that the angular acceleration (the rate of change of velocity) is a.

8.4 Definitions of the circular functions

Let θ be an angle in standard position generated by a radius vector of length r. (Several cases are sketched in Fig. 8.4.1.) Let (x, y) be the coordinates of the end of the radius vector in its terminal position. Then the **sine** and **cosine functions** are defined as follows:

*8.4.1
$$\sin \theta = y/r$$
(Definition)
$$\cos \theta = x/r$$

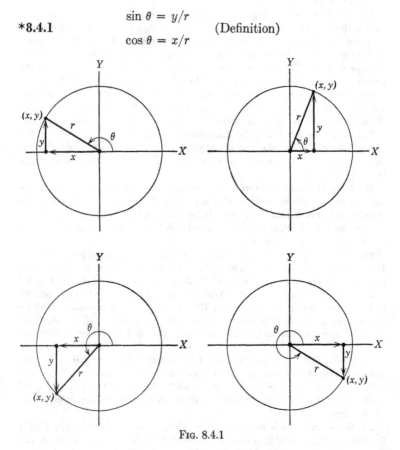

FIG. 8.4.1

We write sin and cos as abbreviations for sine and cosine. The student should note that these symbols have no meaning unless an angle is written after them. Thus cos $= \frac{1}{2}$ is nonsense; cos $(\pi/2)$ $= \frac{1}{2}$ is merely false.

For each real value of θ there is just one value for each of these functions, because the angle uniquely determines the terminal position of the vector, and the ratios are independent of the length of the radius vector. In order to see this, consider two radius vectors r and r' that generate the same angle θ (Fig. 8.4.2). If

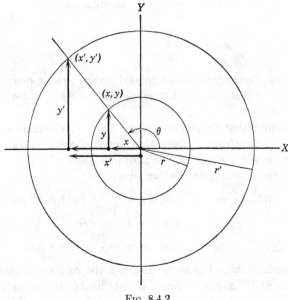

Fig. 8.4.2

(x, y) and (x', y') are the coordinates of their end points in the terminal positions, it is easy to see that the triangles formed by the radius vectors and the vectors x, y, x', y' are similar triangles. Hence their corresponding sides are proportional, and the ratios that give the sine and cosine are equal.

The **tangent function** is defined by

*8.4.2 $$\tan \theta \equiv \frac{\sin \theta}{\cos \theta} \qquad \text{(Definition)}$$

It has a unique value for each θ, unless $\cos \theta$ is zero, in which case it is undefined. Evidently $\tan \theta = y/x$. (Why?)

The **cosecant, secant,** and **cotangent** functions are defined as the reciprocals of the sine, cosine, and tangent, that is,

*8.4.3

$$\csc \theta \equiv \frac{1}{\sin \theta}$$

$$\sec \theta \equiv \frac{1}{\cos \theta} \qquad \text{(Definition)}$$

$$\cot \theta \equiv \frac{1}{\tan \theta}$$

They are evidently uniquely defined except when a denominator is zero. From *8.4.1 and *8.4.2 we see that $\csc \theta = r/y$, $\sec \theta = r/x$, and $\cot \theta = x/y$.

The six functions defined above are called the **circular** or **trigonometric functions.** Since they are independent of r, it is legitimate to use the unit vector and unit circle. Then $r = 1$, and we have the following simple formulas:

*8.4.4   When $r = 1$ $\sin \theta = y$ $\csc \theta = 1/y$

$\cos \theta = x$ $\sec \theta = 1/x$

$\tan \theta = y/x$ $\operatorname{ctn} \theta = x/y$

The student should observe carefully the relation between the definitions and Fig. 8.4.1. Since all coordinates are positive in the first quadrant, and since r is always taken as positive, all functions of angles in the first quadrant are positive. In the second quadrant, since $y > 0$ and $x < 0$, the sine and cosecant are positive and the other functions are negative. The student should work out a chart showing the sign of each function in each quadrant—not for the purpose of memorizing it, but in order to learn how to "figure out" quickly the sign of any function in any quadrant.

For certain angles the circular functions can be calculated exactly by elementary geometry. First, consider the angle 0. Here the terminal position coincides with the initial position of the vector, and the coordinates of the end of the vector are $(r, 0)$. Hence

(1)

$\sin 0 = 0/r = 0$ $\csc 0 = r/0$ (undefined)

$\cos 0 = r/r = 1$ $\sec 0 = r/r = 1$

$\tan 0 = 0/r = 0$ $\cot 0 = r/0$ (undefined)

We can find these results even more easily by considering the unit vector. Then $r = 1$, the coordinates are $(1, 0)$, and (1) follows from *8.4.4. When the denominator is zero in one of the ratios that define the circular functions, the expression is undefined, but this is usually expressed by saying that the undefined expression is "infinite." The symbol ∞ ("infinity") is used, and we may write $\csc 0 = \infty$ and $\cot 0 = \infty$. But ∞ is not a number. The expression $\cot 0 = \infty$ means simply that $|\cot \theta|$ can be made arbitrarily large by taking θ close enough to zero, and that $\cot \theta$ is not defined for $\theta = 0$.

Consider the angle $\pi/3 = 60°$, sketched in a unit circle in Fig. 8.4.3. By the Pythagorean theorem, $x^2 + y^2 = 1$. In plane geometry it is shown that in a right triangle with a 60° angle, the side adjacent to the 60° angle is one-half the hypotenuse. Since $r = 1$, $x = 1/2$, and $y = \sqrt{1 - x^2} = (1/2)\sqrt{3}$. Hence

$$\sin \pi/3 = (1/2)\sqrt{3} \qquad \csc \pi/3 = (2/3)\sqrt{3}$$

(2)
$$\cos \pi/3 = 1/2 \qquad \sec \pi/3 = 2$$

$$\tan \pi/3 = \sqrt{3} \qquad \cot \pi/3 = (1/3)\sqrt{3}$$

Finally, consider $\theta = 135° = 90° + 45°$, sketched in Fig. 8.4.4. Triangle OBA' is evidently an isosceles triangle. (Why?) Hence

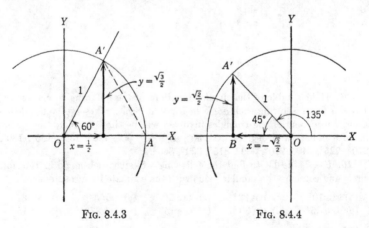

Fig. 8.4.3   Fig. 8.4.4

the two legs are equal in length. However, x is negative and y is positive. Hence $x = -y$. Substituting $-y$ for x in $x^2 + y^2 = 1$, we have $y^2 + y^2 = 2y^2 = 1$ and $y = 1/\sqrt{2} = (1/2)\sqrt{2}$. Then $x = -(1/2)\sqrt{2}$. Now the functions can easily be written down.

EXERCISE 8.4

1. Express in words *8.4.1 through *8.4.4.

2. In each case the coordinates of a point on the terminal side of an angle in standard position are given. Find the circular functions of each angle and sketch:

(a) (1, 5)   (c) (−1, −2)   (e) (0, 2)   (g) (4, −1)
(b) (−2, 3)   (d) (3, −4)   (f) (−3, 0)   (h) (1, −1)

3. Find the angles in prob. 2 to the nearest 30°.

4. For what positions of the terminal side is $x = 0$? Why are sec θ and tan θ undefined in this case? What are the corresponding angles? For what positions of the terminal side are csc θ and cot θ undefined? Why?

5. Why do two congruent angles have identical circular functions?

6. Draw figures similar to Fig. 8.4.2 for angles in quadrants I, III, IV. Give a careful proof of the similarity of the triangles involved and hence of the fact that the circular functions are independent of the length of the radius vector.

7. Draw figures to illustrate *8.4.4, showing angles in each quadrant and labeling the vectors that are equal to the sine and cosine in each case.

8. Complete the following chart showing the signs of the circular functions in each quadrant. Be prepared to justify each entry.

	I	II	III	IV
sin θ	+			
cos θ		−		
tan θ			+	
cot θ				
sec θ	+			
csc θ				−

9. Make a table of the circular functions of the following special angles. Draw a figure showing each angle in the unit circle. Preserve your calculations, since some of the triangles involved will be similar. Show the angles in both radians and degrees. 0°, 30°, 45°, 60°, 90°, 120°, 135°, 150°, 180°, 210°, 225°, 240°, 270°, 300°, 315°, 330°, 360°.

10. Use Table IV to find the following. Sketch each angle in the unit circle and compare the tabular value with the estimate by measurement.

(a) sin 10°   (c) cos 32°   (e) tan 20°   (g) tan 5°   (i) cot 87°
(b) tan 45°   (d) sin 30°   (f) cos 15°   (h) cos 87°   (j) sec 15°

11. What can you say about an angle A less than 360° if:

(a) sin A > 0 and cos A > 0   (c) sin A > 0
(b) sin A > 0 and cos A < 0   (d) cos A > 0

    (e) $\tan A > 0$ (h) $\sin A < 0$ and $\cos A < 0$
    (f) $\tan A < 0$ (i) $\sin A < 0$ and $\sec A > 0$
    (g) $\sin A < 0$ and $\tan A < 0$ (j) $\csc A > 0$ and $\cos A < 0$

    Answers. (a) $0 < A < \pi/2$. (c) $0 < A < \pi$. (e) $0 < A < \pi/2$ or $\pi < A < 3\pi/2$.

    †**12.** Prove that: (a) $|\sin x| \leq 1$. (b) $|\cos x| \leq 1$. (c) $|\sec x| \geq 1$. (d) $|\csc x| \geq 1$. (e) $\tan x$ may take any real value.

    13. Show that the area of an equilateral triangle is given by $A = a^2\sqrt{3}/4$, where a is a side.

    14. Draw a sketch to illustrate the discussion of $\cot 0$ and $\csc 0$. Show several angles near zero with assumed values for y and explain how the functions get larger in absolute value as y approaches zero. Consider both positive and negative y.

    15. Show how the tangent and secant become large in absolute value as the angle approaches $90°$.

    †**16.** It is a fact that when the angle is near zero $\sin \theta$ is approximately equal to θ, provided the angle is measured in radians. Draw a figure and explain why this is plausible. (Compare Ex. 8.3.29.)

    17. In the unit circle the circular functions are represented by the lengths of certain segments as shown in Fig. 8.4.5. (a) Justify the figure. (b) Draw similar figures for angles in the other three quadrants and identify the segments giving the functions. (*Suggestion.* Let the tangent lines remain in

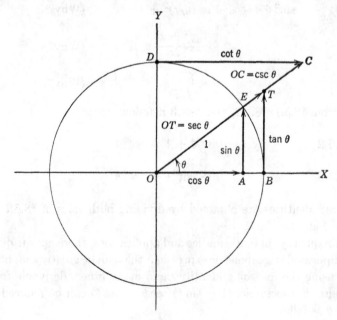

F‌ɪɢ. 8.4.5

the same positions, and let T and C be the points where the terminal side, or its continuation, meets these tangents. Then measure directed distances on the terminal side positive in the direction of the radius vector.)

8.5 Fundamental identities

It is evident from their definitions that the circular functions are interrelated. The fact that $\sin \theta$ and $\cos \theta$ are the sides of a right triangle whose hypotenuse is 1 (*8.4.4 and Fig. 8.4.1) suggests the following identity:

*8.5.1 $$\sin^2 \theta + \cos^2 \theta \equiv 1$$

In this identity $\sin^2 \theta$ and $\cos^2 \theta$ mean $(\sin \theta)^2$ and $(\cos \theta)^2$. It is customary to write exponents in this way to avoid using parentheses. However, the exponent -1 is not put in this position because $\sin^{-1} \theta$ will be assigned a different meaning in 8.9. Hence we write $1/\sin \theta$ as $(\sin \theta)^{-1}$ or $\csc \theta$. A formal proof of *8.5.1 is easy:

(1) $\sin^2 \theta + \cos^2 \theta \equiv (y/r)^2 + (x/r)^2$ (Why?)

(2) $\equiv \dfrac{y^2 + x^2}{r^2}$ (Why?)

(3) $\equiv 1$ (Why?)

From *8.5.1, *8.4.2, and *8.4.3 it follows that:

*8.5.2 $$\tan^2 \theta + 1 \equiv \sec^2 \theta$$

*8.5.3 $$1 + \cot^2 \theta \equiv \csc^2 \theta$$

These identities are obtained by dividing both sides of *8.5.1 by $\cos^2 \theta$ and by $\sin^2 \theta$.

Frequently, in mathematics and applications, there appear quite complicated trigonometric expressions that can be greatly simplified by using the foregoing identities and many others derivable from them. For example, $(1 - \sin \theta)(\sec \theta + \tan \theta)$ can be reduced to $\cos \theta$ as follows:

(4) $(1 - \sin \theta)(\sec \theta + \tan \theta) \equiv (1 - \sin \theta)\left(\dfrac{1}{\cos \theta} + \dfrac{\sin \theta}{\cos \theta}\right)$

$$(\text{*}8.4.3 \text{ and } \text{*}8.4.2)$$

(5) $\equiv (1 - \sin \theta)(1 + \sin \theta)/\cos \theta$

$$(\text{Why?})$$

(6) $\equiv (1 - \sin^2 \theta)/\cos \theta$ (Why?)

(7) $\equiv \cos^2 \theta/\cos \theta$ (Why?)

(8) $\equiv \cos \theta$ (Why?)

Similar methods can be used to find an expression for each function of an angle in terms of any other function alone. For example, from *8.5.1 it follows immediately that $\sin \theta = \pm\sqrt{1 - \cos^2 \theta}$, so that we have a formula for $\sin \theta$ in terms of $\cos \theta$ alone. Whether the plus or the minus sign is appropriate depends on the quadrant in which θ lies. Formulas of this kind can be found in all cases by using the above identities, but the easiest way is to go back to the definitions of the circular functions and choose convenient values of x, y, and r. For example, in order to find all functions in terms of the cosine we consider the angle θ in standard position and choose the radius vector as 1 (Fig. 8.5.1). Then $x = \cos \theta$. By the

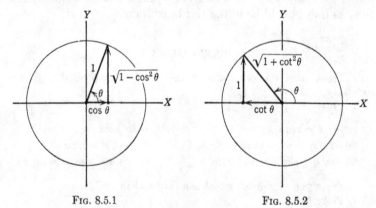

FIG. 8.5.1   FIG. 8.5.2

Pythagorean theorem, $y = \pm\sqrt{1 - \cos^2 \theta}$. The other functions can now be written down by inspection. Similarly, to find all functions in terms of the cotangent, we let $x = \cot \theta$ (Fig. 8.5.2).

Then $y = 1$, since $x/y = \cot \theta$. Hence by the Pythagorean theorem, $r = \sqrt{1 + \cot^2 \theta}$. Using the definitions of the circular functions, we have:

$$\sin \theta = \frac{1}{\pm\sqrt{1 + \cot^2 \theta}} \qquad \csc \theta = \pm\sqrt{1 + \cot^2 \theta}$$

$$(9) \qquad \cos \theta = \pm\frac{\cot \theta}{\sqrt{1 + \cot^2 \theta}} \qquad \sec \theta = \frac{\pm\sqrt{1 + \cot^2 \theta}}{\cot \theta}$$

$$\tan \theta = \frac{1}{\cot \theta} \qquad\qquad \cot \theta = \cot \theta$$

The plus or minus signs are required because in each quadrant one or the other must be chosen to make the identity valid. Thus in the second quadrant we would choose the plus sign for the sine and cosecant since these are positive. We would also choose the plus sign in the foregoing formulas for $\cos \theta$ and $\sec \theta$, because $\cot \theta$ is negative in the second quadrant and we must have $\cos \theta$ and $\sec \theta$ negative also. However, if θ were in the third quadrant, where $\cot \theta$ is positive, we would choose the minus sign in the formulas for $\cos \theta$ and $\sec \theta$ in order to make these functions negative, as they should be in the third quadrant.

EXERCISE 8.5

1. Prove from the definitions of the circular functions that $\cot \theta \equiv \cos \theta/\sin \theta$.

2. Prove:

(a) $\sin \theta \equiv \tan \theta/\sec \theta$ \qquad (d) $\cot^2 x \sec^2 x \equiv 1 + \cot^2 x$

(b) $(\tan x + \cot x) \tan x \cos^2 x \equiv 1$ \qquad (e) $\sec x \cot x \equiv \csc x$

(c) $\cos^2 \theta - \sin^2 \theta \equiv 1 - 2 \sin^2 \theta$ \qquad (f) $\cos x + \sin^2 x \sec x \equiv \sec x$

3. Prove *8.5.2 by dividing both sides of *8.5.1 by $\cos^2 \theta$.

4. Prove *8.5.3.

5. Do *8.5.2 and *8.5.3 hold for *all* real values of θ? (Examine the manner in which you derived them.) Complete the following statement: "A trigonometric identity is an equation that is true for all values of the variables for which its members are. . . ."

6. Make up a table showing a formula for each circular function in terms of every other circular function.

7. Prove the following identities:

(a) $\sin^2 \theta + \cos^2 \theta \equiv \sec^2 \theta - \tan^2 \theta$

(b) $\tan^2 \theta \equiv \sin^2 \theta \tan^2 \theta + \sin^2 \theta$

(c) $\dfrac{\cos A}{1 + \sin A} \equiv \dfrac{1 - \sin A}{\cos A}$

(d) $\dfrac{\sec B}{\tan B + \cot B} \equiv \sin B$

(e) $\cos^2 \theta \cot^2 \theta \equiv (\cos \theta + \cot \theta)(\cot \theta - \cos \theta)$

(f) $\dfrac{\cos A}{1 - \tan A} + \dfrac{\sin A}{1 - \cot A} \equiv \sin A + \cot A$

(g) $\dfrac{\sec^2 \theta + \csc^2 \theta}{\sec \theta \csc \theta} \equiv \tan \theta + \cot \theta$

(h) $\dfrac{\tan \theta - \sin \theta}{\sin^3 \theta} \equiv \dfrac{\sec \theta}{1 + \cos \theta}$

(i) $\dfrac{\cot A + \cot B}{\tan A + \tan B} \equiv \cot A \cot B$

(j) $(\cos x \cos y - \sin x \sin y)^2 + (\sin x \cos y + \cos x \sin y)^2 \equiv 1$

(k) $(1 + \tan \theta)(1 - \tan \theta) + \sec^2 \theta \equiv 2$

(l) $\dfrac{1 + \sin \theta - \cos \theta}{1 + \sin \theta + \cos \theta} \equiv \dfrac{1 - \cos \theta}{\sin \theta}$

8. Prove that $\log (\tan \theta) \equiv \log (\sin \theta) - \log (\cos \theta)$.

9. Simplify:

(a) $\cos x / \cot x$

(b) $\dfrac{1 + \dfrac{\sin \theta}{\sec \theta}}{1 - \dfrac{\cos \theta}{\csc \theta}}$

(c) $\tan A (1 + \tan^2 A)^{-\frac{1}{2}}$

(d) $\sin x \cot^2 x \tan x$

(e) $(\cos^{-2} \theta - 1)^{\frac{1}{2}}$

(f) $\dfrac{\tan x (1 - \cos^2 x)^{\frac{1}{2}}}{(1 - \sin^2 x)^{\frac{1}{2}}}$

(g) $\dfrac{\sin a}{\csc a} + \dfrac{\cos a}{\sec a}$

(h) $\sin A \cos A \tan A + \sin A \cos A \cot A$

(i) $(1 - \cos^2 x) \csc^2 x / \cot x$

(j) $(\sec^2 \theta - 1) \cos^2 \theta$

(k) $(\sin^6 A - \sin^4 A \cos^2 A)^{\frac{1}{2}}$

(l) $(1 + \sin x)(1 - \sin x)^{-1}$

10. If $\sin^2 x + \cos^2 y \neq 1$, what can you say about x and y?

8.6   Identities involving related angles

It is evident from *8.4.1 that, if two angles are congruent (that is, have coincident terminal sides), they have identical circular functions.   Since one complete revolution brings the radius vector to the same position, two angles are congruent if and only if they differ by some multiple of 2π (or 360°). Hence $\sin\theta \equiv \sin(\theta + 2\pi)$ $\equiv \sin(\theta - 2\pi) \equiv \sin(\theta + 4\pi)\cdots$, etc. To sum up, $\sin\theta = \sin(\theta + 2k\pi)$, where k is any integer, positive, zero, or negative. This is true for any of the circular fnctions. Hence

*8.6.1 $f(\theta) \equiv f(\theta + 2k\pi)$ where k is any integer and f is any circular function.

If the angles are measured in degrees, $\theta + 2k\pi$ is replaced by $\theta + 360k$ in the foregoing identities. Thus $\cos 400° = \cos(400 - 360°)$ $= \cos 40°$. Here we took $f(\theta)$ to be $\cos\theta$, $\theta = 400°$, and $k = -1$. This identity means that we can replace any function of an angle greater than 360° by the same function of an angle less than 360°. The property of *8.6.1 is sometimes described by saying that the circular functions are periodic.

In Fig. 8.6.1 we have shown an angle θ in the first quadrant and the angle $-\theta$ in the fourth quadrant. The triangles OAB and OAB' are congruent (why?); hence their corresponding sides are

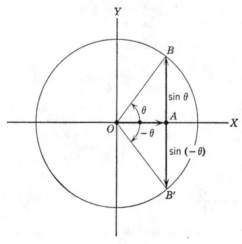

FIG. 8.6.1

equal in length. The vector representing $\cos\theta$ is identical with that representing $\cos(-\theta)$. The vector $\sin(-\theta)$ is the negative of the vector $\sin\theta$. This would be true of any angle θ and its negative, as we shall show in 8.7. Hence

*8.6.2
$$\sin(-\theta) \equiv -\sin\theta$$
$$\cos(-\theta) \equiv \cos\theta$$
$$\tan(-\theta) \equiv -\tan\theta$$

For example, $\sin(-25°) = -\sin 25°$, $\cos(-325°) = \cos 325°$, and $\tan(-10°) = -\tan 10°$.

The student will undoubtedly have noticed that the circular functions have names that can be paired: sine and cosine, secant and cosecant, tangent and cotangent. In a pair, each function is said to be the **cofunction** of the other. Thus the sine is the cofunction of the cosine and vice versa. This terminology is good because any function of an angle is equal to the cofunction of the complement of the angle. Thus $\sin\theta \equiv \cos(\pi/2 - \theta)$, $\cos\theta \equiv \sin(\pi/2 - \theta)$, $\tan\theta \equiv \cot(\pi/2 -)$, etc. If we represent the cofunction of $f(\theta)$ by co-$f(\theta)$, we may summarize this by writing:

*8.6.3
$$f(\theta) \equiv \text{co-}f(\pi/2 - \theta)$$

In 8.7 we shall prove that these identities are true for all values of θ, but it is easy to see that they hold for any acute angle. In

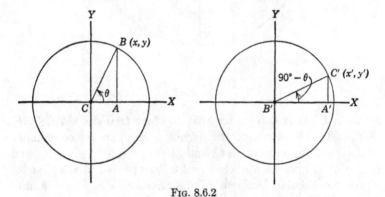

Fɪɢ. 8.6.2

Fig. 8.6.2 are sketched an acute angle θ and its complement $\pi/2 - \theta$, each in standard position in a unit circle. The triangles ABC and $A'B'C'$ are evidently congruent since they are right triangles with the same hypotenuse and equal acute angles. Hence $x' = y$, $y' = x$, and

(1) $\sin \theta = y = x' = \cos (90° - \theta)$

(2) $\tan \theta = y/x = x'/y' = \cot (90° - \theta)$

and similarly for the other functions.

In Fig. 8.6.3 are sketched four angles such that the smallest angle between the terminal side of each one and the positive or negative

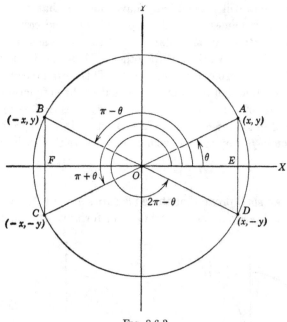

FIG. 8.6.3

x-axis is θ. It is easy to see that the four triangles OAE, OBF, OCF, and ODE are congruent. Hence, if (x, y) are the coordinates of A, the coordinates of B, C, and D are $(-x, y)$, $(-x, -y)$, and $(x, -y)$. It appears plausible, and will be proved in 8.7, that for any θ the circular functions of the angles θ, $\pi - \theta$, $\pi + \theta$, and $2\pi - \theta$ are the same in absolute value, that is,

*8.6.4 $|f(\theta)| \equiv |f(\pi - \theta)| \equiv |f(\pi + \theta)| \equiv |f(2\pi - \theta)|$,

where f is any circular function.

In order to write a true relation not involving the absolute value sign, we have to consider a particular function. For example, since 215° is in the third quadrant, its sine is negative. But 215° = 180° + 35°. Hence sin 215° = sin (180° + 35°) = −sin 35°. On the other hand, since tan 215° is positive, tan 215° = tan 35°.

The theorems of this section enable us to express any function of any angle in terms of some function of an angle not greater than 45°. (It follows that tables of the trigometric functions from 0° to 45° are sufficient.) Given a function of a negative angle, we can by *8.6.2 rewrite it in terms of a positive angle. Then, if the angle is greater than 360°, we can rewrite the function in terms of an angle less than 360° by using *8.6.1. With the aid of *8.6.4 we can rewrite this in terms of an angle less than 90°. If this angle is greater than 45°, we can use *8.6.3 to rewrite in terms of an angle less than 45°. The easiest way to do such problems in practice is not to go through the foregoing steps but to draw a sketch of the angle and then use those properties that give the desired result most quickly. For example, to rewrite sin (−305°) in terms of an angle less than 45°, we sketch as in Fig. 8.6.4a. Then we see that

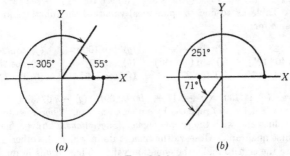

(a)   (b)

FIG. 8.6.4

−305° has the same terminal side as 55°. Hence sin (−305°) = sin 55° = cos 35°. Here we used the identity *8.6.3 and the reasoning behind *8.6.1. To find cos 251° we draw a sketch, such as Fig. 8.6.4b. Then by *8.6.3, cos 251° = −cos 71° = −sin 19°. The minus sign is required because sin 19° and cos 71° are positive (like all functions of acute angles), but cos 251° is negative, as we can see by inspection of the figure.

EXERCISE 8.6

1. Write *8.6.1 in both radians and degrees for the special cases $k = 1, 2, -1, -2$.

2. Sketch the angle θ and the angle $\theta - 2\pi = -(2\pi - \theta)$ for $\theta = 45°$, $120°$, $270°$.

3. Write the relations between the cotangent, secant and cosecant of θ and $-\theta$.

4. Make a sketch like Fig. 8.6.1 for θ in the second quadrant.

5. Write *8.6.3 with f replaced in turn by each circular function.

6. Express in terms of a positive angle between 0 and $\pi/4$. Draw a figure in each case and indicate the angles involved.

(a) $\sin(-10°)$	(i) $\sin 3\pi/2$	(q) $\sec 195°$
(b) $\sin 150°$	(j) $\cos 279°$	(r) $\sin 270°$
(c) $\tan 340°$	(k) $\tan 225°$	(s) $\csc 5$
(d) $\cos 370°$	(l) $\sin 4$	(t) $\tan 240°$
(e) $\cot 88°$	(m) $\cos 328°$	(u) $\sin 1000°$
(f) $\cos 50°$	(n) $\sec 61°$	(v) $\tan(-359°)$
(g) $\tan 3\pi/4$	(o) $\csc 100°$	(w) $\sec(-5\pi)$
(h) $\cos 180°$	(p) $\cot 293°$	(x) $\cos(7\pi - \pi/8)$

Answers. (a) $-\sin 10°$. (c) $-\tan 20°$. (e) $\tan 2°$. (g) $-\tan \pi/4$. (i) $-\sin \pi/2$. (k) $\tan 45°$. (m) $\cos 32°$. (o) $\sec 10°$. (q) $-\sec 15°$.

7. Use Table IV to find the numerical values of the functions in prob. 6. Do the same for:

(a) $\cot 185°$	(d) $\sin(-39°)$	(g) $\tan 1000°$	(j) $\tan 243°$
(b) $\sin 46° 10'$	(e) $\sin(-318°)$	(h) $\csc 32°$	(k) $\cos 382° 40'$
(c) $\cos 99° 50'$	(f) $\tan 82°$	(i) $\sec 116°$	(l) $\sin 277° 15'$

Answers. (a) 11.430. (c) -0.1708. (e) 0.6691. (g) -5.6713.

8. Write *8.6.4 with f replaced by sin, cos, and tan. Then rewrite without absolute values and with the proper signs. [*Suggestions.* Draw a figure with θ in the first quadrant. Observe the quadrants in which the other angles lie and choose the sign to make the equation valid. Thus, if θ is in quadrant I, $\pi - \theta$ is in quadrant II, and its sine is positive. Hence $\sin \theta \equiv \sin(\pi - \theta)$. However, $\pi + \theta$ is in the third quadrant, and its sine is negative. Hence $\sin \theta \equiv -\sin(\pi + \theta)$.]

9. Draw a figure showing θ in the first quadrant and the angles $\pi/2 - \theta$, $\pi/2 + \theta$, $3\pi/2 - \theta$, and $3\pi/2 + \theta$. Show that in this case:

*8.6.5
$$|f(\theta)| \equiv |\text{co-}f(\pi/2 - \theta)| \equiv |\text{co-}f(\pi/2 + \theta)|$$
$$\equiv |\text{co-}f(3\pi/2 - \theta)| \equiv |\text{co-}f(3\pi/2 + \theta)|$$

10. Write out *8.6.5 for sine, cosine, and tangent with the absolute values removed and appropriate signs.

11. Show the plausibility of:

*8.6.6 $\tan \theta \equiv \tan (\pi + \theta)$

*8.6.7 $\cot \theta \equiv \cot (\pi + \theta)$

12. Construct two angles less than 360° that satisfy each of the following equations:

(a) $\tan \theta = 1.5$ (c) $\cos \theta = -0.25$ (e) $\tan \theta = 5$
(b) $\sin \theta = 1/3$ (d) $\sin \theta = 0.3$ (f) $\cot \theta = -3$

Suggestions. The terminal side of θ must be such that the x, y, and r of any point on it satisfies the given relation. In part a, this means that $y/x = 3/2$. Hence the terminal side may be either half of the line $y = 1.5x$. In part b, $y/r = 1/3$, so that points on the terminal side can be found by drawing the line $y = 1$ and the circle $x^2 + y^2 = 3^2$.

13. Draw the angle that satisfies both of each of the following pairs of equations:

(a) $\sin \theta = 0.5; \cos \theta = -\dfrac{\sqrt{3}}{2}$ (c) $\tan \theta = 1.5; \cos \theta = -\dfrac{2}{\sqrt{13}}$

(b) $\sin \theta = -2^{-\frac{1}{2}}; \sin \theta = \cos \theta$ (d) $\cos \theta = -0.25; \sin \theta = -\dfrac{\sqrt{15}}{4}$

14. A function $f(x)$ such that $f(-x) \equiv f(x)$ is called **even,** whereas one for which $f(-x) \equiv -f(x)$ is called **odd.** The graph of an odd function is symmetric with respect to the origin; that of an even function is symmetric with respect to the y-axis. By *8.6.2 the sine is odd and the cosine is even. Classify each of the following as odd, even, or neither, and justify your statements:

(a) $\tan x$ (f) x^3 (k) e^{x^2} (p) $\sin x + \cos x$
(b) $\sec x$ (g) $3x$ (l) e^{-x} (q) $\sin^2 x$
(c) $\cot x$ (h) 10 (m) x^{-1} (r) $\sqrt[3]{x}$
(d) $\csc x$ (i) $x^5 + 1$ (n) $x^3 + x$
(e) x^2 (j) $\log x$ (o) $x^3 + 1$

Answers. (a) Odd. (c) Odd. (e) Even. (g) Odd. (i) Neither.

8.7   The sum formula and related results

The student has probably already observed that:

(1) $\sin (A + B) \not\equiv \sin A + \sin B$

In words, the operation of finding the sine is not distributive over addition. For example, if $A = B = 45°$, then $A + B = 90°$, $\sin (A + B) = 1$, and $\sin A + \sin B = 1/\sqrt{2} + 1/\sqrt{2} = \sqrt{2}$. It would be easy to give other examples, but this one justifies (1).

The correct relation is given by the following identity, which is fundamental to all understanding of the circular functions.

*8.7.1 $\sin(\theta + \phi) \equiv \sin\theta\cos\phi + \cos\theta\sin\phi$

If we compute the right side of *8.7.1 for $\theta = \phi = 45°$ we find $(1/\sqrt{2})(1/\sqrt{2}) + (1/\sqrt{2})(1/\sqrt{2}) \equiv 1$. The companion identity for the cosine is:

*8.7.2 $\cos(\theta + \phi) \equiv \cos\theta\cos\phi - \sin\theta\sin\phi$

In order to prove these theorems we find it convenient first to prove:

*8.7.3 $\cos(\theta - \phi) \equiv \cos\theta\cos\phi + \sin\theta\sin\phi$

Consider any two angles θ and ϕ and the angle $\theta - \phi$ in standard position in a unit circle (Fig. 8.7). The angle from the terminal

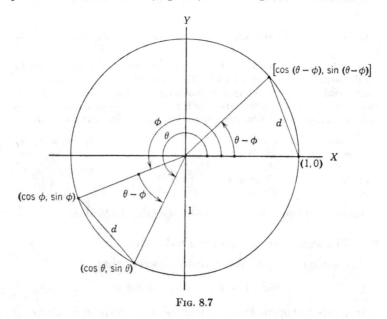

FIG. 8.7

side of ϕ to the terminal side of θ is $\theta - \phi$. Hence the chords labeled d are equal. From *8.4.4, the coordinates of the points where the terminal sides of θ, ϕ, and $\theta - \phi$ meet the unit circle are as indicated in the figure.   It follows that:

(2) $d^2 = (\cos \theta - \cos \phi)^2 + (\sin \theta - \sin \phi)^2$

(3) $= 2 - 2(\cos \theta \cos \phi + \sin \theta \sin \phi)$

and also that:

(4) $d^2 = [\cos (\theta - \phi) - 1]^2 + [\sin (\theta - \phi) - 0]^2$

(5) $= 2 - 2 \cos (\theta - \phi)$

Equating the right members of (3) and (5), we find *8.7.3 as desired.   The proof holds regardless of the values of θ and ϕ.

Now we shall derive the other identities by letting θ and ϕ take special values in *8.7.3.   For $\theta = 90°$,

(6) $\cos (90° - \phi) \equiv \cos 90° \cos \phi + \sin 90° \sin \phi$

(7) $\equiv \sin \phi$

In this result, we replace ϕ by $90° - \phi$ and so find:

(8) $\cos \phi \equiv \sin (90° - \phi)$

Now we can write:

(9) $\sin (\theta + \phi) \equiv \cos [(90° - \theta) - \phi]$ (from 7)

(10) $\equiv \cos (90° - \theta) \cos \phi + \sin (90° - \theta) \sin \phi$

(*8.7.3)

(11) $\equiv \sin \theta \cos \phi + \cos \theta \sin \phi$ (from 7 and 8)

Thus we have proved *8.7.1.

Returning to *8.7.3 with $\theta = 0$, we find:

(12) $\cos (-\phi) \equiv \cos 0 \cos \phi + \sin 0 \sin \phi$

(13) $\equiv \cos \phi$

Now from (7) with ϕ replaced by $-\phi$, we have:

(14) $\sin (-\phi) \equiv \cos [90° - (-\phi)]$

(15) $\equiv \cos [\phi - (-90°)]$ (Why?)

(16) $\equiv \cos \phi \cos (-90°) + \sin \phi \sin (-90°)$ (*8.7.3)

(17) $\equiv - \sin \phi$ (Why?)

Now *8.7.2 follows immediately:

(18) $\cos(\theta + \phi) \equiv \cos[\theta - (-\phi)]$

(19) $\equiv \cos\theta\cos(-\phi) + \sin\theta\sin(-\phi)$ (Why?)

(20) $\equiv \cos\theta\cos\phi - \sin\theta\sin\phi$ (Why?)

From *8.7.1 and *8.7.2 the identities of 8.6 can easily be derived as required in the exercise. The following are also immediate consequences, whose proof we leave to the student.

*8.7.4 $\sin(\theta - \phi) \equiv \sin\theta\cos\phi - \cos\theta\sin\phi$

*8.7.5 $\tan(\theta + \phi) \equiv \dfrac{\tan\theta + \tan\phi}{1 - \tan\theta\tan\phi}$

*8.7.6 $\tan(\theta - \phi) \equiv \dfrac{\tan\theta - \tan\phi}{1 + \tan\theta\tan\phi}$

From *8.7.1 we can also derive formulas that give the functions of twice or half an angle. Thus:

(21) $\sin 2\theta \equiv \sin(\theta + \theta)$

(22) $\equiv 2\sin\theta\cos\theta$ (Why?)

Hence:

*8.7.7 $\sin 2\theta \equiv 2\sin\theta\cos\theta$

Similarly, it can be shown that:

*8.7.8 $\cos 2\theta \equiv \cos^2\theta - \sin^2\theta$

In order to find formulas for functions of half an angle we use this last result "backwards." First we note that *8.7.8 can be written:

(23) $\cos 2\theta \equiv (1 - \sin^2\theta) - \sin^2\theta \equiv 1 - 2\sin^2\theta$

Solving for $\sin\theta$, we find:

(24) $\sin\theta \equiv \pm\sqrt{\dfrac{1 - \cos 2\theta}{2}}$

Now we replace θ by $\phi/2$ in order to find:

*8.7.9 $$\sin \frac{\phi}{2} \equiv \pm \sqrt{\frac{1 - \cos \phi}{2}}$$

Similarly, the student can derive:

*8.7.10 $$\cos \frac{\phi}{2} \equiv \pm \sqrt{\frac{1 + \cos \phi}{2}}$$

EXERCISE 8.7

1. Construct another example to justify (1).

2. Check *8.7.1 for $\theta = 60°$, $\phi = 30°$.

3. Construct examples to show that:

(a) $\cos (A + B) \not\equiv \cos A + \cos B$ (c) $\sin 2A \not\equiv 2 \sin A$

(b) $\tan (A + B) \not\equiv \tan A + \tan B$ (d) $\cos \dfrac{\phi}{2} \not\equiv \dfrac{\cos \phi}{2}$

4. Check the identities *8.7.1 through *8.7.10 for: (a) $\theta = 90°$, $\phi = 60°$. (b) $\theta = 30°$, $\phi = 60°$.

5. Find exact expressions for the following without reference to tables. (Do not evaluate in decimal form.)

(a) $\sin 75°$ (d) $\cos 15°$ (g) $\sin 37.5°$ (j) $\sec 97.5°$
(b) $\cos 75°$ (e) $\sin 7.5°$ (h) $\tan 75°$ (k) $\cos 52.5°$
(c) $\sin 15°$ (f) $\cos 22.5°$ (i) $\cos (-11.25°)$ (l) $\sin 41.25°$

Answers. (a) $0.5\sqrt{2 + \sqrt{3}}$. (c) $0.5\sqrt{2 - \sqrt{3}}$.

6. Prove *8.7.5. [*Hint.* $\tan (\theta + \phi) \equiv \sin (\theta + \phi)/\cos (\theta + \phi)$.]

7. Prove *8.7.4 and *8.7.6. [*Hint.* $\theta - \phi \equiv \theta + (-\phi)$.]

8. Prove *8.7.8.

9. Prove *8.7.10.

10. Prove:

*8.7.11 $$\tan \frac{\phi}{2} \equiv \sqrt{\frac{1 - \cos \phi}{1 + \cos \phi}} \equiv \frac{1 - \cos \phi}{\sin \phi}$$

*8.7.12 $$\tan 2\theta \equiv \frac{2 \tan \theta}{1 - \tan^2 \theta}$$

11. Prove that $\cos 2\theta \equiv 1 - 2 \sin^2 \theta \equiv 2 \cos^2 \theta - 1$.

12. Use *8.7.1 and *8.7.2 to prove *8.6.1 for sine and cosine. Why does it immediately follow that it holds for all functions?

13. Note that in this section we proved the first two equations in *8.6.2. Prove the third one.

14. Prove *8.6.3 for all θ and all circular functions.

15. Derive formulas like *8.6.4 and *8.6.5 for each function without the absolute values.

16. Derive formulas for $\cot(\theta \pm \phi)$ in terms of $\cot\theta$ and $\cot\phi$.

17. Prove by evaulating the right members:

(25) $$2\sin A \cos B \equiv \sin(A+B) + \sin(A-B)$$

(26) $$2\cos A \sin B \equiv \sin(A+B) - \sin(A-B)$$

(27) $$2\cos A \cos B \equiv \cos(A+B) + \cos(A-B)$$

(28) $$-2\sin A \sin B \equiv \cos(A+B) - \cos(A-B)$$

18. Use the results of the previous problem to prove:

*8.7.13 $$\sin\theta + \sin\phi \equiv 2\sin\left(\frac{\theta+\phi}{2}\right)\cos\left(\frac{\theta-\phi}{2}\right)$$

*8.7.14 $$\sin\theta - \sin\phi \equiv 2\cos\left(\frac{\theta+\phi}{2}\right)\sin\left(\frac{\theta-\phi}{2}\right)$$

*8.7.15 $$\cos\theta + \cos\phi \equiv 2\cos\left(\frac{\theta+\phi}{2}\right)\cos\left(\frac{\theta-\phi}{2}\right)$$

*8.7.16 $$\cos\theta - \cos\phi \equiv -2\sin\left(\frac{\theta+\phi}{2}\right)\sin\left(\frac{\theta-\phi}{2}\right)$$

Hint. Let $A + B = \theta$ and $A - B = \phi$ in prob. 17.

19. Find formulas for $\sin 3\theta$ and $\cos 3\theta$ in as simple a form as you can in terms of $\sin\theta$ and $\cos\theta$.

20. Same as prob. 19 for $\sin 4\theta$ and $\cos 4\theta$.

21. Prove the following identities:

(a) $$\cos 2x \equiv \frac{\csc^2 x - 2}{\csc^2 x}$$

(b) $$\tan(\phi/2) \equiv \frac{\sin\phi}{1 + \cos\phi}$$

(c) $$8\sin^4 x + 4\cos 2x - \cos 4x - 3 \equiv 0$$

(d) $$\sin(n+1)A \equiv 2\sin nA \cos A - \sin(n-1)A$$

(e) $$\frac{\sin a + \sin b}{\sin a - \sin b} \equiv \frac{\tan\left(\dfrac{a+b}{2}\right)}{\tan\left(\dfrac{a-b}{2}\right)}$$

(f) $$\tan a + \tan b \equiv \frac{\sin(a+b)}{\cos a \cos b}$$

(g) $\left(1 + \tan a \tan \dfrac{a}{2}\right)^{-1} \equiv \cos a$

(h) $\dfrac{2 \tan \dfrac{x}{2}}{1 - \tan^2 \left(\dfrac{x}{2}\right)} \equiv \tan x$

(i) $2 \sin^2 (45° + 0.5x) - 1 \equiv \sin x$

(j) $\tan A + \cot A \equiv 2 \csc 2A$

22. Simplify:

(a) $2[\tan (0.5B) + \cot (0.5B)]^{-1}$

(b) $\dfrac{\cot (0.5x) - \tan (0.5x)}{\cot (0.5x) + \tan (0.5x)}$

(c) $1 - 2 \sin^2 (\pi/4 - \theta/2)$

(d) $\sin^2 A \cos^2 B - \cos^2 A \sin^2 B$

(e) $\cos (\pi/4 + 0.5x) \cos (\pi/4 - 0.5x)$

(f) $\sin (30° + x) - \sin (30° - x)$

(g) $\cot A - 2 \cot 2A$

(h) $(1 - \tan^2 \theta)(1 + \tan^2 \theta)^{-1}$

(i) $\dfrac{\cos 2\theta - \cos 6\theta}{\sin 6\theta - \sin 2\theta}$

(j) $\dfrac{\sin 3x}{\sin x} + \dfrac{\cos 3x}{\cos x}$

(k) $\dfrac{\cot R + \cot R'}{\cot R - \cot R'}$

†23. Simplify:

$$\frac{\cos \dfrac{20° - A}{2} \cos \dfrac{20° + A}{2} - \sin \dfrac{20° - A}{2} \sin \dfrac{20° + A}{2}}{\cos \dfrac{20° - A}{2} \cos \dfrac{20° + A}{2} + \sin \dfrac{20° - A}{2} \sin \dfrac{20° + A}{2}}$$

This expression is taken from the field of geology.

24. Express $\sqrt{3} \cos \theta - \sin \theta$ in the form $k \sin (\theta + \phi)$ and find ϕ. Do the same for: (a) $\cos \theta - \sin \theta$. (b) $\sin \theta + \cos \theta$. (c) $\cos \theta + \sqrt{3} \sin \theta$. (d) $2 \cos \theta + 3 \sin \theta$. (e) $5 \cos \theta - 1 \sin \theta$.

Solution to first problem. $\sqrt{3} \cos \theta - \sin \theta = 2 \left[\dfrac{\sqrt{3}}{2} \cos \theta - \dfrac{1}{2} \sin \theta \right]$.
But $\sqrt{3}/2 = \sin 120°$ and $-0.5 = \cos 120°$. Hence the expression equals $2 \sin (\theta + 120°)$.

†25. From $n' \sin \theta = (n - n' \cos \theta) \tan \theta'$ and $n - n' = \dfrac{hnn'}{mc^2} (1 - \cos \theta)$

derive $\tan \theta' = \dfrac{\cot (\theta/2)}{1 + \dfrac{hn}{mc^2}}$.

26. Show that:

(a) $\sin^3 x \equiv 0.75 \sin x - 0.25 \sin 3x$

(b) $\sin^4 x \equiv \dfrac{3}{8} - \dfrac{1}{2} \cos 2x + \dfrac{1}{8} \cos 4x$

(c) $\sin^5 x \equiv \dfrac{5}{8} \sin x - \dfrac{5}{16} \sin 3x + \dfrac{1}{16} \sin 5x$

†27. Find $\cos 22.5°$ and $\cos 11.25°$.   Show that $\cos (45°/2^n) \equiv$ $\frac{1}{2}\sqrt{2 + \sqrt{2 + \sqrt{2 + \cdots}}}$ to $n + 1$ radicals.

8.8   Graphs of the circular functions

In graphing the functions, we make use of ideas introduced in Chapters 3 and 4. Angles are measured in radians.

Graph of y = sin x. If we plot the pairs of values found in Ex. 8.4.9 and connect these points by a smooth curve, we get the heavy lined portion of the graph in Fig. 8.8.1. We know from *8.6.1 that,

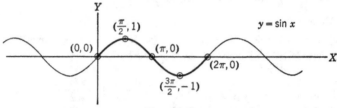

Fig. 8.8.1

if x changes by any multiple of 2π, y remains the same. Hence we can continue the graph in either direction indefinitely. If we divide the x-axis into intervals of length 2π, the curve looks the same in each interval.

We have assumed that it is legitimate to graph $y = \sin x$ by plotting a few points and drawing a smooth curve. The correctness of this procedure is made plausible by returning to the definition of the sine. In a unit circle with a variable angle θ in standard position, $\sin \theta$ is the vertical vector from the x-axis to the terminal point of the radius vector (Fig. 8.8.2). We visualize the radius vector as starting from coincidence with the x-axis and rotating in the positive direction. Then $\sin \theta$ becomes positive, increasing rapidly at first and then more slowly until it comes to a maximum at $\theta = \pi/2$. Then it decreases to zero, repeating in reverse order the values it took in the first quadrant. At $\theta = \pi$, it is zero and then decreases to -1 at $3\pi/2$, passing over the negatives of the values it took in the first quadrant. In the fourth quadrant it repeats the negatives of the values of the second quadrant. (Compare Figs. 8.8.1 and 8.8.2.)

Graph of y = cos x. We could proceed as for the sine, but from *8.7.1:

*8.8.1 $\cos x \equiv \sin (x + \pi/2)$

Hence to plot $y = \sin (x + \pi/2)$, we translate the origin to the point $(-\pi/2, 0)$. (See Fig. 8.8.3.) The new coordinates are

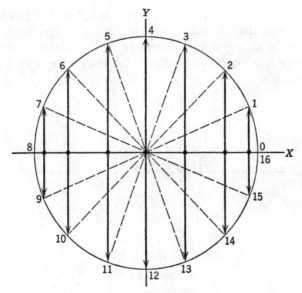

Fig. 8.8.2

$x' = x + \pi/2$, $y' = y$, and the equation of the curve in the new coordinates is $y' = \sin x'$. Hence the graph of $y = \cos x$ is just the graph of $y = \sin x$ displaced $\pi/2$ to the left.

Graph of $y = \tan x$. In order to graph $y = \tan x$, we return to the unit circle of Fig. 8.8.2 and recall that $\tan x = \sin x/\cos x$. When $x = 0$, the numerator is zero and the denominator is 1.

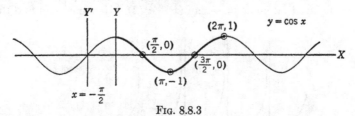

Fig. 8.8.3

Hence the point $(0, 0)$ lies on the curve. As x increases in the first quadrant, $\sin x$ increases and $\cos x$ decreases. Hence $\tan x$ increases and is always positive. As the angle approaches $\pi/2$, $\sin x$ approaches 1 and $\cos x$ approaches zero. Hence $\tan x$ gets larger and larger. In fact, no matter how large a number we choose, we can make $\tan x$ larger than this number by taking x close enough to $\pi/2$. At $x = \pi/2$ tan x is undefined. We plot a few points from Ex. 8.4.9 and draw the heavy curve in Fig. 8.8.4. The line $x = \pi/2$ is evidently an asymptote (compare 6.8), since the curve approaches it as y "approaches infinity." Since tan $(-x) \equiv -$ tan x,

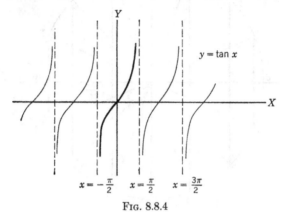

Fig. 8.8.4

we can plot the curve in the interval from $-\pi/2$ to 0 by taking the negatives of the corresponding values in the interval 0 to $\pi/2$. From *8.7.5, tan $x =$ tan $(x + \pi)$. It follows that:

*8.8.2   tan $x \equiv$ tan $(x + k\pi)$, where k is any integer.

This means that the graph of the tangent repeats itself in every interval of length π.

Graph of $y = $ cot x. The easiest way to graph $y = $ cot x is to use the fact that cot $x \equiv -$ tan $(x + \pi/2)$. Hence we can find the

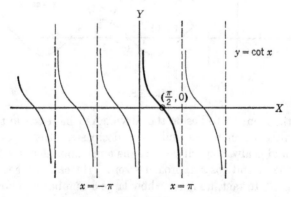

Fig. 8.8.5

graph of cot x by translating the graph of $y = \tan x$ a distance $\pi/2$
and then rotating it about the x-axis.   In the interval $0 < x < \pi$,
cot x has a graph given by the heavy curve in Fig. 8.8.5.   Note
that at $x = 0$, where $\tan x = 0$, cot x is undefined.   On the other
hand, at $x = \pi/2$, where $\tan x$ is undefined, cot $x = 0$.   Whereas
$\tan x$ always increases with increasing x, cot x always decreases
with increasing x.

Graphs of $y = csc\ x$ and $y = sec\ x$.   The reciprocal relationships
csc $x \equiv 1/\sin x$ and sec $x \equiv 1/\cos x$ afford the easiest means of
graphing these functions.   The secant is graphed in Fig. 8.8.6.

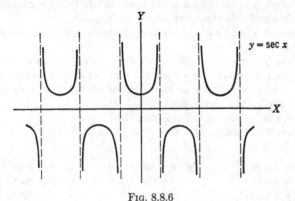

FIG. 8.8.6

The circular functions are the simplest of the large class of
functions that are **periodic.**   The formal definition of this property
is as follows:

*8.8.3 $f(x)$ is periodic $\longleftrightarrow$.   There is a number p such that
for all x, $f(x) \equiv f(x + p)$.   (Definition)

The smallest p such that $f(x) \equiv f(x + p)$ is called the **period**
of the function.   We can sum up the periodic properties of the
circular functions by saying that the sine, cosine, secant, and
cosecant have the period 2π, whereas the tangent and cotangent
have the period π.   The periodic nature of the trigonometric
functions makes them useful for representing repeating or recurring
phenomena such as sound, water waves, light, economic oscilla-
tions, and many other wave-like phenomena.   A tremendous
variety of periodic curves can be obtained by considering various
combinations of these functions.

EXERCISE 8.8

1. Make large graphs for each of the circular functions. Begin by plotting the values found in Ex. 8.4.9. Then draw asymptotes and make use of symmetry and periodicity.

2. Prove that the graphs of the cosine and secant are symmetric with respect to the y-axis, whereas those of the sine, cosecant, tangent, and cotangent are symmetric with respect to the origin.

3. Graph $y = 1 + \sin x$. [*Suggestion.* Translate the origin to $(0, 1)$.]

4. Graph $y = \sin (x - \pi/4)$. [*Suggestion.* Translate the origin to $(\pi/4, 0)$.]

5. Graph $y = \sin x + \cos x$. (*Suggestion.* Graph $y = \sin x$ and $y = \cos x$. Then add ordinates by measuring on the graph.)

6. Graph $y = 2 \sin x$. (*Suggestion.* Graph $y = \sin x$ and double each ordinate.)

7. Graph $y = \sin 2x$. (Note that if we make the change of variable $x' = 2x$, we get the function $y = \sin x'$. Thus $\sin 2x$ runs through the same values as $\sin x$ but "twice as fast." Since $\sin 2 (x + \pi) = \sin (2x + 2\pi)$ $= \sin 2x$, $\sin 2x$ has a period of π.)

8. Graph $y = \sin 3x$. What is its period?

9. Graph $y = \sin (0.5x)$. What is its period?

10. Graph:

(a) $y = \cos 2x$	(d) $y = \cos (x - 1)$	(g) $y = 3 \cos x$
(b) $y = \cos 4x$	(e) $y = \sin (\pi/3 + x)$	(h) $y = \sin \lvert x \rvert$
(c) $y = \sin (x/3)$	(f) $y = \cos 2x - 1$	(i) $y = \lvert \sin x \rvert$

11. Graph:

(a) $y = 2 \sin x + \cos 2x$	(h) $y = x + \tan x$
(b) $y = 3 \cos x + 2 \sin x$	(†i) $y = \dfrac{\sin x}{x}$
(c) $y = 2 \sin 3x - 3 \cos 2x$	
(d) $y = 1 + \sin x + \sin 2x + \sin 3x$	(j) $y = \log \sin x$
(e) $y = \sin 2x - 2 \sin x$	(k) $y = e^{-x} \sin x$
(f) $y = \cos (x/2) + 2 \sin 2x$	(l) $y = (2x + 1) \sin x$
(g) $y = x + \sin x$	

†*12.* Show that a function of the form $y = A \sin (ax + b)$ with $a \neq 0$, $A > 0$ has a period of $2\pi/a$, a maximum A, and a minimum $-A$. Show that its graph can be found from that of $\sin x$ by translating $-b$ in the x-direction, then stretching in the ratio $1/a$ in the x-direction and in the ratio A in the y-direction. Show that $y = B \cos (ax + b)$ has the same graph as $y = B \sin (ax + b + \pi/2)$. Graph:

(a) $y = 2 \sin (x + 30°)$	(d) $y = 2 \cos (x/2 - \pi/4)$
(b) $y = 0.5 \sin (x - \pi/4)$	(e) $y = 10 \sin (\pi/3 - x)$
(c) $y = 3 \sin (2x - 3)$	(f) $y = -5 \cos (x/\pi - \pi/2)$

13. When a curve similar in shape to the graph of $y = \sin x$ is used to represent wave motion, the high points are called *crests*, the low points are called *troughs*, and the points on the horizontal axis are called *nodes*. The period is called the *wavelength*, and half the vertical distance between a crest and a trough is called the *amplitude*. A portion of the curve occupying one wavelength is called a *wave*. Points separated by a horizontal distance equal to a multiple of the period are said to be *in phase*. (See Fig. 8.8.7.) In prob. 12 find the amplitudes, wavelengths, nodes, troughs, and crests.

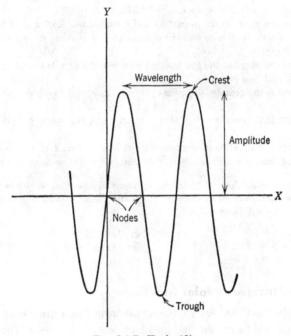

FIG. 8.8.7. (Prob. 13).

14. A body that oscillates on a straight line so that its distance from a fixed point on the line is given by $y = R \sin wt$, where t is the time, is said to describe *simple harmonic motion*. The variable y is called the *displacement*, the fixed point is called the *center* or equilibrium position, R is called the *amplitude*, $2\pi/w$ is called the *period*, and $w/2\pi$ is called the *frequency*. Show that when a point moves in a circle at a constant angular velocity its projection on a diameter (the foot of the perpendicular from the point to the diameter) describes a simple harmonic motion whose amplitude is the radius of the circle and whose period is the time of one revolution.

15. Show by induction that $f(x) \equiv f(x + p) \longrightarrow f(x) \equiv f(x + kp)$ for any integral k.

16. One advantage of radian measure is that when x is in radians:

*8.8.4 $D_x(\sin x) = \cos x$

*8.8.5 $D_x(\cos x) = -\sin x$

These formulas do not hold when the angle is measured in other units.

(a) Find $\Delta y/\Delta x = [\sin(x + \Delta x) - \sin x]/\Delta x$ for $x = 0.5061$, $\Delta x = 0.003$.

(b) Find $\Delta y/\Delta x$ for $y = \cos x$, $x = 0.8930$, $\Delta x = 0.0029$.

(c) How are your results in parts a and b related to *8.8.4 and *8.8.5?

(d) Use the derivative to find the points at which $y = \sin x$ has a maximum or a minimum.

(e) Find the slope of the line tangent to $y = \cos x$ at $x = 0$, $\pi/2$, $\pi/6$, $\pi/4$, and check with the graph.

(f) What is the greatest slope that $y = \sin x$ has, and where does it achieve it?

(g) What is the smallest slope that $y = \sin x$ has, and where does it achieve it?

(h) Write the equation of the tangent line to $y = 2 \sin x$ at $x = 2\pi/3$.

(i) Show that $y = 3 \cos x$ always has three times the slope of $y = \cos x$.

17. For a body describing simple harmonic motion find the times when it is instantaneously at rest (has zero velocity) and when it has maximum velocity.

18. Solve the following:

(a) $\sin x \leq 0$ (d) $\tan x \leq 1$

(b) $\cos x \geq 0$ (e) $|\sin x| = \sin x$

(c) $\sin x \leq 0.5$ (f) $|\cos x| = -\cos x$

8.9 The inverse circular functions

We have seen that each circular function has a unique value for every real value of the independent variable, except for those special values for which it is undefined. Thus for each value of x there is one and only one value of $y = \sin x$. Can we say that for each value of y, there is one and only one value of x? Consider $\sin x = 0.5$. One possibility is $x = \pi/6$, but it is also true that $\sin(5\pi/6) = 0.5$. These are the only two values of x between 0 and 2π, but either of these plus any multiple of 2π gives an x that satisfies $\sin x = 0.5$. We see that when a value of y is given, there is an infinity of values of x. Interchanging x and y so that x will be the independent variable, we may say that $\sin y = x$ defines y as a many-valued function of x. We call this function the **inverse sine** of x or **arcsine** of x. It is written $y = \sin^{-1} x$ or $y = \arcsin x$.

In a similar way we define the inverses of the other circular functions.

$$y = \arcsin x \longleftrightarrow x = \sin y$$

$$y = \arccos x \longleftrightarrow x = \cos y$$

*8.9.1
$$y = \arctan x \longleftrightarrow x = \tan y$$ (Definition)
$$y = \text{arccot } x \longleftrightarrow x = \cot y$$

$$y = \text{arcsec } x \longleftrightarrow x = \sec y$$

$$y = \text{arccsc } x \longleftrightarrow x = \csc y$$

Thus $y = \text{arccot } x$ means that $x = \cot y$. We read arccot x as "arc-cotangent x," "inverse cotangent of x," or "the angle whose cotangent is x." The inverse functions are also written as $\sin^{-1} x$, $\cos^{-1} x$, $\tan^{-1} x$, $\cot^{-1} x$, $\sec^{-1} x$, $\csc^{-1} x$. The student should note that the -1 is not an exponent in the ordinary sense, and $\sin^{-1} x \neq (\sin x)^{-1}$.

Since arcsin x means an angle whose sine is x, it is obvious that $\sin(\arcsin x) = x$. In general, for any circular function f,

*8.9.2
$$f[\text{arc-}f(x)] \equiv x$$

The loci of $y = \sin x$ and $x = \arcsin y$ are identical, so that the graph of the former may be used as a graph of the inverse function. However, if x is to be the independent variable, we consider rather the graph of $y = \arcsin x$ or $x = \sin y$. Points may be found by evaluating x for various values of y, but, because of the relation between the graphs of a function and its inverse, the locus of $y = \arcsin x$ may be obtained by rotating the graph of $y = \sin x$ about the line $y = x$. (See 6.10.) Figure 8.9.1 brings out effectively that $y = \arcsin x$ is a many-valued function of x for $-1 \leq x \leq 1$ and is undefined for x outside this interval. As in the case of previous many-valued functions (see 5.3 and 6.10), we

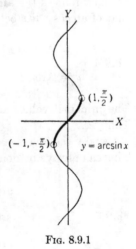

Fɪɢ. 8.9.1

choose a particular value and call it the **principal value** of arcsin x. The formal definition is:

*8.9.3 $y = \text{Arcsin } x \longleftrightarrow y = \arcsin x$ and $-\pi/2 \le y \le \pi/2$

The principal value is drawn with a heavy line in the figure.

The other inverse functions are treated in a similar way. In Fig. 8.9.2 we have marked with a heavy line the portions of the

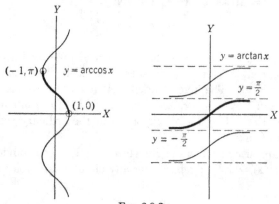

FIG. 8.9.2

curves that give the principal values. The principal value of arctan x follows the same convention as that of arcsin x, whereas that of arccos x lies between 0 and π. Formally,

*8.9.4

$$y = \text{Arccos } x \longleftrightarrow y = \arccos x \text{ and } 0 \le y \le \pi$$

$$y = \text{Arctan } x \longleftrightarrow y = \arctan x \text{ and } -\pi/2 \le y \le \pi/2$$

The principal values are indicated also by $\text{Sin}^{-1} x$, $\text{Cos}^{-1} x$, and $\text{Tan}^{-1} x$.

Usually the inverse cotangent, secant, and cosecant are avoided. This can always be done, since

*8.9.5

$$\text{arccot } x \equiv \arctan (1/x)$$

$$\text{arcsec } x \equiv \arccos (1/x)$$

$$\text{arccsc } x \equiv \arcsin (1/x)$$

To prove the first of these identities, we let arccot $x = y$. Then $x = \cot y$ and

(1) $$\tan y \equiv 1/\cot y \qquad \text{(Why?)}$$

(2) $$\equiv 1/x \qquad \text{(Why?)}$$

or

(3) $$y \equiv \arctan (1/x) \qquad \text{(Why?)}$$

The result holds for all x except $x = 0$. The others are proved similarly. Arccot x is usually taken as the value between 0 and π, inclusive, but there is no universally accepted convention for Arcsec x and Arccsc x.

The student will recall that problems involving logarithms can be referred to exponents since $\log_b x$ is defined as the inverse of b^x. Similarly, problems involving inverse circular functions can be referred to the direct functions as in (1)–(3) above. As an example, we prove:

(4) $$\text{Arcsin } (0.5) + \text{Arccos } (0.5) = \pi/2$$

Let Arcsin $(0.5) = y$. Then $\sin y = 0.5$, and $-\pi/2 \le y \le \pi/2$. The only y satisfying these conditions is $y = \pi/6$. Similarly, Arccos $(0.5) = \pi/3$, and (4) follows immediately.

EXERCISE 8.9

1. Test the following by *8.9.1:

(a) arcsin $(0.5) = 30°$?
(b) arcsin $(0.5) = 150°$?
(c) arccos $0 = \pi/2$?
(d) arccos $0 = 3\pi/2$?
(e) arctan $1 = -135°$?
(f) $\cos^{-1} 1 = 0$?
(g) $\sin^{-1} (0.5) = 45°$?
(h) $\cot^{-1} 1 = \pi/4$?

(i) $\sin^{-1} 0.9730 = 76°$?
(j) $\tan^{-1} (0.5) = 26° 34'$?
(k) arctan $100 = 89° 56'$?
(l) $\sec^{-1} 4 = 31°$?
(m) $\tan^{-1} 1 = 405°$?
(n) $\cos^{-1} 1 = 720°$?
(o) arcsin $(0.5) = 510°$?

Answers. (a) T. (c) T. (e) T. (g) F. (i) F. (k) F.

2. Test each equation in prob. 1 with each inverse function replaced by its principal value. [*Answers.* (a) T. (c) T. (e) F. (g) F. (i) F. (k) F.]

3. Find:

(a) Arctan (-1)	(f) $\sin^{-1} 0.75$	(k) Arctan 0.6703
(b) Arcsin (-1)	(g) $\tan^{-1} 10$	(l) $\cos^{-1} 0.7547$
(c) Arctan (-0.5)	(h) $\tan^{-1}(-0.1)$	(m) Arccos 0.2700
(d) $\cos^{-1}(-0.5)$	(i) $\cos^{-1}(-1/\sqrt{2})$	(n) Arccot 114.59
(e) Arcsin $(0.5\sqrt{3})$	(j) $\sin^{-1}(0.3584)$	(o) Arcsin 0.9899

Answers. (a) $-45°$. (c) $-27°$ approx. (e) $60°$. (g) $84°$ approx.

4. Find all values of the following and indicate which is the principal value:

(a) arcsin $(1/\sqrt{2})$	(e) arctan 1	(i) arcsin 1
(b) arcsin $(-1/\sqrt{2})$	(f) arctan (-1)	(j) arccos (-1)
(c) arccos (-0.5)	(g) arctan $(\sqrt{3})$	(k) arctan 0
(d) arccos $(0.5\sqrt{3})$	(h) arctan $(-1/\sqrt{3})$	(l) arccos 0

Answers. (a) Arcsin $(1/\sqrt{2}) = 45°$; arcsin $(1/\sqrt{2}) = 45° + k360°$, $135° + k360°$. (c) Arccos $(-0.5) = 120°$; arccos $(-0.5) = 120° + k360°$, $240° + k360°$. (e) Arctan $1 = 45°$; arctan $1 = 45° + k180°$. (i) Arcsin $1 = 90°$, arcsin $1 = 90° + k360°$.

5. Draw large graphs of each of the six inverse functions.

6. Explain *8.9.2.

7. Find:

(a) sin (arcsin 1)	(d) cos (arccos 0.5)	(g) csc (arccsc 5)
(b) tan (arctan 2)	(e) tan (arctan 0.5)	(h) sin (arcsin 0.178)
(c) sin (Arcsin 1)	(f) sec (arcsec 10)	(i) cos (arccos 0.399)

8. Why is Arc-$f[f(x)] \not\equiv x$? Illustrate for $x = 225°$ and f replaced by cos.

9. For what values of x does there exist no value of $\sin^{-1} x$? Such values are called **excluded values.** What are the excluded values of x for each of the inverse functions? If $x = a$ is an excluded value for $y = f(x)$, the line $x = a$ cannot cross the graph. If x is excluded from an interval on the x-axis, the curve cannot enter the region directly above or below this interval. Such a region is called an **excluded region.** Shade the excluded regions in your graphs of the inverse functions.

10. Evaluate the following, making a sketch in each case:

(a) sin (Arcsin 0.5)	(f) cos (Arccos 0.7)
(b) Arcsin (sin 540°)	(g) Arctan (tan 135°)
(c) Arcsin (sin 225°)	(h) Arccos (cos 460°)
(d) Arccos (cos 300°)	(i) Arccos (cos 1000°)
(e) Arctan (tan 100°)	(j) Arcsin (sin 330°)

Answers. (a) 0.5. (c) $-45°$. (e) $-80°$. (g) $-45°$.

11. Solve the following equations:

(a) $\sin x = 1$	(d) $\cos 2x = 1$	(g) $\cos x = -0.5163$
(b) $\tan x = 6.3$	(e) $\sin 3x = -1$	(h) $\tan x = -0.8813$
(c) $\cos x = 0.853$	(f) $\tan 2x = 1$	

Answers. (a) $x = (2k + 0.5)\pi$. (c) $x = 31.5° + k360°$, $328.5° + k360°$.

12. Evaluate:

(a) tan (arcsin x)   (d) sec (arcsin x)

(b) sin (arccos x)   (e) sec (arccos x)

(c) cos (arctan x)   (f) cot (arcsin x)

Answers. (a) $\pm x/\sqrt{1 - x^2}$. (c) $\pm x/\sqrt{1 + x^2}$. (e) x^{-1}.

13. If sin x = sin y, what is the relation between x and y?

14. Show that all values of arcsin x are given by Arcsin $x + 2k\pi$ and $(2k + 1)\pi - $ Arcsin x. Derive similar formulas that give all values of arccos x and arctan x.

15. The following formula is used in petrology: $\theta = \tan^{-1}\left(\dfrac{1 - s}{s}\right)^{\frac{1}{2}}$.

Solve for s. Graph s as a function of θ for $45° \leq \theta \leq 90°$.

16. Show that:

(a) Arcsin (0.5) + Arccos (-0.5) = $5\pi/6$

(b) 2 Arcsin (0.5) = Arcsin ($0.5\sqrt{3}$)

(c) $\text{Cos}^{-1} x \equiv \pi/2 - \text{Sin}^{-1} x$

†**17.** In order for a vehicle rounding a curve of radius R at speed V to have no tendency to skid, the curve must be banked at an angle $\theta = \text{Arctan } \dfrac{V^2}{gR}$, where g is the acceleration of gravity. In the light of this formula discuss the common side-show stunt of riding a motorcycle around a vertical circular wall.

†**18.** Solve for x:

(a) sin $3x$ = sin x (c) tan x = tan $4x$ (e) tan $2x$ = cot x

(b) cos $2x$ = cos x (d) sin $3x$ = cos x (f) sin $2x$ = sin x

Answers. (a) $x = k\pi$ or $0.25(2k + 1)\pi$. (c) $x = k\pi/3$.

19. Show that: (a) sin (Arccos x) $\geq$ 0. (b) cos (Arcsin x) $\geq$ 0. (c) cos (Arctan x) $\geq$ 0.

†**20.** Prove the last two parts of *8.9.5.

†**21.** Show that:

(a) arcsin x $\equiv$ arccsc ($1/x$)

(b) arctan x $\equiv$ arccot ($1/x$)

(c) arcsin x $\equiv$ arccos ($\pm \sqrt{1 - x^2}$)

(d) sin ($2 \cos^{-1} x$) $\equiv$ $\pm 2x\sqrt{1 - x^2}$

(e) cos ($\cos^{-1} x + \cos^{-1} y$) = $xy \pm \sqrt{1 - x^2}\sqrt{1 - y^2}$

(f) cos ($2 \cos^{-1} x$) = $2x^2 - 1$

22. Find:

(a) tan ($\tan^{-1} x + \tan^{-1} y$)   (c) sin ($\sin^{-1} x + \cos^{-1} y$)

(b) sin ($0.5 \cos^{-1} x$)   (d) sin ($\cos^{-1} x - \sin^{-1} y$)

Answers. (a) $(x + y)(1 - xy)^{-1}$. (c) $xy \pm \sqrt{1 - x^2}\sqrt{1 - y^2}$.

†**23.** Sketch $y = \sin x$ in the interval $-\pi/2 \leq x \leq \pi/2$ and $y = \text{Arcsin } x$. Show that Arcsin $(\sin x) \equiv x$ for x in this interval. How does this result and *8.9.2 illustrate *6.10.9?

†**24.** Graph the following functions and their inverses and verify that *6.10.9 holds: (a) $y = \cos x$ for $0 \leq x \leq \pi$. (b) $y = \tan x$ for $-\pi/2 \leq x \leq \pi/2$.

†**25.** An examination of Figs. 8.9.1 and 8.9.2 shows that the principal values have been chosen so that they are single-valued functions that yield y as a first quadrant angle when x is positive and that are continuous curves. (a) Use this criterion to define Arccot x and sketch it. (b) Explain why this criterion cannot be satisfied by any definition of Arcsec x and Arccsc x.

8.10 Trigonometry of the right triangle

The simplest application of trigonometry is to the right triangle. The word trigonometry comes from Greek words meaning "measurement of triangles," and trigonometry had its origin in problems of surveying, building, etc.

Consider a right triangle ABC, with sides a, b, and c and angles A, B, and $C = 90°$ (Fig. 8.10.1). Note that angles are labeled with

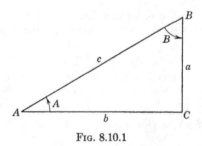

FIG. 8.10.1

capital letters and opposite sides are labeled with the corresponding small letters. Then:

$$\sin A = \frac{a}{c} = \cos B \qquad \csc A = \frac{c}{a} = \sec B$$

*8.10.1 $$\cos A = \frac{b}{c} = \sin B \qquad \sec A = \frac{c}{b} = \csc B$$

$$\tan A = \frac{a}{b} = \cot B \qquad \cot A = \frac{b}{a} = \tan B$$

With these relationships and a table of circular functions, we can find all the sides and angles of a right triangle if we are given one side and either another side or one of the acute angles. The sides and angles of a triangle are called its **elements,** and finding them is called **solving** the triangle. Evidently, we are now in a position to solve any right triangle, if we have sufficient information. We illustrate with examples.

Suppose we wish to find the height of a tree without climbing to its top with a tape measure. We measure out a convenient

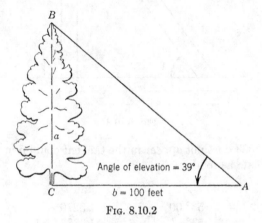

FIG. 8.10.2

distance from its base, say 100 feet. From this point we measure the angle to the top. This angle between the horizontal and the line of sight to the top is called the **angle of elevation** of the top. (See Fig. 8.10.2.) Suppose it is 39°. In the right triangle ABC we know $A = 39°$ and $b = 100$. But $a/b = \tan A$, and hence $a = 100 \tan 39°$. From Table IV, $\tan 39° = 0.8098$. Hence $a = 81$ to the nearest foot. We could also find c and B, since we have $B = 90° - A = 51°$ and $c = b/\sin B$.

As a second example, suppose we wish to brace a vertical post with a diagonal from a point on the floor 3 feet from the base to a point 4 feet above the floor (Fig. 8.10.3). At what angles should the end of the brace be cut, and how long should it be on its outside edge? We have:

(1) $\tan A = a/b = 4/3 = 1.3333$

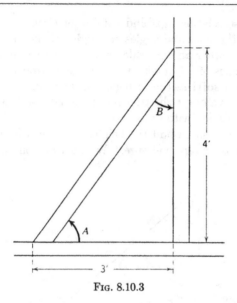

Fig. 8.10.3

Since 1.3333 does not appear in the tangent column in Table IV, we interpolate as follows:

$$
\begin{array}{cc}
x & \tan x \\
10\begin{bmatrix} -53^\circ\ 00' \\ 53^\circ\ \ ? \\ \to 53^\circ\ 10' \end{bmatrix}h & 63\begin{bmatrix} -1.3270 \\ \to 1.3333 \\ 1.3351 \end{bmatrix}81
\end{array}
$$

We have $h/10 = 63/81$ and $h \doteq 8$ to the nearest unit. Hence $A = 53^\circ\ 8'$ and $B = 36^\circ\ 52'$. The Pythagorean theorem may be used to find c, or we may note that:

(2)
$$
c = \frac{a}{\cos B} = \frac{4}{\cos 36^\circ\ 52'}
$$

To find $\cos 36^\circ\ 52'$ we interpolate in Table IV as follows:

$$
\begin{array}{cc}
x & \cos x \\
10\begin{bmatrix} -36^\circ\ 50' \\ 36^\circ\ 52' \\ \to 37^\circ\ 00' \end{bmatrix}2 & h\begin{bmatrix} -0.8004 \\ \to\ ? \\ 0.7986 \end{bmatrix}18
\end{array}
$$

Then $h/18 = 2/10$ and $h \doteq 4$ to the nearest unit. Note that we subtract the 4 from 8004 to get the interpolated result, since the values of cosine are decreasing.   (It is better not to learn rules about whether to add or to subtract, but to use common sense; the interpolated value must obviously be between the tabular values.   If we were to add 4 we would get 0.8008!)   Using cos B = 0.8000, we find c = 5, which is consistent with the Pythagorean theorem.

EXERCISE 8.10

1. Explain *8.10.1 in terms of the definitions of the circular functions. Restate each equation in words, using "hypotenuse," "adjacent side," and "opposite side."

2. Solve the following right triangles, working to two significant digits for length and to the nearest degree for angles: ·

(a) $a = 10$, $A = 60°$

(b) $c = 20$, $A = 45°$

(c) $a = 10$, $b = 20$

(d) $b = 100$, $A = 31°$

Answers.   (a) b = 5.8.   (c) 12, B = 30°.   (e) c = 22, B = 63°.

3. A man 6 feet tall looks toward the top of a building 100 yards away at an angle of 27°.   How high is the building to the nearest foot?

4. Solve the following right triangles, working to the nearest 10 minutes of angle and to three digits:

(a) $c = 3.45$, $a = 1.67$

(b) $A = 15° 20'$, $a = 82.1$

(c) $c = 725$, $A = 45° 42'$

(d) $b = 95$, $A = 78° 57'$

Answers.   (a) $A = 29°$, $B = 61°$, $b = 3.02$.   (c) $a = 519$, $b = 507$, $B =$ 44° 20'.

5. In order to measure the width of a river, an engineer sights straight across to a boulder on the opposite bank.   Then he measures 100 yards in a line perpendicular to this line of sight and again sights to the boulder, noting the angle between this line of sight and the 100-yard base line.   If this angle is 63°, find the width of the river.   (Work to three-digit accuracy.)

6. Solve the following right triangles, working to four significant digits and the nearest minute:

(a) $A = 28° 47'$, $b = 1.934$

(b) $a = 1582$, $b = 1218$

(c) $c = 0.004835$, $b = 0.004129$

(d) $c = 951.2$, $A = 30° 15'$

(e) $B = 27° 54'$, $a = 30.81$

(f) $A = 77° 18'$, $c = 100.8$

Answers.   (a) $a = 1.063$, $c = 2.207$.   (c) $A = 31° 20'$, $a = 0.002514$. (e) $b = 16.31$, $c = 34.86$.

7. A ship sights a certain lighthouse at an angle of 63° 25' with its course. After it has traveled 3.185 miles, the lighthouse is exactly on its beam, that is, the angle is now 90° with its course.   How far away was the lighthouse when first sighted and at the moment of the second observation?

8. A spotter for coastal defense guns stationed in a tower 100 feet high finds that the angle of depression of his line of sight to a hostile ship is 0° 45'. What range should he give to the guns?   (Ignore the curvature of the earth.)

9. Calculate the diameter of the sun by trigonometry from the data of Ex. 8.3.29 and compare with the approximation.

†**10.** Find the angles at which a brace 2 inches thick should be cut to go between two parallel beams 2 feet apart and 12 inches wide as in Fig. 8.10.4. Find the lengths of the edges of the brace.

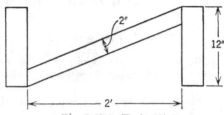

Fɪɢ. 8.10.4. (Prob. 10).

†**11.** The angle of elevation from one point on the ground to the top of a flagpole is 42° 16′. Twenty feet farther away it is 36° 38′. How high is the pole? (*Answer.* 82.24 feet.)

†**12.** Write a general formula for finding the height of an object from angles of elevation A and B measured from points on a line with the base of the object and a distance d apart.

8.11 Trigonometry of the general triangle

A triangle that does not have a right angle can always be divided into two right triangles, as we have done by drawing the altitudes BD in Fig. 8.11.1; hence any triangle may be dealt with by the methods of 8.10. However, there are a number of interesting theorems about the general triangle which make it possible to solve any triangle directly. The first of these is:

*8.11.1 $$\frac{\sin A}{a} = \frac{\sin B}{b} = \frac{\sin C}{c}$$ (Law of Sines)

To prove it, we note in either triangle in Fig. 8.11.1 that

$$\sin A = \frac{h}{c} \quad \text{and} \quad \sin C = \frac{h}{a}$$

Hence

$$\frac{\sin A}{a} = \frac{h}{ac} \quad \text{and} \quad \frac{\sin C}{c} = \frac{h}{ac}$$

From this the equality of the first and third expressions follows. If we draw the altitude from C, we find $\sin A/a = \sin B/b$. The law of sines enables us to find the other sides if two angles and a side are given or to find the other angles if two sides and an angle opposite one of them are given.

A second, and very interesting, relation is:

*8.11.2 $c^2 = a^2 + b^2 - 2ab \cos C$ (Law of Cosines)

This theorem states that in any triangle the square of any side equals the sum of the squares of the other two sides minus twice

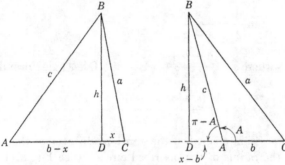

Fig. 8.11.1

the product of the other two sides and the cosine of the included angle.   In the special case where the included angle is 90°, *8.11.2 reduces to the Pythagorean theorem.   In order to prove it, we note in either triangle in Fig. 8.11.1 that:

(1) $x^2 + h^2 = a^2$ and $(b - x)^2 + h^2 = c^2$

Combining these two equations, we find:

(2) $c^2 = a^2 + b^2 - 2bx$

But $x = a \cos C$.   Hence *8.11.2 is proved, since we could follow the same procedure for any side.   The law of cosines enables us to find the third side if two sides and an included angle are given.

The law of sines and the law of cosines are the most important theorems about the general triangle, both for the solution of triangles and in advanced mathematics, but there are other interesting relations that are useful in particular situations.   We list them here, leaving the proofs to the exercise:

*8.11.3 $\dfrac{a + b}{a - b} = \dfrac{\tan \frac{1}{2}(A + B)}{\tan \frac{1}{2}(A - B)}$ (The Law of Tangents)

Of course A and B stand for any angles of a triangle and a and b for the corresponding opposite sides. When two sides and the angle included between them are given, this formula enables us to find a second angle and so solve the triangle. If we are given a, b, and C in a triangle, we know that $A + B = 180° - C$. Hence we can solve *8.11.3 for $\tan \frac{1}{2}(A - B)$ and hence $\frac{1}{2}(A - B)$. From this and $\frac{1}{2}(A + B)$ we can find A and B. The formula is also useful when two angles and a side are known.

$$\tan \frac{A}{2} = \frac{r}{s - a}$$

8.11.4   where $s = \dfrac{a + b + c}{2}$ (Half-Angle Formulas)

$$\text{and } r = \left[\frac{(s - a)(s - b)(s - c)}{s}\right]^{\frac{1}{2}}$$

The expression s is called the **semi-perimeter** of the triangle, and r is the radius of the inscribed circle. (See Fig. 8.11.2.) Of

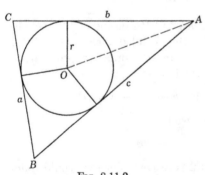

course A stands for any angle in the triangle and a for the opposite side. These formulas are convenient for solving triangles when three sides are given.

The student will recall from plane geometry that the area of a triangle is equal to one half the product of any side and the altitude drawn to that side. For

Fig. 8.11.2

example, in Fig. 8.11.1, the area is given by $S = \frac{1}{2}bh$. We can always use this formula if we have sufficient information to find b and h, but it is convenient to have formulas that give the area directly in terms of the sides and angles of the triangles. The most useful are:

8.11.5 $S = \frac{1}{2}ab \sin C$

8.11.6 $S = \dfrac{a^2 \sin B \sin C}{2 \sin A}$ (Area Formulas)

8.11.7 $S = [s(s - a)(s - b)(s - c)]^{\frac{1}{2}}$

The first formula is useful when two sides and an included angle are given, the second when a side and two angles are given, and the third (called Heron's formula) when three sides are given. Since the order in which we label the elements of a triangle is arbitrary, *8.11.5 and *8.11.6 each stands for three formulas.

The solution of a general triangle is somewhat more complicated than that of a right triangle, because in a right triangle we are always given one angle (the right angle) and hence we need know only two additional sides or a side and another angle. In order to solve a triangle, we must know at least one side. (Why?) If, in addition, we are given two other elements, either sides or angles, we can find all other elements by judicious use of the formulas in this section. It is not worth while to memorize the best ways of doing this in each case unless you are going to be a surveyor, and even then you had better not memorize them just now. The best way is to draw a figure showing what is known and then see what formulas may be used to find unknown elements. In the following exercise we give examples with simple numerical values, leaving to the next section the more accurate and complete solution of the general triangle.

EXERCISE 8.11

(In each problem draw a figure accurate enough to help check the results. Work to two-digit accuracy and the nearest degree, to three digits and the nearest 10', or to four digits and the nearest minute.)

1. Write out the other two forms of the law of cosines. Explain equations (1).

2. State the law of sines in the proportion language. Show that $a/b = \sin A/\sin B$. Explain the statements in the proof.

3. The sides of a triangle are 10, 20, 15. Find the cosines of each of the angles by the law of cosines and check by the relation $A + B + C = 180°$.

4. Two sides and the included angle of a triangle are 10, 30, and 25°. Find the third side by using the law of cosines. (*Answer.* 21.)

5. Two angles of a triangle are 54° and 37°. The side opposite the second is 25. Find the other elements by the law of sines. (*Answer.* 34 and 42.)

6. Two angles and the included side of a triangle are 29°, 42°, and 100. Find the other two sides by the law of sines. Check by the law of cosines.

7. Two sides of a triangle are 10 and 15. The angle opposite 15 is 43°. Solve, using the law of sines. (*Answer.* 21 is the side opposite 110°.)

8. Same as prob. 7 with 15 replaced by 9. Note that there are two solutions. Explain how they appear in both your figure and the algebra. (*Answer.* 13 opposite 88° and 1.4 opposite 6°.)

9. Same with 9 replaced by 6.82. (A right triangle.)

10. Same with 6.82 replaced by 4. Show how the impossibility of the solution appears in the algebra as well as in the figure.

11. Show, using the congruence theorems of plane geometry, that when a side and two other elements are given the triangle is uniquely determined, unless the given elements are two sides and an angle opposite one of them.

12. The case when two sides and an angle opposite one of them are given is called the **ambiguous case.** The various possibilities when $A < 90°$ are

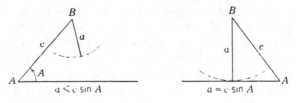

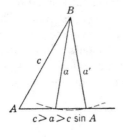

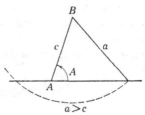

Fig. 8.11.3

illustrated in probs. **7** through **10** above. Let a and c be the given sides and A the given angle. Then explain, using Fig. 8.11.3, the following statements:

$$\text{When } A < 90°: a < c \sin A \longrightarrow \text{ no solution}$$
$$a = c \sin A \longrightarrow \text{ one solution, a right triangle}$$
$$c > a > c \sin A \longrightarrow \text{ two solutions}$$
$$a > c \longrightarrow \text{ one solution}$$
$$\text{When } A \geq 90°: a \leq c \longrightarrow \text{ no solution}$$
$$a > c \longrightarrow \text{ one solution}$$

13. Write out the other two cases of *8.11.3 and *8.11.4.

14. Do prob. 3 by using the half-angle laws. (This method is usually preferable to the law of cosines when the numbers are not very simple.)

15. If C is given, how can we find $\frac{1}{2}(A + B)$? If, in addition, $\frac{1}{2}(A - B)$ is found by using *8.11.3, how can A and B be determined? (*Suggestion.* We have $\frac{1}{2}(A + B) = u$ and $\frac{1}{2}(A - B) = v$.)

16. Do prob. 4 by the law of tangents. (This method is better than the law of cosines for all but the simplest numbers.)

17. Fill in the following table, showing the laws used to solve triangles in the cases indicated:

	Given	Laws
	Three sides	
	Two sides and included angle	
	Two sides and angle opposite	
	Two angles and a side	

18. Show how to construct a triangle with ruler and compass in each of the cases in prob. 17. Pay particular attention to the ambiguous case, considering all four possibilities.

19. Rewrite *8.11.5 and *8.11.6 in as many ways as possible.

20. Prove *8.11.5 and *8.11.6. Draw a clear figure.

21. Find the area of each of the triangles considered in probs. 3 through 10, using just the original data if possible. (*Answers.* (3) 73. (4) 63. (5) 4.2×10^2. (6) 1.7×10^3. (7) 72. (8) 4.8 and 44. (9) 25.)

22. Show that the area of a circular segment (a figure bounded by the arc of a circle and its chord) is given by $\frac{1}{2}r^2(\theta - \sin \theta)$, where θ is the angle subtending the arc and chord. What values of θ are possible?

†**23.** Construct or look up in a text on trigonometry the proofs of *8.11.3 and *8.11.4.

†**24.** Construct a proof of *8.11.7 and show that the radius of the inscribed circle is given by the formula that appears in *8.11.4.

†**25.** Show that the ratios that appear in the law of sines are equal to $1/2R$, where R is the radius of the circle circumscribed about the triangle. (*Suggestion.* It is sufficient to show that $a/\sin A = 2R$. Begin by drawing a careful figure; then draw a diameter through C.)

†**26.** A coordinate system is constructed by drawing two axes at an angle θ and measuring coordinates parallel to these axes. Derive: (*a*) A formula for the distance between two points. (*b*) The equation of a straight line.

27. A geologist records an earthquake on his seismograph. If it occurred at a point 30° away from him (the angle measured at the center of the earth between the radii to the two points), and if the vibrations travel at 5.4 miles a second straight through the earth, when did the quake occur? (Take the radius of the earth as 3960 miles.)

†**28.** Prove that, if two sides of a triangle are unequal, the angle opposite the greater is larger than the angle opposite the smaller. State and prove the converse.

†**29.** Prove that, if two triangles have two sides of one respectively equal to two sides of the other, but the included angle of the first greater than the included angle of the second, the third side of the first is greater than the third side of the second. Also state and prove an obvious converse.

†**30.** Find a formula for the area of a regular polygon of n sides inscribed in a circle of radius r.

8.12 Numerical methods in trigonometry

In order to concentrate attention on fundamental theorems and methods we have avoided the complicated numbers that turn up in science and engineering. In such work precision is limited only by the accuracy of the data and the tables. When dealing with any but the smallest numbers it is convenient to use logarithms. We could, of course, find any circular function desired from a table such as Table IV and then look up its logarithm, but tables have been constructed that give directly the logarithms of the circular functions. Table IV gives these for angles in degrees and minutes for every 10 minutes from 0° to 90°. The student should note that the logarithms of csc x, sec x, and cot x are the negatives of those of sin x, cos x, tan x, respectively. Also it should be noted that the -10 must be appended to most of the tabular values in order to have the appropriate characteristic.

Interpolation in tables of logarithms of circular functions follows familiar methods, but we illustrate the technique by finding log sin 31° 22′:

$$
\begin{array}{cc}
x & \log \sin x
\end{array}
$$

(1) $\quad 10\left[\begin{matrix}-31° 20′ \\ 31° 22′ \\ \rightarrow 31° 30′\end{matrix}\right]2 \qquad h\left[\begin{matrix}-9.7160 - 10 \\ \rightarrow 9. \ ? \ - 10 \\ 9.7181 - 10\end{matrix}\right]21$

Then $h/21 = 2/10$ and $h \doteq 4$. Accordingly the desired logarithm is 9.7164 − 10. On the other hand, to find the x for which log cos $x = 9.4846 - 10$, we have

$$
\begin{array}{cc}
x & \log \cos x
\end{array}
$$

(2) $\quad 10\left[\begin{matrix}-72° 10′ \\ 72° \ ? \\ \rightarrow 72° 20′\end{matrix}\right]h \qquad -15\left[\begin{matrix}-9.4861 - 10 \\ \rightarrow 9.4846 - 10 \\ 9.4821 - 10\end{matrix}\right]-40$

Here $h/10 = 15/40$ and $h \doteq 4$. Hence $x \doteq 72° 14′$. Note that in this case the log cos x is decreasing as x increases. This need add no complications if the student thinks in terms of proportional changes.

The accuracy of interpolation depends upon the function, the size of the angle, and the table. Except for angles near 0° and 90°, we can say roughly that five-digit accuracy in the functions or their logarithms corresponds to accuracy to the nearest ten seconds in the angle. Other correspondences are given in the note at the beginning of Ex. 8.11.

In solving triangles, it is essential to work in a systematic way. The following is a good plan:

1. Draw a diagram showing the data. Make it accurate and large enough to be helpful in planning and checking your computations. Label sides and angles.

2. Write the formulas you plan to use.

3. Make out a form for the complete logarithmic calculation.

4. Look up the logarithms, carry through the computations, and check against your sketch.

5. Check your results against the original conditions of the problem and by alternative methods of solution.

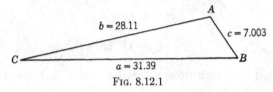

FIG. 8.12.1

We illustrate by doing three examples. Suppose that in a triangle $a = 31.39$, $b = 28.11$, and $c = 7.003$ (Fig. 8.12.1). We plan to use the following formulas for the solution:

$$\tan (A/2) = \frac{r}{s - a} \qquad \text{where } 2s = a + b + c$$

$$\tan (B/2) = \frac{r}{s - b} \qquad r = \left[\frac{(s - a)(s - b)(s - c)}{s} \right]^{1/2}$$

$$C = 180° - B - A$$

$a = 31.39$	$s - a = \ \ 1.862$
$b = 28.11$	$s - b = \ \ 5.142$
$c = \ \ 7.003$	$s - c = 26.249$
$2s = 66.503$	$\log (s - a) = 0.2700$
$s = 33.252$	$\log (s - b) = 0.7112$
	$\log (s - c) = 1.4191$
	$\text{colog } s = 8.4782 - 10$
	$\log r^2 = 0.8785$
	$\div 2$

$$\log r = 10.4392 - 10 \qquad\qquad \log r = 0.4392$$
$$-\log (s - b) = \ \ 0.7112 \qquad\qquad -\log (s - a) = 0.2700$$

$$\log \tan (B/2) = \ \ 9.7280 - 10 \qquad \log \tan (A/2) = 0.1692$$
$$B/2 = 28° 8' \qquad\qquad A/2 = \ \ 55° 53'$$
$$B = 56° 16' \qquad\qquad A = 111° 46'$$

$$C = 180° - (A + B) = 11° 58'$$

$$\text{Check by } \tan (C/2) = \frac{r}{s - c}$$

$$\log r = 10.4392 - 10$$

$$- \log (s - c) = 1.4191$$

$$\overline{\log \tan (C/2) = 9.0201 - 10}$$

$$C/2 = 5° 59'$$

$$C = 11° 58' \text{ as above}$$

Note that, although we are working to only four digits, we kept five in s in order to have four in $s - a$ and $s - b$.

Now consider a triangle with $a = 94.23$, $b = 55.61$, and $B = 27° 44'$. (See Fig. 8.12.2.) The figure suggests that there are two solutions.

$$\sin A = \frac{a \sin B}{b}$$ $$\log 94.23 = 1.9742$$
$$\log \sin 27° 44' = 9.6678 - 10$$
$$C = 180° - (A + B)$$ $$\operatorname{colog} 55.61 = 8.2548 - 10$$
$$c = \frac{a \sin C}{\sin A}$$ $$\overline{\log \sin A = 9.8968 - 10}$$

$$A = 52° 3'$$ $$A' = 180 - A = 127° 57'$$

$$C = 100° 13'$$ $$C' = 24° 19'$$

$$\log 94.23 = 1.9742$$ $$\log 94.23 = 1.9742$$

$$\log \sin 100° 13' = 9.9930 - 10$$ $$\log \sin 24° 19' = 9.6146 - 10$$
$$\operatorname{colog} \sin 52° 3' = 0.1032$$ $$\operatorname{colog} \sin 127° 57' = 0.1032$$

$$\overline{\log c = 2.0704}$$ $$\overline{\log c' = 1.6920}$$

$$c = 117.6$$ $$c' = 49.20$$

Note how the two solutions arise from the fact that there are two angles less than 180° with the same sine.

As our final example, we consider a problem in which the law of tangents is convenient. We have given $b = 15.31$, $c = 20.01$, $A = 61° 35'$. (See Fig. 8.12.3.)

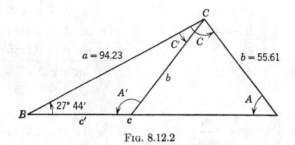

Fig. 8.12.2

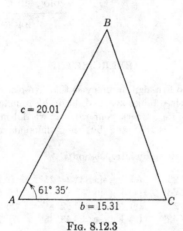

Fig. 8.12.3

$$\tan \frac{C-B}{2} = \frac{c-b}{c+b} \tan \frac{C+B}{2} \qquad a = \frac{b \sin A}{\sin B}$$

$$C = \frac{C+B}{2} + \frac{C-B}{2} \qquad B = \frac{C+B}{2} - \frac{C-B}{2}$$

$$c + b = 34.32 \qquad \log c - b = 0.6721$$

$$c - b = 4.70 \qquad \operatorname{colog} c + b = 8.4520 - 10$$

$$C + B = 180° - A = 118° \, 25' \qquad \log \tan \frac{C+B}{2} = 0.2246$$

$$\frac{C+B}{2} = 59° \, 12'$$

$$\log \tan \frac{C-B}{2} = 9.3487 - 10$$

$$\frac{C-B}{2} = 12° \, 35'$$

$$C = 71° 47'$$
$$B = 46° 37'$$

$$\log b = 1.1850$$
$$\log \sin A = 9.9443 - 10$$
$$\text{colog} \sin B = 0.1386$$

$$\log a = 1.2679$$
$$a = 18.53$$

Check by $a = \dfrac{c \sin A}{\sin C}$

$$\log c = 1.3012$$
$$\log \sin A = 9.9443 - 10$$
$$\text{colog} \sin C = 0.0223$$

$$\log a = 1.2678$$
$$a = 18.53 \text{ as above}$$

EXERCISE 8.12

(Data are given to five-digit accuracy so that five-place tables may be used if desired. If four-place tables are used, begin by rounding off to four digits and the nearest minute. Check your work as we did in the section by the equation $A + B + C = 180°$ and by using different relations among the elements.)

1. Find the following by interpolation:

(a) $\log \sin 31° 45' 25''$ (e) $\log \sin 83° 24' 21''$ (i) $\log \cos 95° 37' 15''$
(b) $\log \cot 15° 18' 56''$ (f) $\log \sin 63° 8' 16''$ (j) $\log \sin 123° 27.6'$
(c) $\log \tan 24° 31.2'$ (g) $\log \sec 36° 52' 8''$ (k) $\log \sin 55° 37' 51''$
(d) $\log \cos 40° 10' 42''$ (h) $\log \csc 22° 43' 48''$ (l) $\log \cos 72° 53''$

Answers. (a) $9.7211 - 10$. (c) $9.6590 - 10$. (e) $9.9971 - 10$. (g) 0.0969. (i) $8.9906 - 10$. (l) $9.4896 - 10$.

2. Solve for x by interpolation:

(a) $\log \sin x = 9.4160 - 10$ (f) $\log \tan x = 9.45416 - 10$
(b) $\log \sin x = 9.98023 - 10$ (g) $\log \tan x = 1.50000$
(c) $\log \cos x = 9.87797 - 10$ (h) $\log \cot x = 0.18558$
(d) $\log \cos x = 9.95142 - 10$ (i) $\log \sin x = 9.97138 - 10$
(e) $\log \cos x = 9.18313 - 10$ (j) $\log \cos x = 9.84102 - 10$

Answers. (a) $15° 6'$. (c) $40° 58'$. (e) $81° 14'$. (f) $15° 53'$. (h) $33° 7'$. (j) $46° 7'$.

3. Solve the following triangles, or show that there is no solution.

(a) $a = 105.27$ $b = 87.141$ $c = 126.43$
(b) $a = 24.153$ $b = 20.996$ $B = 31° 42' 25''$
(c) $b = 20.912$ $c = 13.319$ $A = 50° 41' 47''$
(d) $a = 1.3876$ $B = 22° 35' 10''$ $C = 56° 51' 39''$
(e) $A = 51° 20.2'$ $a = 23.711$ $c = 8.2978$
(f) $a = 18.915$ $A = 105° 10'$ $B = 23° 14' 33''$

(g) $a = 3.0021$ $b = 3.9984$ $c = 5.0095$
(h) $B = 97° 15' 18''$ $a = 5.1932$ $c = 35.914$
(i) $C = 110° 31.6'$ $c = 5482.7$ $b = 3018.5$
(j) $a = 23.911$ $A = 143° 28.1'$ $B = 10° 39' 42''$
(k) $a = 157.83$ $b = 103.29$ $c = 262.83$
(l) $A = 21° 53' 48''$ $a = 1.3926$ $b = 1.0437$
(m) $a = 1000.6$ $c = 99.832$ $B = 7° 56''$
(n) $A = 3° 51' 39''$ $B = 15° 43' 44''$ $c = 75.351$
(o) $a = 33.829$ $b = 36.198$ $A = 71° 10' 52''$
(p) $a = 33.829$ $b = 36.198$ $B = 71° 10' 52''$

4. Find the areas of the following triangles:

(a) $a = 149.38$ $b = 181.16$ $c = 200.92$
(b) $b = 18.837$ $c = 8.2913$ $C = 11° 21' 15''$
(c) $c = 1.8194$ $A = 21.2°$ $B = 84.26°$
(d) $B = 19° 53' 45''$ $a = 45.289$ $c = 45.282$
(e) $a = 25.211$ $b = 5.7613$ $c = 21.667$

(Check by finding the area in a different way.)

5. Find the area of each triangle in prob. 3.

†**6.** Find θ from $2.186 \cos \theta + 1.937 \sin \theta = 3.8991$. (See Ex. 8.7.24.)

†**7.** Show that, when θ is near 0, $\sin \theta \doteq \tan \theta \doteq \theta$; and, when θ is near $\pi/2$, $\cos \theta \doteq \cot \theta \doteq \pi/2 - \theta$. (*Suggestion.* Consider the definitions in a unit circle.) Compare the results of using these formulas with those obtained by interpolating in the table. Which do you think is more accurate?

8. Find the angle subtended by a man 6 feet tall standing 2 miles away.

†**9.** The following formulas, called **Mollweide's equations,** are useful for checking purposes:

(3) $$\frac{a - b}{c} = \frac{\sin \frac{1}{2}(A - B)}{\cos (C/2)} \qquad \frac{a + b}{c} = \frac{\cos \frac{1}{2}(A - B)}{\sin (C/2)}$$

(a) Derive the law of tangents from these formulas.
(b) Derive these formulas.

†**10.** Solve for x, or show that there is no real solution.

(a) $2 \sin^2 x + 3 \sin x + 1 = 0$
(b) $\sin^2 x + 3 \sin x + 2 = 0$
(c) $5 (\log x)^2 - 6 \log x + 1 = 0$
(d) $e^{2x} - 2e^x - 4 = 0$
(e) $3 \tan^2 x + 4 \tan x + 1 = 0$
(†f) $\tan^3 x + 3 \tan^2 x - \tan x - 3 = 0$
(†g) $\sin 2x + \sin x = 1$
(h) $\sin^2 x - \cos^2 x = 1$

9

Analytic Geometry

___

In this chapter we discuss some topics in analytic geometry that had to be postponed until the reader was familiar with circular functions. We use these ideas to complete the discussion of the straight line begun in Chapter 4.

9.1 Rotations

In 4.6 we introduced the idea of a translation, that is, a displacement of the origin of coordinates to a new point without changing the direction of the axes. We found translations useful in dealing with the straight line and various other graphs. It is also helpful at times to rotate the axes.

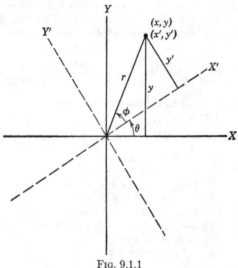

Fig. 9.1.1

Suppose (Fig. 9.1.1) that we rotate the axes through the angle θ. What is the relation between (x, y), the old coordinates of a point, and (x', y'), the new coordinates of the same point? Let ϕ

be the angle between the new x-axis and the radius vector to the point. Then from the definitions of the circular functions:

(1) $\sin(\theta + \phi) = y/r$ $\cos(\theta + \phi) = x/r$

(2) $\sin \phi = y'/r$ $\cos \phi = x'/r$

But we know that

$$\sin(\theta + \phi) \equiv \sin\theta\cos\phi + \cos\theta\sin\phi$$

and

$$\cos(\theta + \phi) \equiv \cos\theta\cos\phi - \sin\theta\sin\phi$$

Hence (1) and (2) imply:

(3) $y/r = (x'/r)\sin\theta + (y'/r)\cos\theta$

(4) $x/r = (x'/r)\cos\theta - (y'/r)\sin\theta$

Multiplying both sides of these equations by r, we have:

$$x = x'\cos\theta - y'\sin\theta$$
*9.1.1   (Rotation Formulas)
$$y = x'\sin\theta + y'\cos\theta$$

These formulas give the old coordinates (x, y) in terms of the new coordinates (x', y') and the angle θ from the old to the new x-axis. In order to illustrate them in a simple case, we consider the line $y = x$, which we know is a line through the origin forming an angle of $45°$ with the x-axis. If we rotate the axes through $45°$, the line should lie along the x'-axis in the new coordinate system and its new equation should be $y' = 0$. Letting $\theta = 45°$ in *9.1.1, we have:

(5)
$$x = x'\cos 45° - y'\sin 45° = x'(\tfrac{1}{2}\sqrt{2}) - y'(\tfrac{1}{2}\sqrt{2})$$
$$y = x'(\tfrac{1}{2}\sqrt{2}) + y'(\tfrac{1}{2}\sqrt{2})$$

Substituting these expressions in $y = x$, we have:

(6) $x'(\tfrac{1}{2}\sqrt{2}) + y'(\tfrac{1}{2}\sqrt{2}) = x'(\tfrac{1}{2}\sqrt{2}) - y'(\tfrac{1}{2}\sqrt{2})$

or

(7) $(\sqrt{2})y' = 0$

or

(8) $y' = 0$

which is, as expected, the equation of the line in the new coordinates.

As a second illustration, consider the circle of radius r and center at the origin:

$$(9) \qquad\qquad x^2 + y^2 = r^2$$

If we rotate the axes through any angle, this should make no difference in the equation, since the radius and center of the circle remain unchanged. Substituting from *9.1.1 in (9), we have:

$$(10) \qquad (x' \cos \theta - y' \sin \theta)^2 + (x' \sin \theta + y' \cos \theta)^2 = r^2$$

which, after carrying out the operations indicated and collecting similar terms, becomes:

$$(11) \qquad x'^2(\cos^2 \theta + \sin^2 \theta) + y'^2(\sin^2 \theta + \cos^2 \theta) = r^2$$

or

$$x'^2 + y'^2 = r^2, \text{ as expected.}$$

We now give an alternative proof of *4.7.4.

*Proof of *4.7.4.* We show that an appropriate translation and rotation make the x-axis coincide with the locus. Then the locus must be a straight line since it will have been shown to be congruent to a straight line. First we translate the origin to $(0, b)$. Then $x = x'$ and $y = y' + b$. The new equation is $y' = mx'$. Now we rotate through an angle $\theta = \arctan m$. Letting (x'', y'') be the coordinates after this rotation, we have:

$$(12) \qquad \begin{aligned} x' &= x'' \cos \theta - y'' \sin \theta \\ y' &= x'' \sin \theta + y'' \cos \theta \end{aligned}$$

The new equation is:

$$(13) \qquad (x'' \sin \theta + y'' \cos \theta) = m(x'' \cos \theta - y'' \sin \theta)$$

or

$$(14) \qquad (\cos \theta + m \sin \theta)y'' = (m \cos \theta - \sin \theta)x''$$

But since $m = \tan \theta = \sin \theta / \cos \theta$, the coefficient of x'' is zero whereas the coefficient of y'' cannot be. (Why?) Hence the equation becomes $y'' = 0$, the equation of the x''-axis, as was to be proved.

EXERCISE 9.1

1. Solve the equations *9.1.1 for x' and y' and so obtain the formulas for the new coordinates in terms of the old:

*9.1.2
$$x' = x \cos \theta + y \sin \theta$$
$$y' = -x \sin \theta + y \cos \theta$$

2. Find the formulas *9.1.2 by expanding $\sin \phi = \sin [(\phi + \theta) - \theta]$ and $\cos \phi = \cos [(\phi + \theta) - \theta]$. (See Fig. 9.1.1.)

3. Consider the points $(0, 0)$, $(3, 4)$, $(0, 5)$, $(-2, 0)$, $(-6, -2)$. Find their coordinates after rotating the axes through the following angles. Make a large sketch for each rotation.

(a) 90°	(c) −90°	(e) 30°	(g) 108°
(b) 45°	(d) 180°	(f) 20°	(h) 210°

4. If the axes are rotated through an angle θ, the new coordinates are (x', y') and the old (x, y). We could look at (x', y') as the old coordinates and (x, y) as the new coordinates after rotating the x', y'-axes through an angle of $-\theta$. Use this idea to derive *9.1.2 from *9.1.1.

5. Prove analytically that the distance of any point from the origin remains unchanged under a rotation. (*Note.* It is required to show that $x^2 + y^2 = x'^2 + y'^2$, where (x, y) and (x', y') are the old and new coordinates of a point.)

6. Consider the equation $y = x^2$. Rotate the axes through 90° and show that the new equation is $y'^2 = x'$. Draw a large sketch.

7. Show that for a rotation through 90°, $x' = y$ and $y' = -x$.

8. Find the new equation corresponding to $y = x^3$ after a rotation of 45°. Draw a large sketch.

9. Find the new equation corresponding to $y = 3x^2$ after translating the origin to $(3, 2)$ and then rotating through 45°.

10. Show that if the axes are translated to (x_0, y_0) and then rotated through θ, the old coordinates are given in terms of the new by:

(15)
$$x = x' \cos \theta - y' \sin \theta + x_0$$
$$y = x' \sin \theta + y' \cos \theta + y_0$$

Suggestion. Let (x'', y'') be the coordinates after the translation.

11. From (15) derive:

(16)
$$x' = (x - x_0) \cos \theta + (y - y_0) \sin \theta$$
$$y' = -(x - x_0) \sin \theta + (y - y_0) \cos \theta$$

†*12.* Show that translating to (x_0, y_0) and then rotating through θ does *not* give the same result as rotating through θ and then translating to (x_0, y_0) in the new system.

†*13.* Show analytically that the distance between two points is not changed by translation or rotation.

†14. We have considered only changes that involve translating or rotating the axes. We might also distort them. Thus if $x' = 2x$ and $y' = 2y$, it means that we have compressed both axes so that a point has coordinates twice as large. Illustrate this change by drawing a figure. Transformations of the type $x' = ax$, $y' = by$ are called **changes of scale**. Interpret and sketch the following: (a) $x' = 3x$, $y' = y$. (b) $x' = 0.5x$, $y' = 0.5y$. (c) $x' = 2x$, $y' = 0.5y$.

†15. Consider $y = 4x^2$. (a) Make the transformation $x' = x$, $y' = y/4$ and graph on the new axes. (b) Make the transformation $x' = 2x$, $y' = y$ and graph on the new axes. (Note that the curve remains unchanged, since we interpret the transformation as distorting the axes.)

†16. So far we have considered the points as fixed and the axes as changing. It is sometimes more convenient to consider the axes fixed and interpret the

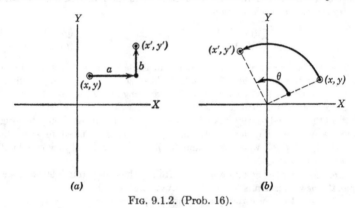

(a)   (b)

Fig. 9.1.2. (Prob. 16).

transformation as moving the points. Thus $x' = x - x_0$, $y' = y - y_0$ is considered as moving each point (x, y) a horizontal distance $-x_0$ and a vertical distance $-y_0$. The points in the plane are translated by $(-x_0, -y_0)$ and the axes remain fixed. Similarly $x' = 2x$ and $y' = 2y$ means moving each point in the plane to a point with double the original coordinates. Interpret the following in this way:

(a) $x' = x + a$, $y' = y + b$

(b) $x' = x \cos \theta - y \sin \theta$
$y' = x \sin \theta + y \cos \theta$

(c) $x' = x + 1$, $y' = y - 2$

(d) $x' = (x + y)/\sqrt{2}$
$y' = (-x + y)/\sqrt{2}$

(e) $x' = 0.5x$, $y' = y$

(f) $x' = (x - 1)/2$, $y' = y$

(g) $x' = x$, $y' = 3y$

(h) $x' = x$, $y' = x + 2y$

(i) $x' = x^2$, $y' = y$

Answers. (a) Each point is shifted a directed distance a horizontally and b vertically. (b) Each point is moved along an arc of a circle with center at the origin through an angle θ. (See Fig. 9.1.2.)

†17. When points in the plane are shifted so that each point changes place with the point symmetric to it with respect to a fixed line, the transformation

is called a **reflection** in this line. Interpret the following reflections from the point of view of prob. 16 and identify the line of reflection in each case:

(a) $x' = -x,\ y' = y$ (c) $x' = y,\ y' = x$ (e) $x' = 2 - x,\ y' = y$
(b) $x' = x,\ y' = -y$ (d) $x' = -y,\ y' = -x$ (f) $y' = -1 - y,\ x' = x$

†**18.** Do probs. 14 and 15 from the point of view of prob. 16.

†**19.** Any pair of functions

(17) $$x' = f(x, y), \qquad y' = g(x, y)$$

may be called a **transformation** or change of variable. It may be interpreted as giving the new coordinates of a fixed point with respect to a different coordinate system. Such was our first interpretation of the equations of translation and rotation. (By what do we replace f and g in order to get the equations of translation and rotation?) Or we may interpret (17) as determining new points in the same coordinate system, so that to each (x, y) there corresponds a (x', y') given by $f(x, y)$ and $g(x, y)$. Letting T stand for (17), we may write $T(x, y) = (x', y')$, meaning that the transformation T carries (x, y) into (x', y'). Then, if T_1 and T_2 stand for two transformations, we say that $T_1 = T_2$ if $T_1(x, y) = T_2(x, y)$ for all (x, y), that is, if they have the same effect on all points. By $T_2 T_1(x, y)$ we mean $T_2[T_1(x, y)]$, that is, the result of transforming by T_1 and then T_2.

(a) Let T_1 stand for $x' = x + 2,\ y' = y + 3$ and T_2 stand for $x' = x - 1$, $y' = y + 4$. Sketch T_1, T_2, $T_1 T_2$, and $T_2 T_1$. (*Suggestion.* Sketch a point and the points into which it goes.)

(b) Let T_1 stand for a rotation through $30°$ and T_2 for one through $60°$. Sketch T_1, T_2, $T_1 T_2$, $T_2 T_1$.

(c) Show that, if T_1 and T_2 are translations, $T_1 T_2 = T_2 T_1$.

(d) Show the same result for two rotations.

(e) Show that $T_1 T_2 \not\equiv T_2 T_1$ for all transformations. (This shows that multiplication of transformations is not commutative.)

(f) Show that, if T_1 and T_2 are translations, so is $T_1 T_2$.

(g) Show the same for two rotations.

†**20.** Do you think that multiplication of transformations is associative? Justify your answer.

†**21.** The transformation I such that $I(x, y) = (x, y)$ for all (x, y) is called the **identity transformation.**

(a) What is its geometric interpretation?

(b) How may I be considered as a special kind of rotation?   translation? change of scale?

†**22.** If T and T' are such that $TT' = T'T = I$, each is called the **inverse** of the other.

(a) To what does T' correspond in the algebra of ordinary numbers?

(b) What is the inverse of a rotation through θ?

(c) What is the inverse of the translation $x' = x + a,\ y' = y + b$?

(d) What is the inverse of $x' = ax,\ y' = by$?

9.2   Angles between lines

Consider a straight line L that meets the x-axis in the point P.
(Two cases are sketched in Fig. 9.2.1.)   The smallest positive

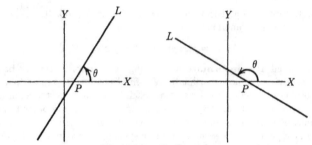

FIG. 9.2.1.   Inclination.

angle θ through which the x-axis must be rotated about P in order
to coincide with the line is called the **inclination** of the line.   If
the line is parallel to the x-axis, its inclination is defined to be zero.
Evidently $0 \leq \theta < \pi$.   If the student recalls the definition of the
slope of a straight line and of the circular functions, he will see that
the slope of any non-vertical straight line is the tangent of its
inclination, that is,

*9.2.1 $m = \tan \theta$

Consider two lines, L_1 and L_2, with inclinations θ_1 and θ_2.   We
define the **angle from L_1 to L_2** as $\phi_{12} = \theta_2 - \theta_1$.   Two cases are

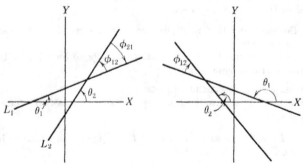

FIG. 9.2.2

sketched in Fig. 9.2.2.   Note that $\phi_{21} = -\phi_{12}$.   When we speak
of the **angle between L_1 and L_2** we mean $|\phi_{12}|$.   From *9.2.1

we can easily derive an expression for the tangent of the angle
from one line to another. We have:

(1) $\tan \phi_{12} = \tan (\theta_2 - \theta_1)$

$$= \frac{\tan \theta_2 - \tan \theta_1}{1 + \tan \theta_2 \tan \theta_1} \quad \text{(Why?)}$$

But $\tan \theta_2 = m_2$ and $\tan \theta_1 = m_1$, where m_2 and m_1 are the slopes
of L_2 and L_1. Hence:

*9.2.2   If ϕ_{12} is the angle from L_1 to L_2,

$$\tan \phi_{12} = \frac{m_2 - m_1}{1 + m_2 m_1}$$

This formula enables us to find the angle between two lines if
we know their slopes. For example, let L_1 be $2x + 4y = 8$ and
L_2 be $5x - 6y = -5$. Then:

(2) $m_1 = -1/2 \quad \text{and} \quad m_2 = 5/6 \quad \text{(Why?)}$

(3) $\tan \phi_{12} = \dfrac{5/6 + 1/2}{1 + (5/6)(-1/2)} = \dfrac{16}{7} = 2.2857$

From Table IV, $\tan 66° 22' = 2.2857$, but a sketch of the lines
indicates that ϕ_{12} is negative. Hence:

(4) $\phi_{12} = -113° 38'$

This is the angle from the first to the second line. The angle
between the lines is $113° 38'$ and this also is the angle from the
second line to the first. The angles involved are sketched in Fig.
9.2.3.

A special case of *9.2.2 arises if the two lines are parallel or coincident. In this case $m_1 = m_2$, $\phi_{12} = 0$ and $\tan \phi = 0$ as we
would expect. (Compare *4.9.2.) Another important special case
arises if we consider two mutually perpendicular lines. Since $\tan$
$90°$ is infinite, we rewrite *9.2.2 as:

(5) $\cot \phi_{12} = \dfrac{1 + m_1 m_2}{m_2 - m_1}$

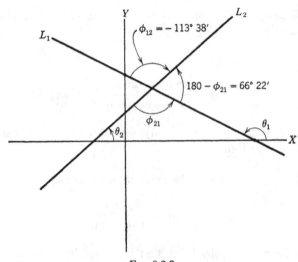

FIG. 9.2.3

Excluding the case when one of the lines is vertical (why?) we have $\phi_{12} = 90°$ and $\cot \phi_{12} = 0$, if, and only if, the numerator of the fraction is zero. Hence:

*9.2.3 Two non-vertical lines are perpendicular if and only if their slopes are negative reciprocals, that is,

$$m_1 m_2 = -1$$

EXERCISE 9.2

1. Find the inclination of a line with each of the following slopes:

(a) 1. (b) −1. (c) $\sqrt{3}$. (d) 2. (e) −4. (f) $\sqrt{2}$. (g) 0.3178.

Answers. (a) 45°. (c) 60°. (e) 104° 2′.

2. Graph each of the following lines and find their inclinations:

(a) $y = 2x - 3$. (b) $4x + 3y = 2$. (c) $7x - 8y = 3$. (d) $x + 4y = 1$.

3. Find the slopes of the lines whose inclinations are:

(a) $\pi/4$	(c) 137° 10′	(e) $\pi/6$	(g) 95°
(b) 25°	(d) $\pi/2$	(f) 179	(h) 19° 43′

Answers. (a) 1. (c) −0.9271. (e) $1/\sqrt{3}$.

4. Find θ_1 and θ_2 for the lines whose slopes are given by (2) and so check the value of ϕ_{12} given in (4).

5. In each of the following pairs of lines find the angle from the first to the second by *9.2.2. Draw and label a large sketch in each case and check by finding θ_1 and θ_2 from the slopes.

(a) $3x - 4y - 6 = 0$, $4x - y = 0$ (d) $x + y + 1 = 0$, $23x - 41y = 80$
(b) $y = 2x$, $y = x$ (e) $2x + 6y = 3$, $x = 4$
(c) $y = 3x + 4$, $3y - 3x = 7$ (f) $x = 2y - 18$, $y = 2x + 18$

Answer for part c. $-26° 34'$.

6. Show that the line $y = x$ bisects the angle formed by the axes.

7. Find the line through the origin that has an inclination of 30°.

8. Find the lines through the point $(0, 5)$ that make an angle of 25° with the y-axis. (*Answer.* $y = \pm 2.145x + 5$.)

9. Prove that the lines $2x - 3y = 1$ and $6x + 4y = 7$ are perpendicular.

10. Find the line through the point $(-1, 3)$ and perpendicular to the line $22x - 14y + 17 = 0$. (*Suggestion.* Find the slope of the required line and then use the point-slope form. *Answer.* $7x + 11y = 26$.)

11. State separately the two theorems combined in *9.2.3.

12. Show that the lines $y = 3$ and $x = -1$ are perpendicular.

13. Find the line through the origin perpendicular to $x - 14y - 3 = 0$.

14. Find the line with y-intercept 6 perpendicular to $2y - 7x - 1 = 0$.

15. Find the line with x-intercept 4 perpendicular to $y = x$.

16. Find the line passing through the point $(2, 1)$ and such that the angle from it to the line $y = x$ is 45°.

17. Find the equation of the perpendicular bisector of the segment determined by $(1, 16)$, $(-3, -4)$.

18. Find the equations of the lines that bisect the angles formed by the intersection of $y = -2x + 3$ and $y = x + 5$.

19. Find the interior angles of the triangle $(-3, 4)$, $(2, 1)$, $(4, -6)$.

20. Do Ex. 8.1.7 another way.

21. Find a formula for the slopes of the bisectors of the angles formed by two lines with slopes m_1 and m_2.

22. Show that $a_1x + b_1y = c_1$ and $a_2x + b_2y = c_2$ are perpendicular if and only if $a_1a_2 + b_1b_2 = 0$.

23. Show that the angles between the lines in prob. 22 are given by $\tan \theta = (a_1b_2 - a_2b_1)(a_1a_2 + b_1b_2)^{-1}$.

9.3 The normal form of the equation of a straight line

Consider any straight line L. Let L' be the line through the origin perpendicular to L and meeting it in A (Fig. 9.3.1). Let p be the distance OA and α the angle XOA, that is, the smallest positive angle from the positive x-axis to the directed line from O to A. We call α the **direction angle** of OA, and we see that it can have any value between 0 and 2π. Note that α is not the same as the inclination of the line OA, although the two are equal when $\alpha < 180°$. We call L' the **normal** to L, and p the **normal**

intercept. If the line passes through the origin, we let $p = 0$ and α be the inclination of the normal.

We wish to find the equation of L in terms of the given constants, α and p. In order to do this we rotate the axes through the angle

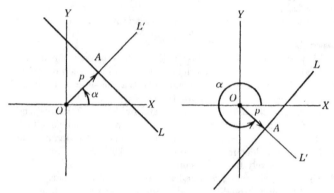

FIG. 9.3.1.　Direction angle and normal intercept.

α. Then the new x-axis coincides with L', and L is parallel to the new y-axis and meets the new x-axis in the point with new coordinates $(p, 0)$ (Fig. 9.3.2). Hence the equation of L in the new co-

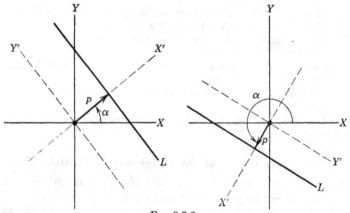

FIG. 9.3.2

ordinates is $x' - p = 0$. But from *9.1.2, $x' = x \cos \alpha + y \sin \alpha$, where x and y are the old coordinates. Hence the equation in the original coordinate system is

(1)　　　　$x \cos \alpha + y \sin \alpha - p = 0$　　　(Normal Form)

Suppose, for example, that a line has a normal intercept of 3 and a direction angle of the normal equal to 45°. Then $p = 3$ and $\alpha = 45°$. Hence the equation is $(1/\sqrt{2})x + (1/\sqrt{2})y = 3$. On the other hand, if $\alpha = 225°$, the equation is $(-1/\sqrt{2})x + (-1/\sqrt{2})y = 3$. (Sketch the two lines.)

Sometimes, when we are given the equation of a straight line, it is convenient to rewrite it in the normal form, and we can always do this by multiplying both sides of the equation by the appropriate number. If the original equation is $Ax + By + C = 0$, we want a number k such that

$$(2) \qquad kAx + kBy + kC = 0$$

is in the form (1), that is, such that

$$kA = \cos \alpha$$

$$(3) \qquad kB = \sin \alpha \quad \text{and}$$

$$kC = -p$$

where p is the normal intercept and α is the direction angle of the normal. In order to find k we square both sides of the first two equations in (3) and add. This gives:

$$(4) \qquad k^2 A^2 + k^2 B^2 = \cos^2 \alpha + \sin^2 \alpha = 1$$

or

$$(5) \qquad k^2 = (A^2 + B^2)^{-1}$$

Now p, the normal intercept, is always positive or zero, since it was taken as the distance OA. Hence kC must be negative, and k must have the opposite sign from C. It follows that

$$(6) \qquad k = \pm(A^2 + B^2)^{-\frac{1}{2}}$$

where the sign is to be taken opposite to that of C. Hence the equation (2) becomes:

$$(7) \qquad \frac{Ax}{\pm\sqrt{A^2 + B^2}} + \frac{By}{\pm\sqrt{A^2 + B^2}} + \frac{C}{\pm\sqrt{A^2 + B^2}} = 0$$

Then $C/\pm\sqrt{A^2 + B^2}$ is p, and α is the angle less than 2π whose cosine and sine are, respectively, equal to the coefficients of x and y in (7). If $C = 0$, then $p = 0$, and the sign of the radical is chosen so that $0 \leq \alpha < \pi$.

Consider, for example, the line $x - 3y + 6 = 0$. Here $A = 1$, $B = -3$, $\sqrt{A^2 + B^2} = \sqrt{10}$, and the equation in normal form is:

(8)
$$\frac{-1}{\sqrt{10}} x + \frac{3}{\sqrt{10}} y - \frac{6}{\sqrt{10}} = 0$$

Note that we have divided both sides of the equation by $-\sqrt{10}$ so that the new constant term is negative. Hence:

(9)
$$p = \frac{6}{\sqrt{10}} \doteq 2$$

and α is an angle with cosine of $-1/\sqrt{10}$ and sine of $3/\sqrt{10}$. The angle lies in the second quadrant. (Why?) Also $3/\sqrt{10} \doteq 0.95$. From Table IV, $\sin 72° \doteq 0.95$ and hence $\alpha \doteq 180° - 72° = 108°$.

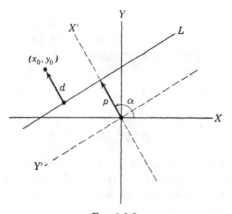

Fig. 9.3.3

The student will recall from plane geometry that the **distance between a point and a line** is defined as the length of the perpendicular from the point to the line. The normal form gives us immediately the distance between the origin and a line. We can also make use of the normal form to get directly the distance between any point and a line. Suppose that we have a straight line L given by $x \cos \alpha + y \sin \alpha = p$ and a point (x_0, y_0) (Fig. 9.3.3). Rotating the axes through α, we find that the line is parallel to the new y-axis, and the distance between the line and the point is $|d| = |x_0' - p|$, where x_0' is the new x-coordinate of the point (x_0, y_0). But $x_0' = x_0 \cos \alpha + y_0 \sin \alpha$. (Why?) Hence:

*9.3.1 The distance between the point (x_0, y_0) and the line $x \cos \alpha + y \sin \alpha = p$ is given by the absolute value of

$$(10) \qquad d = x_0 \cos \alpha + y_0 \sin \alpha - p$$

In *9.3.1 we were not concerned with the direction of the point from the line. If, however, we agree to measure **directed distances from a line** by choosing the positive direction along the normal from the origin to the line, d in (10) gives the directed distance from the line to the point. It is positive or negative according as the point is on the opposite or same side of the line as the origin.

In order to use (10), the equation of the line must be in the normal form. Suppose, for example, that we wish the distance from $3x - 4y + 12 = 0$ to $(-2, 3)$. The equation in normal form is

$$(11) \qquad \frac{-3x}{5} + \frac{4y}{5} - \frac{12}{5} = 0$$

Hence:

$$(12) \qquad d = \frac{-3(-2) + 4(3) - 12}{5} = \frac{6}{5}$$

On the other hand, the distance from this line to $(3, 1)$ is

$$(13) \qquad d' = \frac{-3(3) + 4(1) - 12}{5} = \frac{-17}{5}$$

EXERCISE 9.3

1. Sketch the lines with:

(a) $p = 3$, $\alpha = 45°$

(b) $p = 3$, $\alpha = 225°$

(c) $p = 3$, $\alpha = 150°$

(d) $p = 3$, $\alpha = 300°$

(e) $p = 5$, $\alpha = 90°$

(f) $p = 4$, $\alpha = 120°$

(g) $p = 1$, $\alpha = 180°$

(h) $p = \sqrt{2}$, $\alpha = 125°$

(i) $p = 1.5$, $\alpha = 327° 15'$

(j) $p = 100$, $\alpha = 220° 30'$

2. Write in normal form the equation of each line in prob. 1. Express the circular functions numerically and simplify. [*Answers.* (a) $x + y = 3\sqrt{2}$. (e) $y = 5$. (g) $x = -1$.]

3. Put each of the following equations in normal form, find the direction angle of the normal and the normal intercept, and sketch.

(a) $4x - 3y - 7 = 0$ (e) $3x + 4y + 10 = 0$ (i) $4x + 3y + 7 = 0$
(b) $x + y + \sqrt{2} = 0$ (f) $5x + 4y + 20 = 0$ (j) $x + 2y - 3 = 0$
(c) $2x - y + 3 = 0$ (g) $4x - 3y + 7 = 0$ (k) $\sqrt{3}x - y = 2$
(d) $8x + y - \sqrt{130} = 0$ (h) $4x + 3y - 7 = 0$ (l) $x = 5$

Answers. (a) $\dfrac{4x}{5} + \dfrac{(-3y)}{5} - \dfrac{7}{5} = 0$, $p = 7/5$, $\alpha = 323°\,8'$. (c) $\dfrac{-2x}{\sqrt{5}} + \dfrac{1y}{\sqrt{5}}$
$-\dfrac{3}{\sqrt{5}} = 0$, $p = 3/\sqrt{5}$, $\alpha = 153°\,26'$.

4. Graph (8) and check p and α.

5. Make a sketch for (12) and (13) in the section.

6. Find the directed distance from the line $3x - 4y + 12 = 0$ to each of the following points. Make large sketches showing the distances as vectors: (a) $(-3, 4)$. (b) $(3, 4)$. (c) $(0, 0)$. (d) $(2, -5)$. (e) $(1, 8)$. [Answers. (a) 13/5. (c) $-12/5$. (e) 17/5.]

7. Justify the following theorem:

*9.3.2   The distance between the line $Ax + By + C = 0$ and the point (x_0, y_0) is given by the absolute value of

(14) $$d = \frac{Ax_0 + By_0 + C}{\sqrt{A^2 + B^2}}$$

8. Find the distance from the line $x - 4y = 6$ to the following points: (a) $(0, 0)$. (b) $(5, 10)$. (c) $(1, -3)$. (d) $(-3, 2)$. (e) $(-3, -7)$.

9. How must *9.3.2 be modified so as to give the distance from the line to the point according to the convention mentioned in the section? (Answer. The sign of the square root must be chosen opposite to that of C.)

10. Find the distance from each line to the indicated point and sketch.

(a) $x + 2y = 4$, $(-1, -7)$ (c) $y = 4x - 15$, $(-3, 2)$
(b) $2x + 4y + 5 = 0$, $(1, -1)$ (d) $y = x + 8$, $(4, -15)$

11. Show that the line $3x - 4y + 10 = 0$ is tangent to the circle with center at the origin and radius 2.

12. Find the equations of the lines passing through the point $(4, 0)$ and tangent to the circle with center at the origin and radius 3. (Answers. $3x \pm \sqrt{7}y = 12$.)

13. Find the equations of the lines tangent to the circle $x^2 + y^2 = 4$ and passing through each of the following: (a) $(0, 3)$. (b) $(4, 0)$. (c) $(3, 4)$. (d) $(-5, 6)$.

14. Find the equations of the lines through the origin and tangent to the circle $(x - 2)^2 + (y + 4)^2 = 16$. (Suggestion. Translate the origin to the center of the circle. Solve and translate back.)

15. Find the distance between the lines $6x - y + 3 = 0$ and $6x - y - 7 = 0$.

16. What would you consider to be the distance between a point and a line through it? Show that *9.3.1 gives the expected result.

17. Find the length of the altitudes of the triangles determined by the following triples of points:

(a) $(1, 4)$, $(3, -2)$, $(-1, 7)$

(b) $(2, 8)$, $(-1, -3)$, $(4, 1)$

(c) $(5, 1)$, $(-1, -2)$, $(-8, 1)$

(d) $(5, 6)$, $(-4, 6)$, $(-7, -2)$

18. Fill in the reasons and details in the proof of the following theorem:

*9.3.3   The area of the triangle with vertices (x_1, y_1), (x_2, y_2), (x_3, y_3) is the absolute value of

$$(15) \qquad A = \frac{1}{2} \begin{vmatrix} x_1 & y_1 & 1 \\ x_2 & y_2 & 1 \\ x_3 & y_3 & 1 \end{vmatrix}$$

Proof. The area is given by $\frac{1}{2}ah$, where a is one side and h the altitude on that side. (See Fig. 9.3.4.) The side P_2P_3 has length

$$(16) \qquad a = \sqrt{(x_2 - x_3)^2 + (y_2 - y_3)^2}$$

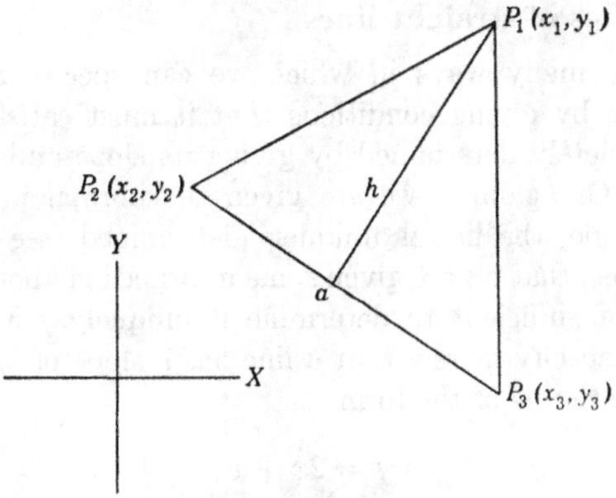

FIG. 9.3.4

The altitude h is the distance between P_2P_3 and P_1. The equation of the line P_2P_3 is

$$(17) \qquad \begin{vmatrix} x & y & 1 \\ x_2 & y_2 & 1 \\ x_3 & y_3 & 1 \end{vmatrix} = 0$$

If we put this in normal form and substitute the coordinates of P_1, we have:

$$(18) \qquad h = \frac{\begin{vmatrix} x_1 & y_1 & 1 \\ x_2 & y_2 & 1 \\ x_3 & y_3 & 1 \end{vmatrix}}{\pm\sqrt{(x_2 - x_3)^2 + (y_2 - y_3)^2}}$$

Substituting these values of a and h in $\frac{1}{2}ah$ gives the desired result.

19. Find the area of each of the following triangles:

  (a) $(0, 0)$, $(1, 1)$, $(2, 3)$ (c) $(-2, 3)$, $(1, -4)$, $(0, 7)$

  (b) $(0, 0)$, $(1, 2)$, $(1, -4)$ (d) $(5, 5)$, $(-5, 5)$, $(0, -3)$

Answers. (a) 0.5. (c) 13.

20. Find the area of each triangle in prob. 17.

21. Use *9.3.3 to show that $(-1, 2)$, $(2, -4)$, and $(1, -2)$ are collinear.

†**22.** Show that in any triangle the sum of the distances of any interior point from the three sides is a constant. (*Suggestion.* Place the axes so that each of the vertices has one coordinate zero.)

23. Find the bisectors of the angles formed by $y = 2x + 3$ and $3x - 4y = 2$. (*Suggestion.* Use the property that the bisector is the locus of points equidistant from the two sides. Note how the two bisectors appear as a result of the ambiguity of the sign of the radical.)

24. Solve Ex. 9.2.18 by the method of this section.

25. Find and sketch the bisectors of the angles formed by each pair of lines in Ex. 4.9.2.

9.4 Families of straight lines

There are many ways in which we can specify a particular straight line by giving conditions that it must satisfy. Thus a line is completely determined by giving its slope and y-intercept. (See 4.8.) Or again, if we are given the normal intercept and direction angle, the line is uniquely determined (see 9.3). Now it may happen that we are given some information about a straight line, but not sufficient to determine it uniquely. For example, suppose we specify merely that a line has a slope of 2. We know that its equation is of the form

$$(1) \qquad\qquad y = 2x + b$$

but we cannot determine b from the given information. For each value of b we get a definite line of slope 2, and all such lines form a set of parallel lines. We call this set a **family** of lines, and say that the equation (1) is its equation. The variable b plays a role different from that played by x and y. It is a constant as far as the graphing of (1) is concerned, but for each of its values we get a different line. We call it a **parameter.**

A set of lines given by

$$A(c)x + B(c)y + C(c) = 0$$

where $A(c)$, $B(c)$, $C(c)$ are functions of a parameter c, is called a

one-parameter family of lines. Consider, for example, the one parameter family

(2)
$$\frac{x}{c} + \frac{y}{\sqrt{r^2 - c^2}} = 1$$

where r is a constant. Here $A(c) = 1/c$, $B(c) = 1/\sqrt{r^2 - c^2}$, and $C(c) = -1$ (Fig. 9.4a). The intercepts are c and $\sqrt{r^2 - c^2}$. Evidently r is the length of the segment cut off between the axes, and since r is constant we see that the family consists of all those lines

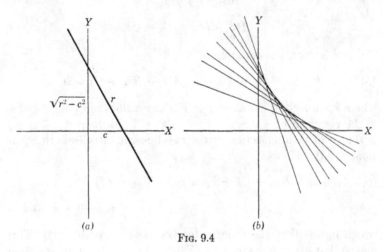

(a)   (b)

FIG. 9.4

on which the axes cut off a segment of length r. In Fig. 9.4b are sketched several members of the family. The student may note that the lines in the figure seem to outline a curve. It often happens, as in this case, that there is a curve to which the lines of a one-parameter family are tangent. Such a curve is called the **envelope** of the family. The exact definition of an envelope and the methods of finding its equation are part of calculus.

We have given examples of two types of problems: (1) Given a description of a family of lines, to find the equation of the family. (2) Given the equation of a family, to graph and describe the lines. In order to solve problems of the first type, we express the given information algebraically. In order to solve problems of the second type, we graph some of the lines and try to see the property that they have in common.

It is often of interest to find the family of lines that pass through the intersection of two given lines. If we know the point of intersection, (x_0, y_0), or can find it, the equation

(3) $$y - y_0 = m(x - x_0)$$

gives all lines of the family except the line $x = x_0$. (Why?) However, there is a way of solving the problem without knowing the point of intersection. Suppose that the two lines are:

(4)
$$Ax + By + C = 0$$
$$A'x + B'y + C' = 0$$

Consider now the equation

(5) $$k_1(Ax + By + C) + k_2(A'x + B'y + C) = 0$$

where k_1 and k_2 are parameters. For any values of k_1 and k_2, this is the equation of a straight line. (Why?) Moreover, if (x_0, y_0) is the point of intersection of the two lines, it satisfies both equations in (4). Hence:

(6) $k_1(Ax_0 + By_0 + C) + k_2(A'x_0 + B'y_0 + C')$

$$= k_1 \cdot 0 + k_2 \cdot 0 = 0$$

which means that the coordinates (x_0, y_0) also satisfy (5). This means that every line given by (5) passes through the intersection of (4).

Not only does every line in the family (5) pass through the point of intersection, but the family (5) includes *all* such lines. If $k_1 = 0$, we have just the second line itself. If $k_1 \neq 0$, we may divide by it and get

(7) $$(Ax + By + C) + c(A'x + B'y + C') = 0$$

where $c = k_2/k_1$. This may be rewritten in either of the following two forms:

(8) $$(A + cA')x + (B + cB')y + (C + cC') = 0$$

(9) $$y = \frac{-(A + cA')x}{(B + cB')} - \frac{(C + cC')}{(B + cB')}$$

From (9) we see that c can be chosen so as to make the line have any desired slope. For, if the slope is to be m, we have:

(10)
$$\frac{-(A + cA')}{(B + cB')} = m$$

which is true if

(11)
$$c = \frac{-(A + mB)}{(A' + mB')}$$

By letting m take all possible values (except $m = -A'/B'$, which is the slope of the second line itself), we get all lines through the point of intersection except the second line and the line parallel to the y-axis. In order to get the latter we take $c = -B/B'$. We have shown that:

*9.4.1 The family of all lines through the intersection of the lines $Ax + By + C = 0$ and $A'x + B'y + C' = 0$ is given by

$$k_1(Ax + By + C) + k_2(A'x + B'y + C') = 0$$

If we are not interested in the line $A'x + B'y + C' = 0$ as part of the family, it is satisfactory to use the equation (7) to represent the family, with the exception of this one line.

The theorem enables us to find the equation of the family of all lines through the intersection of two lines and also to find a particular member satisfying some additional condition. For example, the family of lines through the intersection of

(12)
$$2x - 6y + 3 = 0$$
$$3x + y + 9 = 0$$

is given by

(13)
$$2x - 6y + 3 + c(3x + y + 9) = 0$$

or

(14)
$$(2 + 3c)x + (-6 + c)y + (3 + 9c) = 0$$

The member that passes through the origin is given by a value of c such that $(0, 0)$ satisfies (13). Substituting $(0, 0)$ for (x, y) we find $3 + 9c = 0$ or $c = -1/3$. Substituting this value in (14), we find $x - (19/3)y = 0$ or $19y = 3x$ to be the desired equation.

Or suppose that we wish the line in the family (13), which has a slope of 3. The slope of (14) is $(2 + 3c)/(6 - c)$. Setting this equal to 3 and solving for c, we have:

$$(15) \qquad\qquad 2 + 3c = 3(6 - c)$$

$$(16) \qquad\qquad 6c = 16 \quad \text{or} \quad c = 8/3$$

The desired equation is $10x + (-10/3)y + 27 = 0$, or $30x - 10y + 81 = 0$.

EXERCISE 9.4

1. Find the equations of the lines that pass through $(1, 4)$ and have the slopes $1, -1, 4, 7, m$. [*Answer to last part.* $y - 4 = m(x - 1)$.]

2. Find the equations of the lines that have a slope of 3 and y-intercept of $8, 4, -1, b$. (*Answer to last part.* $y = 3x + b$.)

3. Find the equation of the line whose y-intercept is a and the sum of whose intercepts is 8. $\left(Answer. \ \dfrac{x}{8 - a} + \dfrac{y}{a} = 1. \right)$

4. Graph several members of each of the following families and characterize the family by a property common to all its members:

(a) $y - 1 = mx$

(b) $y = bx + 3$

(c) $x \cos t + y \sin t = 7$

(d) $y = x + b$

(e) $sx + 2y = 6$

(f) $\dfrac{x}{-a} + \dfrac{y}{a} = 1$

Answers. (a) Pass through $(0, 1)$. (c) Tangent to $x^2 + y^2 = 49$. (e) Pass through $(0, 3)$.

5. Write the equation of the one-parameter family of lines satisfying each of the following conditions:

(a) Passing through $(2, 7)$ and not parallel to the y-axis.

(b) Having a y-intercept of -5.

(c) Having a slope of 1.

(d) With x-intercept 3.

(e) With slope of $-1/3$.

Answers. (a) $y - 7 = m(x - 2)$. (c) $y = x + b$. (e) $3y = -x + b$.

6. Graph several members and describe each of the following families:

(a) $cx + 2cy = 7$

(b) $2x + y = 2a$

(c) $y = mx + m^2$

(d) $y = x + b$

(e) $ax + y = a^2$

(f) $(3 - k)x + ky = k(3 - k)$

(g) $|a|x + ay = a|a|$

(h) $a^2x + y = a^3$

(i) $y = mx + \sqrt{m}$

Answers. (a) Lines with slope -0.5. (c) Lines whose y-intercept is the square of the x-intercept. (e) Same family as c.

7. Write the equation of the family of lines:

(a) Having an inclination of 63°.
(b) Having a y-intercept three times the x-intercept.
(c) The sum of whose intercepts is 86.
(d) At a distance 5 from the origin.
(e) Parallel to the line $2x - 14y = 3$.
(f) Perpendicular to the line $y = 4x - 2$.
(g) At a distance 5 from the point $(-3, 1)$.
(h) The distance between whose intercepts is 3.
(i) Having an x-intercept twice the y-intercept.
(j) Tangent to the circle $x^2 + y^2 = 16$.

8. Graph several members of each family in probs. 5 and 7.
†9. Which of the families considered above seem to have envelopes? Can you identify these envelopes?
10. For the examples beginning with (12) in the section, graph each of the lines involved and check the algebra. Find the line of the family with slope -2.
11. Carry through the previous problem by solving for the intersection of the lines.
12. Find the line that passes through the intersection of the lines $x + 7y - 6 = 0$ and $5x = y - 15$ and the point $(1, -3)$. Do not solve for the intersection. (*Answer.* $17x + 15y + 28 = 0$.)
13. Find the line that passes through the intersection of the lines $223x - 159y - 301 = 0$ and $99x + 183y - 211 = 0$ and is parallel to the y-axis. Solve for the intersection if you wish!
14. Find the line that passes through the intersection of $x + y - 1 = 0$ and $3x = 14y - 2$ and is: (a) Parallel to the x-axis. (b) Parallel to the y-axis. (c) Passes through the origin. (d) Has slope of 8. (e) Passes through $(10, 15)$.
15. The word family is used to describe any set of curves. A family may be determined by an equation involving one or more parameters. Graph several members of each of the following families by assigning values to c:

(a) $y = cx^2$	(f) $y = c \sin x$
(b) $y = (x - c)^2$	(g) $x^2 + (y - c)^2 = c^2$
(c) $y - c = (x - c)^2$	(h) $xy = c$
(d) $(x - c)^2 + (y - c)^2 = 1$	(i) $y = x^c$
(e) $y - c^2 = (x - c)^2$	

9.5 Equation of a locus

When we study equations by finding their loci or graphs, we are using geometric ideas to help us study algebraic and functional relations. In such cases the problem is: Given an equation, find its locus. On the other hand, when we find the equation of a given locus (for example, of a straight line or of a circle), we are using algebra to help us study a geometric figure. The problem in such cases is: Given a locus, find its equation. It is this problem that we wish to illustrate further in this section.

A **locus** or graph is just a set of points that satisfy some condition. If the condition is expressed by an equation, this is the equation of the locus. (Compare 4.6.) We can easily find the equation if the locus is described by giving directly the condition that must be satisfied by a point on the locus. For example, to find the locus consisting of those points that lie at a distance 3 units to the right of the y-axis, we note that this condition means $x = 3$. Hence this is the desired equation of the locus. Moreover, we see that the locus is a straight line parallel to the y-axis. This is a very simple example of a case in which we can identify the nature of a locus by finding its equation.

The equation may be found also from a geometric description of the locus. Consider, for example, the problem of finding the equation of the perpendicular bisector of the segment joining the points $(3, 5)$ and $(2, -4)$. The line must pass through the midpoint of the segment, that is, through (x_0, y_0) given by

$$x_0 = \frac{3 + 2}{2} = \frac{5}{2}$$

(1)   (Why?)

$$y_0 = \frac{5 - 4}{2} = \frac{1}{2}$$

Also its slope must be the negative reciprocal of the slope of the segment. The latter is:

(2) $$m = \frac{5 - (-4)}{3 - 2} = 9$$

Hence the desired slope is $m' = -1/9$, and the equation of the locus is:

(3) $$y - 1/2 = (-1/9)(x - 5/2)$$

In order to illustrate the way in which we may use the equation of a locus to identify the locus, we find the equation of the locus of a point that moves so that its distance from the point $(0, 1)$

is equal to its distance above the line $y = -1$. Let (x, y) be the coordinates of a point on the locus. Then the distance from $(0, 1)$ is $\sqrt{x^2 + (y - 1)^2}$, and the distance above $y = -1$ is $y + 1$ (Fig. 9.5.1). Hence the equation of the locus is:

$$(4) \qquad \sqrt{x^2 + (y - 1)^2} = y + 1$$

This equation does not suggest any curve with which we are familiar, but, if we square both sides and simplify, we find

$$(5) \qquad y = (1/4)x^2$$

Often a geometric theorem whose proof by elementary geometry is not easy can be proved with dispatch by using the equations of

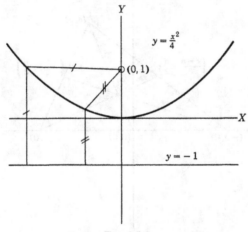

FIG. 9.5.1

the loci involved. Methods of this sort that use algebra and co-ordinate systems are called **analytic,** whereas the methods that deal only with the geometric figures themselves are called **synthetic.** To illustrate, we give an analytic proof of the familiar geometric theorem that the medians of a triangle meet in a point. (This point is called the **centroid** or center of gravity.) In order to simplify the proof, we place the axes so that the x-axis lies along one side of the triangle and the origin lies at the midpoint of this side. Then the vertices of the triangle and the midpoints are as

indicated in Fig. 9.5.2. (Why?) The equations of the medians
are:

$$y = \left(\frac{d}{c}\right) x$$

(6)
$$y = \left(\frac{d}{c - 3a}\right) (x - a)$$

$$y = \left(\frac{d}{c + 3a}\right) (x + a)$$

These lines are proved concurrent by solving any pair for their
intersection and substituting in the third, or by applying the result

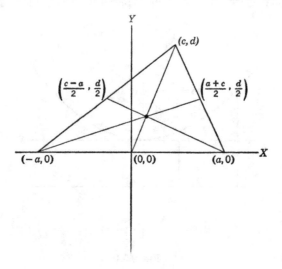

Fig. 9.5.2

stated in Ex. 4.11.12. Incidentally, we can see from Fig. 9.5.2 an
easy way to prove the well-known theorem that a line joining the
midpoints of two sides of a triangle is parallel to the third side.
The equation of the line joining the midpoints, $([c - a]/2, d/2)$
and $([a + c]/2, d/2)$, is $y = d/2$ (why?), and this is the equation
of a line parallel to the x-axis and hence to the third side.

In the simplest cases, the condition determining the locus is
immediately expressible as a single equation involving x and y.
It may happen, however, that the coordinates of a point on the

locus must satisfy two or more equations simultaneously. This would be the case, for example, if we want the locus of a point lying on *both* of two lines. Then the locus would consist of a single point determined by the two equations of the lines or by the two equations $x = a$ and $y = b$, if (a, b) is the intersection. It may happen also that there are no points that satisfy the given condition. This would be the case, for example, if we required that the locus consist of points common to two different but parallel lines. In this case we say that the locus is the null set. In still other cases the point lies on the locus if it satisfies *at least one* of several conditions. For example, the locus of points at a distance 4 from the y-axis is the locus of points for which $x = 4$ or $x = -4$. This means that the condition on the point is that $x - 4 = 0$ or $x + 4 = 0$. Hence the equation of the locus is $(x + 4)(x - 4) = 0$, since this is satisfied if and only if one of the factors is zero. This technique can always be used to write as a single equation a condition consisting in the satisfying of at least one of several equations. We write each equation with the right member zero and then set the product of the left members equal to zero.

The following check list of steps may be helpful to the student in solving locus problems:

1. List the given information about the locus.

2. Draw a sketch showing as much of the information as possible.

3. Choose a coordinate system that appears to simplify the problem as much as possible.

4. State the given information in terms of the coordinates of an arbitrary point on the locus.

5. Simplify the resulting equation(s).

6. Draw an accurate graph and check against the given information.

EXERCISE 9.5

(Where proofs are requested, they are to be analytic.)

1. Find the equation of the locus of points that satisfy each of the following conditions. Graph and describe.

   (a) At a distance 3 from the origin.
   (b) At a distance 2 to the left of the y-axis.
   (c) Equidistant from the points $(1, 0)$ and $(-1, 0)$.
   (d) Equidistant from the points $(-4, 7)$ and $(-1, 0)$.
   (e) At a distance 5 from the point $(-2, -3)$.

(*f*) At a distance 2 from the *y*-axis.
(*g*) Equidistant from the axes.
(*h*) On either the line $x = 3$ or the line $y = 2$.

Answers. (a) $x^2 + y^2 = 9$. (c) $x = 0$. (e) $x^2 + y^2 + 4x + 6y = 12$.

2. Same as prob. 1 for:

(*a*) On both the lines $2x - 5y - 10 = 0$ and $x + 4y - 5 = 0$.
(*b*) On both the lines $2y - 7x + 6 = 0$ and $7x - 2y - 8 = 0$.
(*c*) Equidistant from $(2, 0)$ and the line $y = -2$.
(*d*) On either of the lines in part *b*.

Answers. (c) $4y = x^2 - 4x$. (d) $(2y - 7x + 6)(7x - 2y - 8) = 0$.

3. Find the equation of the locus of a point:

(*a*) whose distance from the origin is twice its distance from $(2, 4)$.
(*b*) whose directed distance from $3x + 4y + 10 = 0$ is equal to its directed distance from $4x - 3y - 15 = 0$. (Draw a careful sketch.)
(*c*) whose directed distance from $3x + 4y + 10 = 0$ is the negative of its directed distance from $4x - 3y - 15 = 0$.
(*d*) the product of whose distances from $y = -2$ and $(0, 1)$ is 2.
(*e*) the sum of whose distances from $(1, 0)$ and $(-1, 0)$ is 4.

Answers. (a) $3x^2 + 3y^2 - 16x - 32y + 80 = 0$. (e) $3x^2 + 4y^2 = 12$.

4. Find the equations of the bisectors of the angles formed by $x - y + 2 = 0$ and $2x + y - 5 = 0$. Draw a careful figure and explain the correspondence between bisectors and equations.

5. Prove that, if two sides of a quadrilateral are equal and parallel, so are the other two sides. [*Suggestion.* Let the vertices be at $(0, 0)$, $(d, 0)$, (a, b), $(a + d, b)$, where *d* is the length of the equal sides.]

6. Prove that a segment joining the midpoints of two sides of a triangle equals one half the third side.

7. Prove that the median of a trapezoid is parallel to the bases and equal to one half their sum.

8. Prove that, if the chord of a circle is bisected by a radius, it is perpendicular to it. [*Suggestion.* Place the center of the circle at the origin and the radius along the *x*-axis. Then the coordinates of the ends of the chord and of the intersection of the chord and radius are (a, b), (c, d), and $(e, 0)$. It is sufficient to prove that $a = c$. (Why?)]

9. Prove that the locus of points equidistant from the end points of a segment is the perpendicular bisector.

10. Find (3) by using the distance formula and the locus property of prob. 9.

11. Prove that the altitudes of a triangle meet in a point. This point is called the **orthocenter**. [*Suggestion.* Place the axes so that the vertices are at $(a, 0)$, $(b, 0)$, and $(0, c)$.]

12. Show that the perpendicular bisectors of the sides of a triangle meet in a point. It is called the **circumcenter**.

13. Prove that the bisectors of the interior angles of a triangle meet in a point. (Use the same axes as in prob. 11.) This point is called the **incenter**.

14. Show that the centroid is two-thirds of the distance from any vertex to the midpoint of the opposite side.

15. Show that a line that divides two sides of a triangle proportionately is parallel to the third side. State and prove a converse.

16. Show that the centroid of any triangle is at $(\bar{x}, \bar{y})$, where $\bar{x}$ and $\bar{y}$ are the averages of the coordinates of the vertices.

17. Show that the diagonals of a parallelogram bisect each other at a point whose coordinates are the averages of the coordinates of the vertices.

18. Show that the perimeter of the quadrilateral formed by joining the midpoints of the sides of a given quadrilateral is equal to the sum of the diagonals of the original quadrilateral.

†**19.** Show that in any triangle the following nine points lie on the same circle (it is called the **nine-point circle**): the feet of the altitudes, the midpoints of the sides, and the points on the altitudes half-way between each vertex and the point of intersection of the altitudes. (*Suggestion.* Use the set-up of prob. 11. Where must the center be?)

†**20.** Show that the centroid, circumcenter, and orthocenter of a triangle are collinear. Show that the centroid trisects the segment joining the other two.

†**21.** Show that the center of the nine-point circle also is collinear with the points in prob. 20 and bisects the segment joining the circumcenter and orthocenter.

†**22.** Sometimes the condition that defines a locus includes inequalities as well as equations. Give examples from previous sections and write the relations that characterize the following loci:

(a) The points within a distance 3 of the origin.

(b) The points on the segment joining $(2, -1)$ and $(5, 6)$.

(c) The points in the third quadrant.

(d) The points within a distance 2 of the x-axis.

(e) The points within a distance 3 of the line $y = 4x - 2$.

(f) The points that are within a distance 1 of the x-axis *and* within a distance 2 of the y-axis.

(g) The points that are within a distance 1 of the x-axis *or* within a distance 2 of the y-axis.

†**23.** Find the condition that determines the set of points: (a) Within a circle of radius 2 and center $(-3, 6)$. (b) Within a square three of whose vertices are $(0, 0)$, $(0, 2)$, $(2, 0)$. (c) Within the square determined by $y = 2x + 3$, $y = 2x - 1$, $2y = -x + 3$, $x + 2y = 5$.

9.6 Parametric methods

Imagine an airplane headed due north with an airspeed of 200 mph. If the wind is blowing due east at 50 mph, the plane is flying north in a mass of air that is moving east. What is the path of the plane? Let us visualize the plane at the origin at a certain time, $t = 0$, and place the positive x-axis toward the east and the positive y-axis to the north. (Draw a figure.) As time passes, the x-coordinate of the plane increases by 50 per hour. The y-coordinate increases by 200 per hour. These statements are based on physical theories that are confirmed by experience. Hence after t hours, the coordinates are given by

(1)
$$x = 50t$$
$$y = 200t$$

These equations give the coordinates of a point on the path at a time t. We call t a parameter and the equations parametric equations. Note that t is a variable, but it is not one of the variables to be plotted; it is an *auxiliary variable*. By substituting different values of t we can plot as many points on the locus as we like, but what is the nature of the locus? In order to see this we manipulate the equations so as to eliminate t. This is legitimate since (1) gives the pair of values (x, y) corresponding to any t. Solving the first equation for t and substituting in the second, we find

(2) $$y = 4x$$

which we recognize as a straight line through the origin with slope of 4.

Two equations of the form

(3)
$$x = F(t)$$
$$y = G(t)$$

determine a locus upon which points may be located by substituting values of t. The equations (3) are called the **parametric equations** of the locus, and t is called the **parameter.** If t can be eliminated, the resulting equation in x and y is satisfied everywhere on the locus. However, it is not always the equation of the locus since it may be satisfied by points not given by the parametric equations. (See Ex. 9.6.12.)

In the example with which we opened this section the functions F and G were linear, and the locus turned out to be a straight line. This relation is not accidental:

*9.6.1 $$x = mt + c$$

$$y = nt + d$$

are the parametric equations of a straight line through the point (c, d) and having a slope n/m.

To prove this we solve the first equation for t and substitute in the second, finding:

(4) $$y - d = (n/m)(x - c)$$

It is often convenient to deal with a locus in terms of its parametric equations without eliminating the parameter. For example, we can see from the equations in *9.6.1, without finding (4), that when $t = 0$, $x = c$ and $y = d$. Also we can see that when x increases by m units, y increases by n units. Thus we could easily plot the locus directly from the parametric equations. As a second example, consider:

(5) $$\begin{aligned} x &= t \cos t \\ y &= t \sin t \end{aligned} \qquad \text{(With } t \geq 0\text{)}$$

We could simply start with $t = 0$ and calculate the x and y corresponding to various values. If the student does this for a considerable number of values of t and then links up the resulting points *in order according to increasing t,* he will get some idea of the locus. (See Fig. 9.6.1.) But it is easier to see the nature of the locus by the following considerations. If we square the equations and add, we get:

(6) $$x^2 + y^2 = t^2$$

This means that t is just the distance from the origin to the point on the locus. Also, if we divide the second equation by the first, we get:

(7) $$\frac{y}{x} = \tan t$$

In order to see the significance of these results, we show in the figure a point (x, y) on the locus and the vector from the origin. We see that t is both the length of this vector and also the angle that the vector makes with the x-axis. Thus we may think of the curve as swept out by the terminal point of a vector that is increasing in length so that its length is equal to the angle (in radians, of course) that it has swept out. Such a curve is one example of a large class of curves called spirals. In 9.8 we will find an easier way to deal with such curves by the use of polar coordinates. We could combine (6) and (7) to eliminate t, but the result

$$y = x \tan \sqrt{x^2 + y^2}$$

is not very helpful.

One of the most important uses of parametric methods is in finding a locus when it is difficult to get a relation between x and y directly, but easy (or at least not as difficult) to express x and y separately in terms of some auxiliary variable or parameter. We began this section with a very simple example of this kind. We now consider a slightly more difficult example: What is the locus

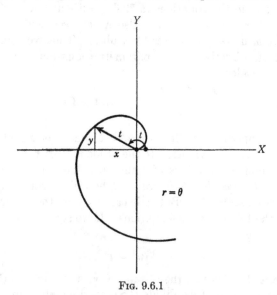

Fig. 9.6.1

of the vertices of all right triangles having a given segment as hypotenuse? (The student may recall the result from plane geometry, but we wish here to derive it analytically.) The situation is sketched in Fig. 9.6.2, where we let $2a$ be the length of the segment and place it on the x-axis with its center at the origin. We take as our auxillary variable θ, the angle between one side of the right triangle and the hypotenuse. Then the right angle is at P and $AP = 2a \cos \theta$, $PB = 2a \sin \theta$. It follows that

$$x = 2a \cos^2 \theta - a = a(2 \cos^2 \theta - 1) = a \cos 2\theta \qquad \text{(Why?)}$$
(8)
$$y = 2a \sin \theta \cos \theta = a \sin 2\theta \qquad \text{(Why?)}$$

The final form of the parametric equations is easily seen to give a circle with center at the origin and radius a, as the student may verify by squaring both sides of each equation and adding.

In problems where it is necessary to introduce a parameter, it is usually possible to proceed in different ways by using different parameters. No rule can be given to tell in advance what parameter should be used. The proper procedure is to draw a figure that includes as many of the conditions of the problem as possible and then consider whether there is a variable related to both x and y

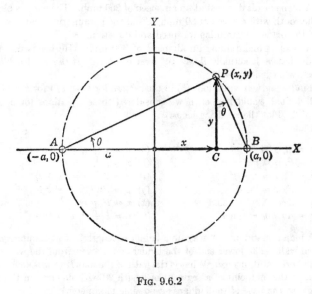

FIG. 9.6.2

in some way. This is a matter for heuristic thinking. It may help to try to construct some points on the locus. If one "hunch" doesn't work, try another.

EXERCISE 9.6

1. A boat is heading due west at a speed of 20 knots. The current is moving north at a speed of 3 knots. Find the parametric equations of the path, using time as a parameter.

2. Plot the following lines without eliminating the parameter. Make a table of values of t, x, and y.

(a) $x = t, y = 2t$ (c) $x = 3t - 4, y = t + 2$

(b) $x = -t - 2, y = 3t + 1$ (d) $x = 0.5t - 13, y = -31t + 1$

3. Eliminate t in each part of prob. 2 and check against the previous graph. *Answers.* (a) $y = 2x$. (c) $x - 3y + 10 = 0$.

4. Show that $x = (\cos\theta)t + x_0$, $y = (\sin\theta)t + y_0$, where t is the parameter, are the parametric equations of a straight line through (x_0, y_0). What is θ?

5. Show that $x = r\cos\theta$, $y = r\sin\theta$ are the equations of a circle with center $(0, 0)$ and radius r. What is θ? (*Hint.* Square and add.)

6. Show that $x = r\sin\theta$, $y = r\cos\theta$ also represent a circle. What is the interpretation of θ here?

7. A plane heads due east at an air speed of 260 mph. The wind is blowing from the south with a speed of 30 mph. Find the parametric equations of the path of the plane. Determine its speed and direction.

8. A pilot is maintaining an airspeed of 300 mph. He heads due north but finds that he is actually flying 10° west of north. If the wind is blowing due west, what is its speed?

9. Find x and y in equations (5) in the section for $t = k\pi/4$ for $k = 0, 1, 2, \cdots, 12$. What would happen if we considered these equations for negative values of t? Plot the resulting locus.

10. Plot:

(a) $x = 2t, y = t^2$

(b) $x = t^2, y = t^3$

(c) $x = t^2, y = 1 - t^3$

(d) $x = t^2, y = \log t$

(e) $x = 2\cos\theta, y = 4\sin\theta$

(f) $x = \sin\theta, y = \tan\theta$

(g) $x = 1 + t, y = t^2 + t + 1$

(h) $x = t^{-1}, y = t^{-2}$

(i) $x = t^{-1}, y = (1 + t)^{-1}$

(j) $x = \sin t, y = \sin^2 t$

(k) $x = t^2, y = t^{-2}$

(l) $x = 1 + t^2, y = 1 - t^2$

11. A man is standing half way up a ladder of length 20 feet leaning against a vertical wall. The lower end of the ladder slides away from the wall along a smooth horizontal surface. What is the path of the man? (*Suggestion.* Find the locus of the midpoint of a segment of length 20 with its ends on the axes. The angle at the base of the ladder is a possible parameter.)

12. Plot each of the following and compare the result with the graph of the equation obtained by eliminating the parameter:

(a) $x = t^2, y = t^4$

(b) $x = e^t, y = 2e^t - 1$

(c) $x = \sin t, y = \csc t$

(d) $x = \sin^2 t, y = \sin^3 t$

13. Graph the locus determined by the following:

$$x = t, y = 0 \text{ when } 0 \le t < 1$$
$$x = 1, y = t - 1 \text{ when } 1 \le t < 2$$
$$x = 3 - t, y = 1 \text{ when } 2 \le t < 3$$
$$x = 0, y = 4 - t \text{ when } 3 \le t \le 4$$
No points when $t < 0$ or $t > 4$.

14. Write parametric equations of the segment joining $(1, 3)$ and $(-2, 12)$. (*Suggestion.* Write parametric equations of the line and then limit the parameter.)

15. Plot $x = t, y = 4t$ for $0 \le t \le 3$.

16. Write the parametric equations of the boundary of the triangle with vertices $(3, 1)$, $(0, 0)$, $(0, 6)$.

17. If a projectile is fired with a muzzle velocity v_0 from a weapon whose barrel makes an angle α with the horizontal, its horizontal and vertical distances from the muzzle after t units of time are given by

$$x = (v_0 \cos \alpha)t$$

$$y = (v_0 \sin \alpha)t - \tfrac{1}{2}gt^2$$

Find the equation of the path. Find the highest point and the time to reach it. Find the time of flight and the range. [*Answer.* $y = (\tan \alpha)x - \left(\dfrac{g \sec^2 \alpha}{2v_0^2}\right) x^2.$]

9.7 Vectors

We have made frequent use of the one-to-one correspondence between real numbers on an axis and vectors. A vector has a length equal to the absolute value of the corresponding number and points toward the right or left on the axis according to whether the number is positive or negative. Moreover, we consider vectors as being movable along the axis and interpret addition, subtraction, multiplication, and division in terms of them. In particular, if a vector has its initial point at x_1 and its terminal point at x_2, it has a length of $| x_2 - x_1 |$, and the number $x_2 - x_1$ indicates both its length and direction. We have used vectors in this fashion to represent directed distances measured parallel to the x-axis, y-axis, or some other line upon which a unit of distance and a positive direction had been established. We have also considered a vector rotating about the origin in order to sweep out an angle. In this section we shall discuss vectors more generally in the plane.

Consider the following very simple problem. A speedboat is going through the water at 20 mph directly against a current of 6 mph. What is the resulting speed in relation to the land? Obviously the answer is $20 - 6$ in the direction the boat is heading. The vector picture is given in Fig. 9.7.1. Note that we lay down

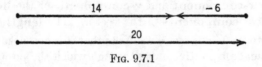

Fig. 9.7.1

a vector 20, then add a vector -6 by laying the -6 vector with its initial point on the terminal point of the 20 vector. Then the vector required to join the initial point of the 20 vector to the terminal point of the −6 vector gives the answer both in magnitude and direction.

Suppose now that the current is the same but that the boat heads at an angle of 45° to the current. Let us apply the same technique and see what happens. In Fig. 9.7.2 we lay out a vector of length 20 pointing at an angle of 45° with the current. Then we lay the −6 vector with its initial point on the terminal point of the 20 vector. The vector joining the initial point of 20 to the final

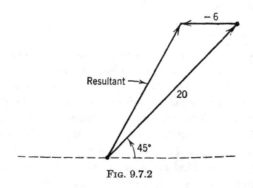

Fig. 9.7.2

point of −6 is a vector that might give the direction and magnitude of the resulting motion. It is a fact (established by experience and consistent with basic physical principles) that this is the case. Velocities, forces, and many other quantities representable by vectors can be handled in this way.

In order to develop the fundamental ideas on vectors, we first consider those with initial points at the origin. Every such vector determines a unique point (its terminal point), and, conversely, every point determines a unique vector (the vector from the origin to this point). Thus we may represent such a vector uniquely by the coordinates (x, y) of its terminal point P. These numbers are called the **x-component** and **y-component**, or the **horizontal** and **vertical components** of the vector. The **length** or **norm** of the vector is $\sqrt{x^2 + y^2} = r$. Note that it is non-negative. Let θ be the smallest positive angle through which the positive x-axis would have to rotate in order to coincide with the vector. We call θ the **direction angle** of the vector. (Compare 9.3 and Fig. 9.3.1.) Evidently the vector is determined as well by giving (r, θ) as by giving (x, y).

We designate the vector itself by (x, y), $\overrightarrow{OP}$, or $\mathbf{V}$. *Note that* $\mathbf{V}$ *stands for an ordered pair of numbers, not a single number.* The components are numbers, but they may also be thought of as vectors on the axes. In fact, x may be thought of as a vector with components x and 0, and y as one with components 0 and y. When we wish to talk of these component vectors, we will write $(x, 0)$ and $(0, y)$ or $\mathbf{V}_x$ and $\mathbf{V}_y$. We note from the definitions of the circular functions that

*9.7.1
$$x = r \cos \theta$$
$$y = r \sin \theta$$

Just as we found it convenient to allow vectors to move about on a line without changing their length and direction, so we now

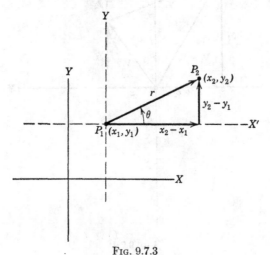

Fig. 9.7.3

find it convenient to let vectors move about in the plane without changing length or direction. Suppose a vector has its initial point P_1 at (x_1, y_1) and its terminal point P_2 at (x_2, y_2). We set axes at (x_1, y_1) parallel to the old axes and define the components, the length, and the direction angle as before. Evidently the components are $x_2 - x_1$ and $y_2 - y_1$. Hence $\overrightarrow{P_1P_2} = \mathbf{V} = (x_2 - x_1, y_2 - y_1)$. (See Fig. 9.7.3.) We call two vectors **equal** if their components are equal, regardless of their location.

We now wish to define what is meant by the **sum,** or **resultant,** of two vectors:

*9.7.2 $(x_1, y_1) + (x_2, y_2) = (x_1 + x_2, y_1 + y_2)$ (Definition)

This means that we define the sum of two vectors to be a vector with components equal to the sums of the corresponding components. It is easy to see that this definition corresponds exactly with the process of adding vectors geometrically that we illustrate in Fig. 9.7.2. In Fig. 9.7.4, we have drawn $V_2 = (x_2, y_2)$

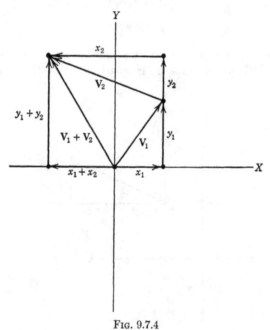

FIG. 9.7.4

with its initial point at the terminal point of $V_1 = (x_1, y_1)$. Then the vector joining the origin to the terminal point of V_2 has as its components just the sums of the components of V_1 and V_2 as indicated by the definition.

The correspondence between the algebraic definition of vector addition and the geometrical and physical picture enables us to solve vector problems by algebra alone. For example, in the problem of Fig. 9.7.2 one vector has $r = 20$ and $\theta = 45°$ and the other $r = 6$ and $\theta = 180°$. Hence these vectors are given by V_1

$$= (20 \cos 45°, \ 20 \sin 45°) = (10\sqrt{2}, \ 10\sqrt{2}) \text{ and } \mathbf{V}_2 = (-6, 0).$$
Hence:

(1) $$\mathbf{V}_1 + \mathbf{V}_2 = (10\sqrt{2}, 10\sqrt{2}) + (-6, 0)$$

$$= (10\sqrt{2} - 6, 10\sqrt{2}) = \mathbf{V}$$

The vector $\mathbf{V}$ has length

$$r = \sqrt{436 - 120\sqrt{2}}$$

and its direction angle may be found from

(2) $$\tan \theta = \frac{10\sqrt{2}}{10\sqrt{2} - 6}$$

It is sometimes more convenient to find the resultant by making use of the geometrical idea that the resultant is the third side of a triangle formed by the given vectors. Then its length and direction can be found by trigonometry. For example, in Fig. 9.7.2 we know two sides and the included angle in the triangle and so can find the other side and angles.

Instead of wanting the resultant of two or more vectors we may be interested in the inverse problem, that is, to find two vectors that have a given resultant. The process is called **resolving the resultant into components.** Every pair of vectors that adds to the given vector is called a pair of components. Evidently the horizontal and vertical components are one such pair, and there are an infinite number of other possibilities. In practical applications it is usually required to find the pair of components that have given directions, and the problem is easily solved by trigonometry, since in all such problems we are given a side and the angles of a triangle.

For example, a pilot is headed due west. The wind is blowing north, and he finds that he is actually traveling at an angle of 21° north of west and at a ground speed of 250 mph. What are the air speed and the velocity of the wind? It is a question of resolving his resultant velocity (a vector of length 250 and angle 159°) into horizontal and vertical components. (See Fig. 9.7.5.) From *9.7.1 we have immediately for these components: $x = 250 \cos 159°$ and $y = 250 \sin 159°$, which is the desired solution. Suppose now that the pilot heads 10° south of west and the wind is blowing 20°

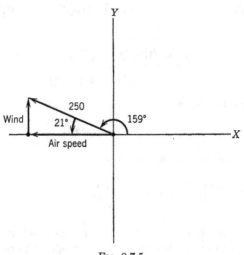

FIG. 9.7.5

west of north, but that the resultant is the same as before (Fig. 9.7.6). What are the wind and air speeds? Now the desired components are not parallel to the axes, but we know a side and two angles of the triangle and can therefore solve it. (How?)

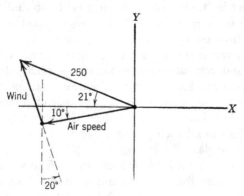

FIG. 9.7.6

EXERCISE 9.7

1. Sketch the following vectors with initial points at the origin:

(a) (0, 3)   (c) (2, 7)   (e) (2, −7)   (g) (0, −5)
(b) (−4, 0)   (d) (−2, 7)   (f) (−2, −7)   (h) (−1, −8)

2. Find the length and direction angle of each of the vectors in prob. 1.
[*Answers.* (a) $r = 3$, $\theta = 90°$. (c) $r = \sqrt{53}$, $\theta = 74°\,3'$.]

3. Find the components, length, and direction angle of each of the following vectors determined by their initial and terminal points, the initial point being given first. Sketch each vector.

(a) (0, 0), (2, 1)   (c) (−1, 3), (−4, 8)   (e) (−4, 4), (0, −4)
(b) (3, 4), (1, 7)   (d) (1, 3), (−4, −2)   (f) (3, 0), (0, 4)

Answers. (c) (−3, 5), $r = \sqrt{34}$, $\theta = 120°\,58'$.

4. Perform the following vector additions, drawing a diagram in each case:

(a) (3, 6) + (8, 1)   (d) (3, 0) + (0, 4)   (g) (3, 5) + (−1, −7)
(b) (0, 1) + (2, −4)   (e) (−1, −4) + (−3, 2)   (h) (−1, 4) + (5, −2)
(c) (0, 0) + (2, 1)   (f) (2, −5) + (2, −8)   (i) (2, −3) + (5, −1)

Answers. (a) (11, 7). (c) (2, 1). (e) (−4, −2).

5. Find to three digits and minutes the length and direction of the resultant vector in (1) and check against a large diagram to scale.

6. Find the components of each vector determined by the length and direction angle:

(a) $r = 1$, $\theta = 30°$ (d) $r = 1.874$, $\theta = 343°$
(b) $r = 3$, $\theta = 225°$ (e) $r = 1000$, $\theta = 27°\,43'$
(c) $r = 0.35$, $\theta = 95°$ (f) $r = 0.0031$, $\theta = 145°\,16'$

7. Show that any vector may be considered as the sum of its vertical and horizontal component vectors, that is, $\mathbf{V} = \mathbf{V}_x + \mathbf{V}_y$.

8. A vector of length 200 has direction angle of 200°. Resolve it into components: (a) Parallel to the axes. (b) Lying along lines with direction angles 150° and 240°. (Note that these components are perpendicular and hence are "projections" of the original vector.) (c) Lying along lines with direction angles of 160° and 240°.

9. Show that the sum (resultant) of two vectors is the diagonal of the parallelogram whose sides are the vectors. This means that finding the resultant amounts to finding the diagonal of a given parallelogram, and resolving a vector into components means finding a parallelogram with a given diagonal. (*Suggestion.* Draw the two sums $\mathbf{V}_1 + \mathbf{V}_2$ and $\mathbf{V}_2 + \mathbf{V}_1$.)

10. The straight line distance from A to B is 1000 feet. A man wishes to walk from A to B by going first in a direction at an angle of 25° with AB and then at an angle of 45° with AB. Sketch and find the distances he must walk.

11. A man is rowing directly toward the bank of a river that is flowing at 2 miles an hour. He is rowing 3 miles an hour in the water. What are his speed and direction with respect to the land? (*Answer.* $\sqrt{13}$ mph, 33°\,41' downstream from line across current.)

12. Directions on the earth are often given in terms of the **azimuth,** the angle measured clockwise from north. Thus a vector with azimuth of 60° would have a direction angle of 30° and one having an azimuth of 180°

would have a direction angle of 270°. Suppose a man in the jungle walks 500 yards on an azimuth of 47° and then 600 yards on an azimuth of 125°. Where will he be, that is, how far from his starting point and in what direction?

13. A boat heads on an azimuth of 100°. The current is drifting on an azimuth of 300°. After one hour the boat is 10 miles east and 1 mile north of its original position. What are the resultant speed and direction? What is the speed of the boat in the water, and what is the velocity of the current?

14. More than two vectors may be added by successive application of *9.7.2. Suppose that a ship navigating by "dead reckoning" travels 100 miles on an azimuth of 25°, then 200 miles on an azimuth of 80°, then 100 miles on an azimuth of 100°. What is its final position? (*Hint.* Begin by finding the horizontal and vertical components.)

15. Forces may be treated like vectors. Suppose that the following forces act on a particle at the origin: 20 pounds at an angle of 40°, 100 pounds at an angle of 200°, 50 pounds at an angle of 95°. Find the single force (a vector) that is equivalent to these three. Find the force that will balance the three so that the resultant is zero.

16. The projection of a point on a line is the foot of the perpendicular from the point to the line. The projection of a vector on a line is the vector formed by the projections of its points. (*a*) Show that the components of a vector are its projections on the axes. (*b*) Show that the length of the projection of a vector $\mathbf{V}$ on a line is $|\mathbf{V}| \cos \theta$, where θ is the angle between the vector and the line and $|\mathbf{V}|$ is its length.

†**17.** Show that if P, Q, R are three points, the projections on any line of $PQ + QR$ equals the sum of the projections of PQ and QR.

18. Show that $(-x, -y)$ is a vector with the same length as (x, y) and the opposite direction. We call $(-x, -y)$ the **negative** of (x, y) and designate it by $-(x, y)$, i.e.,

$$*9.7.3 \qquad -(x, y) \equiv (-x, -y) \qquad \text{(Definition of Negative)}$$

19. The vector $(0, 0)$ is called the **null vector.** It has zero length, and its direction is not defined. Show that the sum of any vector and its negative equals the null vector. Show that the addition of the null vector has no effect on any vector. (What identities in Chapter 3 correspond to these results?)

20. Show that (kx, ky) is a vector of length k times the length of (x, y) and having the same or opposite direction according as $k > 0$ or $k < 0$. This vector is called the **scalar product** of (x, y) by k, and

$$*9.7.4 \qquad k(x, y) \equiv (kx, ky) \qquad \text{(Definition)}$$

21. Show that $(-1)(x, y) \equiv -(x, y)$.

†**22.** Show that for any vectors $\mathbf{V}$, $\mathbf{V}_1$, $\mathbf{V}_2$, and $\mathbf{V}_3$:

$$*9.7.5 \qquad\qquad \mathbf{V}_1 + \mathbf{V}_2 \equiv \mathbf{V}_2 + \mathbf{V}_1$$

$$*9.7.6 \qquad\qquad \mathbf{V}_1 + (\mathbf{V}_2 + \mathbf{V}_3) \equiv (\mathbf{V}_1 + \mathbf{V}_2) + \mathbf{V}_3$$

*9.7.7 $$k(\mathbf{V}_1 + \mathbf{V}_2) \equiv k\mathbf{V}_1 + k\mathbf{V}_2$$

*9.7.8 $$(k + k')\mathbf{V} \equiv k\mathbf{V} + k'\mathbf{V}$$

*9.7.9 $$k(k'\mathbf{V}) \equiv (kk')\mathbf{V}$$

23. Illustrate *9.7.3 to *9.7.9 with vector diagrams. Suggest appropriate names for the first three.

24. Sometimes the length or norm of a vector V is represented by $|\,\mathbf{V}\,|$. The symbol is the same as the absolute value sign, but no confusion can arise, because the absolute value of a number is just the length of the corresponding vector. Show that:

*9.7.10 $$|\,\mathbf{V}\,| \geq 0$$

*9.7.11 $$|\,\mathbf{V}\,| = 0 \longleftrightarrow \mathbf{V} = (0, 0)$$

*9.7.12 $$|\,k\mathbf{V}\,| = |\,k\,| \cdot |\,\mathbf{V}\,|$$

25. If we define the **difference** of two vectors by

*9.7.13 $\quad \mathbf{V} = \mathbf{V}_1 - \mathbf{V}_2 \longleftrightarrow \mathbf{V}_2 + \mathbf{V} = \mathbf{V}_1 \qquad$ (Definition)

show that

*9.7.14 $$(x_1, y_1) - (x_2, y_2) \equiv (x_1 - x_2, y_1 - y_2)$$

*9.7.15 $$\mathbf{V}_1 - \mathbf{V}_2 \equiv \mathbf{V}_1 + (-\mathbf{V}_2)$$

26. Show that any vector may be regarded as the difference between the vectors from the origin to its initial and terminal points.

27. Show that the sum of three vectors is a vector that closes the polygon formed by laying the vectors end to end.

†28. Show by induction that the sum of n vectors is a vector that closes the polygon formed by laying the vectors end to end.

9.8 Polar coordinates

The correspondence between points and pairs of rectangular coordinates is not the only one-to-one correspondence that can be set up between points and ordered number pairs. A most useful alternative is the polar coordinate system. Consider a point (x, y), and let (r, θ) be the length and direction angle of the vector from the origin to the point. (See Fig. 8.4.1 and compare 9.7 and 9.3.) Then we call r and θ the **polar coordinates** of the point. We call r the **radius vector** or polar distance and θ the **direction angle** or vectorial angle. We call the origin the **pole**. It is the

point from which r is measured. We call the positive x-axis the
polar axis. It is from this directed line that the angle θ is measured. With this scheme there is a one-to-one correspondence
between points and pairs (r, θ), where r is positive and θ lies between 0 and 2π.

However, we wish to be able to interpret (r, θ) when r is negative
and when θ takes any value. This is easily done. Given any r
and θ, we place θ in standard position and then go a distance $| r |$
from the origin in the direction of the terminal side if $r > 0$ and in
the opposite direction if $r < 0$. This process yields a unique
point. Hence for every ordered pair (r, θ) there is just one corresponding point. On the other hand, many different pairs yield
the same point, and hence the correspondence is no longer one-to-
one. In fact, there are an infinite number of pairs of coordinates
that yield the same point, since a change in θ of any integral
multiple of 2π has no effect on the point, that is,

*9.8.1 $(r, \theta) \cong (r, \theta + 2k\pi)$

where the congruence sign, $\cong$, means that the points are the same.
Also

*9.8.2 $(r, \theta) \cong (-r, \theta + \pi)$

This means that rotation through 180° coupled with changing the
direction of r yields the same point.

The relations between the rectangular and polar coordinates of
a point follow immediately from the definitions of the circular
functions and the distance formula. They are:

*9.8.3
$$x = r \cos \theta$$
$$y = r \sin \theta$$
(Rectangular in Terms of Polar)

$$r = \pm\sqrt{x^2 + y^2}$$

*9.8.4
$$\theta = \arctan (y/x)$$
$$= \arcsin (y/r)$$
$$= \arccos (x/r)$$
(Polar in Terms of Rectangular)

Either sign may be taken for r, but then θ must be chosen so that
$\sin \theta = y/r$ and $\cos \theta = x/r$.

We begin by considering some simple equations in polar coordinates. Obviously $r = 0$ and $\theta = 0$ are the equations of the pole and polar axis, respectively. (Why?) The equation $r = a$ is the equation of a circle with center at the origin and radius a. The equation $\theta = b$ is the equation of a straight line through the origin making an angle of b with the positive x-axis. Thus we may think of the plane as being covered by lines through the origin and circles with centers at the origin, and these coordinate curves are printed on "polar coordinate paper."

Just as for rectangular coordinates, we may graph an equation in polar coordinates by substituting values for θ or r and calculating the corresponding values of r or θ; but it pays to find certain key points and to observe special properties of the locus such as symmetry, excluded values of the variables, infinite values of the variables, and periodicity. It may help also to transform the equation into rectangular coordinates, where it may be recognized as the equation of a familiar curve. The intercepts of a curve in polar coordinates may be found by letting θ equal $k\pi$ (for the x-intercepts) and $(2k + 1)\pi/2$ (for the y-intercepts). Often these values for $k = 0$ and 1 plus a few simple angles such as $\pi/4$, $3\pi/4$, $5\pi/4$, $7\pi/4$, etc., are sufficient to plot the curve. If the equation remains unchanged when r is replaced by $-r$, the locus is symmetric with respect to the origin. (Why?) The same is true if it remains unchanged when θ is replaced by $\theta + \pi$ or $\theta - \pi$. If the equation remains unchanged when θ is replaced by $\pi - \theta$ (by $-\theta$) the locus is symmetric with respect to the y-axis (x-axis). (Draw a sketch and explain why this is so.) Values of θ for which r is imaginary or infinite should be noted.

We illustrate these techniques with several examples.

$$(1) \qquad\qquad r = \frac{3}{\cos \theta}$$

Multiplying both sides by $\cos \theta$ we find $r \cos \theta = 3$. If we transform to rectangular coordinates this becomes $x = 3$, which is the equation of a familiar straight line. Of course we could plot the locus directly from (1). Since r is a periodic function with period 2π, we need consider only $0 \leq \theta < 2\pi$. Since the equation remains unchanged when θ is replaced by $-\theta$, the locus is symmetric with respect to the x-axis. When $\theta = 0$, $r = 3$. Also when

$\theta = \pi/2$, r is infinite. This means that the locus does not meet the y-axis. Plotting a few more points would be sufficient to suggest that the locus is a straight line, but the easiest way to prove that this appearance is correct is by transforming to rectangular coordinates.

(2) $$r = \cos 2\theta$$

Here r is a periodic function but the period is π, since $\cos 2(\theta + \pi)$ $\equiv \cos (2\theta + 2\pi) \equiv \cos 2\theta$. Also we note that the curve is symmetric with respect to both axes and with respect to the origin. Hence it would be sufficient to plot the curve for $0 \le \theta \le \pi/2$. We find the following values:

(3)

θ	0	$\pi/8$	$\pi/4$	$3\pi/8$	$\pi/2$
2θ	0	$\pi/4$	$\pi/2$	$3\pi/4$	π
r	1	$(1/2)\sqrt{2}$	0	$(-1/2)\sqrt{2}$	-1

If we plot these and a few intermediate points and connect them *in order according to* θ by a smooth curve, we find the heavy line in Fig. 9.8.1. The rest of the curve is found by symmetry. Because

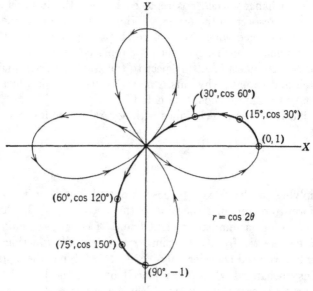

FIG. 9.8.1

of the periodicity of cos 2θ we know that we have found all the locus. The student should note how convenient polar coordinates are for plotting periodic functions. The function $y = \cos 2x$ yields an infinite wave, whereas the same function in polar coordinates yields a finite and equally decorative curve.

$$(4) \qquad\qquad r = \sqrt{\theta}$$

First we note that $\theta \geq 0$, for otherwise r is imaginary. We see that as θ increases, r increases. We find the following values:

(5)

θ	0	1	4	16	$\pi/2$	$3\pi/2$	2π
r	0	1	2	4	$(1/2)\sqrt{2\pi}$	$(1/2)\sqrt{6\pi}$	$\sqrt{2\pi}$

Plotting these and a few other points and connecting them in order, we get what is evidently a spiral. (See Fig. 9.8.2.) We

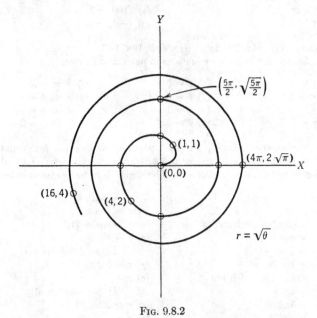

Fig. 9.8.2

could think of this locus as generated by a vector rotating about the origin and having a length equal to the square root of the angle through which it has moved.

EXERCISE 9.8

1. Plot the points whose polar coordinates (r, θ) are given. Show both rectangular axes, the pole, and the polar axis.

(a) $(5, 0)$, $(-5, 0)$, $(10, \pi/2)$, $(5, 3\pi/2)$, $(5, \pi)$
(b) $(1, 45°)$, $(1, 60°)$, $(1, 270°)$, $(1, 360°)$, $(1, 1000°)$
(c) $(1, 30°)$, $(3, 30°)$, $(7, 30°)$, $(-2, 30°)$, $(-2, 210°)$, $(5, 210°)$
(d) $(3, 146°)$, $(-4, 178°)$, $(7, 215°)$, $(10, 313°)$

2. Make up several illustrations of *9.8.1 and *9.8.2.
3. Transform to rectangular coordinates each of the points in prob. 1.
[*Answer to part a.* $(5, 0)$, $(-5, 0)$, $(0, 10)$, $(0, -5)$, $(-5, 0)$.]
4. Find the polar coordinates of the following points whose rectangular coordinates are given. Choose θ between 0 and 2π and r positive. Plot, showing both coordinate pairs:

(a) $(2, 1)$ (f) $(-5, 5)$
(b) $(0, -7)$ (g) $(4, -6)$
(c) $(-3, -1)$ (h) $(6, \sqrt{7})$
(d) $(5, 0)$ (i) $(100, -2)$
(e) $(-5, 0)$ (j) $(-5, -5)$

Answers. (a) $(\sqrt{5}, 26° 34')$. (c) $(\sqrt{10}, 198° 26')$.
5. Sketch the following equations in polar coordinates:

(a) $r = 1$ (e) $\theta = 0$ (i) $\theta = -45°$
(b) $r = 5$ (f) $\theta = 3\pi/2$ (j) $r \cos \theta = 4$
(c) $r = -9$ (g) $\theta = 1$ (k) $r \cos \theta = -5$
(d) $r = 9$ (h) $\theta = 45°$ (l) $r \sin \theta = 2$

6. Sketch the following curves. Begin by finding the intercepts and other key points. Indicate special properties such as symmetry and find each equation in rectangular coordinates. The angle θ is to be measured in radians.

(a) $\theta = r$ (l) $r^2 = 3 \sin \theta$
(b) $r = 2\theta$ (m) $r = \tan \theta$
    (Spiral of Archimedes)   (n) $r = \cos (\theta/2)$
(c) $r = \csc \theta$ (o) $r = e^{\theta}$
(d) $r = 3 \sec \theta$ (Logarithmic Spiral)
(e) $r = \cos 2\theta$ (p) $r = \log \theta$
    (Four-Leaved Rose)   (q) $r = \theta^{-1}$
(f) $r = \sin \theta$ (Hyperbolic Spiral)
    (Circle)   (r) $r = 2 \cos 4\theta$
(g) $r = \sin 2\theta$ (s) $r = 2 \cos 3\theta$
(h) $r = \cos \theta$ (t) $r^2 = \cos 2\theta$
(i) $r = \sin 3\theta$ (u) $r = \sin^2 \theta$
(j) $r^2 = \cos \theta$ (v) $r = (1 - \theta)^{-1}$
(k) $r^2 = \sin \theta$

7. Show that if the polar axis is rotated through an angle θ_0, the new coordinates will be related to the old by

(6)
$$r' = r$$
$$\theta' = \theta - \theta_0$$

8. Sketch the following loci:

(a) $\theta < 45°$ (c) $-\pi/2 \le \theta \le \pi/2$ (e) $r > \sin\theta$
(b) $r \ge 2$ (d) $1 \le r \le 2\pi$ (f) $r \le \sin 3\theta$

9. The locus of $r = a\cos\theta + b$ is called a **limaçon**. Graph it for:
(a) $a = 2, b = 1$. (b) $a = 1, b = 2$. (c) $a = -1, b = 1$. (d) $a = 2, b = -3$.

10. When $a = b$, the limaçon is called a **cardioid**. Graph several examples.

11. The graph of $r^2 = a^2 \cos 2\theta$ is called a **lemniscate**. Graph several examples.

12. The class of curves given by $r = a\cos n\theta$ are called **roses**. Graph examples for various values of n.

13. Graph several examples of $r^n\theta = a$.

14. Finding analytically the points of intersection of two curves whose equations are given in polar coordinates is complicated by the fact that they may meet in points for which no single set of coordinates satisfies *both* equations. This arises from the fact that a point has infinitely many different pairs of polar coordinates, of which one may satisfy one equation and a different one the other. The best procedure is to graph both equations and investigate the possibilities. Evidently all intersections may be obtained by replacing r and θ in *one* of the equations by $(-1)^n r$ and $\theta + n\pi$, where n is any integer, and then solving simultaneously. (Why?) Solve the following in order to find the points of intersection of their loci:

(a) $r = 2, \theta = \pi/4$ (d) $r = 2 + 4\cos\theta, r = 1$
(b) $r = 2\cos\theta, r = \sin\theta$ (e) $r = \cos\theta, r = \sin\theta$
(c) $r = \sin 2\theta, r = \cos\theta$ (f) $r\theta = 1, r = \tan\theta$

15. Find the equation in polar coordinates of the following: (a) The line whose equation in normal form is $x\cos\alpha + y\sin\alpha = p$. (b) The circle whose equation in rectangular coordinates is $(x - h)^2 + y^2 = a^2$.

†16. Find in polar coordinates the equation of the circle with center at $\theta = \theta_0$, $r = r_0$ and radius a. (*Suggestion.* Draw a figure and consider the triangle determined by the origin, the center, and a point on the circle.)

10

The Complex
Number System

In this chapter we make the extension of the number system that is suggested by the fact that $x^2 = -2$ has no real solution. As for the previous extensions to negatives (3.9), rationals (3.11 and 3.12), and reals (3.13), the new numbers and operations on them are defined in such a way that the laws of algebra still hold and the enlarged number system includes the numbers previously defined.

It might begin to look as if there would be no end to these extensions of the number system, but such is not the case. With the inclusion of the imaginary numbers, no further extensions are necessary! Each of the extensions is dictated by the fact that when we operate on numbers by addition, subtraction, multiplication, division, taking powers, and root extraction we find in some cases that the result is not among the numbers previously defined. But it turns out that when we perform these operations on real and imaginary numbers we always get another real or imaginary number! We say that the real and imaginary numbers form a number system that is **closed** under these operations.

In the first section of this chapter we proceed heuristically and experimentally to manipulate complex numbers without defining them or their properties. The second section gives a definition of complex numbers and the operations upon them and justifies the manipulations of the first section. The third section is devoted to geometric interpretation. The remainder of the chapter introduces the student to the most important elementary theory and applications of complex numbers.

10.1 Manipulation of complex numbers

We know that there is no real solution of $x^2 = -2$, that is, no real number equal to $\sqrt{-2}$. However, if $\sqrt{-2}$ is a number of some sort, by definition we have:

(1) $$(\sqrt{-2}\,)^2 = -2$$

Now we manipulate the left member of (1) according to the rules worked out in 5.3, without taking into account that these were proved only for real radicals, that is, square roots whose radicands are positive. We write accordingly:

(2) $\quad (\sqrt{-2}\,)^2 = (\sqrt{-2}\,)(\sqrt{-2}\,) \stackrel{?}{=} \sqrt{(-2)(-2)}$ (*5.3.4)

(3) $\qquad\qquad\qquad\qquad\quad \stackrel{?}{=} \sqrt{4} = 2$ (!!?)

The result (3) is in flat contradiction with (1) and would enable us to prove that $-2 = 2$!! What is wrong? The difficulty is that the theorems of 5.3 were proved for real numbers, but they apparently do not hold when the radicands are negative.

Suppose now that we use *5.3.4 to rewrite $\sqrt{-2}$ as $\sqrt{(-1)2}$ $= (\sqrt{-1}\,)\sqrt{2}$; then we find:

(4) $\quad (\sqrt{-2}\,)^2 = (\sqrt{-1}\,)(\sqrt{2}\,)(\sqrt{-1}\,)(\sqrt{2}\,)$

(5) $\qquad\qquad = (\sqrt{-1}\,)(\sqrt{-1}\,)(\sqrt{2}\,)(\sqrt{2}\,)$

(6) $\qquad\qquad = (-1)2 = -2$ (Why?)

The final result is consistent with (1). We are still using some results of 5.3 without knowing that they apply to imaginaries, but it now looks as though we have found a way to use them that does not immediately lead to a contradiction. The device we used was to rewrite the imaginary square root as $\sqrt{-1}$ times a real square root. This is in fact the device that enables us to handle complex numbers by the usual rules of algebra.

Since we are going to write $\sqrt{-1}\sqrt{2}$ in place of $\sqrt{-2}$, and in general $\sqrt{-1}\sqrt{a}$ in place of $\sqrt{-a}$ when $a > 0$, it is convenient to adopt a special symbol for $\sqrt{-1}$. We use the letter i, to which we assign the property

(7) $$i^2 = -1$$

Then we always express the square root of a negative real number by writing i times the square root of its absolute value. Thus:

$$(8) \qquad \sqrt{-3} = \sqrt{-1}\sqrt{3} = i\sqrt{3}$$

$$(9) \qquad \sqrt{-4} = i\sqrt{4} = 2i$$

$$(10) \qquad -\sqrt{-5} = -i\sqrt{5}$$

When the square root of a negative number has been rewritten in this way, the coefficient of i is a real number. We then manipulate the result according to the usual rules that apply to real numbers, treating i as we would any variable and using (7). Thus $\sqrt{-2}\sqrt{-3} = (i\sqrt{2})(i\sqrt{3}) = i^2\sqrt{6} = -\sqrt{6}$. *We shall see in the next section that this manipulative procedure is simply a convenient way of dealing with complex numbers, and that (7) can be derived from appropriate definitions of complex numbers and operations upon them.*

In doing some manipulations we need to know higher powers of i. From (7) we find:

$$(11) \qquad i^3 = i \cdot i^2 = i(-1) = -i$$

$$(12) \qquad i^4 = i \cdot i^3 = i(-i) = -i^2 = -(-1) = 1$$

More generally, for n an integer,

$$(13) \qquad i^{4n} \equiv 1 \qquad i^{4n+2} \equiv -1$$

$$(14) \qquad i^{4n+1} \equiv i \qquad i^{4n+3} \equiv -i$$

The first result is derived as follows:

$$(15) \qquad i^{4n} \equiv (i^4)^n \equiv 1^n \equiv 1$$

The student will note that we are using algebraic laws, such as the laws of exponents, without having justified their application to complex numbers. This will be done in the next section.

So far we have dealt with square roots of negatives alone, but they often appear in combination with real numbers. Thus the quadratic formula gives $2 + i$ as a root of $x^2 - 4x + 5 = 0$. We shall manipulate such expressions by the usual rules of algebra, taking care always to eliminate negative radicands by rewriting in terms of i. Thus

$$(16) \quad (2 + \sqrt{-1})^2 = (2 + i)^2 = 4 + 4i + i^2$$

$$= 4 + 4i - 1 = 3 + 4i$$

We can use this result to check by substitution that $2 + i$ is a root of $x^2 - 4x + 5 = 0$. We have $(2 + i)^2 - 4(2 + i) + 5 = 3 - 8 + 5 + i(4 - 4) = 0$. Here we write $0 \cdot i = 0$ in accordance with the usual property of zero. We give further examples of manipulation:

$$(17) \qquad (2 + i)(2 - i) = 2^2 - i^2 = 4 - (-1) = 5$$

This is expected since the product of the roots of $x^2 - 4x + 5 = 0$ should equal the constant term.

$$(18) \qquad (3 + 4i)(1 - i) = 3 - 3i + 4i - 4i^2$$
$$= 3 + (-3 + 4)i + 4 = 7 + i$$

$$(19) \qquad 1/i = i/i^2 = i/(-1) = -i$$

$$(20) \qquad \frac{2}{3 + i\sqrt{3}} = \frac{2(3 - i\sqrt{3})}{(3 + i\sqrt{3})(3 - i\sqrt{3})}$$
$$= \frac{2(3 - i\sqrt{3})}{9 + 3} = \frac{3 - i\sqrt{3}}{6}$$

In all these cases we proceeded as though i were a real number except that we made use of $i^2 = -1$. The student should note that in every case the result reduces to the form $a + ib$, where a and b are real.

EXERCISE 10.1

1. Rewrite in terms of i:

   (a) $\sqrt{-3}$, $\sqrt{-16}$, $\sqrt{-7}$, $\sqrt{-a^2}$, $1 - \sqrt{-5}$, $\sqrt{-8}$, $3 + \sqrt{-1}$
   (b) $3 - \sqrt{-9}$, $\sqrt{-a^4}$, $1 + 2\sqrt{-6}$, $\sqrt{2} + 3\sqrt{-2}$, $3 - 2\sqrt{-9}$

Answers. (a) $i\sqrt{3}$, $4i$, $i\sqrt{7}$, ia, $1 - i\sqrt{5}$, $i2\sqrt{2}$, $3 + i$.

2. Find from (7) by successive multiplication by i: i^3, i^4, i^5, i^6, i^7, i^8.

3. Complete the derivation of (13) and (14).

4. Find the squares of the following: (a) $2i$. (b) $i\sqrt{3}$. (c) $4i\sqrt{2}$. (d) $\sqrt{-3}$. (e) $-3i\sqrt{5}$. (f) $\sqrt{2} + i\sqrt{3}$. [*Answers.* (a) -4. (c) -32. (e) -45.]

5. Find the cube and fourth power of each number in prob. 4.

6. Rewrite the following expressions in the form $a + ib$, where a and b are real:

   (a) $2 - 6i + 3\sqrt{-2} - 1$
   (b) $2\sqrt{-4} + 3i - \sqrt{-3} + 1 - \sqrt{3}$
   (c) $(1 + i)^2$
   (d) $(1 - i)^2$
   (e) $(3 + 2i)^2$
   (f) $(\sqrt{3} - i)^2$

(g) $(2 + 3i)(2 - 3i)$

(h) $(\sqrt{5} + i\sqrt{2})(\sqrt{5} - i\sqrt{2})$

(i) $(1 + i)^3$

(j) $(2 - 8i)(1 + i)^{-1}$

(k) $1 + 2i - (\sqrt{3} + i)i$

(l) $(3 + 4i)(2 - i)$

(m) $(3 - i)(i + 1)$

(n) $(2 + i)^3 - (2 + i)^2$

(o) $(1 + i)^4$

(p) $2/i$

(q) i^{-3}

(r) $(3i - 1)/i$

(s) $i^{-4} - 2i^{-2} + i + 3$

(t) $(1 + i)^{-1}$

(u) $(1 + 2i)^{-2}$

(v) $3(1 - 2i)^{-1}$

(w) $(5i + 1)(1 - i)^{-1}$

(x) $(1+i)^2 + 3i(1+i) + 2 + (1+i)^{-1}$

(y) $(2 + 3i)(1 - i)^2$

(z) $(i^6 - i^4 + i^2)(1 + i)$

Answers. (a) $1 + i(3\sqrt{2} - 6)$. (c) $2i$. (e) $5 + 12i$. (g) 13. (i) $-2 + 2i$. (k) $2 + i(2 - \sqrt{3})$. (m) $4 + 2i$. (o) -4. (q) i. (s) $6 + i$. (u) $(-3/25) + (-4/25)i$. (w) $3i - 2$.

7. Find the first three positive and negative powers of $2 + 3i$.

8. Find:

(a) $(a + ib) + (a' + ib')$

(b) $(a + ib) - (a' + ib')$

(c) $(a + ib)(a' + ib')$

(d) $(a + ib)/(a' + ib')$

9. Find $(a + ib) + (a - ib)$ and $(a + ib)(a - ib)$.

10. Show by substitution in the equation that the quadratic formula gives the roots of the quadratic.

11. Find the roots in the form $a + ib$ of the following and check by substitution:

(a) $3x^2 + x + 2 = 0$

(b) $x^2 - 2x + 4 = 0$

(c) $5x^2 - 7x + 5 = 0$

(d) $x^2 + x + 1 = 0$

12. Show that $-0.5 + 0.5\sqrt{-3}$ and $-0.5 - 0.5\sqrt{-3}$ are cube roots of one and that each equals the square of the other.

13. Show that $1, -1, i, -i$ are fourth roots of one.

14. Indicate what algebraic laws were used in the manipulations in the text.

15. Show that $\sqrt{i} = \dfrac{1 + i}{\sqrt{2}}$.

10.2 Complex numbers

We call an expression of the form $x + iy$ or $x + yi$, where x and y are real numbers, a **complex number**. If $y = 0$, we call it a real number. If $y \neq 0$, we call it an **imaginary number**. If $y \neq 0$ and $x = 0$, we call it a **pure imaginary**. Thus 2 is a real number, $2 + 3i$ is an imaginary number, $3i$ is a pure imaginary, and all three are complex numbers. We call x the **real part** or real component and y the **imaginary part** or imaginary component. *Note that both the real and imaginary parts are real numbers.* The imaginary part is distinguished by the i that is written preceding or following it. Thus in $2 + 5i$, the real part is 2 and the imaginary part is 5. In $3 - i\sqrt{2}$, the real part is 3 and the imaginary part is $-\sqrt{2}$. We say nothing yet about the meaning of i, except that it distinguishes the real and imaginary components.

We now define equality for complex numbers. (Compare 3.11.)

*10.2.1 $x + iy = x' + iy' \longleftrightarrow x = x'$ and $y = y'$

(Definition)

Also we identify $x + i \cdot 0$ with the real number x, that is,

*10.2.2 $x + i \cdot 0 \equiv x$ (Definition)

It follows from this that $0 + i \cdot 0 = 0$, and hence that a necessary and sufficient condition for a complex number to equal zero is that both its real and imaginary parts are zero, that is,

*10.2.3 $x + iy = 0 \longleftrightarrow x = 0$ and $y = 0$

It is easy to show that equality defined in the above manner has the properties discussed in 3.3.

We next define the sum and product of two complex numbers as follows:

*10.2.4 $(x + iy) + (x' + iy') \equiv (x + x') + i(y + y')$

(Definition)

*10.2.5 $(x + iy)(x' + iy') \equiv (xx' - yy') + i(xy' + yx')$

(Definition)

The reasonableness of these definitions will be apparent to the student in terms of the results of Ex. 10.1.8. In fact these definitions give the same results as would be found by treating the sum and product as though they represented real numbers with $i^2 = -1$. However, they are definitions and cannot be proved by reference to algebraic identities that have been proved only for real numbers.

It is not hard to show that the commutative, associative, and distributive laws hold for complex numbers. For example, to prove that addition is commutative we must show that

(1) $(x + iy) + (x' + iy') \equiv (x' + iy') + (x + iy)$

Now

(2) $\quad (x + iy) + (x' + iy') \equiv (x + x') + i(y + y')$ $\qquad$ (*10.2.4)

(3) $\qquad\qquad\qquad\qquad \equiv (x' + x) + i(y' + y)$ $\qquad$ (*3.4.2)

(4) $\qquad\qquad\qquad\qquad \equiv (x' + iy') + (x + iy)$ $\qquad$ (*10.2.4)

The other laws are proved in a similar way by using the definitions of addition and multiplication and then applying the theorems of Chapter 3 to the real and imaginary parts—which are real numbers and hence are already known to satisfy these laws.

We define subtraction of complex numbers exactly as we did for real numbers. Thus as in *3.7.1, $(x + iy) - (x' + iy')$ is the number $a + ib$ such that $(a + ib) + (x' + iy') = (x + iy)$. But $(a + ib) + (x' + iy') = (a + x') + i(b + y')$. Hence, from *10.2.1, $a + x' = x$ and $b + y' = y$. It follows that $a = x - x'$ and $b = y - y'$, and

*10.2.6 $\quad (x + iy) - (x' + iy') \equiv (x - x') + i(y - y')$

Similarly, as in *3.7.4, we define the quotient of two complex numbers, $(x + iy)/(x' + iy')$, as the number $a + ib$ such that $(x' + iy')(a + ib) = (x + iy)$. But $(x' + iy')(a + ib) \equiv (x'a - y'b) + i(x'b + y'a)$. Hence

(5) $\qquad\qquad x'a - y'b = x \quad \text{and} \quad x'b + y'a = y$

These are two linear equations in a and b whose solutions are:

(6) $\qquad a = \dfrac{xx' + yy'}{x'^2 + y'^2} \quad \text{and} \quad b = \dfrac{x'y - xy'}{x'^2 + y'^2}$

Hence

*10.2.7 $\qquad \dfrac{x + iy}{x' + iy'} \equiv \dfrac{(xx' + yy') + i(x'y - xy')}{(x'^2 + y'^2)}$

As for real numbers, we write $-(x + iy)$ in place of $(-1)(x + iy)$ and $-iy$ for $i(-y)$. Then it is easy to show that

*10.2.8 $\qquad\qquad -(x + iy) \equiv -x - iy$

*10.2.9 $\quad (x + iy) + (-1)(x + iy) \equiv 0 \qquad$ (Compare *3.10.2.)

The following can also be proved without difficulty:

*10.2.10 $(x + iy)^{-1} \equiv \dfrac{x - iy}{x^2 + y^2}$

*10.2.11 $(x + iy)(x + iy)^{-1} \equiv 1$ (Compare *3.11.5.)

Also, using z's to stand for complex numbers,

*10.2.12 $\dfrac{zz_1}{zz_2} \equiv \dfrac{z_1}{z_2}$, where $z \neq 0, z_2 \neq 0$ (Compare *3.7.10.)

We could continue to list identities for complex numbers that correspond to identities for real numbers given in Chapter 3. However, this is not necessary, because all the identities given in Chapter 3 and in later chapters follow from the fundamental laws, that is, the commutative and associative laws of addition and multiplication, the distributive law, the definition of negative and reciprocal, and the properties *3.10.2 (*10.2.9) and *3.11.5 (*10.2.11). Once we have shown that complex numbers satisfy these fundamental laws we may with confidence treat them according to the same rules as real numbers, just as we did in the previous section. Of course we extend the definitions of exponents and other operations to apply to complex as well as real numbers. For example, in *10.2.10 we write $(x + iy)^{-1}$ to mean $1/(x + iy)$.

So far in this section we have said nothing about the properties of i or indicated its nature in any way. It simply appeared as a symbol to distinguish the real and imaginary parts of a complex number. However, we can now easily prove that

*10.2.13 $i^2 = -1$

In fact,

(7) $i^2 = (0 + i \cdot 1)^2 = (0 + i \cdot 1)(0 + i \cdot 1)$ (Why?)

(8) $= (0 \cdot 0 - 1 \cdot 1) + i(0 \cdot 1 + 1 \cdot 0)$ (*10.2.5)

(9) $= -1 + 0i$ (Why?)

(10) $= -1$ (*10.2.2)

It is almost obvious that

*10.2.14 $x + iy \equiv (x + i \cdot 0) + (0 + iy)$

Thus the plus sign that connects the real and imaginary parts of a complex number has a meaning consistent with our definition of addition. A complex number is the sum of two complex numbers, one of which is real and the other a pure imaginary.

The results in this section together with those in the following exercise provide a complete justification of the manipulative methods used in the previous section. Complex numbers have been shown to behave under algebraic operations exactly like the real numbers, and hence we may manipulate them according to the familiar rules of ordinary algebra.

EXERCISE 10.2

1. Indicate which of the following complex numbers are real, imaginary, or pure imaginary:

(a) $\sqrt{2}$ (e) $2i - \sqrt{3}$ (i) $(1 + i)(1 - i)$
(b) $3i$ (f) $-\sqrt{7}$ (j) i^2
(c) $2 - i$ (g) $i\sqrt{5}$
(d) $1 - \sqrt{2}$ (h) $-1 - 0.5i$

2. Indicate the real and imaginary parts in each of the numbers in prob. 1 and in the following:

(a) $2 + 3i$ (e) $6i$ (i) $-i$
(b) $1 - i$ (f) $-3i$ (j) $-\sqrt{-12}$
(c) $\sqrt{3} - i\sqrt{5}$ (g) $2i\sqrt{10}$ (k) $-\sqrt{3} + \sqrt{-3}$
(d) 4 (h) $3\sqrt{11} - 2i\sqrt{13}$

Answers. (a) 2 and 3. (c) $\sqrt{3}$ and $-\sqrt{5}$. (e) 0 and 6. (g) 0 and $2\sqrt{10}$.

3. Explain the derivation of *10.2.6 and verify steps (5) and (6) in the derivation of *10.2.7.

4. Prove *10.2.14.

5. Show that equality as defined in *10.2.1 is reflexive, symmetric, and transitive. (Compare 3.3.)

6. Show that for any complex numbers u, v, and w:

(11) $u + v \equiv v + u$

(12) $u + (v + w) \equiv (u + v) + w$

(13) $uv \equiv vu$

(14) $u(vw) \equiv (uv)w$

(15) $u(v + w) \equiv uv + uw$

7. Suggest names for (11) through (15).

8. Show that *10.2.7 can be obtained by multiplying both numerator and denominator of the left member by $x' - iy'$.

9. Derive *10.2.8 through *10.2.11 by using the formulas for addition, subtraction, multiplication, and division.

10. Prove *10.2.12.

11. Show that

$$*\textbf{10.2.15} \qquad k(x + iy) \equiv kx + iky, \text{ when } k \text{ is real}$$

12. Each of the two numbers $x + iy$ and $x - iy$ is called the **conjugate** of the other. Find their sum and product.

13. Show from *10.2.3 that a necessary and sufficient condition that $x + iy = 0$ is $x^2 + y^2 = 0$.

14. Show that for any complex numbers u and v, $u - v \equiv u + (-v)$.

15. If $\bar{z}$ represents the conjugate of z, show that

$$(16) \qquad \overline{uv} \equiv \bar{u} \cdot \bar{v}$$

$$(17) \qquad \overline{u + v} \equiv \bar{u} + \bar{v}$$

$$(18) \qquad \overline{z^n} \equiv \bar{z}^n$$

†**16.** Show that the complex numbers form a field.

†**17.** Show that complex numbers with rational components form a field.

†**18.** Show that complex numbers with integral coefficients do not form a field, but do form an integral domain.

10.3 Geometric interpretation of complex numbers

From the last two sections we are familiar with the idea that a complex number is determined by two real numbers, the real and imaginary parts. Two complex numbers are equal if and only if their real and imaginary parts are equal respectively. It follows that there is a one-to-one correspondence between complex numbers and ordered pairs of real numbers. Given the complex number $x + iy$ we have the corresponding pair (x, y), and given a pair (x, y) we have the corresponding complex number $x + iy$. Thus $2 - 3i$ corresponds to $(2, -3)$ and conversely. Since we have already established a one-to-one correspondence between ordered pairs of real numbers and points in the plane and between ordered pairs of real numbers and vectors, it would seem appropriate to represent a complex number $x + iy$ by the point (x, y) in a rectangular coordinate system or by the vector (x, y) with components x and y. Thus our use of the notation $x + iy$ and of the names real component for x and imaginary component for y were not accidental.

In Fig. 10.3.1 are plotted the complex numbers 3, $2i$, $-1 + 3i$, and $1 - 2i$. When complex numbers are represented geometrically in this way we speak of the x-axis as the **axis of reals** and the y-axis as the **axis of imaginaries**. However, it should be noted that we are plotting pairs of real numbers. Nothing like "imaginary distances" are involved!

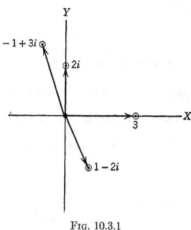

FIG. 10.3.1

A comparison of *10.2.4 with the definition of the sum of two vectors given in *9.7.2 suggests that the vector corresponding to the sum of two complex numbers is just the sum of the vectors corresponding to the two numbers being added. As shown for $(2 + 3i) + (-4 - i)$ in Fig. 10.3.2, it is the diagonal of the parallelogram built on the vectors corresponding to the terms in the sum.

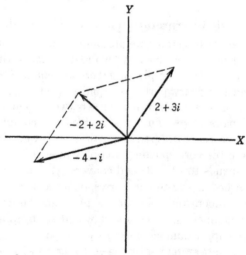

FIG. 10.3.2.   Addition.

Multiplication of a complex number by a real number may be interpreted exactly as the scalar multiplication of the corresponding vector by the number. (See Ex. 9.7.20.) In Fig. 10.3.3 we have sketched $3(2 + i)$ and $-2(-1 + 2i)$. The geometric interpretation of multiplication will be given in the next section.

It follows from the foregoing remarks that we may use and speak of points, vectors, and complex numbers interchangeably. Thus we may think of $z = x + iy$ as the point (x, y) or the vector (x, y).

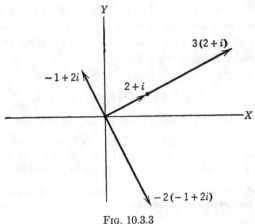

Fɪɢ. 10.3.3

Conversely, we may represent the point or vector (x, y) by the complex number $x + iy$. Actually it would be possible to dispense with the letter i altogether and merely represent a complex number by its real and imaginary parts written as an ordered number pair (x, y). Then the basic identities of the previous section could be rewritten:

*10.2.1 $(x, y) = (x', y') \longleftrightarrow x = x', y = y'$ (Definition)

*10.2.2 $(x, 0) = x$ (Definition)

*10.2.3 $(x, y) = 0 \longleftrightarrow x = 0$ and $y = 0$

*10.2.4 $(x, y) + (x', y') = (x + x', y + y')$ (Definition)

*10.2.5 $(x, y)(x', y') = (xx' - yy', xy' + x'y)$ (Definition)

*10.2.15 $k(x, y) = (kx, ky)$

The student should compare these results with the corresponding vector statements in 9.7. (Of course there is nothing in 9.7 corresponding to *10.2.5, since we did not define a vector product.) It appears from this discussion that *the important and essential thing about a complex number is that it is an ordered pair of real numbers* following *10.2.1 to *10.2.5. We could use any notation that indicated which was the real and which the imaginary part, for instance $x:y$, x/y, or $x?y$. The notation $x + iy$ is convenient because we can then operate with complex numbers according to the usual rules of algebra and $i^2 = -1$, without having to remember special rules for sums and products.

EXERCISE 10.3

1. Plot the following complex numbers, showing both points and vectors:

(a) $1, i, -1, -i$

(b) $0, 1 - i, 2 - 2i$

(c) $1 + i, 1 - i, -1 - i$

(d) $-\sqrt{2} + 4i, -3 - 5i$

(e) $3 - ie, -2 + 6i$

(f) $\cos 30° + i \sin 30°$

2. Sketch the following pairs of numbers and their sums. Draw the appropriate vectors and parallelograms.

(a) $2 + 3i, 3 + i$

(b) $1 + 2i, -3 + 5i$

(c) $-3 - 2i, -1 + i$

(d) $3 + i, 3 - i$

(e) $-1 + 4i, -1 - 4i$

(f) $-2 + 5i, 2 - 5i$

(g) $6 - 9i, -(2 + i)$

(h) $-3 - 5i, 1 + 4i$

(i) $i, 3 + 2i$

(j) $1 - 2i, 3$

(k) $-5i, -1 + 3i$

(l) $-2 - 7i, 1 - 4i$

(m) $3 + 0.5i, 5 + 25i$

(n) $\sqrt{5} - i\sqrt{3}, \log_3 9 + i$

(o) $-6, 2i - 18$

3. Sketch:

(a) 3 and $2(3)$

(b) $2i$ and $2(2i)$

(c) $-2 + i$ and $2(-2 + i)$

(d) $2 - 5i$ and $-(2 - 5i)$

(e) $-5 - 1.5i$ and $-(-5 - 1.5i)$

(f) $3 + 7i$ and $-2(3 + 7i)$

(g) $3 + 2i$ and $4(3 + 2i)$

(h) $-1 - i$ and $10(-1 - i)$

(i) $3i$ and $(-1)(3i)$

(j) $15 + i$ and $-2(15 + i)$

4. Sketch $z = u - v$, where:

(a) $u = 5, v = 2$

(b) $u = 5i, v = 2i$

(c) $u = 2i, v = 1 + i$

(d) $u = 3 + 4i, v = 2 + i$

(e) $u = 5 - 6i, v = -3 - 2i$

(f) $u = -3 - 4i, v = 6 - i$

5. Give a geometric interpretation of Ex. 10.2.13.

6. Give a geometric interpretation of the commutative law of addition.

7. Sketch $1 + i$, $1 + 2i$, $4 + i$, and their sum. Indicate how the sum is the vector that closes the polygon made up of the three vectors laid end to end.

8. Let $u = 3 - 2i$, $v = 1 + i$, and $w = -2 + 2i$. Diagram $u + (v + w)$ and $(u + v) + w$. Give a geometric interpretation of the associative law of addition.

9. Make sketches of $[(1 + 2i) + (-3 + i)] + (-2i)$, $(1 + 2i) + [(-3 + i) + (-2i)]$, and $[(1 + 2i) + (-2i)] + (-3 + i)$.

10. Plot z and z^{-1} for $z = 1, 2, i, -i, 2 + i$.

11. Consider the function $w = z^2 - 1$. Evaluate this function for $z = 1$, i, $2 + i$, $2i$, $-1 - i$, $-i$. Plot the values of z on one coordinate plane and the corresponding values of w on another. Note that it takes two planes to plot a function of a complex variable.

†12. Plot $w = z + (1 + i)$, using two planes as in prob. 11 and plotting z and w for $z = 1, -1, i, 2i - 1, -4 + 2i$. If z lies along the line $y = 3x - 1$, along what curve does w lie?

10.4 Polar form of complex numbers

Let (x, y) be the point corresponding to a complex number $x + iy$, and let $r \geq 0$ and θ be the polar coordinates of this point. Then from *9.8.3 and *9.8.4,

*10.4.1 $x + iy = r(\cos \theta + i \sin \theta)$, where $r = \sqrt{x^2 + y^2}$ and $\theta = \arccos(x/r) = \arcsin(y/r)$.

When a complex number is expressed in this way it is said to be in **polar form.** The positive quantity r is called the **modulus** or absolute value, and is sometimes written $|x + iy|$. The angle θ is called the **amplitude,** argument, or phase. A complex number may be put in this form merely by dividing and multiplying by $\sqrt{x^2 + y^2}$. For example,

$$(1) \qquad 2 + 3i = \sqrt{13}\left[\frac{2}{\sqrt{13}} + i\frac{3}{\sqrt{13}}\right]$$

This form does not display θ explicitly, but it can be found from tables. Thus $56.3° \doteq \arcos(2/\sqrt{13}) = \arcsin(3/\sqrt{13})$ and

$$(2) \qquad 2 + 3i \doteq \sqrt{13}(\cos 56.3° + i \sin 56.3°)$$

The angle θ must satisfy *both* $\cos \theta = x/r$ and $\sin \theta = y/r$. For example,

$$(3) \qquad 2 - 3i \doteq \sqrt{13}(\cos 303.7° + i \sin 303.7°)$$

The operation of multiplication has a simple interpretation in terms of polar coordinates. Consider:

(4) $$z_1 = x_1 + iy_1 = r_1(\cos \theta_1 + i \sin \theta_1)$$

(5) $$z_2 = x_2 + iy_2 = r_2(\cos \theta_2 + i \sin \theta_2)$$

The product is:

(6) $\quad z_1 z_2 \equiv r_1 r_2 (\cos \theta_1 + i \sin \theta_1)(\cos \theta_2 + i \sin \theta_2)$

(7) $\quad\quad \equiv r_1 r_2 [\cos \theta_1 \cos \theta_2 - \sin \theta_1 \sin \theta_2$

$$+ i(\cos \theta_1 \sin \theta_2 + \sin \theta_1 \cos \theta_2)] \quad \text{(Why?)}$$

(8) $\quad\quad \equiv r_1 r_2 [\cos (\theta_1 + \theta_2) + i \sin (\theta_1 + \theta_2)] \quad \text{(Why?)}$

The right member of (8) is in polar form. Hence:

*10.4.2   The modulus of the product of two complex numbers is the product of the moduli and the amplitude is the sum of the amplitudes, that is,

$$z_1 z_2 = r_1 r_2 [\cos (\theta_1 + \theta_2) + i \sin (\theta_1 + \theta_2)]$$

This result might be symbolized in terms of the polar coordinates of the complex numbers as follows: $(r_1, \theta_1)(r_2, \theta_2) \equiv (r_1 r_2, \theta_1 + \theta_2)$.

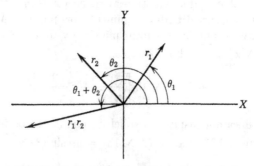

FIG. 10.4.   Multiplication of complex numbers.

It means that when we multiply z_1 by z_2 we rotate the vector z_1 through an angle equal to the amplitude of z_2 and multiply the length of this vector by the modulus of z_2. (See Fig. 10.4.)

We leave it to the reader to prove the following:

*10.4.3 $z_1/z_2 \equiv (r_1/r_2)[\cos(\theta_1 - \theta_2) + i \sin(\theta_1 - \theta_2)]$

*10.4.4 $1/z \equiv (1/r)(\cos\theta - i\sin\theta)$

*10.4.5 $z^2 \equiv r^2(\cos 2\theta + i \sin 2\theta)$

Thus when we multiply, divide, take the reciprocal, or square we do the same operations on the moduli (the r's), but we add, subtract, take the negative, or double the amplitudes (the θ's). Thus taking the reciprocal means to take the reciprocal of the modulus and the negative of the amplitude, and to square a complex number means to square the modulus and double the angle. It begins to look as though the amplitude of a complex number behaves as a logarithm or exponent! (See Ex. 10.4.12.)

We now state a famous theorem:

*10.4.6 $(\cos\theta + i\sin\theta)^n \equiv \cos n\theta + i \sin n\theta$

(De Moivre's Theorem)

We begin by proving this theorem by induction for all positive integral values of n. Obviously it holds when $n = 1$. If

(9) $(\cos\theta + i\sin\theta)^k \equiv \cos k\theta + i \sin k\theta$

then

(10) $(\cos\theta + i\sin\theta)^{k+1} \equiv (\cos\theta + i\sin\theta)(\cos k\theta + i \sin k\theta)$

(Why?)

(11) $\equiv \cos(\theta + k\theta) + i \sin(\theta + k\theta)$

(*10.4.2)

(12) $\equiv \cos(k+1)\theta + i \sin(k+1)\theta$

(Why?)

We have shown that the theorem is true for $n = 1$ and that the theorem for $n = k$ implies the theorem for $n = k + 1$. Hence by the principle of finite induction it holds for all positive integral values of n. It is now easy to extend the theorem to negative values

of n. We know that *10.4.6 holds for all positive integral values of
n. Taking the reciprocal of both sides, we have:

(13) $(\cos \theta + i \sin \theta)^{-n} \equiv (\cos n\theta + i \sin n\theta)^{-1}$

(14) $\equiv \cos n\theta - i \sin n\theta$ (*10.4.4)

(15) $\equiv \cos(-n\theta) + i \sin(-n\theta)$ (Why?)

But this final equation is just the theorem with n replaced by $-n$.
Hence the theorem holds for all negative integral values of n. Incidentally, this argument shows that if the theorem holds for any
value of n, it holds also for the negative of that value. De Moivre's
theorem holds for rational, irrational, and even complex values of
the exponent, but integral values are sufficient for our purposes.

To illustrate the theorem, consider:

(16) $\dfrac{\sqrt{3}}{2} + i\dfrac{1}{2} = \cos 30° + i \sin 30°$

Squaring the left side by direct multiplication, we find:

(17) $\dfrac{3}{4} + i\dfrac{\sqrt{3}}{2} - \dfrac{1}{4} = \dfrac{1}{2} + i\dfrac{\sqrt{3}}{2}$

De Moivre's theorem applied to the square of the right member of
(16) gives

(18) $\cos 60° + i \sin 60°$

which is equal to the right member of (17).

EXERCISE 10.4

1. Put each of the following in polar form and sketch, showing r and θ.
Check against the graph.

(a) $3 + 2i$ (e) 1 (i) $-4 - 8i$
(b) $-1 + 4i$ (f) -1 (j) $-5 + 2i$
(c) $1 + i$ (g) $0.5 - (0.5\sqrt{3})i$ (k) $100 + 5i$
(d) i (h) $0.5\sqrt{2}(1 + i)$ (l) $1.51 + \sqrt{-2.3}$

2. Write in the form $x + iy$:

(a) $2(\cos 13° + i \sin 13°)$ (d) $3(\cos \pi/2 + i \sin \pi/2)$
(b) $\sqrt{2}(\cos 45° + i \sin 45°)$ (e) $5(\cos \pi + i \sin \pi)$
(c) $\sqrt{57}(\cos 32° + i \sin 32°)$ (f) $\cos 42° - i \sin 42°$

Answers. (a) $1.9488 + i(0.4500)$. (e) -5.

3. Plot the complex numbers (2) and (3) in the section.

4. Diagram the following multiplications, showing each factor and the product:

(a) $(\cos 30° + i \sin 30°)(\cos 60° + i \sin 60°)$

(b) $(\cos 20° + i \sin 20°)(\cos 40° + i \sin 40°)$

(c) $(\cos 210° + i \sin 210°)(\cos 50° + i \sin 50°)$

(d) $(\cos 193° + i \sin 193°)^2$

(e) $(\cos 95° + i \sin 95°)^3$

(f) $(\cos 5° + i \sin 5°)^{18}$

5. Diagram each multiplication in Ex. 10.1.6, showing each factor and the product. Find the moduli and amplitudes and verify that *10.4.2 holds.

6. Calculate the product $(3 + 4i)(4 - 3i)$ directly by the definition of multiplication and by changing to polar form and using *10.4.2. Draw a large sketch and compare the two results.

7. Find $(0.5 + 0.5\sqrt{3}\,i)^2$ by direct multiplication and by De Moivre's theorem, and sketch.

8. In the following problems we represent the complex number by its polar coordinates, (r, θ). Rewrite each number in the form $r\,(\cos \theta + i \sin \theta)$, perform the indicated operations, and sketch.

(a) $(2, 45°)(3, 45°)$ (d) $(1, 120°)^3$ (g) $(1, 30°)^{-2}$

(b) $(3, 154°)(1, 100°)$ (e) $(1, 90°)^4$ (h) $(1, 90°)^{0.5}$

(c) $(1, 120°)^2$ (f) $(4, 150°)(2, 50°)^{-1}$ (i) $(\sqrt{3}, 1)(2\sqrt{3}, 2)$

9. Use De Moivre's theorem to derive the formulas for $\sin 2\theta$ and $\cos 2\theta$. [*Suggestion.* Expand $(\cos \theta + i \sin \theta)^2$ in two ways and equate real and imaginary parts.]

†10. Use De Moivre's theorem to derive formulas for $\sin 3\theta$ and $\cos 3\theta$.

11. Prove *10.4.3, 4, 5.

†12. If we define

$$(19) \qquad e^{x+iy} = e^x\,(\cos y + i \sin y)$$

it turns out that imaginary exponents obey the usual exponential laws. Show that:

*10.4.7 $e^{iy} \equiv \cos y + i \sin y$

*10.4.8 $e^{z_1}e^{z_2} \equiv e^{z_1+z_2}$ for any complex z_1, z_2

*10.4.9 $(e^{z_1})^x \equiv e^{xz_1}$ for any complex z_1 and real x

*10.4.10 $e^{\pi i} = -1$

†13. Show that

$$(20) \qquad \sin x = \frac{e^{ix} - e^{-ix}}{2i}$$

$$(21) \qquad \cos x = \frac{e^{ix} + e^{-ix}}{2}$$

10.5 Roots of complex numbers

We stated in 6.10 that $y^n = x$ defines y as an n-valued function of x. To justify this statement and, at the same time, to show how to find all the nth roots of any complex number, we begin by considering the nth roots of unity. These are the solutions of the equation $z^n - 1 = 0$. If we write z and 1 in the polar form, the equation becomes:

$$(1) \qquad r^n(\cos n\theta + i \sin n\theta) = 1(\cos 0 + i \sin 0)$$

Since r must be positive and real (why?), $r = 1$. Also, since $\cos n\theta = \cos 0 = 1$ and $\sin n\theta = \sin 0 = 0$ (why?), $n\theta$ is an angle with the same terminal side as the angle 0. There are an infinite number of such angles, given by $2k\pi$, where $k = 0$, ± 1, ± 2, ± 3, $\cdots$. Hence $n\theta = 2k\pi$, for any integral k, gives a solution of (1). The solutions are therefore given by

$$(2) \qquad z = \cos \frac{2k\pi}{n} + i \sin \frac{2k\pi}{n}$$

However, only n of the solutions (2) are distinct. Taking $k = 0$, 1, 2, 3, $\cdots$, $n - 1$ we get n values of θ, which represent angles with different terminal sides, since each is $2\pi/n$ greater than the preceding one. If we continue by setting $k = n$, $n + 1$, $\cdots$, we get angles congruent to those found previously, and hence we repeat the cycle of values of $\cos \theta$ and $\sin \theta$. Negative values of k yield congruent angles also. For example, $2(-1)\pi/n \cong 2(n - 1)\pi/n$. It follows that there are just n nth roots.

To illustrate, we find the cube roots of unity:

$$(3) \quad z^3 = \cos 3\theta + i \sin 3\theta = \cos 2k\pi + i \sin 2k\pi = 1$$

$$(4) \quad z = \cos (2k\pi/3) + i \sin (2k\pi/3)$$

$$= \cos 0 + i \sin 0 = 1, \text{ when } k = 0$$

$$= \cos (2\pi/3) + i \sin (2\pi/3) = -1/2 + i(1/2)\sqrt{3}, \text{ when } k = 1$$

$$= \cos (4\pi/3) + i \sin (4\pi/3) = -1/2 - i(1/2)\sqrt{3}, \text{ when } k = 2$$

As a second example, we find the fourth roots of unity:

$$(5) \qquad z^4 = \cos 4\theta + i \sin 4\theta = \cos 2k\pi + i \sin 2k\pi = 1$$

(6) $z = \cos (2k\pi/4) + i \sin (2k\pi/4)$

$= \cos 0 + i \sin 0 = 1$, when $k = 0$

$= \cos (\pi/2) + i \sin (\pi/2) = i$, when $k = 1$

$= \cos \pi + i \sin \pi = -1$, when $k = 2$

$= \cos (3\pi/2) + i \sin (3\pi/2) = -i$, when $k = 3$

In this manner we can find the n nth roots of unity for any positive integral n. The process has a simple geometric interpretation that can be used to write down the results without doing the algebra. Since $r = 1$ for all the roots, it follows that they all lie on the unit circle. Referring to (2), we see that one root is always the real number 1, which lies at the point where the unit circle crosses the positive x-axis. The remaining roots are spaced evenly on the unit circle, neighboring roots being separated by an angle (or arc) of $2\pi/n$. Hence we can locate the roots graphically by dividing the unit circle into n equal arcs, beginning on the positive x-axis. The amplitudes can then be read off as multiples of $2\pi/n$, and the components of the corresponding complex roots can be read graphically or computed by trigonometry. The cube and fifth roots of unity are plotted in Fig. 10.5.

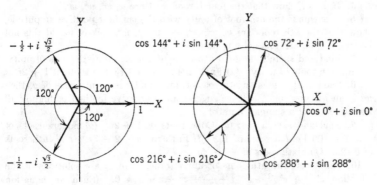

Fig. 10.5.   Cube and fifth roots of unity.

The foregoing method applies to any complex number. The solutions of $z^n = a = r(\cos \phi + i \sin \phi)$ are given by

(7) $z = \sqrt[n]{r} \left(\cos \dfrac{\phi + 2k\pi}{n} + i \sin \dfrac{\phi + 2k\pi}{n} \right)$

where k takes the values $0, 1, 2, 3, \cdots, n-1$. For example, to find the cube roots of $2i$, we proceed as follows:

$$(8) \qquad z^3 = 2i = 2(\cos \pi/2 + i \sin \pi/2)$$

$$(9) \qquad z = \sqrt[3]{2}\left(\cos \frac{\pi/2 + 2k\pi}{3} + i \sin \frac{\pi/2 + 2k\pi}{3}\right)$$

$$= \sqrt[3]{2}\,(\cos 30° + i \sin 30°) \qquad \text{for } k = 0$$

$$= \sqrt[3]{2}\,(\cos 150° + i \sin 150°) \qquad \text{for } k = 1$$

$$= \sqrt[3]{2}\,(\cos 270° + i \sin 270°) \qquad \text{for } k = 2$$

Note that each root may be obtained by adding $120° = 2\pi/3$ to the amplitude of the previous one. More generally, once an nth root has been found, the others may be written down by successive additions of $2\pi/n$ to the angle.

EXERCISE 10.5

1. Find the cube roots of: (a) 8. (b) −1. (c) 20. (d) $3 - 2i$.

2. If $w = (1/2)(1 + i\sqrt{3}\,)$, show that $w^3 = -1$.

3. Write the roots (9) in the form $x + iy$.

4. Find the fourth roots of: (a) −1. (b) 16. (c) i. (d) $-4 - 3i$.

5. If $w = i$, show that the fourth roots of 1 are w, w^2, w^3, w^4.

6. If w equals the nth root of unity with the smallest positive amplitude, show that the nth roots are $w, w^2, w^3, \cdots w^{n-1}, w^n$. (Why would this not be true for the nth roots of any number?)

7. Find and sketch the: (a) Fifth roots of unity. (b) Sixth roots of unity. (c) Seventh roots of unity. (d) Fifth roots of i. (e) Eighth roots of $1 + 2i$.

8. Show that if w is an nth root of unity and z is an nth root of a, then wz is an nth root of a. Hence suggest an alternative method of finding all the nth roots of a complex number.

9. Find and sketch the: (a) Cube roots of $1 + 2i$. (b) Square roots of $-3 + i$. (c) Fourth roots of 16. (d) Fifth roots of $2 - 3i$. (e) Tenth roots of 6.129. (f) Sixth roots of $-4 - 5i$.

10. Show that the nth roots of unity (other than 1) are solutions of the equation $z^{n-1} + z^{n-2} + z^{n-3} + \cdots + z^2 + z + 1 = 0$. (Such an equation is called **cyclotomic.**)

11. Show that the sum of the n nth roots of unity for any n is equal to zero.

12. Solve for x: $x^n - x^{n-1} + \cdots + (-1)^n x^2 + (-1)^{n-1}x + (-1)^n = 0$.

13. Find all the pairs of numbers such that the square of each is equal to the other.

14. Indicate how the discussion of this chapter justifies our statement that the complex numbers are closed under addition, multiplication, raising to a power, and their inverse operations.

Since an equation of the form $Ax + By + C = 0$ always represents a straight line, it would be natural to ask what sort of curve is represented by an equation of the form

(I) $$Ax^2 + Bxy + Cy^2 + Dx + Ey + F = 0$$

where at least one of A, B, C is not zero. We have already studied special cases. The locus of (I) is a parabola when $B = C = 0$ (5.5), a hyperbola when $A = C = D = E = 0$ (6.8), and a circle when $B = D = E = 0$ and $A = C$ (8.2).

An equation of type (I) is called an **equation of the second degree** in x and y. Its left member is evidently a function of x and y, and is called a **quadratic function** or **polynomial of the second degree**. The word "degree" refers to the powers of x and y that appear. "Second" indicates that the sum of the exponents is two in at least one term and no higher in any term. Equation (I) is called the **general equation of the second degree**.

In this chapter we study the set of curves represented by (I). Its members are called **conics** or conic sections. The name goes back to the Greeks, who studied these curves without the benefit of algebra or analytic geometry by considering them as the intersections of planes with a right circular cone. It can indeed be proved that the set of curves represented by (I) is the same as the set of curves obtained by letting a plane cut a right circular cone, but we are not particularly interested in this because it is much easier to study the curves directly by more modern methods. We begin by considering special cases: the ellipse, parabola, and hyperbola. In 11.4 we show that (I) always represents one of these unless it "degenerates" into two straight lines, a single point, or an imaginary locus. The remainder of the chapter is devoted to the algebraic and geometric properties of the conics as a whole.

11.1   The ellipse

A special case of (I) is

(1)
$$\frac{x^2}{a^2} + \frac{y^2}{b^2} = 1$$

A locus whose equation can be put in this form is called an **ellipse**. It is assumed that $a > 0$, $b > 0$. We begin by graphing (1). Setting $x = 0$ and $y = 0$, we find the intercepts: $(0, b)$, $(0, -b)$, $(a, 0)$, $(-a, 0)$. We note that the curve is symmetric with respect to the origin. (Why?) Solving for x and y, we find:

(2)
$$x = \pm \frac{a}{b} \sqrt{b^2 - y^2}$$

(3)
$$y = \pm \frac{b}{a} \sqrt{a^2 - x^2}$$

Evidently $|x| > a$ and $|y| > b$ are excluded regions. (Why?) Also as x increases from 0 to a, y decreases from b to 0. With these results and a few plotted points, we have Fig. 11.1.1.

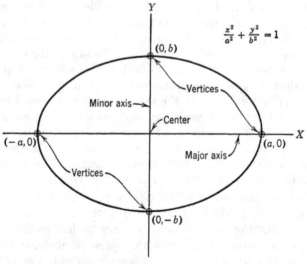

FIG. 11.1.1

In the locus (1), the origin is called the **center**. It is the center of symmetry of the curve. Segments cut off by the curve on

lines through the center are called **diameters**. Those that lie on
the axes are called the **axes** of the ellipse. The longer one is called
the **major axis** or principal axis; the other one is called the **minor
axis**. The ends of the axes are called **vertices**. We may describe
(1) by saying that it is the equation of an ellipse with center at the
origin, and axes of length $2a$ and $2b$ lying along the x-axis and y-
axis, respectively. The lengths a and b are often called the **semi-
axes,** the larger being called the semi-major axis and the smaller
the semi-minor axis. The equation (1) is called the **standard
equation** of an ellipse, and its locus is said to be in **standard
position.**

A special case of (1) arises when $a = b$. In that case we have
$x^2 + y^2 = a^2$, a circle with radius a and center at the origin.
Hence the circle is a special kind of ellipse.

What is the equation of an ellipse when the center is not at the
origin? When the axes of the ellipse are parallel to the coordinate
axes, the answer is given by the following theorem:

*11.1.1 The locus of

$$(4) \qquad \frac{(x - h)^2}{a^2} + \frac{(y - k)^2}{b^2} = 1$$

is an ellipse with center at (h, k) and axes parallel to
the coordinate axes.

The proof is easy. If we translate the origin to the point (h, k),
$x' = x - h$, $y' = y - k$; then the equation reduces to the form of
(1) and hence is an ellipse with the stated properties. When
$a = b$, we have a circle with center at (h, k) and radius a. (Compare *8.2.1.)

If we expand (4) and collect about powers of x and y, we find:

$$(5) \quad b^2x^2 + a^2y^2 + (-2hb^2)x + (-2ka^2)y$$
$$+ (b^2h^2 + a^2k^2 - a^2b^2) = 0$$

This is in the form (I), and we are led to the guess that (I) represents an ellipse if $A > 0$, $B = 0$, $C > 0$. Under these conditions
we can rewrite (I) as follows:

$$(6) \qquad A[x^2 + (D/A)x \qquad] + C[y^2 + (E/C)y \qquad] = -F$$

Completing the square yields

(7) $A(x - h)^2 + C(y - k)^2 = S$

where $h = -D/2A$, $k = -E/2C$, and $S = -F + h^2A + k^2C$.
Then we write (8) as

(8) $\dfrac{(x - h)^2}{(S/A)} + \dfrac{(y - k)^2}{(S/C)} = 1$ (Why?)

This is in the form (4) if we can replace the denominators by a^2
and b^2, where a and b are real. Since $A > 0$ and $C > 0$, this is
possible if and only if $S > 0$. (Why?) If $S < 0$, we can rewrite
the equation:

(9) $\dfrac{(x - h)^2}{(-S/A)} + \dfrac{(y - k)^2}{(-S/C)} = -1$

which is in the form

(10) $\dfrac{(x - h)^2}{a^2} + \dfrac{(y - k)^2}{b^2} = -1$

Evidently (10) is satisfied by no points (why?), and we call the
locus an **imaginary ellipse.** If $S = 0$, the locus consists of only
the one point (h, k) and is called a **point ellipse,** or a **degenerate
ellipse.** With this understanding, we have proved the following
theorem:

*11.1.2   When A and C have the same sign

(11) $Ax^2 + Cy^2 + Dx + Ey + F = 0$

is the equation of an ellipse with center at $(-D/2A,
-E/2C)$, and axes parallel to the coordinate axes.

It is sufficient to state that A and C have the same sign because
we can then always make them both positive. (How?) In case
$A = C$, the locus reduces to a circle. Moreover, we can then divide
both sides of equation (11) by $A = C$. Hence we have the special
case:

*11.1.3   The locus of

(12) $x^2 + y^2 + Dx + Ey + F = 0$

is a circle with center $(-D/2, -E/2)$.

As in (6)–(8), the center may be found by completing the square in both x and y, and so putting (12) in the form

(13) $$(x - h)^2 + (y - k)^2 = r^2$$

where $h = -D/2$, $k = -E/2$, and $r^2 = -F + D^2/4 + E^2/4$. If the right member is zero, we have a "circle of zero radius" or **point circle** consisting of the point (h, k). If the right member is negative, there are no points whose coordinates satisfy the equation and we call the locus an **imaginary circle**.

EXERCISE 11.1

1. Plot the following ellipses, in each case finding several points in the first quadrant and making use of symmetry to find others:

(a) $\dfrac{x^2}{16} + \dfrac{y^2}{4} = 1$ (b) $\dfrac{x^2}{4} + \dfrac{y^2}{9} = 1$ (c) $\dfrac{x^2}{2} + \dfrac{y^2}{4} = 1$

2. Put in the form (1) and plot, identifying in each case the center, major and minor axes, semi-axes, and vertices. Where a translation is required, show both sets of axes.

(a) $25x^2 + 9y^2 = 225$ (g) $16(x - 1)^2 = 32 - 2(y + 4)^2$
(b) $x^2 + 4y^2 - 4 = 0$ (h) $2(x + 5)^2 + 6(y - 3)^2 = 9$
(c) $2x^2 + 8y^2 - 6 = 0$ (i) $x^2 + 100y^2 = 100$
(d) $3x^2 + 3y^2 - 7 = 0$ (j) $100x^2 + 121y^2 = 12100$
(e) $9(x - 3)^2 + 4(y - 1)^2 = 36$ (k) $3x^2 + 5y^2 = 0$
(f) $(x + 2)^2 + (y + 5)^2 = 4$ (l) $100x^2 + 18y^2 + 7 = 0$

3. In the following complete the square and put in the form (4), or identify as a point or imaginary ellipse. By an appropriate translation put in the form (1) and plot.

(a) $x^2 + y^2 + 2x + 2y - 1 = 0$ (g) $x^2 + 4y^2 + 2x - 24y + 41 = 0$
(b) $3x^2 + 3y^2 - 6x - 12y - 3 = 0$ (h) $x^2 + y^2 + 3x - 3y = 0$
(c) $4x^2 + 9y^2 - 8x + 36y + 4 = 0$ (i) $81x^2 + y^2 + 2y - 77 = 0$
(d) $4x^2 + y^2 + 8y - 3 = 0$ (j) $2x^2 + 5y^2 - x + 8y - 3 = 0$
(e) $x^2 + 2y^2 = -1$ (k) $2x^2 + 2y^2 - x + 3y - 5 = 0$
(f) $x^2 + y^2 = 0$ (l) $5x^2 + 7y^2 + 3x - 4y = 0$

4. Show that the equations:

$$x^2 + y^2 + Dx + F = 0 \quad \text{and} \quad x^2 + y^2 + Ey + F = 0$$

are the equations of circles having their centers on the x-axis and y-axis, respectively.

5. Find the equation of:

(a) A circle with center at (2, 3) and radius 3.

(b) A circle with center at (−1, 4) and diameter 5.

(c) An ellipse with center at the origin, horizontal axis 4, vertical axis 6.

(d) A circle of radius 2, tangent to both axes and having its center in the second quadrant.

(e) An ellipse with center at (0, 3), one vertex at (0, 7), and another at (−2, 3).

Answers. (a) $x^2 + y^2 - 4x - 6y + 4 = 0$. (c) $9x^2 + 4y^2 = 36$.

6. The student will observe from equation (12), and from *8.2.1, that there are three parameters in the general equation of a circle. Hence we need three conditions to determine a circle. For example, if (12) is to pass through the three points (1, 0), (0, 5), (2, 2) we must have (why?):

$$1^2 + 0^2 + D + 0 \cdot E + F = 0$$

(14)
$$0^2 + 5^2 + 0 \cdot D + 5E + F = 0$$

$$2^2 + 2^2 + 2D + 2E + F = 0$$

or

$$D \quad\quad + F = -1$$

(15)
$$5E + F = -25$$

$$2D + 2E + F = -8$$

These equations can be solved for D, E, and F. Similar methods apply when other conditions are given. One of the forms of the equation is chosen (*8.2.1 may be more convenient in some cases), the conditions are expressed as equations involving the parameters, and these equations are then solved. Similar remarks apply to determining the equation of an ellipse. Since four essential parameters are involved in (4) or (11) after dividing by a non-zero parameter, four conditions are required to determine an ellipse that has its axes parallel to the coordinate axes. Determine the equations of the following loci:

(a) The circle passing through the origin, (1, 0), and (1, 4).

(b) The circle passing through (2, −4), (0, 7), (−3, 6).

(c) The ellipse passing through (2, −4), (0, 0), (8, 0), (4, 2) and having axes parallel to the coordinate axes.

(d) The ellipse having center at (3, 7) and semi-axes of 3 and 2 parallel respectively to the x- and y-axes.

(e) The ellipse having vertices at (3, 14), (3, 6), (−2, 10), and (8, 10).

(f) The ellipse having center at (3, 5), axes parallel to the coordinate axes, and two vertices on the coordinate axes.

(g) The ellipse in standard position passing through (1, 2) and (5, −1). [*Hint.* Use (1) and solve for $1/a^2$ and $1/b^2$.]

(h) The circle with center at the origin and passing through (2, −3).

(i) The circle having as diameter the segment joining the origin to (−6, −4).

(j) The circle with center at $(-3, 2)$ and tangent to the x-axis.

(†k) The circle with center at $(5, 5)$ and tangent to $y = 2x$.

(†l) The inscribed and circumscribed circles of the triangle $(-1, -3)$, $(1, 5)$, $(3, 1)$.

Answers. (a) $x^2 + y^2 - x - 4y = 0$. (c) $6x^2 + 11y^2 - 48x + 26y = 0$.
(e) $16(x - 3)^2 + 25(y - 10)^2 = 400$. (g) $x^2 + 8y^2 = 33$. (i) $x^2 + 6x + y^2 + 4y = 0$.

†7. Show that the equation of the circle through (x_1, y_1), (x_2, y_2), (x_3, y_3) is given by:

$$\begin{vmatrix} x^2 + y^2 & x & y & 1 \\ x_1^2 + y_1^2 & x_1 & y_1 & 1 \\ x_2^2 + y_2^2 & x_2 & y_2 & 1 \\ x_3^2 + y_3^2 & x_3 & y_3 & 1 \end{vmatrix} = 0$$

†8. Show that $x = a \sin \theta$, $y = b \cos \theta$ are parametric equations of an ellipse. How do you interpret θ?

Simultaneous Equations

9. Finding the points where a straight line meets an ellipse is equivalent to solving simultaneous equations of the form $ax + by + c = 0$ and $Ax^2 + By^2 + Dx + Ey + F = 0$. The easiest method is to solve the linear equation for one of the variables and substitute in the other. If the line is tangent, there is only one solution. If it does not meet the ellipse, the solutions are imaginary. Find and sketch the intersections of the following pairs of curves:

(a) $x^2 + y^2 = 1$
$2x + y + 1 = 0$

(b) $x^2 + y^2 = 2$
$x - y = 1$

(c) $4x^2 + y^2 = 4$
$x = 1$

(d) $9x^2 + 16y^2 = 144$
$2x + 4y = 12$

(e) $x^2 + 4y^2 - 12x + y + 39 = 0$
$3x - y = 0$

(f) $x^2 + y^2 = 2$
$x + y = 2$

Answers. (a) $(0, -1)$, $(-4/5, 3/5)$. (c) $(1, 0)$.

10. Finding the intersection of two circles is equivalent to solving simultaneous equations of the form $x^2 + y^2 + Dx + Ey + F = 0$ and $x^2 + y^2 + D'x + E'y + F' = 0$. The best procedure is to subtract one equation from the other to get a linear equation and then work as in prob. 9. Solve and sketch the following:

(a) $x^2 + y^2 - 4 = 0$
$x^2 + y^2 - 2x - 3 = 0$

(b) $x^2 - 4x + y^2 = 0$
$x^2 - 4y + y^2 = 0$

(c) $x^2 + y^2 - 12x - 6y + 41 = 0$
$x^2 + y^2 - 6x - 6y = 17$

(d) $x^2 + 4x + y^2 + 2y + 4 = 0$
$x^2 + y^2 + 4x + 6y + 4 = 0$

(e) $x^2 + y^2 - 3x + 4y - 1 = 0$
$(x + 2)^2 + (y + 2)^2 = 1$

(f) $x^2 + y^2 = 100$
$x^2 + y^2 - 2x + 3y = 0$

11. Find and sketch the intersections of the following ellipses. (These are special cases in which the equations are linear in x^2 and y^2. In general two ellipses meet in four points and give rise to a fourth-degree equation.)

(a) $x^2 + 2y^2 = 4$
  $3x^2 + y^2 = 2$

(b) $x^2 + y^2 = 16$
  $81x^2 + y^2 = 81$

(c) $5x^2 + y^2 = 10$
  $x^2 + 5y^2 = 10$

(d) $100x^2 + y^2 = 100$
  $121x^2 + y^2 = 121$

12. Show by algebra that each of the following pairs of curves have no points in common. Check by sketching.

(a) $x^2 + y^2 = 2$
  $x^2 + y^2 = 4$

(b) $9x^2 + 4y^2 = 36$
  $4x^2 + y^2 = 4$

(c) $x^2 + y^2 = 1$
  $y = x - 3$

(d) $3x^2 = 48 - 16y^2$
  $x + y = 6$

13. Find the value(s) of k for which the following pairs of curves have only one common point.

(a) $2kx + y + 3 = 0$
  $x^2 + y^2 = 1$

(b) $x^2 + y^2 = 3$
  $x + ky + 2 = 0$

Answer for a. $\pm\sqrt{2}$.

14. Show analytically that two **concentric circles** (circles with the same center) either do not meet or are identical.

15. Show that the line $y = mx$ meets the ellipse (1) in two points equidistant from the origin, and hence that every diameter is bisected by the center of the ellipse.

Focal Properties

16. Prove:

*11.1.4 The locus of points the sum of whose distances from two fixed points is constant is an ellipse.

Suggestions. Place the axes so that the points lie at $(c, 0)$ and $(-c, 0)$. Let the sum be $2a$. (See Fig. 11.1.2.) Then a necessary and sufficient condition that (x, y) lies on the locus is:

(16) $\sqrt{(x - c)^2 + y^2} + \sqrt{(x + c)^2 + y^2} = 2a$

This is equivalent to

(17) $(x - c)^2 + y^2 = [2a - \sqrt{(x + c)^2 + y^2}]^2$

which reduces to

(18) $a^2 + cx = a\sqrt{(x + c)^2 + y^2}$

Squaring both sides of this and letting $b^2 = a^2 - c^2$ yields (1) as required.

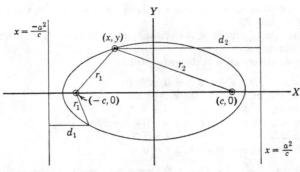

$$x = \frac{-a^2}{c}$$

(x, y)

d_2

r_1

r_2

r_1 $(-c, 0)$

$(c, 0)$

X

d_1

$$x = \frac{a^2}{c}$$

Fɪɢ. 11.1.2

17. The two points involved in *11.1.4 are called the **foci** of the ellipse, and their distance from the center is called the **focal distance.** The segments joining the foci to a point on the curve are called **focal radii.** Show that the foci always lie on the major axis and that $c = \sqrt{a^2 - b^2}$, where a and b are the semi-major and semi-minor axes. Find the foci of several of the ellipses considered above.

18. When the ellipse specializes to a circle, what happens to the foci? What is the focal distance for a circle?

19. Light (or sound) issuing from one focus of an ellipse is reflected by the curve to the other focus. This property is responsible for such effects as being able to hear at one focus of a large elliptically shaped room what is whispered at the other focus. The student may test it by constructing an elliptically shaped box with flat bottom and vertical sides. A marble projected from one focus will go to the other.

20. The property *11.1.4 can be used to construct an ellipse as follows. Place two pegs at a distance apart of $2c$. Construct a loop of string of length $2a + 2c$. Place the loop about the pegs and extend it as far as possible with the point of a pencil. Move the pencil, keeping the string taut, and so trace out an ellipse. Experiment with this method and explain why it works.

21. Show that if r_1 represents the distance from any point on the ellipse (1) to the focus $(-c, 0)$, and d_1 represents the distance between this point and the line $x = -a^2/c$, then $r_1/d_1 = e$, where $e = c/a$. The constant e is called the **eccentricity.** Note that $e < 1$. The line is called the **directrix.** (See Fig. 11.1.2.)

†**22.** Show that:

*11.1.5   The locus of a point whose distance from a fixed point divided by its distance from a fixed line is a constant less than 1 is an ellipse.

Suggestion. Place the axes so that the point is at $(-c, 0)$ and the line at $x = -d$, where $d > c > 0$. Then choose a so that $a^2/c = d$ and express the condition defining the locus. (See Fig. 11.1.2.)

†23. Show that the other focus $(c, 0)$ and the line $x = a^2/c$ also have the property of prob. 22.

24. In each of several of the ellipses plotted in probs. 1 to 3 draw the foci and directrices.

25. Find the equation of the ellipse with:

(a) Focus $(-3, 0)$, $e = 1/2$, center at $(0, 0)$.
(b) Focus $(5, 0)$, $e = 5/6$, center at $(0, 0)$.
(c) Focus at $(0, 2)$, directrix $y = 4$, $e = 2/3$.
(d) Focus at $(0, -3)$, directrix $y = 16/3$, $e = 3/4$.
(e) Focus $(2, 3)$, directrix $x = 4$, $e = 1/3$.
(f) Focus $(1, 4)$, directrix $y = 2$, $e = 0.9$.

Suggestions. See probs. 17 and 22 for relations among a, b, c, e. In c, use *11.1.5 and prob. 22 to write:

$$(19) \qquad \frac{\sqrt{(x - 0)^2 + (y - 2)^2}}{4 - y} = \frac{2}{3}$$

Then simplify and put in the form (11).

Answers. (a) $3x^2 + 4y^2 = 108$. (e) $8x^2 - 28x + 9y^2 - 54y + 101 = 0$.

26. Find the equations of the following loci. Note that they must be ellipses (why?), yet they do not fit into the cases so far considered.

(a) A point whose distance from the origin divided by its distance from the line $x - y + 1 = 0$ equals $\frac{1}{2}$.
(b) Points the sum of whose distances from $(3, 5)$ and $(-1, 4)$ is 6.
(c) A point whose distance from the point $(-4, -3)$ divided by its distance from the line $2x + y - 3 = 0$ is 0.3.
(d) Points the sum of whose distances from $(-1, 5)$ and $(3, -4)$ is 11.

†27. How does the ellipse change its shape as we hold a fixed and let e vary from 0 to 1? [*Suggestion.* For a convenient numerical value of a, plot various examples of (1) starting with $b = a$ and letting b go toward zero. In each case find e and the foci. Then consider c fixed and let a and b vary.]

28. A segment joining two points on an ellipse is called a **chord**. Diameters are special cases. A chord through a focus and perpendicular to the major axis is called a **latus rectum**. Find the length of the latus rectum of each of several ellipses in probs. 1 through 3.

29. Find a formula for the length of the latus rectum of (1). (*Answer.* $2b^2/a$, where a is the semi-major axis.)

30. Collect the formulas connecting a, b, c, e, the directrices, and the latus rectum of (1). How many independent relations are there? How many of the parameters have to be known to determine the ellipse? Find the equations of the following ellipses in standard position and sketch, showing foci, directrices, and latus rectum.

(a) Focus $(3, 0)$, semi-major axis 4.
(b) Focus $(-2, 0)$, semi-minor axis 2.
(c) Focus $(1, 0)$, eccentricity $1/2$.

(d) Distance between foci 5, eccentricity 2/3.

(e) Major axis 8, directrix $x = 5$.

(f) Focus $(4, 0)$, latus rectum 6.

(g) Directrix $x = -6$, eccentricity 1/3.

Answers. (a) $7x^2 + 16y^2 = 112$. (c) $3x^2 + 4y^2 = 12$. (e) $9x^2 + 25y^2 = 144$.

Families of Ellipses

31. Just as we considered families of lines in 9.4, so we may consider families of conics. For example, (1) is the equation of all ellipses in standard position. It contains two parameters and hence determines a two-parameter family. Write the equations of the following families:

(a) Circles with center at the origin.

(b) Circles with center at $(2, -0.1)$.

(c) Ellipses with centers along the line $x = -3$, one vertex on the y-axis, and a vertical semi-axis of 2.

(d) Ellipses with center at $(3, 2)$ and axes parallel to the coordinate axes.

(e) Circles through the points $(3, 1)$ and $(0, 2)$.

(f) Circles tangent to both axes and having their centers in the first and third quadrants.

(g) All circles tangent to both axes.

(h) Circles with centers on the line $3y + x - 6 = 0$ and having radii of 2.

(i) Ellipses in standard position with centers at $(0, 0)$ and major axis 6.

Suggestions for part e. Make use of the given conditions to try to determine the parameters in (12). From these express all parameters in terms of one of them.

Answers. (a) $x^2 + y^2 = r^2$. (c) $4(x + 3)^2 + 9(y - c)^2 = 36$. (g) $(x - c)^2 + (y \pm c)^2 = c^2$. (i) $b^2x^2 + 9y^2$ and $9x^2 + b^2y^2 = 9b^2$ with $b \leq 3$.

32. Show that:

*11.1.6 The equation

(20) $$k_1(x^2 + y^2 + Dx + Ey + F)$$
$$+ k_2(x^2 + y^2 + D'x + E'y + F') = 0$$

is the equation of the family of all circles and the one straight line through the intersections of the circles $x^2 + y^2 + Dx + Ey + F = 0$ and $x^2 + y^2 + D'x + E'y + F' = 0$.

Hint. See *9.4.1. The line is obtained by taking $k_2 = -k_1$. Note that all but the second circle itself are obtained with $k_1 = 1$ and all but the first with $k_2 = 1$.

33. Find, if possible, the circle(s) that passes through the intersection of $x^2 + y^2 - 9 = 0$ and $x^2 + y^2 - 2x + 4y - 11 = 0$ and: (a) Passes through the origin. (b) Passes through $(-6, -8)$. (c) Has its center on $x = 2$. (d) Has radius 10. (*e*) Has its center on $y = 6$. (*f*) Is tangent to the x-axis. (*g*) Is tangent to the y-axis. (*h*) Is tangent to $x = -7$. (*i*) Is tangent to $x - y = 8$.

34. Same as prob. 33 for each pair of circles in prob. 10.

35. Ellipses with the same foci are called **confocal.** Find the equation of the family of ellipses with foci at $(\pm c, 0)$, and draw several members for $c = 3$.

Radical Axes and Tangents

36. The common chord of two intersecting circles is called their **radical axis.** It is given by $f(x, y) - g(x, y) = 0$, where $f = 0$ and $g = 0$ are the equations of the circles in the form (12). Even when the circles do not intersect, the line obtained in this way is called the radical axis. (*a*) Find the radical axis of each pair of circles in prob. 10 and sketch. (†*b*) Show that the radical axis is perpendicular to the line of centers of the circles. (†*c*) Show that if the circles intersect, their line of centers is the perpendicular bisector of the segment joining their points of intersection.

†**37.** To find the line with a slope of 2 that is tangent to the circle $x^2 + y^2 = 4$, we consider the lines $y = 2x + b$, which intersect the circle in points whose x-coordinates are given by $5x^2 + 4bx + b^2 - 4 = 0$, whose roots are $x = (-2b \pm \sqrt{20 - b^2})/5$. A necessary and sufficient condition for tangency is that these roots should be equal, that is, that $b^2 = 20$. (Why?) Hence the two tangent lines are $y = 2x \pm 2\sqrt{5}$. In a similar way, prove:

*11.1.7   The lines with slope m that are tangent to the ellipse (1) are given by

$$(21) \qquad y = mx \pm \sqrt{a^2m^2 + b^2}$$

*11.1.8   The line tangent to the ellipse (1) at the point (x_0, y_0) is given by

$$(22) \qquad \frac{xx_0}{a^2} + \frac{yy_0}{b^2} = 1$$

38. Find the indicated tangents to the following ellipses and sketch. (*a*) To the unit circle with slope $\frac{1}{2}$. (*b*) To $25x^2 + 9y^2 = 225$ with slope 1. (*c*) To $x^2 + 4y^2 - 4 = 0$ at the points where $x = 1$. (†*d*) To $(x + 2)^2 + (y + 5)^2 = 4$ where $x = -2$. (†*e*) To $x^2 + y^2 + 4x + 4y - 1 = 0$ where $x = -1$.

Answers. (*a*) $2y = x \pm \sqrt{5}$. (*c*) $x \pm 2\sqrt{3}y = 4$.

39. Find the slopes of the tangent lines to the ellipse $x^2 + 4y^2 = 4$ at the points where $x = 1$.

40. The **normal** to a curve at a point is defined to be the line perpendicular to the tangent to the curve at the point. Show from *11.1.8 that the normal to the ellipse (1) at (x_0, y_0) is given by

$$(23) \qquad b^2x_0(y - y_0) = a^2y_0(x - x_0)$$

41. Find the normal line at the point of tangency of each tangent found in prob. 38.

†**42.** Prove that the radical axes determined by three circles are concurrent.

†**43.** Show that the length of the tangent lines from (x_1, y_1) to the circle $x^2 + y^2 + Dx + Ey + F = 0$ is given by $d^2 = x_1{}^2 + y_1{}^2 + Dx_1 + Ey_1 + F$. (See Fig. 11.1.3.)

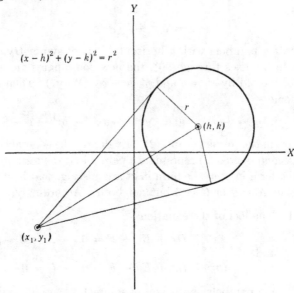

$$(x - h)^2 + (y - k)^2 = r^2$$

(h, k)

(x_1, y_1)

FIG. 11.1.3. (Prob. 43).

44. Use the result of prob. 43 to find the length of a tangent line from the following points to each of the circles in prob. 10. Sketch and interpret your results: (a) $(4, 5)$. (b) $(2, 0)$. (c) $(-1, -1)$. (d) $(0, 10)$.

†**45.** Prove that the radical axis is the locus of points from which the tangents to the two circles have equal lengths.

11.2 The parabola

A locus whose equation can be put in the form

$$(1) \qquad\qquad y = ax^2$$

is called a **parabola.** We graphed parabolas in 5.1 and showed in 5.5 that any equation of the form

$$(2) \qquad\qquad y = ax^2 + bx + c \qquad (a \neq 0)$$

can be put in the form (1). We call equation (1) the **standard form** and say that its locus is in **standard position.**

An equation of types (1) or (2) represents a parabola with vertical axis of symmetry. It is easy to see that an equation of the form

(3) $$x = ay^2$$

or

(4) $$x = ay^2 + by + c$$

represents a parabola with a horizontal axis of symmetry. For if we rotate the axes through 90°, the new coordinates are given in terms of the old by $x' = y$ and $y' = -x$. (Why?) Then (3) and (4) become:

(5) $$y' = -ax'^2 \quad \text{and} \quad y' = -ax'^2 - bx' - c$$

which are in the forms (1) and (2). Evidently the general equation of the second degree (I) represents a parabola if the term in xy *and either* the term in y^2 *or* the term in x^2 are missing, that is, if $B = 0$, and either $A = 0$ or $C = 0$ but not both. Accordingly,

*11.2.1   The loci of the equations

$$Ax^2 + Dx + Ey + F = 0 \qquad A \neq 0$$

and

$$Cy^2 + Dx + Ey + F = 0 \qquad C \neq 0$$

are parabolas with axes of symmetry parallel to the y- and x-axes, respectively.

An equation of the form $Ax^2 + Dx + Ey + F = 0$ may be treated as in 5.5 after solving for y. An equation of the form $Cy^2 + Dx + Ey + F = 0$ may be put in the form $Ax^2 + Dx + Ey + F = 0$ by rotating the axes, but it is easier to solve for x and then complete the square in y. The procedure is the same as for equations of type (2), but with x and y interchanged. Consider the equation:

(6) $$6y^2 - 3x + 24y + 18 = 0$$

or

(7) $$x = 2y^2 + 8y + 6$$

(8) $$x = 2(y^2 + 4y \quad) + 6$$

(9) $$= 2(y^2 + 4y + 4) + 6 - 8$$

(10) $$x + 2 = 2(y + 2)^2$$

A translation of the axes to the point $(-2, -2)$ yields the new equation:

(11) $$x' = 2y'^2$$

This is a parabola with vertex at the new origin. The x'-axis is its axis of symmetry and it extends only to the right of the y'-axis. In Fig. 11.2.1 we have graphed the locus, showing both old and new axes.

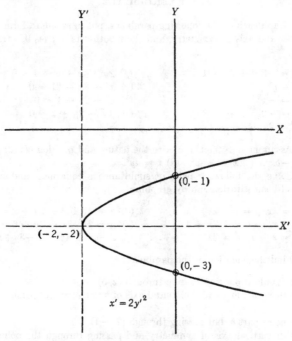

FIG. 11.2.1

It may happen that $E = 0$ in $Ax^2 + Dx + Ey + F = 0$ or that $D = 0$ in $Cy^2 + Dx + Ey + F = 0$. In that case the equation reduces to

(12) $$Ax^2 + Dx + F = 0 \quad \text{or} \quad Cy^2 + Ey + F = 0$$

and may always be written as

(13) $$A(x - x_1)(x - x_2) = 0 \quad \text{or} \quad C(y - y_1)(y - y_2) = 0$$

where x_1, x_2, y_1, y_2 are roots of the respective equations. The first equation in (13) represents the two straight lines $x = x_1$ and $x = x_2$, parallel to the y-axis. These lines are real and different, coincident, or imaginary according to the nature of the roots. Similar remarks apply to the second equation in (13). Thus when all terms in one variable are missing, the conic (I) "degenerates" into two parallel straight lines that we call a **degenerate parabola.**

EXERCISE 11.2

1. Graph each of the following parabolas, put in standard form, indicate the vertex and axis of symmetry, and show both sets of axes if a translation is used.

(a) $y = -3x^2 + 2x - 1$ (f) $2y^2 - x + y - 4 = 0$
(b) $x = y^2$ (g) $x^2 - 3x - 21 = 0$
(c) $x = -y^2$ (h) $y^2 - 2y + 1 = 0$
(d) $x = -2y^2 + 4y - 8$ (i) $3x^2 - 2y + x = 0$
(e) $y^2 = 2x + 3$

2. Assuming $a > 0$, characterize the nature and position of: (a) $y = ax^2$. (b) $y = -ax^2$. (c) $x = -ay^2$. (d) $x = ay^2$.

3. Solve the following pairs of simultaneous equations, and check the solutions by substitution and by graphing.

(a) $y = 3x$, $y = x^2$ (d) $y = 2x$, $3x^2 - 2y + x = 0$
(b) $y = x$, $y = x^2 + 2x + 1$ (e) $y = 2x^2$, $3x + 2y + 6 = 0$
(c) $y = x - 1$, $y^2 = 3x$ (f) $x - y = 2$, $4y^2 + 7x - 3y + 11 = 0$

4. Find the equation of the parabola:

(a) In standard position passing through $(2, 5)$.
(b) With vertex at $(0, 0)$, horizontal axis of symmetry and passing through $(2, 5)$.
(c) Same as part b, but passing through $(2, -1)$.
(d) With vertical axis of symmetry and passing through the points $(0, 1)$, $(-6, 3)$, and $(2, 4)$.

Answers. (a) $5x^2 - 4y = 0$. (c) $2y^2 - x = 0$.

5. Solve:

(a) $y = x^2 + 2$ (b) $y^2 = 2x$ (†c) $8y - x^2 = 0$
    $y = -2x^2$ $y^2 - x - 2 = 0$ $x - y^2 = 0$

Focal Properties

6. Prove:

*11.2.2   The locus of points equidistant from a fixed line and a fixed point is a parabola.

Suggestion. Place the axes so that the point lies at $(0, c)$ and the line is $y = -c$. (See Fig. 11.2.2.)

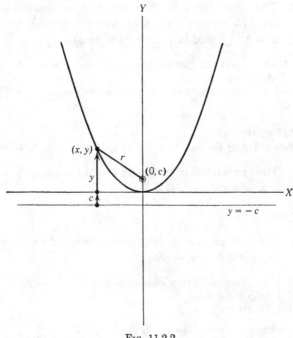

Fig. 11.2.2

7. The proof of *11.2.2 shows that $y = ax^2$ is the locus of points equidistant from $(0, 1/4a)$ and $y = -1/4a$. The point is called the **focus,** and the line is called the **directrix.** Find and sketch the focus and directrix of each of the parabolas in prob. 1.

8. Find the equation of the parabola with the given focus and directrix:

(a) $(0, 2)$, $y = -2$.
(b) $(0, 3)$, $y = -3$
(c) $(2, 0)$, $x = -2$
(d) $(0, -4)$, $y = 4$

(e) $(-5, 0)$, $x = 5$
(f) $(4, 3)$, $y = -2$
(g) $(2, -3)$, $x = 3$
(h) $(1, 4)$, $y = 2$

Answers. (a) $x^2 = 8y$. (c) $y^2 = 8x$. (e) $y^2 = -20x$. (g) $y^2 + 6y + 2x + 4 = 0$.

9. Find the equation of the parabola with: (a) focus $(0, 4)$, vertex $(0, 0)$; (b) focus $(3, 5)$, vertex $(1, 5)$.

10. The segment cut off by a parabola on a line through the focus perpendicular to the axis of symmetry is called the **latus rectum.** Find the length of this segment for each parabola in prob. 1. Show that the latus rectum of $y = ax^2$ has length $|a|^{-1} = 4|c|$, where c is the distance from the focus to the vertex.

11. Prove that:

*11.2.3   The equation $(y - k)^2 = 4c(x - h)$ represents a parabola with vertex at (h, k), focus $(h + c, k)$, directrix $x = h - c$, and latus rectum of length $4|c|$.

Note. Many authors call this and the special case $y^2 = 4cx$ the standard forms in place of our form (1).

12. Find the locus of points equidistant from the line $2x - y = 7$ and the point $(2, 2)$.

Tangents

13. Show either by the method of Ex. 11.1.37 or by using *5.4.3 that:

*11.2.4   The line with slope m tangent to the parabola $y = ax^2$ is $y = mx - (m^2/4a)$.

14. Prove:

*11.2.5   The line tangent to the parabola $y = ax^2$ at the point (x_0, y_0) is $y + y_0 = 2ax_0x$.

15. Find the equation of the normal to $y = ax^2$ at (x_0, y_0). [*Answer.* $x_0(x - x_0) + 2y_0(y - y_0) = 0$.]

16. Find and sketch:

(a) The tangent and normal to $y = 2x^2$ at $(2, 8)$.

(b) The tangent to $x^2 = 3y$ at the point where $x = 3$.

(c) The tangents to $y^2 = x$ at the points where $x = 4$.

(d) The tangent to $y = x^2 - 3x + 6$ where $x = 0$.

(e) The tangent to $y = 5x^2 - x + 1$ having a slope of 4.

(f) The tangent and normal to $y^2 + 3x - y + 3 = 0$ at $(-1, 0)$.

(g) The tangent to $y = 4x^2$ with slope of 4.

(h) The tangent to $x - y^2 + 2 = 0$ with slope of -1.

Answers. (a) $y = 8x - 8$, $x + 8y = 6$. (c) $x + 4y + 4 = 0$.

17. A line tangent to the parabola $y + 3x^2 = 0$ has slope of -0.25. Where does it meet the curve?

†18. Find the condition on the coefficients of $Ax + By + C = 0$ so that it is tangent to $y = ax^2$. (*Answer.* $A^2 = 4aBC$.)

19. Plot several members of the following families:

(a) $2ax + a^2y + 1 = 0$ (b) $c^2x + cy + 1 = 0$ (c) $y = x^2 - 2cx + 2c^2$

†20. In prob. 19 try to find a curve to which all members of the family appear to be tangent.

†21. The focus of a parabola has the interesting property that light from it is reflected by the parabola parallel to the axis of symmetry, and light coming into the parabola parallel to the axis of symmetry is reflected to the

focus. This property underlies the use of parabolic reflectors. To prove it
we must make use of the physical law that light is reflected from a smooth
surface so that the "angle of reflection" equals the "angle of incidence." (See
Fig. 11.2.3a.) We must show that a line from the focus to any point on the
parabola makes an angle with the normal at the point equal to the angle made
by a line parallel to the axis of symmetry. In Fig. 11.2.3b we have drawn
such a line from the focus $(c, 0)$ of the parabola $y = (1/4c)x^2$. The line QP
is parallel to the y-axis, BPR is the tangent, and PN is the normal. We wish
to show that $\angle CPN = \angle QPN$. This would be as good as done if we could
show that $\angle BPC = \angle QPR$. (Why?) But we know that $\angle QPR = \angle CBP$.

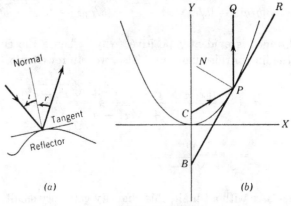

Fig. 11.2.3. (Prob. 21).

(Why?) Hence we need only find a way of showing that $\angle CBP = \angle BPC$,
or in other words that the triangle BPC is . . . (?) . . . . In order to do this
we may find the equation of the tangent line and discover where it meets the
y-axis. With these hints construct the proof.

11.3 The hyperbola

A locus whose equation can be put in the form

(1)
$$\frac{x^2}{a^2} - \frac{y^2}{b^2} = 1$$

is called a **hyperbola.** The x-intercepts are $(a, 0)$ and $(-a, 0)$,
but there are no y-intercepts. (Why?) We observe that the curve
is symmetric with respect to both axes and the origin. (Why?)
We find:

(2)
$$x = \pm(a/b)\sqrt{y^2 + b^2}$$
$$y = \pm(b/a)\sqrt{x^2 - a^2}$$

These equations show that there are two values of x for every value of y and two values of y for values of x such that $|x| > a$, whereas $|x| < a$ is an excluded region. We are going to show that the curve lies between the lines $y = \pm(b/a)x$ and approaches them as x and y approach infinity. (See Fig. 11.3.1.) Consider the line $y = (b/a)x$ and the branch of the curve in the first quadrant given by $y = (b/a)\sqrt{x^2 - a^2}$. For any value of x, the vertical directed distance from the curve to the line is:

$$(3) \quad d = [b/a]x - [b/a]\sqrt{x^2 - a^2} = [b/a][x - \sqrt{x^2 - a^2}\,]$$

This distance is evidently positive (why?); hence the curve lies below the line in this quadrant. Also, we can rewrite d as:

$$(4) \qquad d = \frac{[b/a][x - \sqrt{x^2 - a^2}\,][x + \sqrt{x^2 - a^2}\,]}{x + \sqrt{x^2 - a^2}}$$

$$(5) \qquad\quad = \frac{ab}{x + \sqrt{x^2 - a^2}}$$

As x increases without limit, this quantity gets very small; in fact, we can make it as small as we like by taking x large enough. This means that the vertical distance between the curve and the line approaches zero as x gets large. The other branches can be treated in a similar way.

In the hyperbola (1), the origin is called the **center,** the x-axis is called the **principal axis** or **transverse axis,** the y-axis is called the **conjugate axis,** the x-intercepts are called the **vertices.** The quantities $2a$ and $2b$ are also called the transverse and conjugate axes, and the quantities a and b are called the **semi-transverse** and **semi-conjugate axes.** The student should note that a may be greater or less than b; the conjugate axis is the one associated with the minus sign. The lines $y = \pm(b/a)x$ are called the asymptotes. (Compare 6.8.) We call (1) the **standard form** of the hyperbola and speak of the corresponding curve as being in **standard position.**

The treatment of the hyperbola is very similar to that of the ellipse. They are called **central conics,** since they possess a

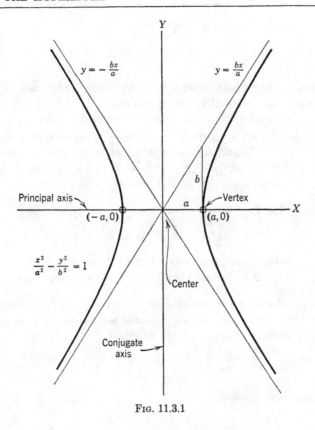

$$\text{F{\scriptsize IG}. } 11.3.1$$

center of symmetry. The student should note similarities and differences in order to avoid confusion. We have, corresponding to *11.1.1:

*11.3.1 The locus of

(6)
$$\frac{(x-h)^2}{a^2} - \frac{(y-k)^2}{b^2} = 1$$

is a hyperbola with center at (h, k) and axes parallel to the coordinate axes.

We leave it to the student to show that this is true and that the transverse and conjugate axes are $2a$ and $2b$.

It is easy to see that the equation

(7)
$$\frac{y^2}{a^2} - \frac{x^2}{b^2} = 1$$

represents a hyperbola, since a rotation through 90° reduces it to the form (1). Note that the transverse axis (which meets the curve) is associated with the positive term and the conjugate axis (perpendicular to it through the center) is associated with the negative term. We are led to suspect that an equation of the form $Ax^2 + Cy^2 + Dx + Ey + F = 0$ represents a hyperbola if A and C are of opposite sign, that is, if $A/C < 0$. The theorem is:

*11.3.2 If $A/C < 0$,

(8)
$$Ax^2 + Cy^2 + Dx + Ey + F = 0$$

is the equation of a hyperbola with center at $(-D/2A, -E/2C)$ and axes parallel to the coordinate axes.

The proof consists in showing as in 11.1 that the equation can be put in the form (6). If the constant term reduces to zero, the equation takes the form

(9)
$$\frac{(x-h)^2}{a^2} - \frac{(y-k)^2}{b^2} = 0$$

which can be written as

(10)
$$\left(\frac{x-h}{a} + \frac{y-k}{b}\right)\left(\frac{x-h}{a} - \frac{y-k}{b}\right) = 0$$

and hence represents two straight lines through the point (h, k) and having slopes $\pm b/a$. In this case we speak of a **degenerate hyperbola** consisting of two straight lines.

EXERCISE 11.3

1. Graph the following hyperbolas. (*Suggestion.* Draw the asymptotes, plot a few points in the first quadrant, and use symmetry to find other points.)

(a) $x^2 - y^2 = 1$

(b) $\dfrac{x^2}{4} - \dfrac{y^2}{9} = 1$

(c) $x^2 - 4y^2 = 0$

(d) $16x^2 - y^2 = 16$

(e) $y^2 - x^2 = 1$

(f) $25x^2 - 9y^2 = -1$

(g) $(x-3)^2 - 4(y+2)^2 = 4$

(h) $4(x+1)^2 - 9(y-4)^2 = 124$

2. Put the following equations in standard form and plot, showing both old and new axes.

(a) $x^2 - y^2 + 2x + 2y - 1 = 0$ (e) $5x - 2x^2 + y^2 - 3y = 4$
(b) $x^2 - 4y^2 - 8x - 32y - 3 = 0$ (f) $y^2 - 2x^2 + 16x + 2y - 35 = 0$
(c) $x^2 - y^2 - 2y + 4x + 3 = 0$ (g) $x^2 - y^2 + 2y = 1$
(d) $(x + 2)^2 - x - y^2 = 1$ (h) $x^2 - 4y^2 - 2x - 16y + 17 = 0$

Partial solution for part b:

(11) $(x^2 - 8x + \quad) - 4(y^2 + 8y + \quad) = 3$
(12) $(x^2 - 8x + 16) - 4(y^2 + 8y + 16) = -45$
(13) $4(y + 4)^2 - (x - 4)^2 = 45$ (See Fig. 11.3.2.)

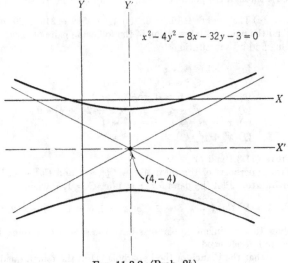

$$x^2 - 4y^2 - 8x - 32y - 3 = 0$$

$(4, -4)$

Fig. 11.3.2. (Prob. 2*b*).

3. Show that the asymptotes of (1) or (6) are given by the equation that results from replacing 1 by zero in the right member.

4. A hyperbola in which $a = b$ is called **equilateral** or **rectangular**. It occupies a position among hyperbolas somewhat similar to that occupied by the circle among ellipses. (Why?) Plot rectangular hyperbolas for $a = 1, 2, 4$.

5. Find and graph the equation of the hyperbola that has its axes parallel to the coordinate axes and:

(a) Its center at the origin, horizontal transverse axis of length 4 and conjugate axis of length 8.

(b) Is rectangular, has center at the origin, and a vertex at $(2, 0)$.

(c) Center at $(3, -5)$, one vertex at $(0, -5)$, and semi-conjugate axis of 2.

(d) Center at $(-2, -6)$, one vertex at $(-2, -3)$, and one asymptote with slope 2.

Answers. (a) $4x^2 - y^2 = 16$. (c) $4x^2 - 9y^2 - 24x - 90y = 225$.

6. The student will observe that there are four essential parameters in (6) or (8). Hence four conditions are required to determine a hyperbola whose axes are horizontal and vertical. Determine the hyperbola that is so oriented and:

(a) Has center at the origin and passes through the points $(1, 3)$ and $(-4, -8)$.

(b) Same as part a but center at $(-1, -1)$.

(c) Passes through the points $(0, 1)$, $(0, 5)$, $(2, 0)$, $(-2, 6)$.

(d) Has center at $(2, -5)$, vertical principal axis of length 6, and passes through the origin.

(e) Passes through the points $(0, -2)$, $(1, -3)$, $(1, 4)$, $(-2, 6)$.

Answers. (a) $11x^2 - 3y^2 + 16 = 0$. (c) $4y^2 - 5x^2 - 24y + 20 = 0$.

7. Find the points of intersection of the following pairs of curves. Check by graphing and by substitution.

(a) $y = 3x$, $x^2 - y^2 = 4$

(b) $3x^2 + 2y^2 = 6$, $x^2 - y^2 = 1$

(c) $x + 2y - 6 = 0$, $x^2 - y^2 = 0$

(d) $x^2 - 4y^2 - 2x - y = 0$, $x + 2y - 4 = 0$

(e) $x^2 - 4y^2 = 4$, $y^2 - 4x^2 = 4$

(f) $x^2 + y^2 = 9$, $2x^2 - 9y^2 = 18$

8. Prove *11.3.1 and *11.3.2.

9. Two hyperbolas of the form $Ax^2 - Cy^2 = 1$ and $Cy^2 - Ax^2 = 1$ are called **conjugate**. Plot the pairs for $A = 1/a^2$, $C = 1/b^2$ when:

(a) $a = 1 = b$　(b) $a = 1, b = 2$　(c) $a = 3, b = 1$　(d) $a = 2, b = 3$

10. Show that conjugate hyperbolas have the same asymptotes, but that their axes are interchanged.

11. Show that the branch of the hyperbola (1) in the fourth quadrant lies above the asymptote there and approaches it as x approaches infinity.

12. Show that the asymptotes of a rectangular hyperbola are perpendicular.

13. Find in terms of a and b a formula for the angle between the asymptotes of a hyperbola. (*Answer.* $\tan^{-1}[2ab/(a^2 - b^2)]$.)

†**14.** Find the equations of the hyperbolas with asymptotes $y - 2x = 4$ and $y + 2x = 0$ and having a transverse axis of 8. [*Answer.* $4(x + 1)^2 - (y - 2)^2 = 64$ and $(y - 2)^2 - 4(x + 1)^2 = 16$.]

15. Show that $x = a \sec \theta$, $y = b \tan \theta$ are parametric equations of a hyperbola. Suggest another pair giving the same curve.

16. Find the equation and draw a few members of the families of hyperbolas with horizontal principal axes and:

(a) With center at $(0, 0)$ and vertices at $(\pm 3, 0)$.

(b) With center at the origin and perpendicular asymptotes.

(c) With centers on $y = x$, perpendicular asymptotes, and one asymptote passing through the origin.

(d) Passing through the origin and the point $(1, 1)$.

17. Show that the equation of the rectangular hyperbola $x^2 - y^2 = a^2$ goes into $x'y' = 0.5a^2$ when the axes are rotated through $45°$.

Focal Properties

†**18.** Prove the following theorem, which corresponds to *11.1.4.

*11.3.3 The locus of points the difference of whose distances from two fixed points is a constant is a hyperbola.

Suggestion. Place the axes so that the points are at $(c, 0)$ and $(-c, 0)$, and take the difference to be $2a$. Compare Ex. 11.1.16.

19. The two points in *11.3.3 are called the **foci**, and the segment joining a focus to a point on the curve is called a **focal radius**. Show that the foci always lie on the transverse axis and that $c^2 = a^2 + b^2$. Compare with the relation for the ellipse. Find the foci of several of the hyperbolas already graphed.

20. The chord through a focus perpendicular to the transverse axis is called a **latus rectum**. Find a formula for its length when the equation is in standard form (1). Find its length for several hyperbolas and sketch. (*Answer.* $2b^2/a$.)

21. Show that if r_1 is the distance from any point on the hyperbola (1) to the focus $(-c, 0)$ and d_1 is the distance from this point to the line $x = -a^2/c$, then $r_1/d_1 = e$, where $e = c/a$. (Compare Ex. 11.1.21.) The constant e is called the **eccentricity**. Note that $e > 1$. (Why?) The line is called the **directrix.**

†**22.** Show that:

*11.3.4 The locus of a point whose distance from a fixed point divided by its distance from a fixed line is a constant greater than 1 is a hyperbola.

Suggestion: Choose the axes so that the point is at $(-c, 0)$ and the line is given by $x = -d$, where $c > 0$ and $d > 0$. Then choose a so that $d = a^2/c$. Compare Ex. 11.1.22.

†**23.** Show that the other focus $(c, 0)$ and the line $x = a^2/c$ serve equally well as focus and directrix.

24. In each of several hyperbolas already plotted locate the foci and directrices.

25. Find the equation of the locus of points:

(*a*) the difference of whose distances from the points $(3, 0)$ and $(-3, 0)$ is 4.

(*b*) the difference of whose distances from the points $(1, 1)$ and $(3, 6)$ is 6.

(*c*) such that the distance to the point $(0, 4)$ equals twice the distance from the line $x = 2$.

(*d*) such that the distance from the point $(-1, -2)$ is three times the distance from the line $3x - y - 2 = 0$.

Answers. (*a*) $5x^2 - 4y^2 = 20$. (*c*) $3x^2 - y^2 - 16x + 8y = 0$.

26. Collect the formulas connecting a, b, c, e, the directrices, asymptotes, and latus rectum of the hyperbola in standard form. Compare with the corresponding relations for the ellipse. Find the equations of the hyperbolas in standard position with the following properties:

(a) Focus (4, 0), transverse axis 6.
(b) Directrix $x = 2$, focus (3, 0).
(c) Eccentricity 2, focus (1, 0).
(d) Conjugate axis 6, asymptotes $x^2 - 9y^2 = 0$.
(e) Latus rectum 4, eccentricity 3.
(f) Passing through (1, 4) and (2, −6).
(g) $e = 2$, vertex at (5, 0).

Answers. (a) $7x^2 - 9y^2 = 63$. (c) $12x^2 - 4y^2 = 27$. (e) $16x^2 - 2y^2 = 1$. (g) $3x^2 - y^2 = 75$.

27. Show that a circle of radius c with center at the center of a hyperbola passes through the vertices of a rectangle with sides $2a$ and $2b$, center at the origin, and sides parallel to the axes.

Tangents

†**28.** Prove that:

*11.3.5** The tangent line to (1) at the point (x_0, y_0) is given by

$$(14) \qquad \frac{xx_0}{a^2} - \frac{yy_0}{b^2} = 1$$

29. Find the normal to (1) at (x_0, y_0).
30. Prove:

*11.3.6** The lines with slope m that are tangent to (1) are given by

$$(15) \qquad y = mx \pm \sqrt{a^2 m^2 - b^2}$$

31. Show that, if $m < b/a$, there is no tangent of the form (15). Explain this in terms of Fig. 11.3.1.
32. Find the following lines: (a) Tangent to $x^2 - y^2 = 19$ at $(3, \sqrt{10})$. (b) Tangent to $5x^2 - 2y^2 = 50$ at the vertex where $x > 0$. (c) Tangent to $x^2 - 4y^2 - 1 = 0$ at the point where $x = 2$, $y > 0$. (d) Tangent to $x^2 - y^2 = 1$ with slope of 2. (e) Tangent to $16x^2 - 9y^2 = 1$ with slope 5. (f) Tangent to $x^2 - 36y^2 - 2x + 124y - 159 = 0$ with slope of 1. (†g) Tangent to $x^2 - y^2 = 4$ and passing through the origin.

Answers. (a) $3x - \sqrt{10}y = 1$. (c) $2x - 2\sqrt{3}y = 1$.

11.4   The general conic

In this section we prove and apply the following theorem:

*11.4.1** Every conic is an ellipse, parabola, hyperbola, or one of their degenerate forms.

As the student will recall from the discussion at the beginning of the chapter, a conic is simply a graph of an equation of the form

(I) $$Ax^2 + Bxy + Cy^2 + Dx + Ey + F = 0$$

where we do not have $A = B = C = 0$. We have already proved the theorem for the cases in which $B = 0$, since we have considered all these possibilities in the last three sections. (See *11.1.2, *11.2.1, *11.3.2 and the discussion of degenerate cases.)

A simple example of the case $B \neq 0$ is

(1) $$xy = 1$$

Its graph (Fig. 6.8) looks something like one of the hyperbolas of 11.3 tilted at an angle of $45°$. We conjecture that the equation (1) reduces to standard form if we rotate the axes through $45°$. Under this rotation, new and old coordinates are related by

(2)
$$x = x'(1/\sqrt{2}) - y'(1/\sqrt{2}) = \frac{(x' - y')}{\sqrt{2}}$$
$$y = \frac{(x' + y')}{\sqrt{2}}$$

Substituting in (1) and simplifying, we find:

(3) $$\frac{x'^2}{2} - \frac{y'^2}{2} = 1$$

which we recognize as the equation of a hyperbola in standard form relative to the new axes.

We now prove that the xy term may be eliminated from the equation of *any* conic by an appropriate rotation. This will complete the proof of *11.4.1. (Why?) If the axes are rotated through an angle θ, (I) becomes

(II) $$A'x'^2 + B'x'y' + C'y'^2 + D'x' + E'y' + F' = 0$$

where

(4) $$B' = -2A \sin \theta \cos \theta + B(\cos^2 \theta - \sin^2 \theta) + 2C \sin \theta \cos \theta$$

This expression can be verified by substituting *9.1.1 in (I). Since

(5) $$B' = (C - A)2 \sin \theta \cos \theta + B(\cos^2 \theta - \sin^2 \theta)$$

(6) $$= (C - A) \sin 2\theta + B \cos 2\theta$$

it is zero if

(7) $$(A - C) \sin 2\theta = B \cos 2\theta$$

or

(8) $$\tan 2\theta = \frac{B}{A - C}$$

For all possible values of A, B, and C there is a value of 2θ between $0°$ and $180°$ and hence a value of θ between $0°$ and $90°$ that satisfies this equation. (Why?) Hence by rotating through this angle we can eliminate the xy term.

As an example, consider the conic

(9) $$2x^2 + xy + y^2 - 1 = 0$$

Here $B/(A - C) = 1$. Hence $\tan 2\theta = 1$, $2\theta = 45°$, and $\theta = 22.5°$. We could look up in the table approximate values of $\sin \theta$ and $\cos \theta$, but since $\tan 2\theta = 1$, $\cos 2\theta = 1/\sqrt{2}$ (why?), and from *8.7.9 and *8.7.10

(10)
$$\sin \theta = \sqrt{\frac{1 - \frac{1}{\sqrt{2}}}{2}} = \frac{\sqrt{2 - \sqrt{2}}}{2}$$
$$\cos \theta = \frac{\sqrt{2 + \sqrt{2}}}{2}$$

Replacing x and y in (9) by their values given by *9.1.1, we find

(11) $$(3 + \sqrt{2})x'^2 + (3 - \sqrt{2})y'^2 = 2$$

which represents an ellipse in standard position with respect to the new axes at an angle of $22.5°$ with the old.

It is clear from the example that the computations involved in simplifying the general conic may be tiresome. Hence it is sometimes convenient to eliminate linear terms by translation before trying to eliminate the xy term by rotation. In order to show how this is done, we subject (I) to the general translation $x = x' + h$, $y = y' + k$. The new equation is

(III) $Ax'^2 + Bx'y' + Cy'^2$

$\qquad + (2hA + Bk + D)x' + (2kC + Bh + E)y'$

$\qquad\qquad + Ah^2 + Bhk + Ck^2 + Dh + Ek + F = 0$

We can eliminate the linear terms by choosing h and k so that

(12)
$$2Ah + Bk = -D$$
$$Bh + 2Ck = -E$$

These equations have a solution in h and k unless $B^2 - 4AC = 0$.
(Why?) We shall show in the next section that $B^2 - 4AC = 0$ if
and only if the conic is a parabola, and in that case we do not
expect to get rid of the linear terms. (Why?) Hence for central
conics it is possible first to eliminate the linear terms by translation,
then the xy term by rotation. After translation to (h, k) given by
(12), the equation (III) takes the form

(13) $$Ax'^2 + Bx'y' + Cy'^2 + F'' = 0$$

It can now be put in standard form by rotating about the new
origin, (h, k), since rotation can eliminate the xy term but cannot
introduce linear terms. It follows that (h, k) given by (12) is the
center of the conic.

EXERCISE 11.4

1. Carry out the manipulations to find equations (2), (3), (4), (8), (10),
(11), and (III). Explain why $B^2 - 4AC \neq 0$ is a necessary and sufficient
condition for the equations (12) to be solvable for a unique h, k. Graph (9)
by using (11).

2. Treat (1) by rotating the axes through $-45°$. Draw old and new
axes and sketch the curve.

3. Treat (9) by rotating through $-67.5°$ and graph.

4. Find the center and eliminate linear terms from the following:

(a) $2x^2 + 3xy + y^2 - 4x - y - 6 = 0$
(b) $x^2 - xy - 5y^2 + x - 4y + 3 = 0$
(c) $5x^2 + xy - 2y^2 + x + 2y = 0$
(d) $3x^2 + 2x + y + y^2 - 5xy + 10 = 0$
(e) $x^2 + 2ax + y^2 + 2by + c = 0$
(f) $x^2 - y^2 + 3xy + x = 5$
(g) $xy - 2y^2 + x^2 - x + 10 = 2y$
(h) $2x^2 - 3axy + Sy^2 - 2x + by - c = 0$

Partial answers. (a) $F'' = 0$. (c) $F'' = 656/41^2$. (e) $F'' = c - a^2 - b^2$.

5. Eliminate the xy term from the following:

(a) $xy = -2$ (d) $5x^2 + 4xy + 2y^2 = 0$
(b) $x^2 + 3xy + y^2 = 0$ (e) $x^2 - 3xy - 3y^2 = 5$
(c) $x^2 + 2xy + y^2 - x + y = 0$ (f) $6x^2 + 12xy + 7y^2 = 0$

Answers. (a) $y'^2 - x'^2 = 4$. (c) $\sqrt{2}x'^2 + y' = 0$. (e) $3y'^2 - 7x'^2 = 10$.

6. Reduce to standard form and graph. Show all three sets of axes.

(a) $5y^2 + 2xy + 5x^2 - 12y - 12x = 6$

(b) $2y^2 - 4xy + 2y^2 + 8y = 10$

(c) $x^2 + 2xy - y^2 + 4x + y - 5 = 0$

(d) $x^2 - 2xy + 3y^2 + 6x - 4 = 0$

Answers. (a) $3x''^2 + 2y''^2 = 9$. (c) $x''^2 - y''^2 = 63/8\sqrt{2}$.

7. Find D', E', F' in (II).

8. Graph conics in probs. 4, 5, and 6 by solving the equation for one of the variables in terms of the others and substituting convenient values. (Plotting a few points in this way is a good method of checking the graph plotted by other means.)

9. Solve simultaneously and sketch. (Use your ingenuity to eliminate one of the variables by manipulating the two equations. Check your results by comparison with the graph.)

(a) $3x - 2y = 5$
  $x^2 - xy = 5 - y$

(b) $9x^2 + xy + 8y^2 = 0$
  $2y + 4x - 1 = 0$

(c) $9x^2 + 4y^2 = 36$
  $4x^2 + 9y^2 = 36$

(d) $x^2 - y^2 = 1$
  $x^2 + y^2 = 4$

(e) $4x^2 + y^2 = 1$
  $x^2 - 3y^2 = 6$

(f) $xy = 4$
  $x^2 - y^2 = 2$

(†g) $2x^2 + 3xy + y^2 = 2$
  $x^2 + 2xy = y^2 + 1$

(h) $x^2 + y^2 = 9$
  $xy - 1 = 0$

10. Find the conic through the following points: $(1, 0)$, $(3, 5)$, $(-2, 1)$, $(5, -10)$, $(-4, -6)$. What is it?

11. The sum of the squares of two numbers is 100 and their product is -6. Find them and sketch the equations and solutions.

12. From (8) find a formula for $\tan \theta$ in terms of A, B, and C. (*Answers.* $[C - A \pm \sqrt{(C - A)^2 + B^2}]B^{-1}$.)

13. If an expression of the form $a + b\sqrt{c}$, where a, b, and c are rational, has a square root of the form $x + y\sqrt{c}$, where x and y are rational, it may be found as follows. Let $x + y\sqrt{c} = \sqrt{a + b\sqrt{c}}$. Then $x^2 + cy^2 + 2xy\sqrt{c} = a + b\sqrt{c}$. Hence $x^2 + cy^2 = a$ and $2xy = b$. (Why?) These equations either yield the desired result or show that there is none of the desired form. Find, if possible, the square roots of the following. Check by squaring.

(a) $9 + 4\sqrt{2}$

(b) $12 - 6\sqrt{3}$

(c) $4 + \sqrt{15}$

(d) $5 - 2\sqrt{6}$

(e) $7 - 2\sqrt{6}$

(f) $(21/4) - \sqrt{5}$

†14. Express the following in the form $\sqrt{x} \pm \sqrt{y}$, where x and y are rational, or show that this is impossible.

(a) $\sqrt{8 - \sqrt{15}}$

(b) $\sqrt{6 + 2\sqrt{5}}$

(c) $\sqrt{3 + 2\sqrt{6}}$

(d) $\sqrt{1 - \sqrt{2/3}}$

(e) $\sqrt{1.25 + \sqrt{0.5}}$

(f) $\sqrt{2\sqrt{10} + 7}$

†15. Investigate the family:

$$k_1(Ax^2 + Bxy + Cy^2 + Dx + Ey + F)$$
$$+ k_2(A'x^2 + B'xy + C'y^2 + D'x + E'y + F') = 0$$

†16. Find a formula for the center of the general conic (I) in terms of the coefficients.

11.5 Invariants

We have seen how translation and rotation of the axes change the equation of a curve, but, since the curve itself is not modified, we would expect to find that some characteristics of the equation are unaltered. The equations before and after the shifts in the axes are:

(I) $\qquad Ax^2 + Bxy + Cy^2 + Dx + Ey + F = 0$

(II) $\quad [A \cos^2 \theta + B \sin \theta \cos \theta + C \sin^2 \theta]x'^2$

$\qquad + [(C - A) \sin 2\theta + B \cos 2\theta]x'y'$

$\qquad + (A \sin^2 \theta - B \sin \theta \cos \theta + C \cos^2 \theta)y'^2$

$\qquad + (D \cos \theta + E \sin \theta)x' + (-D \sin \theta + E \cos \theta)y' + F = 0$

(III) $\quad Ax'^2 + Bx'y' + Cy'^2 + (2Ah + Bk + D)x'$

$\qquad + (2Ck + Bh + E)y' + Ah^2 + Bhk$

$\qquad + Ck^2 + Dh + Ek + F = 0$

We are looking for some features of (I) that remain in (II) and (III). Such features we call **invariants** under translation and rotation. The first and most obvious invariant is the degree of the equation. Next we note that the coefficients of x^2, xy, and y^2 are invariant under translation. They do change under rotation, but, denoting the coefficients in (II) by primed letters, we note that:

(1) $\qquad A' + C' = A(\cos^2 \theta + \sin^2 \theta) + C(\sin^2 \theta + \cos^2 \theta)$

$\qquad\qquad = A + C$

Thus, $A + C$, the sum of the coefficients of x^2 and y^2, is an invariant!

Two other important invariants are:

(2) $\Delta = B^2 - 4AC$

(3) $R = \frac{1}{2} \begin{vmatrix} 2A & B & D \\ B & 2C & E \\ D & E & 2F \end{vmatrix}$

The invariant Δ (read "delta") is called the **discriminant**. Note
that $-\Delta$ is a minor of the determinant that gives R. It is a good
exercise in algebraic manipulation to prove that these are invariants
by substituting the coefficients from (II) and (III) in (2) and (3)
and showing that the resulting expressions reduced again to (2)
and (3). (These invariants would not be easy to discover from
scratch! The methods for finding invariants and finding all of
them for any particular curve are known, but they involve algebraic
theories as yet unfamiliar to the student.)

Since we are interested in the relation between these invariants
and the nature of the conic, we calculate their values for the conics
in standard form. We find for the ellipse given by 11.1(1):

(4) $A + C = 1/a^2 + 1/b^2;$ $\Delta = -4/a^2b^2;$ $R = -4/a^2b^2$

For the parabola, we find from 11.2(1) written as $ax^2 - y = 0$:

(5) $A + C = a;$ $\Delta = 0;$ $R = -a$

For the hyperbola, we find from 11.3(1):

(6) $A + C = 1/a^2 - 1/b^2;$ $\Delta = 4/a^2b^2;$ $R = 4/a^2b^2$

For the degenerate cases, the values of $A + C$ and Δ are the
same, but R is $4/a^2b^2$ for an imaginary ellipse, 0 for a point ellipse,
0 for a degenerate parabola, and 0 for a degenerate hyperbola.
The student should check these statements by references to 11.1,
11.2, and 11.3.

Remembering that these functions of the coefficients are invariants, we see that:

*11.5.1   The invariant R is zero if and only if the conic consists
            of a point or two distinct or coincident straight lines.

In order to prove this we need only observe that R cannot be
zero in the non-degenerate cases listed in (4)–(6) and that it is

zero in the cases listed in the theorem. Also if $R > 0$ for an ellipse, the locus is imaginary. Hence by calculating R for any conic, whether in standard form or not, we can tell whether it is degenerate.

We observe that Δ is negative, zero, or positive according to whether the curve is an ellipse, parabola, or hyperbola. Hence we can tell the kind of curve represented by a second-degree equation without plotting or simplifying. We have:

*11.5.2 The invariant Δ is negative, zero, or positive according as the conic is an ellipse, parabola, or hyperbola.

Now we note that if $A + C$ and Δ are known, we can find a and b from (4), (5), and (6). Hence we may be able to graph a conic without going to the trouble of making any of the awkward substitutions involved in rotating the axes! We introduce the method with an example.

Consider:

$$(7) \qquad x^2 + 3xy + 3y^2 + 10x - 15y + 24 = 0$$

We find that $A + C = 4$, $\Delta = -3 < 0$, and $R < 0$. Hence the equation represents a non-degenerate ellipse. Now we translate so as to get rid of the terms in x and y. The equations 11.4(12) are:

$$(8) \qquad \begin{aligned} 2h + 3k &= -10 \\ 3h + 6k &= 15 \end{aligned}$$

The solution $(-35, 20)$ gives the center. To find the new equation after translation, we do not need actually to substitute, for a glance at (III) indicates that the first three coefficients do not change and the constant term is just the result of substituting (h, k) for (x, y) in the original equation. (See also Ex. 11.5.4.) Hence the new equation with respect to the center as origin is:

$$(9) \qquad x'^2 + 3x'y' + 3y'^2 - 301 = 0$$

The angle through which these axes must now be rotated to put the equation in standard form is given by

$$(10) \qquad 2\theta = \arctan(-3/2) = 123°\,41'$$

$$(11) \qquad \theta = 61°\,51'$$

We do not need to carry out the substitution involved in the rotation, for we notice in (II) that rotation does not change the constant term. Also, if we let primed letters stand for the coefficients after rotation, we have $B' = 0$, and

(12)
$$A' + C' = 4 = A + C$$
$$-4A'C' = -3 = \Delta$$

We can solve these two equations by solving the first for C' and substituting in the second. The result is:

(13) $$4A'^2 - 16A' + 3 = 0$$

The roots are:

(14) $$A' = \frac{4 \pm \sqrt{13}}{2} \qquad C' = \frac{4 \mp \sqrt{13}}{2}$$

The roots are paired as indicated by the signs before the radicals. The fact that we get two pairs of values is not surprising since we could eliminate the xy term by rotating through either θ or $\theta + 90°$.

One of the two pairs corresponds to our rotation, but which one? In order to decide we must make use of the fact that we choose θ as the solution less than $90°$ of the equation $\tan 2\theta = B/(A - C)$. Looking at (II), we note that

(15) $$A' - C' = (A - C) \cos 2\theta + B \sin 2\theta$$

$$= B \left[\frac{(A - C) \cos 2\theta}{B} + \sin 2\theta \right]$$

(16) $$= B \left[\frac{\cos^2 2\theta}{\sin 2\theta} + \sin 2\theta \right] \qquad \text{(Why?)}$$

But since $0 < 2\theta < \pi$, $\sin 2\theta > 0$ and the coefficient of B is positive. Hence B and $A' - C'$ must have the same sign, and we choose the pair of solutions with this feature. Here $B = 3 > 0$ and

$$\left[\frac{4 + \sqrt{13}}{2} \right] - \left[\frac{4 - \sqrt{13}}{2} \right] = \sqrt{13} > 0 \quad \text{whereas} \quad \left[\frac{4 - \sqrt{13}}{2} \right]$$

$-\left[\dfrac{2+\sqrt{13}}{2}\right] = -\sqrt{13} < 0.$ Hence the proper pair is $A' =$
$\dfrac{4+\sqrt{13}}{2}$ and $C' = \dfrac{4-\sqrt{13}}{2}$, and the equation becomes

(17) $\qquad (4+\sqrt{13})x''^2 + (4-\sqrt{13})y''^2 = 602$

This equation is in the standard form with $a^2 = 602/(4+\sqrt{13})$, $b^2 = 602/(4-\sqrt{13})$; $a \doteq 8.9$ and $b \doteq 39$.

To graph the original equation (7) we locate the new origin at $(-35, 20)$ and the new axes at an angle of $61° 50'$ with the original

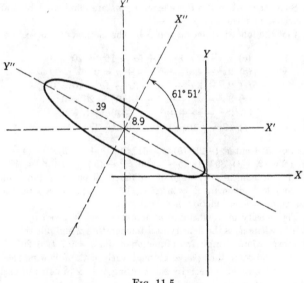

Fig. 11.5

axes. Then on these axes we plot an ellipse with x-intercepts $(\pm 8.9, 0)$ and y-intercepts $(0, \pm 39)$ (Fig. 11.5).

The foregoing method is suitable only for central conics. For a parabola, when the xy-term is eliminated by rotation, either the x^2 or y^2 term drops out also, and the computations are not unreasonably tedious.

EXERCISE 11.5

1. Calculate $A + C$, and R for several conics in the previous four sections, and verify that *11.5.1 and *11.5.2 hold.

2. Verify (II) and (III).

3. In each of the following show that $R = 0$ and graph the locus. (*Note.* Degenerate hyperbolas and parabolas may be dealt with by factoring. For the point ellipse, the center is the only point that lies on the locus.)

(a) $x^2 + 2xy + y^2 - x - y = 0$ (e) $3x^2 + 4xy + x + y^2 + y = 0$
(b) $x^2 + 3xy + 2y^2 + x - 2 = 0$ (f) $x^2 - y^2 + 2(x + y) = 0$
(c) $x^2 + y^2 - 2x + 4y + 5 = 0$ (g) $y^2 + 4y + 3 = 0$
(d) $x^2 - y^2 - 2x - 4y - 3 = 0$ (h) $xy + y - x = 1$

Answers.   (a) $(x + y)(x + y - 1) = 0$.   (c) $(1, -2)$.   (e) $(x + y)$ $(3x + y + 1) = 0$.

†**4.** Show that when the first-degree terms are eliminated by translation the new constant term is given by $-R/\Delta$.

5. Plot the following central conics by the method of this section:

(a) $x^2 + 3xy + 3y^2 + 5x - 15y + 20 = 0$
(b) $2x^2 - 4xy - y^2 - x + y = 0$
(c) $2x^2 + 3y^2 - 5xy + 3x - 4y + 10 = 0$
(d) $8y^2 - xy + 6x^2 - x + 8y - 20 = 0$
(e) $3x^2 - xy + 4y^2 - x - 10y + 3 = 0$
(f) $10x^2 - 10xy - y^2 + 14x - 3y = 0$

Answers. (a) Center: $(-25, 15)$, $\theta = 61° 51'$, $(4 + \sqrt{13})x''^2 + (4 - \sqrt{13})y''^2 = 310$. (c) $(-2, -1)$, $39° 21'$, $(5 - \sqrt{26})x''^2 + (5 + \sqrt{26})y''^2 = 18$.

6. The mechanical advantage of a differential pulley of inner and outer radius r and R is given by $A = 2R(R - r)^{-1}$. Graph A as a function of R for fixed r, and A as a function of r for fixed R.

7. The density of a substance at temperature t is given by $d = d_0[1 + \beta(t - t_0)]^{-1}$, where d_0 is the density at a temperature t_0 and β is the coefficient of volume expansion. Graph d as a function of t for $\beta = 0.2$, $t_0 = 20°$, $d_0 = 1.3$.

8. Find the conic that passes through each of the following sets of five points. Identify each. Check by substituting the coordinates of each point in the final equation.

(a) $(0, 0)$, $(2, 0)$, $(0, 5)$, $(1, 8)$, $(3, 4)$
(b) $(1, 1)$, $(5, 5)$, $(10, -5)$, $(3, 5)$, $(8, 6)$
(c) $(0, 5)$, $(0, -5)$, $(-3, 0)$, $(3, 7)$, $(4, -9)$
(d) $(1, 1)$, $(0, 0)$, $(8, 7)$, $(-1, -3)$, $(7, 3)$
(e) $(8, 0)$, $(-3, 2)$, $(-1, -5)$, $(3, 4)$, $(10, 7)$
(f) $(5, 7)$, $(-3, 4)$, $(-5, -1)$, $(2, -3)$, $(1, 2)$

9. The student probably used the method of Ex. 11.1.6 to solve the previous problem. However, the conic through five points, P_1, P_2, P_3, P_4, P_5.

may be found in the following simpler way. Let $f_{12} = 0$ and $f_{34} = 0$ be the equations of the straight lines through P_1 and P_2, P_3 and P_4. Then $f_{12}f_{34} = 0$ is the equation of a degenerate conic passing through the first four points. Similarly $f_{13}f_{24} = 0$ is the equation of a degenerate conic passing through the same points. Then $f_{12}f_{34} + kf_{13}f_{24} = 0$ is the equation of a family of conics passing through these points. By substituting the coordinates of the fifth point, k may be determined so that the desired member of the family is found. Use this method to do prob. 8, and compare with previous results.

   10. Graph each locus found in prob. 8.

   †11. Show that Δ is invariant under translation and rotation, i.e., show that $B'^2 - 4A'C' = B^2 - 4AC$, where the primed letters represent the new coefficients.

   †12. Show the same for R.

   †13. The conic consists of two straight lines if and only if its equation factors into two linear factors. Solving (I) for x as a function of y, we find:

$$(18)\quad x = \frac{1}{2A}\left[-(By + D) \pm \sqrt{(B^2 - 4AC)y^2 + 2(BD - 2AE)y + D^2 - 4AF}\,\right]$$

For (I) to factor, the radical must be a perfect square. Verify (18) and show that $R = 0$ is a necessary and sufficient condition for factorization. Why does it follow that R is invariant under translation and rotation?

11.6 Geometric properties of conics

   The ellipse, parabola, and hyperbola have in common the property of being the locus of a point that moves so that the ratio of its distance from a fixed point (the focus) to its distance from a fixed line (the directrix) is a constant (the eccentricity) (Ex. 11.1.21, 22, Ex. 11.2.6, and Ex. 11.3.21, 22). This property is the geometrical counterpart of the fact that the equation of a conic is of the second degree. Now we state that any locus having this property is a conic, that is,

*11.6.1   The locus of a point that moves so that the ratio of its distances from a fixed point and a fixed line is constant is a conic.

   The proof of this is easy. We place the axes so that the point is at $(c, 0)$ and the line is given by $x = k$. (See Fig. 11.6.1.) Then the distance of a point (x, y) on the locus from the point and the line are:

$$(1)\qquad \begin{aligned} d_1 &= \sqrt{(x - c)^2 + y^2} \\ d_2 &= |\, x - k \,| \end{aligned}$$

Let the constant ratio be e. Then we have as a necessary and sufficient condition for a point to lie on the locus:

(2) $$\sqrt{(x-c)^2 + y^2} = e \mid x - k \mid$$

This simplifies to

(3) $$(1 - e^2)x^2 + 2(e^2k - c)x + y^2 = k^2e^2 - c^2$$

which is of the form (I) and hence represents a conic.

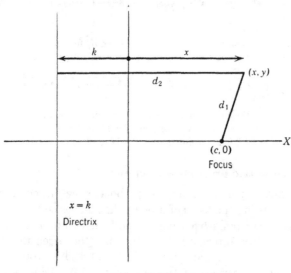

Fig. 11.6.1

It is instructive to see how the equation (3) reduces to the special types of conics, according to the value of e. There are three possible mutually exclusive cases: $e < 1$, $e = 1$, and $e > 1$. If $e = 1$, (3) becomes:

(4) $$2(k - c)x + y^2 = k^2 - c^2$$

which is evidently a parabola. (Why?) Now if we place the y-axis so that it is halfway between the directrix $x = k$ and the focus $(c, 0)$, we have $k = -c$, and (4) reduces to

(5) $$y^2 = 4cx$$

which is the form considered in Ex. 11.2.11. (See Fig. 11.6.2a.) Of course a rotation through $-90°$ would reduce (5) to $y' = (1/4c)x'^2$.

If the focus is on the directrix, the equation reduces to $y^2 = 0$, the equation of the x-axis. (Why?) If we let k approach c and at the same time let k and c get large so that $k^2 - c^2 = (k - c)(k + c)$ remains constant, the locus of (4) approaches the two straight lines $y^2 = k^2 - c^2$. Thus two parallel straight lines may be thought of as a limiting case of the parabola as focus and directrix move off in opposite directions.

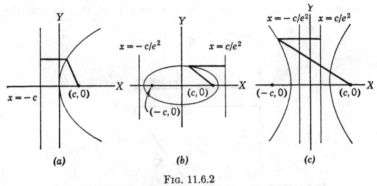

FIG. 11.6.2

If $e \neq 1$, we place the y-axis so that $c = e^2 k$. (Why is this always possible?) Then (3) becomes:

(6) $$(1 - e^2)x^2 + y^2 = k^2 e^2 (1 - e^2)$$

which may be written

(7) $$\frac{x^2}{k^2 e^2} + \frac{y^2}{k^2 e^2 (1 - e^2)} = 1$$

If $e < 1$, the denominators are positive, and this equation is the standard form of the ellipse with $a = ke$, $b = ke\sqrt{1 - e^2}$, $c = ae = \sqrt{a^2 - b^2}$ (Fig. 11.6.2b). If $e > 1$, we rewrite (7) as

(8) $$\frac{x^2}{k^2 e^2} - \frac{y^2}{k^2 e^2 (e^2 - 1)} = 1$$

This is the standard form of the hyperbola, with $a = ke$, $b = ke\sqrt{e^2 - 1}$, $c = ae = \sqrt{a^2 + b^2}$ (Fig. 11.6.2c). Equations (6)–(8) give rise to the degenerate central conics as limiting cases. Suppose e approaches 0 while $a = ke$ is constant. (This means that the focus approaches the origin and the directrix "moves away toward infinity.") Then (7) approaches the equation of the circle $x^2 + y^2 = a^2$. If we let c and a approach 0 so that e remains constant and hence k approaches 0, the equation approaches that of a point ellipse. If in (8) k approaches 0 while e remains constant, the locus approaches a degenerate hyperbola consisting of two straight lines.

Polar coordinates are convenient for developing the equation of the general conic. We place the pole and polar axis so that the

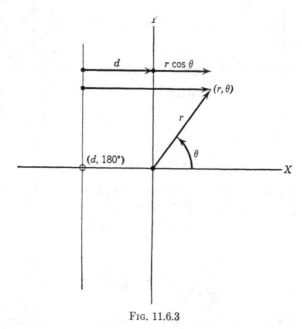

Fig. 11.6.3

focus is at the pole and the directrix is perpendicular to the polar axis and crosses its extension at the point $r = d$, $\theta = 180°$ (Fig. 11.6.3). Then the distance to the focus of a point on the locus is $d_1 = r$. The distance to the directrix is $d_2 = d + r \cos \theta$. Hence the equation of the conic is:

(9) $$r = e(d + r \cos \theta)$$

or

(10) $$r = \frac{ed}{1 - e \cos \theta}$$

Thus we have:

*11.6.2 The equation in polar coordinates of a conic with eccentricity e, focus at the origin, and directrix $x = -d$ is:

$$(11) \qquad r = \frac{ed}{1 - e \cos \theta}$$

Of course this theorem can be proved also by transforming (11) into rectangular coordinates and showing that it is the equation of a conic with the required properties.

A conic whose equation is given in polar coordinates may be plotted by using *11.6.2 or by transforming to rectangular coordinates. Of course (11) covers only conics in the special position considered. Other conics with foci at the origin can be put in the desired form by rotating the axes. For example, the equation

$$(12) \qquad r = \frac{2}{1 - 2 \sin \theta}$$

can be put in the form (11) by rotating the axes through 90°. Then the equation becomes (see Ex. 9.8.7):

$$(13) \qquad r' = \frac{2}{1 - 2 \cos \theta'}$$

which is the equation of a hyperbola with transverse axis along the new x-axis or old y-axis.

EXERCISE 11.6

1. Carry through the derivation of the conic with focus at $(3, 0)$, directrix at $x = -2$, and eccentricity 1. Where should the axes be placed so that this reduces to the form (5)? Make this change two ways: (*a*) by translating the axes and finding the new equation from the one already derived, and (*b*) by restating the problem with axes properly placed.

2. Check equations (1) through (8) by carrying through the algebra. Explain the formulas for a, b, and c.

3. Find the equations of each conic determined by its focus, directrix, and eccentricity:

(*a*) $(-2, 0)$, $x = -4$, $e = 0.5$ (*c*) $(-2, 0)$, $x = -3$, $e = 1$

(*b*) $(-2, 0)$, $x = -1$, $e = 3$ (*d*) $(3, -1)$, $x + y = 4$, $e = 2$

(e) $(0, 3)$, $y = 1$, $e = 1$ (h) $(0, 0)$, $x - y = 10$, $e = 0.2$

(f) $(-1, -1)$, $y = 2x - 1$, $e = 1$ (i) $(0, -5)$, $3x - y = 4$, $e = 1$

(g) $(4, -3)$, $y + 4x - 8 = 0$, $e = 4$ (j) $(2, 7)$, $y = 4$, $e = 3$

Suggestion. In part d the distance from the line to (x, y) on the locus is, by *9.3.1, $d_1 = (x + y - 4)/\sqrt{2}$, and the distance from $(3, -1)$ is $d_2 = \sqrt{(x - 3)^2 + (y + 1)^2}$. The desired equation results from simplifying $d_2 = 2d_1$.

Answers. (a) $3x^2 + 8x + 4y^2 = 0$. (b) $8x^2 - y^2 + 14x + 5 = 0$. (c) $y^2 - 2x - 5 = 0$. (e) $x^2 - 4y + 8 = 0$. (g) $239x^2 + 128xy - y^2 - 888x - 358y + 599 = 0$.

4. Graph the equations found in prob. 3. Show foci and directrices.

5. Prove that the locus of a point that moves so that the ratio of its distances from two fixed points is a constant greater than 1 is a circle. What is the locus if the constant is 1?

6. Put in the form (11) and identify:

(a) $r(2 - 3 \cos \theta) = 4$ (b) $2r(1 - \cos \theta) = 5$ (c) $r(4 - \cos \theta) = 1$

7. Put in the form (11) by rotating the axes and identify:

(a) $r(1 - 2 \sin \theta) = 3$ (b) $r(1 + \cos \theta) = 1$ (c) $r(2 + \sin \theta) = 3$

†**8.** Prove that the following represent conics with eccentricity e and focus at the origin. Identify the orientation of each. What is p? (We assume $e \geq 0$ and $p > 0$.)

(14) $r = \dfrac{p}{1 \pm e \cos \theta}$ (16) $r = \dfrac{p}{1 \pm e \sin \theta}$

(15) $r = \dfrac{-p}{1 \pm e \cos \theta}$ (17) $r = \dfrac{-p}{1 \pm e \sin \theta}$

9. Plot the following:

(a) $r = 2(1 + \cos \theta)^{-1}$ (d) $r[2 - 4 \cos (\theta + 30°)] = 8$

(b) $r(3 - \sin \theta) = 6$ (e) $r - 3 = 2r \cos \theta$

(c) $r = (\cos \theta - 2)^{-1}$ (f) $r = 2(1 + \sin \theta)^{-1}$

†**10.** Prove *11.6.2 by transforming (11) to rectangular coordinates.

A **polynomial of the nth degree** in x is an expression of the form

$$A_n x^n + A_{n-1} x^{n-1} + \cdots + A_2 x^2 + A_1 x + A_0$$

where n is a positive integer or zero, the A's are constants, and $A_n \neq 0$. The coefficient of x^n is called the **leading coefficient.** By "constant" we mean in this chapter any number, real or complex, although the A's will usually be taken as integers for simplicity. When $n = 0$, 1, 2, 3, or 4 the polynomial is called a **constant,** a **linear function,** a **quadratic,** a **cubic,** or a **quartic,** respectively. If $P(x)$ stands for a polynomial, $y = P(x)$ determines y as a single-valued function of x. The equation $P(x) = 0$ is called an **algebraic equation** or sometimes a polynomial equation.

We introduce polynomials of higher degree by considering a special case:

(1) $$y = 2(x - 1)(x + 2)(x + 3)$$

(2) $$= 2x^3 + 8x^2 + 2x - 12$$

We begin, as always, by finding the intercepts $(0, -12)$, $(1, 0)$, $(-2, 0)$, and $(-3, 0)$. We can continue by calculating the y's corresponding to a number of values of x, and joining the resulting points by a smooth curve. (See Fig. 12.) This procedure is simple, but for several reasons it is not satisfactory. In the first place, we found the x-intercepts so easily only because the function was conveniently factored in (1). To find the intercepts in all cases, we need to be able to find the zeros of polynomials, that is, to solve algebraic equations. Also, the process of substituting a large number of values of x in a polynomial is tedious. Hence it would be convenient to have some simple method of calculating values of a polynomial. Finally, it appears plausible that if we find a number of points and join them with a smooth curve we get a fairly accurate graph. But how can we be sure of the location of the high and low points? There is no reason to think that these points are located exactly at points that we happen to plot. We need a method of finding exactly the location of the high and low points, or bend points, of a polynomial.

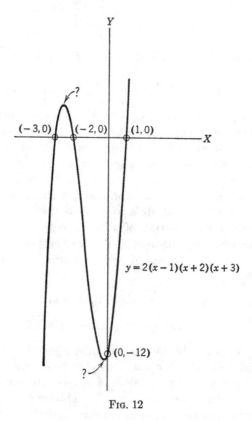

$$y = 2(x-1)(x+2)(x+3)$$

FIG. 12

This chapter will supply the student with the techniques indicated in the foregoing paragraph and at the same time familiarize him with the corresponding algebraic theory.

12.1 Division by $x - a$

We are concerned in this section with the problem of dividing a polynomial $P(x)$ by a binomial of the form $x - a$. By definition, $P(x)/(x - a)$ means a function $f(x)$ such that $(x - a)f(x) \equiv P(x)$.

If $f(x)$ is a polynomial we say that $P(x)$ is **divisible** by $x - a$ and that $x - a$ is a **factor** of $P(x)$. For example,

(1) $$\frac{x^2 - 1}{x - 1} \equiv x + 1$$

and $x^2 - 1$ is divisible by $x - 1$. However,

(2) $$\frac{x^2 + 1}{x - 1} \equiv x + 1 + \frac{2}{x - 1}$$

since

(3) $$(x - 1)\left[x + 1 + \frac{2}{x - 1}\right] \equiv x^2 - 1 + 2 \equiv x^2 + 1$$

and $x^2 + 1$ is not divisible by $x - 1$. Equation (2) illustrates the fact that the fraction always equals a polynomial plus a constant divided by the denominator, that is,

(4) $$\frac{P(x)}{x - a} \equiv Q(x) + \frac{R}{x - a}$$

where $Q(x)$ is a polynomial and R is a constant. The polynomial $Q(x)$ is called the **quotient,** and the constant R is called the **remainder.** The remainder is zero if and only if $P(x)$ is divisible by $x - a$. Since (4) does not hold when $x = a$ we prefer to state this idea in the following form:

*12.1.1   For any polynomial $P(x)$ and constant a, there exist a polynomial $Q(x)$ and a constant R such that

(5) $$P(x) \equiv (x - a)Q(x) + R$$

Proof of *12.1.1.   Consider any polynomial

(6) $$P(x) \equiv A_n x^n + A_{n-1} x^{n-1} + \cdots + A_2 x^2 + A_1 x + A_0$$

The student can easily verify that

(7) $$P(x) \equiv (x - a)A_n x^{n-1} + (A_{n-1} + aA_n)x^{n-1}$$
$$+ A_{n-2} x^{n-2} + \cdots + A_1 x + A_0$$

This means that

(8) $$P(x) \equiv (x - a)Q_1(x) + R_1(x)$$

where $Q_1(x)$ is a polynomial (equal to $A_n x^{n-1}$) and $R_1(x)$ is a polynomial of degree less than n. But we can rewrite R_1 according to the same procedure, that is,

(9) $$R_1(x) \equiv (x - a)Q_2(x) + R_2(x)$$

where the degree of R_2 is less than that of R_1. Hence, after n steps

(10) $$P(x) = (x - a)(Q_1 + Q_2 + \cdots Q_n) + R_n$$

where $Q_1 + Q_2 + \cdots + Q_n$ is a polynomial and R_n is of degree $n - n$, that is, a constant.

The procedure that we used to prove *12.1.1 is just the familiar process of long division, sometimes called the **division algorithm.** For example, consider the fraction

(11) $$\frac{2x^3 + 7x^2 - x - 3}{x - 2} \equiv 2x^2 + 11x + 21 + \frac{39}{x - 2}$$

Here $n = 3$, $a = 2$, $A_3 = 2$, $A_2 = 7$, $A_1 = -1$, and $A_0 = -3$. The work of long division is arranged as follows:

$$\begin{array}{r} 2x^2 + 11x + 21 \\ x - 2\overline{\smash{)}2x^3 + 7x^2 - x - 3} \end{array}$$

(12) $2x^3 - 4x^2$

(13) $11x^2$

(14) $11x^2 - 22x$

(15) $21x$

(16) $21x - 42$

            39

The first step in the process is to divide x into $2x^3$ and write the result $2x^2$ above. "Multiplying back" $2x^2$ by $x - 2$, we write the product in line (12), and subtract to get (13). At this stage we have:

(17) $$2x^3 + 7x^2 - x - 3 \equiv (x - 2)2x^2 + 11x^2 - x - 3$$

This corresponds to equation (7) in the general process. Note that the subtraction of (12) to get (13) is what gives us the polynomial of lower degree as remainder at this stage. Now we repeat the process with line (13), that is, we divide x into $11x^2$ to get $11x$, "multiply back" to get line (14), and subtract to get line (15).

Then we have:

(18) $\quad 2x^3 + 7x^2 - x - 3 \equiv (x - 2)(2x^2 + 11x) + 21x - 3$

which corresponds to equation (9). Repeating the process gives us line (16) and the final constant remainder. The final result corresponding to (10) is:

(19) $\quad 2x^3 + 7x^2 - x - 3 \equiv (x - 2)(2x^2 + 11x + 21) + 39$

which is the same as (11). Of course, in actually carrying through a process of division we do not need to go through all this reasoning, which is given here to illustrate the proof of *12.1.1 and to show the student that the process of long division is based on fundamental algebraic identities.

We shall now develop methods of carrying through division by $x - a$ that are easier than those used in equations (12) through (16), where we really did a lot of needless writing. Since we are always dividing by x, the next term in the quotient always has the same coefficient as the term into which we divide. We really do not need to write x at all, since it is only the coefficients that are involved in the computations. Thus we could write:

$$
(20) \qquad
\begin{array}{r}
2 \quad 11 \quad 21 \\
\hline
-2/2 \quad\ 7 \quad -1 \quad -3 \\
2 \quad -4 \\
\hline
11 \\
11 \quad -22 \\
\hline
21 \\
21 \quad -42 \\
\hline
39
\end{array}
$$

But still we have written 2, 11, and 21 three times each when it would have been enough to write them twice each, since we know that when we multiply back we will get the same thing. If we leave out these repetitions and push together the remaining numbers, we have:

$$
(21) \qquad
\begin{array}{r}
2 \quad\ 11 \quad\ 21 \\
\hline
-2/2 \quad\ 7 \quad -1 \quad\ -3 \\
-4 \quad -22 \quad -42 \\
\hline
11 \quad\ 21 \quad\ 39
\end{array}
$$

But we can do better by omitting the top line altogether, and copying the 2 down below:

(22)
$$-2 \quad \begin{array}{cccc} 2 & 7 & -1 & -3 \\ & -4 & -22 & -42 \\ \hline 2 & 11 & 21 & 39 \end{array}$$

Here the bottom row gives us the coefficients in the quotient and the constant remainder. We proceed by copying down 2, writing down $-2(2) = -4$, subtracting to get 11, writing down $-2(11)$ $= -22$, subtracting to get 21, writing down $-2(21) = -42$, and subtracting to get 39. But since we multiply each time by -2 and subtract, why not multiply each time by 2 and add? Then the computation would look like this:

(23)
$$2 \begin{array}{|cccc} 2 & 7 & -1 & -3 \\ & 4 & 22 & 42 \\ \hline 2 & 11 & 21 & 39 \end{array}$$

We simply copy down 2. Then $2(2) = 4$, $4 + 7 = 11$, $2(11)$ $= 22$, $22 + (-1) = 21$, $2(21) = 42$, $42 + (-3) = 39$.

In general terms the procedure is as follows. To divide $P(x)$ by $x - a$, write down the coefficients of $P(x)$ in descending order of powers of x, writing zeros for missing terms, and place a at the left. Then proceed as follows:

(24)
$$a \begin{array}{|ccccccc} A_n & A_{n-1} & A_{n-2} & A_{n-3} & \cdots & A_1 & A_0 \\ & aA_n & aB_1 & aB_2 & \cdots & aB_{n-2} & aB_{n-1} \\ \hline A_n & B_1 & B_2 & B_3 & \cdots & B_{n-1} & R \end{array}$$

The process is called **synthetic division.** It is useful when dividing by $x - a$, and also when we wish to evaluate a polynomial at $x = a$, as we shall see in the next section.

We illustrate the method once more by dividing $3x^5 - 2x^3 + x^2 - 6x - 2$ by $x + 2$:

$$-2 \begin{array}{|cccccc} 3 & 0 & -2 & 1 & -6 & -2 \\ & -6 & 12 & -20 & 38 & -64 \\ \hline 3 & -6 & 10 & -19 & 32 & -66 \end{array}$$

Hence $3x^5 - 2x^3 + x^2 - 6x - 2 \equiv (x + 2)(3x^4 - 6x^3 + 10x^2 - 19x + 32) + (-66)$.

EXERCISE 12.1

1. Evaluate the following by long division:

(a) $(x^3 - 3x^2 + x - 3)(x - 1)^{-1}$ (c) $(x^4 - 1)(x + 1)^{-1}$
(b) $(4x^4 - 2x^2 + x - 8)(x - 2)^{-1}$ (d) $(x^3 + x)(x - 1)^{-1}$

Answers. (a) $Q = x^2 - 2x - 1$, $R = -4$. (c) $Q = x^3 - x^2 + x - 1$, $R = 0$.

2. Use synthetic division to do prob. 1 and compare the two methods.
3. Find the quotient and remainder:

(a) $(5x^4 - 2x^2 - x + 7)(x + 1)^{-1}$
(b) $(8x^6 - 7x^5 - 3x + 2)(x - 1)^{-1}$
(c) $(4x^3 + 6x^2 - 2x)(x + 2)^{-1}$
(d) $(7x^4 + 5x - 2x^3 - 8)(x + 3)^{-1}$
(e) $(3x^7 - 2x^4 + x^2 - 2x + 1)(x + 4)^{-1}$
(f) $(3x^6 + x^4 - x^2 + 3)(x^2 - 1)^{-1}$
(g) $(3x^3 - 2x^2 + 1 + 6x)(x - 0.5)^{-1}$
(h) $(x^3 - x^4 + x^5 + x - x^2)(x - 1)^{-1}$
(i) $(ax^6 + a^2x^4 + a^3x^3 + 2)(x - a)^{-1}$
(j) $(x^3 + 3x^2y - xy^2 + y^4)(x - y)^{-1}$
(k) $(A_0 + A_1x + A_2x^2)(x - a)^{-1}$
(l) $(A_3x^3 + A_2x^2 + A_1x + A_0)(x - a)^{-1}$

Answers. (a) $Q = 5x^3 - 5x^2 + 3x - 4$, $R = 11$. (c) $Q = 4x^2 - 2x + 2$, $R = -4$. (e) $R = -49639$. (g) $Q = 3x^2 - 0.5x + 5.75$, $R = 3.875$.

4. Find:

(a) $(x^5 + 2x - 1)(2x + 6)^{-1}$
(b) $(x^4 + 3x^2 - 2x)(3x - 1)^{-1}$
(c) $(1 - 15x^4 + x^2 - 3x^3)(2x - 3)^{-1}$
(d) $(4 - 3x^2 + 18x^4 - 2x^6)(6x^2 + 7)^{-1}$

Suggestions. (a) $2x + 6 = 2(x + 3)$. (c) $2x - 3 = 2(x - 1.5)$.

5. How is finite induction involved implicitly in the proof of *12.1.1? Rewrite the proof so as to include it explicitly.

12.2 The remainder theorem and its consequences

The following remarkable proposition is known as the **remainder theorem:**

*12.2.1 If R is the constant remainder when a polynomial $P(x)$ is divided by $x - a$, then $R = P(a)$.

Since finding the remainder by division is often easier than substitution, this theorem provides a good method of evaluating a polynomial. For example, to evaluate

(1) $$P(x) = x^5 - 6x^4 + 10x^2 - x + 3$$

when $x = 2$, we write

(2)

$$
\begin{array}{r|rrrrr}
2 & 1 & -6 & 0 & 10 & -1 & 3 \\
  &   & 2 & -8 & -16 & -12 & -26 \\
\hline
  & 1 & -4 & -8 & -6 & -13 & -23
\end{array}
$$

Hence $P(2) = -23$. When synthetic division is used in this way it is called **synthetic substitution**.

Proof of *12.2.1. From *12.1.1 we know that for any polynomial $P(x)$ and constant a there is a polynomial $Q(x)$ and a constant R such that $P(x) \equiv (x - a)Q(x) + R$. If we let $x = a$, we get $P(a) = (a - a)Q(a) + R = R$, and the theorem is proved. We may combine *12.2.1 and *12.1.1 in the following theorem:

*12.2.2   For any polynomial $P(x)$ and constant a, there exists a polynomial $Q(x)$ such that

$$P(x) \equiv (x - a)Q(x) + P(a)$$

It is evident from *12.2.2 that $x - a$ is a factor if and only if the remainder, that is, $P(a)$, is zero. Hence we have proved the **factor theorem** and its converse:

*12.2.3   If a is a zero of a polynomial $P(x)$, then $x - a$ is a factor of $P(x)$.

*12.2.4   If $x - a$ is a factor of a polynomial $P(x)$, then a is a zero of $P(x)$.

The factor theorem is of considerable importance, both in the numerical solution of equations and in studying the principles of solving algebraic equations. It means that if we can find one root r of an algebraic equation, we may factor the polynomial into $x - r$ and a polynomial of lower degree. Thus to solve

(3) $$x^4 + 2x^3 - x - 2 = 0$$

we note that $x = 1$ is a solution. (Why?) Hence the left member must be divisible by $x - 1$. By synthetic division we find:

(4) $$(x - 1)(x^3 + 3x^2 + 3x + 2) = 0$$

Next we note that $x = -2$ is a zero of the second factor. (Why?) By synthetic division we find:

$$(5) \qquad (x - 1)(x + 2)(x^2 + x + 1) = 0$$

The roots of the original equation are the values of x that make at least one of these factors zero. They are $x = 1, -2, (-1 \pm i\sqrt{3})/2$. (Why?)

Our main purpose in giving the foregoing illustration is to emphasize the fact that, if we can find one zero of a polynomial, we can factor it into a linear factor and a polynomial of lower degree. The existence of such a zero is guaranteed by the **fundamental theorem of algebra:**

*12.2.5 Every algebraic equation with complex coefficients has at least one root in the field of complex numbers.

Accepting this theorem without proof,[1] we may state:

*12.2.6 Every algebraic equation of degree n has exactly n roots.

Proof. Let the equation be $P(x) = 0$. Then *12.2.5 implies that there is a root r_1. Hence from *12.2.3, $P(x) = (x - r_1)P_1(x)$, where $P_1(x)$ is a polynomial of degree $n - 1$. By the same reasoning there is a root r_2 of $P_1(x)$ and hence

$$(6) \qquad P(x) \equiv (x - r_1)(x - r_2)P_2(x)$$

where $P_2(x)$ is of degree $n - 2$. We continue in this way until the quotient is a constant, that is,

$$(7) \quad P(x) = C(x - r_1)(x - r_2)(x - r_3) \cdots (x - r_{n-1})(x - r_n)$$

There are exactly n factors, since each time we find a root and a factor we reduce the degree of the quotient by 1. Some of the r's may be the same. In this case we speak of **multiple roots.** With this understanding, it is clear from (7) that the roots of $P(x) = 0$ are just the n numbers $r_1, \cdots r_n$, and hence the theorem is proved.

[1] The proof involves ideas beyond the scope of this book. One of the simplest is given in *What Is Mathematics?* by Richard Courant and Herbert Robbins, Oxford University Press, New York, 1941, pp. 269 ff.

EXERCISE 12.2

1. Use the remainder theorem to find the indicated values and check by substitution.

(a) $f(x) = 4x^3 - x^2 + 7x + 13; f(3), f(3/8), f(-2)$
(b) $f(x) = x^4 + 13x^3 + 2x^2 + 5x - 4; f(-1), f(1.1), f(4)$
(c) $f(x) = 5x^6 - 3x^5 + 8x^4 - x + 6; f(-2), f(0.5), f(3)$

2. Plot the following functions by calculating at least 10 values. Where the shape of the curve is doubtful, consider fractional values of the variable.

(a) $y = 2x^4 - 3x^3 + x^2 - 6x + 1$ (d) $y = x^5 + x^4 + 3x^3 - x + 2$
(b) $y = x^3 + x^2 - 5x + 3$ (e) $s = 2t - 3t^4 + t^2 - 15t^3$
(c) $y = x^4 - 8x^2 + 18$ (f) $u = v(v^3 - 2v + 8 + 7v^2) - 3$

3. Show that the equation $x^4 - 2x^3 - 9x^2 + 2x + 8 = 0$ has roots equal to -2, -1, 1 and 4.

Solution:

$$
\begin{array}{rrrrrr}
1 : & 1 & -2 & -9 & 2 & 8 \\
    &   & 1 & -1 & -10 & -8 \\
\hline
-1 : & 1 & -1 & -10 & -8 & 0 \\
     &   & -1 & 2 & 8 & \\
\hline
-2 : & 1 & -2 & -8 & 0 & \\
     &   & -2 & 8 & & \\
\hline
4 : & 1 & -4 & 0 & & \\
    &   & 4 & & & \\
\hline
    & 1 & 0 & & & \\
\end{array}
$$

4. Find a function with the indicated zeros, write as a polynomial; then check the zeros.

(a) $2, 1, -2, 3$ (d) $0.1, 2, 6, -0.5$ (g) $1/\sqrt{2}, 1/\sqrt{3}, 2$
(b) $-1, 5, \sqrt{2}$ (e) $1, -3, 2, 2$ (h) $0.3, 0.2, 5, -0.2$
(c) $2m - 1, 3, 4, -0.5$ (f) $3, -1, -1, 5$ (i) $1, 18, 0, 6$

5. Find all the roots of the following and write the left member in factored form.

(a) $x(x^3 - 8) = 0$ (d) $x^3 + 3x^2 + 5x + 3 = 0$
(b) $x^6 - 1 = 0$ (e) $x^3 - 6x^2 + 3x + 10 = 0$
(c) $x^4 + x = 0$ (f) $x^4 + 4x^3 + 6x^2 + 4x + 1 = 0$

6. Show that 1 is a triple root of $x^4 - x^3 - 3x^2 + 5x - 2 = 0$.

7. Give as many examples as you can from previous work to illustrate *12.2.6. How many roots does the equation $\sin x = 1$ have? Why does this not contradict *12.2.6?

†8. Show that, if two polynomials of degree n are equal for more than n values of the variable, they are identical, that is, equal for all values of the variable and have the same coefficients. (*Hint.* Form the polynomial of their difference.)

9. Sketch the regions determined by the following inequalities:

(a) $y < x^3 + 2x - 1$
(b) $y \geq x^3 + 7x^2 + 16x$
(c) $y < x^4 - 6x^3 + 9x^2$
(d) $0 < y < x^2$ and $0 < x < 2$
(e) $0 < y < (x - 1)(2 - x)(x - 3)(x - 4)$
(f) $x < y < x^2(x + 4)$

10. Explain the confusion involved in the following "proof" of the remainder theorem. "We show that the remainder is $f(r)$ directly by long division as follows:

$$x - r \overline{\smash{\big)}\,\begin{matrix} f \\ f(x) \\ \underline{f(x) - f(r)} \\ f(r)! \end{matrix}}$$

"(?)

†11. Use the factor theorem to evaluate the determinant

$$\begin{vmatrix} x & y & z \\ x^2 & y^2 & z^2 \\ x^3 & y^3 & z^3 \end{vmatrix}$$

†12. Show that, if a complex number z is a root of an algebraic equation with real coefficients, its conjugate $\bar{z}$ is also a root.

†13. Show that, if $\bar{z}$ is the conjugate of the complex number z, then $P(\bar{z}) = \overline{P(z)}$ for any polynomial $P(z)$. (*Suggestion.* See Ex. 10.2.15.)

†14. Look up the proof of *12.2.5 in *What Is Mathematics?* and rewrite in your own words. (See footnote on page 497.)

12.3 Rational roots

The student is familiar with the fact that an equation with integral coefficients may not have any rational roots or even any real roots. However, it does sometimes happen that there are rational roots (especially in textbook problems!), and hence we discuss a method of finding them. It depends on the following theorem:

*12.3.1 If $a_n x^n + a_{n-1} x^{n-1} + \cdots + a_2 x^2 + a_1 x + a_0 = 0$, where the a's are integers, has a rational root given in lowest terms by p/q, then p is a divisor of a_0 and q is a divisor of a_n.

Proof. If p/q is a root,

(1) $a_n(p/q)^n + a_{n-1}(p/q)^{n-1} + \cdots$
$$+ a_2(p/q)^2 + a_1(p/q) + a_0 = 0$$

(2) $a_n p^n + a_{n-1}p^{n-1}q + a_{n-2}p^{n-2}q^2 + \cdots$
$$+ a_2 p^2 q^{n-2} + a_1 p q^{n-1} + a_0 q^n = 0$$

This can be written in either of the following forms:

(3) $a_{n-1}p^{n-1}q + a_{n-2}p^{n-2}q^2 + \cdots + a_2 p^2 q^{n-2} + a_1 p q^{n-1}$
$$+ a_0 q^n = -a_n p^n$$

(4) $a_n p^n + a_{n-1}p^{n-1}q + a_{n-2}p^{n-2}q^2 + \cdots + a_2 p^2 q^{n-2}$
$$+ a_1 p q^{n-1} = -a_0 q^n$$

The hypothesis that p/q is in lowest terms means that p and q have no common factor other than 1. In (3), since the left member is divisible by q, $-a_n p^n$ must be also. Since q cannot divide p, it must divide a_n. Similarly, from (4) p must be a divisor of $-a_0 q^n$ and hence of a_0.

To illustrate the use of this theorem we consider:

(5) $6x^4 + x^3 + 11x^2 + 2x - 2 = 0$

The divisors of 6 are ± 1, ± 2, ± 3, ± 6, and the divisors of -2 are ± 1, ± 2. Hence, if there are rational roots, they must be among the following: ± 1, $\pm 1/2$, $\pm 1/3$, $\pm 1/6$, ± 2, $\pm 2/3$. Synthetic division shows that $-1/2$ is a root:

(6)
$$
-\tfrac{1}{2} \,\big|\;
\begin{array}{rrrrr}
6 & 1 & 11 & 2 & -2 \\
  & -3 & 1 & -6 & 2 \\
\hline
6 & -2 & 12 & -4 & 0
\end{array}
$$

Now we consider:

(7) $6x^3 - 2x^2 + 12x - 4 = 0$

or

(8) $3x^3 - x^2 + 6x - 2 = 0$

Here possible values of p are ± 1, ± 2, and possible values of q are

± 1, ± 3. The possible rational roots are ± 1, ± 2, $\pm 1/3$, $\pm 2/3$. It turns out that $1/3$ is a root:

(9)
$$\frac{1}{3} \begin{array}{|rrrr} 3 & -1 & 6 & -2 \\ & 1 & 0 & 2 \\ \hline 3 & 0 & 6 & 0 \end{array}$$

Now we have the equation

(10)
$$x^2 + 2 = 0$$

whose roots are $\pm i\sqrt{2}$. Hence the roots of (5) are $-1/2$, $1/3$, and $\pm i\sqrt{2}$.

The method is evidently quite handy when the number of possibilities is not large, but consider:

(11)
$$y^4 + y^3 + 66y^2 + 72y - 432 = 0$$

The possible divisors of the last term are ± 1, ± 2, ± 3, ± 4, ± 6, ± 8, ± 9, ± 12, ± 16, ± 18, ± 24, ± 27, ± 36, ± 48, ± 54, ± 72, ± 108, ± 144, ± 216, and ± 432! In this particular case, it is easy to see that large values of y cannot be roots (why?), and it turns out that 2 and -3 are solutions. But things might not go so well for

(12)
$$35y^4 - y^3 - 66y^2 - 72y - 432 = 0$$

Here there are 160 possible rational roots to test! Numerous more refined methods of solving equations have been worked out and are included in the branch of mathematics called theory of equations. In the next section we shall consider one very useful technique of finding roots approximately.

EXERCISE 12.3

1. Show that 2 and -3 are roots of (11), and find the other two roots.

2. Verify that (12) has 160 possible rational roots.

3. Show that, if the leading coefficient is one, any rational root must be integral.

4. In each of the following equations, find all the rational roots or establish that there are none. (*Note.* When a root r is found, you need work only with the **depressed equation**, found by dividing the left member by $x - r$. The trials should be done neatly and systematically so that the work may be checked.)

(a) $x^3 - 2x^2 - 2x + 1 = 0$
(b) $x^3 - 4x + 2 = 0$
(c) $x^4 - x^3 - x^2 - x - 2 = 0$
(d) $x^4 + 2x^2 + 3 = 0$
(e) $x^6 - 7x^4 + 14x^2 + 8 = 0$
(f) $x^4 + 10x^3 + 20x^2 - 10x - 21 = 0$

Answers. (a) -1. (c) $-1, 2$. (e) None.

5. Find all the rational roots of the following:

(a) $2x^3 + x^2 - 8x - 4 = 0$
(b) $3x^4 + 9x^3 + x + 3 = 0$
(c) $2x^3 - 5x^2 - 9 = 0$
(d) $4x^3 - x - 3 = 0$
(e) $3x^4 - x^3 - 5x^2 + x + 2 = 0$
(f) $6x^6 + 5x^5 + 5x^4 - 9x^3 - 14x^2 + 4x - 8 = 0$
(g) $5x^4 + 3x^3 + 3x^2 + 13x - 6 = 0$
(h) $6x^5 - 41x^4 - 3x^3 - 27x^2 - 9x + 14 = 0$

Answers. (a) $2, -2, -0.5$. (c) 3. (e) $1, 1, -1, -2/3$.

6. Find all the roots of:

(a) $x^3 + 3x^2 - 2 = 0$
(b) $x^3 - 7x - 6 = 0$
(c) $2x^3 + 5x^2 + 9 = 0$
(d) $x^3 - x^2 - 12x + 12 = 0$
(e) $x^3 + 2ax^2 + 5a^2x + 4a^3 = 0$
(f) $x^6 + 4x^5 - 6x^4 - 40x^3 - 31x^2 + 36x + 36 = 0$
(g) $3x^3 + 5x^2 + 20x + 12 = 0$
(h) $30x^3 - 3x^2 + 10x - 1 = 0$

7. Find some of the integral solutions (x, y) to the equation $(x + y)^{x-y} - (xy)^y - (x - y)^x = 0$. (*Answer.* One pair is 0, 1.)

12.4 Approximate solution of equations

To introduce the technique of **successive approximations** we consider an equation whose roots we can already find:

$$(1) \qquad\qquad x^2 + 3x - 1 = 0$$

The roots are:

$$(2) \qquad\qquad x = \frac{-3 \pm \sqrt{13}}{2} \doteq 0.303 \text{ and } -3.303$$

to three decimal places. We shall find these roots by approximate methods.

Our first step is to plot $y = f(x) = x^2 + 3x - 1$ for a number of successive integral values of x (Fig. 12.4.1). We note that $f(-4) > 0$, $f(-3) < 0$, $f(0) < 0$, and $f(1) > 0$, that is, the curve must cross the axis somewhere between -4 and -3 and again somewhere between 0 and 1. These points of crossing are the only possible zeros of $f(x)$. (Why?)

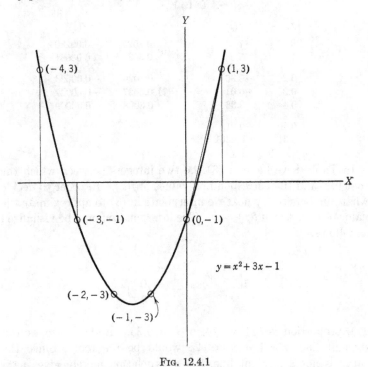

Fɪɢ. 12.4.1

We concentrate our attention on the root between 0 and 1. We could proceed by calculating $f(x)$ for $x = 0.1, 0.2, \cdots$, etc., until we found two successive values between which $f(x)$ changed from negative to positive, but we can save ourselves some trouble by noting on the graph that the curve seems to cross somewhere past a quarter of the way from 0 to 1. We find $f(0.2) = -0.36$, $f(0.3) = -0.01$, $f(0.4) = 0.36$. Hence the root lies between 0.3 and 0.4, and to one decimal place it is $x = 0.3$. Again we could try $x = 0.31, 0.32$, etc. However, we see that $f(0.3)$ is much closer to zero than $f(0.4)$, so we expect that we will not have to make many trials. In fact $f(0.31) = 0.0261$, so that 0 is the next digit in the root.

Now we are getting to the point where the graph does not help much. Let us look back on our computations and see how we could do them without the graph. We make a table of values of x and $f(x)$:

TABLE 1

	x	$f(x)$		x	$f(x)$
(3)	0	−1	(6)	0.302	−0.00280
	1	3		0.303	0.00081
	0.2	−0.36		0.3026	−0.00063
(4)	0.3	−0.01	(7)	0.3027	−0.00027
	0.4	0.36		0.3028	0.00009
(5)	0.30	−0.01			
	0.31	0.0261			

In Table 1 we have in (3) the two integers between which the root lies, and the corresponding values of $f(x)$. In order to decide what numbers to try next we interpolate in (3) to approximate the value of x for which $f(x) = 0$. The interpolation may be visualized as follows:

$$x \qquad\qquad f(x)$$

(8)

$$1\begin{bmatrix} 0 \\ ?\leftarrow \\ \rightarrow 1 \end{bmatrix}h \qquad 1\begin{bmatrix} -1 \\ \rightarrow 0 \\ 3 \end{bmatrix}4$$

The proportion is $h/1 = 1/4$, or $h = 0.25$. If the curve were a straight line, $x = 0 + h = 0.25$ would be the root. Since the curve is not a straight line, the interpolation merely gives us a clue. (See 5.2.) Hence in (4) we try values beginning with 0.2 and find that the root lies between 0.3 and 0.4. Now we interpolate in (4) to find the next number to try. The total difference of the f's is 0.37. The partial difference of the f's is 0.01. Hence the partial difference of the x's is given by $h/0.1 = 0.01/37$ or $h = 1/370 < 0.01$. Hence we begin with $x = 0.30$ and find that the root lies between 0.30 and 0.31. To interpolate in (5), we ignore the decimal points and remember that we are interested only in *estimating* the next digit. The partial over the total difference of the f's is $1/(3.61) \doteq 0.3$. Hence we start with $x = 0.302$ and get

(6), which shows that the root lies between 0.302 and 0.303. Note that it was wise to *start one digit below the estimate* by interpolation. If we had started with 0.303 we should have had a positive value of $f(x)$ already. At this stage we know that the root lies between 0.302 and 0.303, but we do not know which is closer. Hence we cannot say which is the root to three-figure accuracy. We might guess that it would be closer to 0.303 (why?), but to be sure we must calculate the next digit in the decimal representation. An interpolation in (6) gives a guess of 0.7. Hence we begin with 0.3026 and get (7). Actually we could have stopped as soon as we saw that $f(0.3026) < 0$, since we knew then that the root was closer to 0.303 than to 0.302, but we see from (7) that the root lies between 0.3027 and 0.3028. Evidently we can continue the process as long as time, paper, and patience last.

The above procedure of successive approximations is applicable to any equation whose left member is a continuous function, that is, has a graph without breaks. To solve $f(x) = 0$ we try successive values of x until we find consecutive integers which "bracket" a root, that is, yield values of $f(x)$ of different sign. Then the continuity of the graph guarantees at least one zero between these integers. We then interpolate for an estimate of the first decimal place and try successive values beginning below the estimate. When we "bracket" the root to the first decimal place, we interpolate again and proceed with the second decimal place. It is important to use common sense in doing the interpolation, since the function may be going from negative to positive or the other way around. Also it must be remembered that to find the root accurate to n places we have to find at least the first $n + 1$ places. It may also happen that there is more than one root in an interval where the function changes sign. A careful graph usually helps to minimize the work, and a calculating machine lightens the labor.

In order to illustrate the method in a simple case we find the roots of

$$(9) \qquad\qquad x^3 + 8x - 1 = 0$$

It is easy to see that there are no rational roots. (Why?) The function is sketched in Fig. 12.4.2, and the computations are outlined in Table 2.

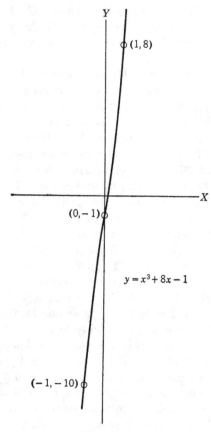

FIG. 12.4.2

TABLE 2

x	$f(x)$	x	$f(x)$
-2	-25	0.123	-0.0141
-1	-10	0.124	-0.0061
0	-1	0.125	0.0020
1	8		
2	23	0.1247	-0.00046
		0.1248	0.00034
0.1	-0.199		
0.2	0.608	0.12475	-0.000059
		0.12476	0.000022
0.12	-0.0383		
0.13	0.0422		

The table indicates that one of the roots of the equation is $x \doteq 0.1248$ to four decimal places. The sketch suggests that there are no other real roots. If the student performs the synthetic division by $(x - 0.1248)$ he will find that the resulting quadratic equation has imaginary roots that can be found by the quadratic formula and that are approximate roots of (9).

EXERCISE 12.4

1. Find by successive approximations the root of (1) between -4 and -3.

2. Factor (9) approximately by dividing by $(x - 0.125)$, and approximate the two imaginary roots.

3. Find by the quadratic formula and by successive approximations the roots of $2x^2 + 3x - 4 = 0$.

4. Find by successive approximations the real roots of:

$$(a) \quad x^3 + 7x - 1 = 0$$
$$(b) \quad x^2 - 2 = 0$$
$$(c) \quad x^3 - 3 = 0$$
$$(d) \quad x^4 - 10 = 0$$
$$(e) \quad x^5 - 15 = 0$$
$$(f) \quad x^6 - 2 = 0$$
$$(g) \quad x^3 - 2x + 2 = 0$$
$$(h) \quad x^5 - 10 = 0$$
$$(i) \quad x^4 - 8x^3 - 18x + 2 = 0$$
$$(j) \quad x^4 - x^2 - 111 = 0$$
$$(k) \quad x^8 + 11x^6 + 41x^4 + 61x^2 - 30 = 0$$
$$(l) \quad x^3 - 10x^2 + 35x + 50 = 0$$

5. The following equation is similar to one that appears in a recent textbook on genetics. Find the smallest real root.

$$283291 - 587519y + 467111y^2 - 105102y^3 = 0$$

Suggestion. To locate the root approximately, divide by 10^5 and round off the coefficients drastically.

6. The method of successive approximations can be used to solve equations that are not algebraic. For example, to solve $x^2 - \sin x = 0$, we can estimate the root by graphing $y = x^2$ and $y = \sin x$ and noting where the curves cross. Then we can carry on the approximation as for algebraic equations. Approximate the roots of the following:

(a) $x^2 - \sin x = 0$	(d) $x^2 - \cos x = 0$	(g) $\log_{10} x - \tan x = 0$
(b) $e^{-x} = 0.5$	(e) $e^x - 2x = 0$	(h) $\sin (x^2 - 1) = x$
(c) $x - 2 \sin x = 0$	(f) $10^{2x - x^2} = 2$	(i) $e^x(1 - 0.1x) = 1$

7. Solve $(1 + 1/t)^t = 2.72$ for t to three significant figures.

8. Sometimes equations may be solved approximately by neglecting "small terms." For example, when a is small the solutions of $ax^2 + bx + c = 0$ and of $bx + c = 0$ are not very different. Find a formula for the error. Check your work by solving $0.01x^2 + 2x - 5 = 0$ exactly and approximately.

9. The **method of false position** is an alternative way of solving algebraic equations. It goes as follows. Let x be a first estimate. Then substitute $x + h$ in the equation, ignore all powers of h higher than one, and solve for h. Repeat with $x + h$ as the estimate. Solve several of the foregoing equations by this method.

10. Solve the following pairs of simultaneous equations:

(a) $y - x^3 = 0$ (c) $y^2 + x^2 = 1$
    $y^2 + y = x$ $y - x^4 = 0$

(b) $x^2 - y^2 = 1$ (d) $y - x^3 = 0$
    $x^2 y = 1$ $y^3 + 2y + x = 0$

Suggestion. You may derive a single equation and solve it approximately, but a more convenient method is to use both equations by substituting a guess for x in one equation, solving that for y, substituting this y in the other, etc.

†11. Show that in any polynomial, $P(x) = A_n x^n + A_{n-1} x^{n-1} + \cdots + A_1 x + A_0$, the next to highest power may be eliminated by a transformation $x' = x - h$, where h is appropriately chosen. What is the geometric interpretation of this change of variable? Apply the method to solve the general quadratic.

12.5 Bend points

Having seen how to find the x-intercepts of a polynomial function, we turn to the problem of finding the points at which the curve reaches a maximum or minimum, that is, changes from going up to going down, or vice versa. In order to find these points, which are called **bend points,** we make use of the idea of the tangent line to the curve at a point. (See 5.4.) The tangent line has the same direction as the curve at the point, and an object moving along the curve, if suddenly allowed to move freely, would move off along the tangent. It appears (Fig. 12.5) that where the tangent line is horizontal, there may be a maximum or minimum (Q and R in the figure). It is also possible that there may be a point where the curve levels off without having either maximum or minimum (S in the figure). Such a point is called a **horizontal inflexion.** Also the curve may have a maximum or minimum without having a horizontal tangent (T in the figure). Since points like T do not occur for polynomials, if we can find the points where the tangent line is horizontal we will have found all the places where the curve has a maximum, minimum, or horizontal inflexion. Since the tangent line is horizontal if and only if its slope is zero, and since the slope of the tangent line is given by the derivative, the problem reduces to finding the derivative of a polynomial and determining its zeros.

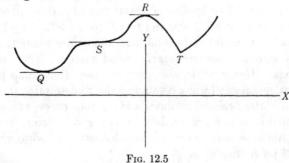

FIG. 12.5

Finding the derivative of any polynomial is easy, thanks to the following generalization of *5.5.2, which we state without proof:

*12.5.1 $D_x(f_1 + f_2 + \cdots + f_n) = D_x f_1 + D_x f_2 + \cdots + D_x f_n$

This theorem states that the derivative of the sum of n functions is the sum of the derivatives. Since a polynomial is a sum of power functions, *12.5.1 and *6.1.3 together make it easy to find the derivative of any polynomial. For example,

$$D_x(3x^2 + 4x - 1) = D_x(3x^2) + D_x(4x) + D_x(-1) = 6x + 4 + 0$$

and

$$D_x(5x^7 - 2x^6 + 3x^4 + x^2) = 35x^6 - 12x^5 + 12x^3 + 2x$$

We now find the bend points of the polynomial

(1) $$y = 2x^3 + 8x^2 + 2x - 12$$

which we graphed in Fig. 12. Taking the derivative of each term and adding, we have

(2) $$D_x y = 6x^2 + 16x + 2$$

which is zero when

(3) $$x = \frac{-4 \pm \sqrt{13}}{3} \doteq -2.54, \ -0.13$$

The corresponding values of y are 1.76 and -12.13. Thus the two key points that are labeled with question marks in Fig. 12 are seen to be approximately $(-2.54, 1.76)$ and $(-0.13, -12.13)$. Note also that the coordinates are irrational, and hence no amount of plotting additional rational points would lead us to their exact locations. Since the derivative has only two zeros, we know that we have found all bend points of the curve and that it will continue to rise to the right of $(2, 40)$ and to fall to the left of $(-4, -20)$.

The student should understand that the points where the derivative is zero are not necessarily bend points. They may be inflections. Hence it is necessary to use discretion, plotting additional points and using common sense to determine the actual situation. More refined methods and rigorous proofs of the statements of this section are available in any good calculus textbook, but the methods here presented are adequate for finding the bend points of polynomials.

Although *12.5.1 is sufficient for finding the derivative of any polynomial, two additional theorems are often convenient both for polynomials and other functions. The first theorem gives a formula for the derivative of a product:

*12.5.2 $D_x(yz) = yD_xz + zD_xy$

This theorem is handy where a function is given in factored form. For example,

(4) $D_x\{(x^2 - 1)(x^2 + 4x - 1)\} \equiv (x^2 - 1)D_x(x^2 + 4x - 1)$
$$+ (x^2 + 4x - 1)D_x(x^2 - 1)$$

(5) $\equiv (x^2 - 1)(2x + 4)$
$$+ (x^2 + 4x - 1)(2x)$$

(6) $\equiv 4(x^3 + 3x^2 - x - 1)$

The second theorem gives a method for differentiating composite functions (Ex. 4.3.14):

*12.5.3   If $y = f(u)$ and $u = g(x)$, then $D_xy = (D_uy)(D_xu)$.

For example, suppose $y = u^2$ and $u = (1 - x^3)$ so that $y = (1 - x^3)^2$. Then $D_uy = 2u = 2(1 - x^3)$ and $D_xu = -3x^2$.

Hence

(7) $$D_x(1 - x^3)^2 \equiv 2(1 - x^3)(-3x^2)$$

(8) $$\equiv 6x^5 - 6x^2$$

We can check this by expanding y as a polynomial and using *12.5.1:

(9) $$D_x(1 - x^3)^2 \equiv D_x(1 - 2x^3 + x^6)$$

(10) $$\equiv -6x^2 + 6x^5$$

EXERCISE 12.5

1. Find, using *12.5.1, the derivatives of the following:

(a) $3x^2 - 2x + 5$

(b) $5x^3 - 2x^2 + 1$

(c) $3 + 32t^2$

(d) $x^5 + 4x^3 - 2x + 15$

(e) $3 - 27x^4 + x$

(f) $3 + 10t + 2t^2$

(g) $2(x^2 - 1) - x^3$

(h) $4x^4 + 2x^3 - x^2 + 3x - 7$

(i) $(x - 3)(2x + 4)$

(j) $1 + 2x - 14x^3 + x^7$

(k) $1 - x - x^2 - x^3 - x^4 - x^5$

(l) $x^5 - 2x^2$

(m) $0.5x^2 + 0.4x^5 - 1 - x$

(n) $x^7 - x$

(o) $x^{2n+1} - x^{2n-1}$

Answers. (a) $2(3x - 1)$. (c) $64t$. (e) $1 - 108x^3$. (g) $x(4 - 3x)$. (i) $2(2x - 1)$. (k) $-(1 + 2x + 3x^2 + 4x^3 + 5x^4)$. (m) $2x^4 + x - 1$. (o) $(2n + 1)x^{2n} - (2n - 1)x^{2n-2}$.

2. Graph the function $y = x^3 + x^2 - 9x - 9$ by finding its zeros and bend points. Find the slope of the tangent line to the curve at each of its intercepts.

3. Graph the following polynomials, indicating zeros and bend points:

(a) $y = 5x^3 - 2x^2 + x - 4$

(b) $y = 6x^2 - x^3$

(c) $y = x(x^2 - 1)$

(d) $y = x^3(x^2 + 2x + 1)$

(e) $y = x^4 - x^2 + 1$

(f) $y = x^5 + 3x^3 + x^2 - 10$

(g) $y = x(x - a)^2$

(h) $y = (x + 1)(x - 2)^2(x - 5)^3$

4. Graph the polynomials of prob. 1.

5. Since we can now find the slope of the tangent line to a polynomial curve at any point, we can find the equation of the tangent there by using the point-slope form. Find the equations of the tangent lines as indicated and sketch the curve and tangent in the neighborhood:

(a) to $y = 3x^4$ at $x = 1$

(b) to $y = 2x^3 - 3x + 4x^2 - 3$ at $x = -1$

(c) to $y = (x - 1)^2 + 1$ at $x = 3$

(d) to $y = (x - 1)^4$ at $x = 1$ and $x = 0$

6. Find y as a function of x if:

(a) $y = u^2,\ u = x^2 + 1$ (d) $y = 2u^4 + u^2 + u,\ u = 1 - x^2$
(b) $y = u^3,\ u = 1 - x$ (e) $y = u^3 + 2,\ u = 1 - 3x$
(c) $y = 2u^2 + au,\ u = x + a$ (f) $y = 1 - 3u^2,\ u = 3x^2 - 4$

Answers. (a) $y = x^4 + 2x^2 + 1$. (c) $y = 2x^2 + 5ax + 3a^2$.

7. In each part of prob. 6 find $D_u y$, $D_x u$, and $D_x y$ by *12.5.3 and directly by using *12.5.1 on the expression for y in terms of x.

8. Find by using *12.5.2 and *12.5.3:

(a) $D_x(Ax^5 - 2Cx^2)$ (e) $D_x(x^2 + 1)^3$ (i) $D_t(1 - 2t)^8$
(b) $D_t(1 - t^3)$ (f) $D_x 2(x^5 + 3x)$ (j) $D_x(1 - 3x)^5$
(c) $D_x x(x + 1)$ (g) $D_x(1 - 4x)^2$ (k) $D_x(ax + b)^n$
(d) $D_x x^2(x^3 + 2)$ (h) $D_x(x + a)^{10}$ (l) $D_u(p + qu)^n$

Answers. (a) $5Ax^4 - 4Cx$. (c) $2x + 1$. (e) $6x(x^2 + 1)^2$. (g) $-8(1 - 4x)$. (i) $-16(1 - 2t)^7$. (k) $na(ax + b)^{n-1}$.

9. Find the derivatives of:

(a) $(1 - x^2)(x^3 + 2x + 1)$ (d) $(3x + 2)^2(1 - x^2)^3$
(b) $(3x + 2)(1 - x^2)$ (e) $(1 - x)^2 + 2(1 - x)$
(c) $(3x + 2)^2(1 - x^2)$ (f) $(4x - 2)(x^2 + 3)(2x - 5)$

10. Find the derivatives of:

(a) $\sin x^2$ (e) e^{1-x} (i) $\sin 2x$
(b) $x \sin x$ (f) $\log_e x^2$ (j) $\cos^2 x - \sin^2 x$
(c) $e^x(x^2 - 1)$ (g) $\log_e(1 + 2x)$ (k) $e^x - e^{-x}$
(d) e^{x^2} (h) $2 \sin x \cos x$ (l) $\cos(Ax + B)$

Answers. (a) $2x \cos x^2$. (c) $e^x(x^2 + 2x - 1)$. (e) $-e^{1-x}$. (g) $2(1 + 2x)^{-1}$. (i) $2 \cos 2x$. (k) $e^x + e^{-x}$.

11. Graph the functions appearing in probs. 8 and 9.

12. Prove that the graph of a polynomial of degree n cannot have more than $n - 1$ bend points.

13. Find the polynomial of third degree that passes through $(0, 0)$, $(1, 1)$, $(2, -2)$, $(-3, 2)$. Graph.

†**14.** Show that *5.4.4 is a special case of *12.5.2.

†**15.** Prove *6.1.3 by induction, using *12.5.2. (*Hint.* $x^n = x \cdot x^{n-1}$.)

16. Show that $D_x(yzw) = yzD_x w + ywD_x z + zwD_x y$.

12.6 The antiderivative

We are now able to find the derivative of any polynomial, and hence the exact rate of change and the slope of the tangent line to its graph at any point. For example, since velocity is the rate of change of distance, we can find the velocity of a body at any time if we know distance as a function of time. Also, since acceleration is the rate of change of velocity, we can find the acceleration if we know velocity as a function of time. It would be natural to

ask whether we could go the other way. Given the velocity as a
function of time, could we find the distance as a function of time?
Given the acceleration, could we find the velocity? More generally,
given a derivative could we find the function from which it came?

Suppose we have a function $F(x)$ and suppose that its derivative
is $f(x)$, that is,

(1) $$D_x F(x) = f(x)$$

When two functions are related as in (1) we call $f(x)$ the derivative
of $F(x)$ and we call $F(x)$ the **antiderivative** or **indefinite integral**
of $f(x)$. The following symbolism is used to represent the antiderivative:

*12.6.1 $F(x) = \int f(x)\,dx \quad \longleftrightarrow \quad D_x F(x) = f(x)$ (Definition)

Thus the elongated S followed by $f(x)\,dx$ means "a function whose
derivative is $f(x)$." The dx merely indicates the independent
variable. Thus

(2) $$\int 3x^2\,dx = x^3, \text{ since } D_x x^3 = 3x^2$$

The antiderivative is analogous to the antilogarithm and other
inverse functions. The student will recall that inverse functions
are often not single-valued. (See 5.3, 6.10, 7.2, and 8.9.) Hence,
he will not be surprised to find that the antiderivative is many-
valued even though the derivative is single-valued. Given $F(x)$
there is only one function $f(x)$ such that $D_x F(x) = f(x)$, but given
$f(x)$ there are infinitely many functions $F(x)$ whose derivatives all
equal $f(x)$. In fact,

*12.6.2 $D_x[F(x) + C] \equiv D_x F(x)$, for any constant C

since $D_x C = 0$. It follows that if $D_x F(x) = f(x)$, then $\int f(x)\,dx$
$= F(x) + C$ where C is an arbitrary constant. Thus $x^3 + 3$,
$x^3 - 15$, $x^3 + 0.01$, and $x^3 + C$, where C is any constant, are all
values of (2). For this reason an arbitrary constant is always
written after the antiderivative. Thus we write:

(3) $$\int 3x^2\,dx = x^3 + C$$

In order to show the usefulness of these ideas, we solve the classic problem of a freely falling body. We start with a single assumption, namely, that a freely falling body has a constant acceleration denoted by g. Now the acceleration is the rate of change of velocity. Thus if $v(t)$ gives velocity as a function of time, we must have:

(4) $$D_t v(t) = g \quad \text{or} \quad v(t) = \int g \, dt$$

We want a function whose derivative is the constant g. We know from experience with polynomials that one such function is gt. Hence we have:

(5) $$v(t) = gt + C$$

where C is some unknown constant. Now velocity is the rate of change of distance. Hence if we represent distance by $s(t)$, we have:

(6) $$D_t s(t) = v(t) \quad \text{or} \quad s(t) = \int (gt + C) \, dt$$

We want a function whose derivative is $gt + C$. The student can easily verify that $gt^2/2 + Ct$ is such a function, hence

(7) $$s(t) = \tfrac{1}{2}gt^2 + Ct + C'$$

where C and C' are unspecified constants. Since $D_t v = g$, and $D_t s = v$, we could have written (4) as $D_t(D_t s) = g$ or $D^2_t s = g$ and (5) as $D_t s = gt + C$. (See Ex. 5.4.21.)

Suppose now that we have additional information about the position and velocity at a certain time. For example, we may know that when $t = 0$, $v = 0$, and $s = 0$. In that case we have from (5):

(8) $$0 = g \cdot 0 + C \quad \text{or} \quad C = 0$$

and then, from (7),

(9) $$0 = g \cdot 0 + C' \quad \text{or} \quad C' = 0$$

Hence the velocity and distance are given by

(10) $$v = gt \quad \text{and} \quad s = \tfrac{1}{2}gt^2$$

More generally, suppose that when $t = t_0$, $v = v_0$, and $s = s_0$.

Then substitution in (5) and (7) yields $C = v_0 - gt_0$, $C' = s_0 - v_0t_0 - gt_0^2/2$, and

(11) $\quad v = v_0 + g(t - t_0) \quad$ and $\quad s = s_0 + v_0(t - t_0) + \frac{1}{2}g(t - t_0)^2$

These are the classic equations of the motion of a freely falling body.

It follows from *12.6.1 and *12.5.1 that

*12.6.3 $\displaystyle\int [f_1(x) + f_2(x) + \cdots + f_n(x)]\, dx = \int f_1(x)\, dx$

$$+ \int f_2(x)\, dx + \cdots + \int f_n(x)\, dx$$

Hence to find the antiderivative of a polynomial we need only find the antiderivatives of power functions. Since finding the derivative of a power function decreases the power by one, we might guess that the antiderivative of a power function would have a power one greater. Thus we might expect the antiderivative of ax^n to be some constant times x^{n+1}. We can easily find this constant. Suppose

(12) $$Cx^{n+1} = \int ax^n\, dx$$

This means $D(Cx^{n+1}) = ax^n$. Hence

(13) $$(n + 1)Cx^n = ax^n$$

and $C = a/(n + 1)$. Thus to find the antiderivative of a power function, we *raise the power by one and divide by the new power.* This is reasonable enough, since it is just the inverse of what we do for finding the derivative of a power function. Now to find the antiderivative of a polynomial we add the antiderivatives of each term and an arbitrary constant. Thus

(14) $$\int (1 + 2x)\, dx = x + x^2 + C$$

(15) $$\int (3 - x + 2x^2)\, dx = 3x - \frac{x^2}{2} + \frac{2x^3}{3} + C$$

EXERCISE 12.6

(Check all antiderivatives by differentiation.)

1. Find the antiderivative of:

(a) 1	(e) $5x^4$	(i) $2x^2$	(m) $4x$	(q) $ABx/2$
(b) $2x$	(f) $(n+1)x^n$	(j) $3x^3$	(n) $2x^2/3$	(r) $10x^{10}$
(c) $3x^2$	(g) 3	(k) x^3	(o) $2cx^3$	
(d) $4x^3$	(h) $4x$	(l) $2x^5$	(p) $2ax$	

Answers. (a) $x + C$. (c) $x^3 + C$. (e) $x^5 + C$. (g) $3x + C$. (i) $2x^3/3 + C$.
(k) $x^4/4 + C$. (m) $2x^2 + C$. (o) $cx^4/2 + C$. (q) $ABx^2/4 + C$.

2. Verify the following by finding the derivatives of the right members:

(a) $\int 2x^2\, dx = 2x^3/3 + C$

(b) $\int x\, dx = x^2/2 + C$

(c) $\int dx = x + C$

(d) $\int (2ax + 3)\, dx = ax^2 + 3x + C$

(e) $\int (1 - 2x)\, dx = x - x^2 + C$

(f) $\int (4x^2 - 5x^3 + x^7 - 1)\, dx = \dfrac{4x^3}{3} - \dfrac{5x^4}{4} + \dfrac{x^8}{8} - x + C$

3. Find the indefinite integrals of each of the following polynomials:

(a) $5x^3 + 7x^6 + 2x - 31$
(b) $2Ax + B$
(c) $1 + 2x + 3x^2 + 4x^3$
(d) $x + 2x^2 + 3x^2 + 4x^4$
(e) $x^2(2 - 9x^5)$
(f) $3cx^2 - 6sx^3 + C$
(g) $(x - 1)^2$
(h) $(1 - x)^2$
(i) $(1 - 3x)^3$

4. Find the function that satisfies the given conditions:

(a) $D_x y = 2x - 6$, $y = 2$ when $x = 0$
(b) $D_x y = -x^2 + 3x$, $y = 0$ when $x = 1$
(c) $D_x y = 2$, $y = 1$ when $x = 2$
(d) $D_x y = m$, $y = y_1$ when $x = x_1$
(e) $D_x y = 2ax + b$, $y = y_1$ when $x = x_1$
(f) $D_x y = Ax^3 - Bx + C$, $y = S$ when $x = 0$
(g) $D_x s = 980t$, $s = 3$ when $t = 0$
(h) $D_x y = 2Mx^2 - Ex$, $y = m$ when $x = n$
(i) $D_x^2 y = 1$, $D_x y = 0$, $y = 0$ when $x = 0$
(j) $D_x^2 y = 1 - x$, $D_x y = 0 = y$ when $x = 0$

Answers. (a) $y = x^2 - 6x + 2$. (c) $y = 2x - 3$. (e) $y - y_1 = a(x^2 - x_1^2)$
$+ b(x - x_1)$.

5. A body is dropping according to the law $v = 32t$, where v is the speed and t is the time. (a) Find the acceleration. (b) Find the distance traveled since the time $t = 0$. Graph s and v as functions of t. [*Answers.* (a) 32. (b) $s = 16t^2$.]

6. The marginal cost of producing a certain commodity is given by $D_x y = 0.1x + 2$, where x is the output in thousands. Find the cost function if the overhead is 1.5.

7. The slope of a tangent line to a certain curve is given by $D_x y = 36x^2 - 5x + 7$, where x is the x-coordinate of a point on the curve. Find the equation of the curve if it passes through the origin.

8. Taking $g = 32$ feet per second per second or 980 centimeters per second per second:

(a) How long does it take a body falling freely from rest to attain a speed of 100 miles per hour? How far has it fallen?

(b) How long does it take an object to fall from the top of a building 100 feet high?

(c) With what initial velocity is a body thrown downward if it attains a speed of 980 centimeters per second in one-half second?

(d) A body falls freely with an initial velocity of 100 centimeters per second. After 3 seconds how far has it fallen, and what is its velocity?

(e) When $t = 0$, $s = 0$, and $v = -100$ feet per second. Thereafter the body is acted upon only by gravity. Find and interpret s when $t = 3$.

(f) Find and interpret s when $v = 0$ in part e.

(g) If a projectile is thrown upward from the ground with an initial velocity of 500 feet per second, how high will it rise and how long will it remain in the air?

(h) If the initial velocity is cut in half in part g, what effect does this have on the maximum height and the time in the air?

Answers. (a) 4.58 seconds, 336 feet. (c) 490 centimeters per second.

†9. Find the antiderivatives of:

(a) $y = x^{-1}$

(b) $y = x^{-2}$

(c) $y = x^{-3}$

(d) $y = \cos x$

(e) $y = e^x$

(f) $y = 3(1 - x)^2$

(g) $y = 2e^{-x}$

(h) $y = -\sin x$

(i) $y = xe^{x^2}$

(j) $y = 2e^{2x}$

(k) $y = x^{0.5}$

(l) $y = x^{-0.5}$

(m) $3 \cos 4x$

(n) $(x^2 + 1)(2x)$

(o) $\dfrac{2 \log_e x}{x}$

12.7 The definite integral

In the previous sections we have introduced one of the fundamental ideas of calculus—the derivative and its inverse, the antiderivative. In this section, we discuss a second important idea, the definite integral. As before, we consider only polynomials and the simplest applications, leaving precise statement and proof to a course in calculus.

The definite integral is conveniently introduced by considering the problem of finding areas. The student is able to find the area of a circle or figure bounded by straight lines. But how could one find the area bounded by the curve $y = x^2$, the x-axis, and the lines $x = 1$ and $x = 2$? (Fig. 12.7.1.) This region does not divide

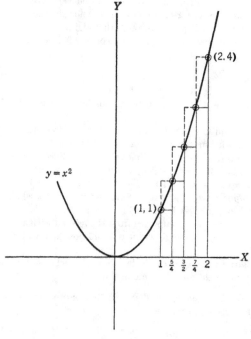

FIG. 12.7.1

up into figures whose areas we have learned to compute. More-over, just what do we *mean* by the area of such a figure? A plausible answer is that the area is the number of square units in the region, but we cannot exactly fill the region with squares no matter how small we choose them!

In order to get an idea of the area of a region bounded by curves, we approximate the region by one whose area we *can* calculate. (Compare 8.2.) In Fig. 12.7.1 we draw vertical lines dividing the interval from $x = 1$ to $x = 2$ into several equal parts. From points where these lines meet the curve, we draw horizontal lines to the adjacent verticals. In this way we have, above and below the curve, broken lines that meet at each point where a vertical meets the curve. The area under each of these broken lines is just the sum of the areas of rectangles, each of which is the product of its base and altitude. For the division into four parts in the figure the area under the upper broken line is:

(1) $\bar{A} = (1/4)(5/4)^2 + (1/4)(3/2)^2 + (1/4)(7/4)^2$

$$+ (1/4)(2)^2 \doteq 2.72$$

The area under the lower broken line is:

(2) $\underline{A} = (1/4)(1)^2 + (1/4)(5/4)^2 + (1/4)(3/2)^2$

$$+ (1/4)(7/4)^2 \doteq 1.97$$

Now the curve $y = x^2$, which bounds the region whose area we wish to define, lies always between the upper and lower broken lines, no matter how many divisions we make. The area evidently lies between $\bar{A}$ and $\underline{A}$, and we can make $\bar{A}$ and $\underline{A}$ very near together by taking a large number of divisions. In fact, each time we increase n, $\bar{A}$ gets smaller and $\underline{A}$ gets larger. (Why?) In the case above for $n = 4$, $\bar{A} - \underline{A} = (1/4)(2^2 - 1^2) = 3/4$. If we divide the interval into n subdivisions, we have:

(3) $\bar{A} = (1/n)\left[\left(\dfrac{n+1}{n}\right)^2 + \left(\dfrac{n+2}{n}\right)^2 + \cdots + \left(\dfrac{2n-1}{n}\right)^2 + 2^2\right]$

(4) $\underline{A} = (1/n)\left[1^2 + \left(\dfrac{n+1}{n}\right)^2 + \cdots + \left(\dfrac{2n-1}{n}\right)^2\right]$

(5) $\bar{A} - \underline{A} = 3/n$

Thus the difference between $\bar{A}$ and $\underline{A}$ is positive, and as n increases it approaches zero. (Why?) It seems plausible that $\bar{A}$ and $\underline{A}$ approach a limiting value as n gets large. We define the area to be this limit and designate it by the symbol

(6) $$A = \int_1^2 x^2\, dx$$

which is read "the definite integral of x^2 between 1 and 2" or "the area under $y = x^2$ between 1 and 2." The same elongated S is used as for the antiderivative, but the meaning is quite different. The definite integral is a number, not a function of the variable

that appears. The definite integral is the limit of sums of the type
(3) or (4), as the number of the terms becomes large. The elongated
S was originally adopted to suggest this sum, and the dx was included to suggest the base of the rectangle whose height is the
ordinate of the curve.

We now generalize the above argument. Suppose we have any
curve given by $y = f(x)$. (See Fig. 12.7.2.) We are interested in

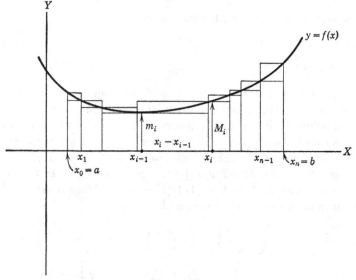

Fig. 12.7.2

the area bounded by this curve and the x-axis between the lines
$x = a$ and $x = b$. As before, we divide up the interval into n subdivisions by taking $a = x_0 < x_1 < x_2 < x_3 \cdots < x_{n-1} < x_n = b$.
We form the sums

(7)
$$\bar{A} = \sum_{i=1}^{i=n} M_i(x_i - x_{i-1})$$

(8)
$$\underline{A} = \sum_{i=1}^{i=n} m_i(x_i - x_{i-1})$$

where M_i is the largest value of $f(x)$ in the interval from x_{i-1} to
x_i and m_i is the smallest value of $f(x)$ in this interval. Thus each
term in a sum is the area of a rectangle of height M_i or m_i and base

$(x_i - x_{i-1})$. The sum $\overline{A}$ is the area under a broken line that lies above the curve. The sum $\underline{A}$ is the area below a broken line lying below the curve. If, as we let the number of intervals n get large and the size of each interval get small, $\overline{A}$ and $\underline{A}$ approach the same limiting value, we call this limit the **definite integral** of $f(x)$ between a and b. It is written:

$$(9) \qquad A = \int_a^b f(x)\,dx$$

We may think of this as meaning the area under the curve $y = f(x)$ between $x = a$ and $x = b$. It is actually the limit of the sums (7) and (8). The numbers a and b are called the **lower** and **upper limits of integration,** $f(x)$ is called the **integrand,** and dx serves to indicate the **variable of integration.** The limit always exists when $f(x)$ is a polynomial.

The discussion above may give us a more precise idea of what is meant by area, but it does not seem to give us any way of actually calculating one! Of course we can approximate an area by calculating $\overline{A}$ and $\underline{A}$. We know that the area lies between them and we can make their difference small by taking n large. However, there is a remarkable connection between the definite integral and the antiderivative. It is given by the **fundamental theorem of integral calculus,** which we state here for polynomials without proof:

*12.7.1 If $F(x)$ is a polynomial such that $D_x F(x) = f(x)$, then

$$\int_a^b f(x)\,dx = F(b) - F(a)$$

This theorem states that in order to evaluate a definite integral we need only find an antiderivative of the integrand and then take the difference between its values at the upper and lower limits of integration! Thus, to find the area under $y = f(x)$, we do not have to do any computations more difficult than finding an antiderivative. Moreover, this method gives us the exact area. The proof is beyond the scope of this course, but the previous discussion should enable the student to understand the meaning of definite integrals and to evaluate them when the integrand is a polynomial. (The fundamental theorem applies to other functions. In particular it holds whenever $f(x)$ is continuous, that is, has an unbroken graph.)

To return to our example (6), the integrand is x^2. An antiderivative of this is $x^3/3$. Hence

$$(10) \qquad A = \int_1^2 x^2 \, dx = \left[\frac{x^3}{3} \right]_1^2$$

$$(11) \qquad\qquad = 2^3/3 - 1^3/3 = 7/3 \doteq 2.33$$

The symbol $[x^3/3]_1^2$ is a way of indicating the antiderivative and the values of x for which it is to be evaluated. Note that we get a fair approximation by averaging $\overline{A}$ and $\underline{A}$, even for $n = 4$. For larger n, the approximation would be much better.

Theorem *12.7.1 enables us to find the area bounded by any curves whose equations are known, provided we can find the required antiderivatives. If we are interested in a region lying between two curves, say $y = f_1(x)$ and $y = f_2(x)$, as in Fig. 12.7.3,

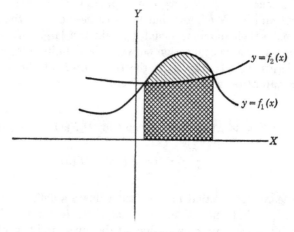

FIG. 12.7.3

we can find the area under each and subtract. If we are interested in the area between the x-axis and a curve below it, as in Fig. 12.7.4, *12.7.1 gives a negative result, and the desired area is given by the absolute value. A complicated region can be broken up into simpler ones. With these indications, the student can handle a large number of situations involving definite integrals. Of course, we cannot here do more than hint at the tremendous power and beauty of the calculus as a mathematical subject and tool of application.

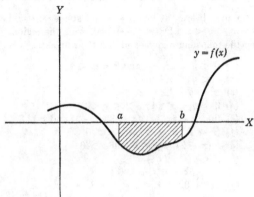

FIG. 12.7.4

EXERCISE 12.7

1. Draw a large graph of $y = x^2$ similar to Fig. 12.7.1, but involving division of the interval from 1 to 2 into 8 subdivisions. Draw the broken lines above and below the curve. Evaluate $\bar{A}$ and $\underline{A}$. Compare their average with the exact area given by (10). (*Answers.* $\bar{A} = 2.52$, $\underline{A} = 2.15$.)

2. Draw a large figure as in prob. 1 with 4 and 8 subdivisions and illustrate how $\bar{A}$ gets smaller and $\underline{A}$ gets larger as n increases.

3. Find the area under $y = x^2$ between $x = 1$ and $x = 3$ by using *12.7.1. Find $\underline{A}$ and $\bar{A}$ for $n = 4$ and compare $(\underline{A} + \bar{A})/2$ with the exact area. (*Answers.* $A = 8.6\dot{6}$, $(\bar{A} + \underline{A})/2 = 8.75$.)

4. Show that $\int_a^b c\,dx = c(b - a)$. Sketch $y = c$ and indicate the area.

5. Show that $\int_0^b mx\,dx = \dfrac{mb^2}{2}$. Sketch $y = mx$ and indicate the area.

6. Evaluate the following definite integrals by using *12.7.1:

(a) $\displaystyle\int_0^1 x\,dx$ (d) $\displaystyle\int_{-1}^3 x^4\,dx$ (g) $\displaystyle\int_2^5 (25 - x^2)\,dx$

(b) $\displaystyle\int_{-1}^1 x^2\,dx$ (e) $\displaystyle\int_{-2}^2 (1 + x + x^2)\,dx$ (h) $\displaystyle\int_{-2}^{-1} dx$

(c) $\displaystyle\int_0^2 2x^2\,dx$ (f) $\displaystyle\int_1^3 (x-1)(x-3)\,dx$ (i) $\displaystyle\int_{-3}^{-1} (1 - x^2)\,dx$

Answers. (a) 0.5. (c) 16/3. (e) 28/3.

7. Draw a sketch for each part of prob. 6 and indicate the region whose area is to be found.

8. Find $\int_{-1}^{0} x^3\, dx$. Draw a sketch and interpret the result.

9. Find the area below the line $y = x$ and above $y = x^2$, i.e., the region determined by $x^2 < y < x$. Draw a sketch showing the region.

10. Sketch the following regions and find their areas:

(a) $0 < y < x^2 + 1$ and $2 < x < 4$
(b) $0 < y < 3 + x - x^2$
(c) $x^2 < y < 3x - 2$
(d) $0 < y < x^4$ and $-2 < x < 2$
(e) $0 < y < (x - 1)(x - 3)(x - 5)$ and $x < 3$
(f) $(x - 1)^3 < y < x$ and $x > 0$
(g) $x^2(x - 2) < y < 0$ and $x > 0$
(h) $0 < y < x^2(4 - x^2)$
(i) $x^2 - 5x - 6 < y < 0$
(j) $2x^2 + 3x + 1 < y < 1 + 8x - 3x^2$

Answers. (a) 62/3. (c) 1/6. (e) 4. (g) 4/3. (i) 343/6.

11. Find:

(a) $\int_0^s x^2\, dx$

(b) $\int_{-a}^{a} x^3\, dx$

(c) $\int_0^a (x^2 - a^2)\, dx$

(d) $\int_0^t x^3\, dx$

(e) $\int_0^t t^3\, dt$

(f) $\int_a^b (Ax^2 + Bx + C)\, dx$

Answers. (a) $s^3/3$. (c) $-2a^3/3$. (e) $t^4/4$.

†12. Show that $\int_a^b f(x)\, dx = \int_a^b f(t)\, dt$, and hence that the value of a definite integral depends only upon the function that is the integrand and the limits of integration and not upon the letter used for the variable in the function. (For this reason, the variable of integration is called a "dummy variable.")

13. Find the area under one loop of the cosine curve. Show that it is the same as the area under one loop of the sine curve.

14. Find the area between the hyperbola $1/x$, the x-axis, and the lines $x = 1$ and $x = 3$. (*Answer.* 0.4771.)

15. Find $\int_1^2 e^x$ and sketch the corresponding area. [*Answer.* $e(e - 1)$.]

16. Read the article on *Calculus* in the *Encyclopaedia Britannica*, and write a discussion of one of the examples given there.

12.8 Arithmetic and number bases

The familiar operations of decimal arithmetic are actually based on the algebra of polynomials, and it is for this reason that we have postponed their discussion until now. We give no formal proofs, but indicate by examples the lines that they would take.

The techniques of arithmetic calculation depend upon our method of writing numbers in decimal form. (If the student doubts this, let him try to multiply XXVI by XIV as they stand according to the usual rules of decimal arithmetic!) In the decimal system, the meaning of a digit depends upon its position. Thus the 3 in 143 stands for just 3, the 3 in 134 stands for 30, and 3 in 314 stands for 300. For this reason the system is called a **positional system.** Thus 247 means $200 + 40 + 7$, 51.03 means $50 + 1 + 3/100$. It is more instructive to write these numbers as follows:

(1) $$247 = 2 \cdot 10^2 + 4 \cdot 10^1 + 7 \cdot 10^0$$

(2) $$51.03 = 5 \cdot 10^1 + 1 \cdot 10^0 + 0 \cdot 10^{-1} + 3 \cdot 10^{-2}$$

We see that *a number written in decimal form is a polynomial in powers of ten.* In general, if $a_0, a_1, a_2, a_3 \cdots$ stand for digits chosen from the set 0, 1, 2, $\cdots$ 9, then an expression of the form $a_n a_{n-1} a_{n-2} \cdots a_3 a_2 a_1 a_0$ stands for

(3) $$a_n 10^n + a_{n-1} 10^{n-1} + \cdots a_3 10^3 + a_2 10^2 + a_1 10 + a_0 10^0$$

Consider a simple addition of two numbers according to the usual algorithm:

(4)
$$
\begin{array}{r}
376 \\
819 \\
\hline
1195
\end{array}
$$

We have added from the right, one digit at a time, "carrying" when the sum exceeds 9. This would be clearer if we wrote it as follows:

(5)
$$
\begin{array}{r}
376 \\
819 \\
\hline
15 \\
8 \\
11 \\
\hline
1195
\end{array}
$$

Now we write the numbers as the polynomials for which they stand and carry out the addition according to the rules of algebra. We have:

(6) $376 + 819 = (3 \cdot 10^2 + 7 \cdot 10 + 6) + (8 \cdot 10^2 + 1 \cdot 10 + 9)$

(7) $= (3 \cdot 10^2 + 8 \cdot 10^2) + (7 \cdot 10 + 1 \cdot 10) + 6 + 9$

(8) $= (3 \cdot 10^2 + 8 \cdot 10^2) + (7 \cdot 10 + 1 \cdot 10) + 15$

(9) $= (3 \cdot 10^2 + 8 \cdot 10^2) + (7 \cdot 10 + 1 \cdot 10 + 1 \cdot 10) + 5$

(10) $= (3 \cdot 10^2 + 8 \cdot 10^2) + 9 \cdot 10 + 5$

(11) $= 11 \cdot 10^2 + 9 \cdot 10 + 5$

(12) $= (10 + 1)10^2 + 9 \cdot 10 + 5$

(13) $= 1 \cdot 10^3 + 1 \cdot 10^2 + 9 \cdot 10 + 5 = 1195$

Of course it would be ridiculous to go through all these steps to do an addition, but the student should examine the steps in (6) through (13) in order to see that they are justified by the fundamental theorems of Chapter 3. Then he should compare with (5) and (4) in order to see that these are merely abbreviations.

Now consider the multiplication algorithm:

$$
\begin{array}{r}
429 \\
37 \\
\hline
3003 \\
1287 \\
\hline
15873
\end{array}
$$

(14)

The process would be clearer if we wrote it as follows:

$$
\begin{array}{r}
429 \\
37 \\
\hline
63 \\
14 \\
28 \\
27 \\
6 \\
12 \\
\hline
15873
\end{array}
$$

(15)

We multiply each digit in the first number by each digit in the second, moving each product one space to the left for each space we have moved in the factors. Then we add. In (14) we "carry" the tens digit in any product and so do not have to write as much as in (15). Thus (14) is an abbreviation of (15). It is easy to see that (15) is just an abbreviation of the work of multiplying the two numbers expressed as polynomials in 10. Letting b stand for 10, we have:

(16) $(37)(429) = (3b + 7)(4b^2 + 2b + 9)$

(17) $\qquad = 3b(4b^2 + 2b + 9) + 7(4b^2 + 2b + 9)$

(18) $\qquad = 12b^3 + 6b^2 + 27b + 28b^2 + 14b + 63$

From here on, the problem is one of addition. Note that the co-efficients of the various powers of b are just the products in (15), where the power of b is indicated by position. The actual numbers that are being added in (15) would be stated explicitly if zeros were placed at the right of each number to make the right-hand margin of the computation even.

The discussion above illustrates the fact that the usual techniques of arithmetic depend upon the fundamental rules of algebra given in Chapter 3 and upon the way in which we write out numbers, namely, as polynomials in 10. If we wrote out numbers in terms of special symbols for one, five, ten, fifty, one hundred, etc., as did the Romans, or by means of letters of the alphabet, as did the Greeks, the techniques of arithmetic manipulation would be different. However, the basic properties of numbers, given in Chapter 3, would still be valid.

Among the many possible ways of representing numbers, there is a whole group of methods that are similar to our decimal system. They make use of zero and several digits, whose values depend upon their positions in the number. In all these systems a number is written as a polynomial in some number b called the **base**. In the decimal system $b = 10$. The system with $b = 2$ is of particular importance in modern computing machines and in theoretical logic, and is called the **binary system**. We describe some of its properties.

Any number can be expressed as a polynomial in 2 with coefficients either 0 or 1. Thus $7 = 1 \cdot 2^2 + 1 \cdot 2^1 + 1 \cdot 2^0$ and $53 = 1 \cdot 2^5 + 1 \cdot 2^4 + 0 \cdot 2^3 + 1 \cdot 2^2 + 0 \cdot 2^1 + 1 \cdot 2^0$. It is easy to evaluate a

polynomial in 2. To express a number as a polynomial in 2, we
find the highest power of 2 that is less than or equal to the number
and subtract this power from the number. Then we find the
highest power that is less than or equal to the remainder and
repeat. In this way we get a polynomial in powers of 2, starting
with the highest power. Each power has a coefficient of either
0 or 1. In order to write the number in the binary system we simply
write down the coefficients in order. The position of a 0 or 1
indicates the power of 2 it represents. Thus we write 7 as 111
and 53 as 110101. The student should verify that 10 (decimal)
= 1010 (binary) and 2 (decimal) = 10 (binary).

Looking back on arithmetic, the student will recall that all that
is required is a knowledge of the addition and multiplication tables
for the digits and the rules of procedure illustrated earlier in this
section. For the binary system, the same holds, but the tables
are very simple. They are as follows:

(19)   Addition   Multiplication

                        0   1   0 1
                    0   0   1   0   0 0
                    1   1   10   1   0 1

The addition table means that $0 + 0 = 0, 0 + 1 = 1 + 0 = 1$,
and $1 + 1 = 10$. For multiplication, any product is zero except
$1 \cdot 1 = 1$. With these tables we can carry out the operations on
binary numbers according to the usual rules. For example, the
multiplication of 7 by 9 becomes in binary:

(20)   111
                                 1001

                                  111
                              111

                               111111

Multiplication is extremely simple. To multiply by 1 we simply
recopy the number. Thus we need to do no real multiplications
in the usual sense. It is just a question of recopying or not. Then
to add, we write down 0 or 1 according to whether the number of
1's to be added is even or odd, and carry over a 1 to the next

column for every two 1's. Thus the addition of 10, 5, 14, and 17 becomes in binary:

$$
\begin{array}{r}
(21) \qquad\qquad 1010 \\
101 \\
1110 \\
10001 \\
\hline
101110
\end{array}
$$

The simplicity of the operations with binary numbers makes them very adaptable to computation by means of electronic devices. Machines have been built to translate numbers from one system to another. The trouble to do this is more than compensated for by the speed with which the computations can be handled by the binary computers—several thousand times as fast as the speediest human calculator!

EXERCISE 12.8

In order to make clear what number system is being used, place the base in parentheses after each number, unless the base is 10. Thus 10(2) means 2.

1. Express the following numbers as polynomials in 10, with coefficients less than 10: (a) 3582. (b) 1980. (c) 1000346. (d) 18.36. (e) 0.01512. (f) 258.003.

2. Explain the rules of algebra that justify each of the steps (6)–(13) and (16)–(18). Explain the connection with the corresponding manipulations in (4)–(5) and (14)–(15).

3. Write a discussion of your own to illustrate the correctness of the addition and multiplication algorithms, using $495 + 187$ and $29 \cdot 698$.

†4. In a certain 3-digit number, the digits are reversed and the smaller of the two numbers subtracted from the larger. The result is then added to the result of reversing the order of its digits. Under what conditions is the final result 1089? (*Suggestion:* Try several numbers, including 185 and 483.)

5. Express the following numbers as polynomials in b with coefficients less than b, for b as indicated:

(a) 4, $b = 2$　　(d) 30, $b = 2$　　(g) 8, $b = 4$　　(j) 40, $b = 3$
(b) 6, $b = 2$　　(e) 4, $b = 3$　　(h) 24, $b = 4$　　(k) 72, $b = 7$
(c) 19, $b = 2$　　(f) 9, $b = 3$　　(i) 38, $b = 2$　　(l) 150, $b = 8$

Answers. (a) $1 \cdot 2^2$. (c) $1 \cdot 2^4 + 1 \cdot 2^1 + 1 \cdot 2^0$. (e) $1 \cdot 3 + 1 \cdot 3^0$. (g) $2 \cdot 4^1$. (i) $1 \cdot 2^5 + 1 \cdot 2^2 + 1 \cdot 2$. (k) $1 \cdot 7^2 + 3 \cdot 7 + 2 \cdot 7^0$.

6. Write in the binary notation: (a) 4. (b) 6. (c) 19. (d) 30. (e) 27. (f) 100. (g) 37. (h) 65. (i) 50. (j) 100. (k) 144. [*Answers.* (a) 100. (c) 10011. (e) 11011. (g) 100101.]

7. Translate into the decimal system the following numbers in the binary notation:

(a) 110   (c) 10101   (e) 111   (g) 101010   (i) 101001
(b) 10011   (d) 1010   (f) 1011011 (h) 110011   (j) 100001

Answers. (a) 6. (c) 21. (e) 7. (g) 42. (i) 41.

8. Carry out the indicated computations on the following numbers written in binary notation:

(a) 10101 + 10001   (d) (11001)(101011)
(b) 10 + 110   (e) (10)(10011)
(c) 10001 + 11001 + 10111 + 10   (f) (1110)(1001)

Answers. (a) 100110. (c) 1000011. (e) 100110.

9. Translate into decimal form the numbers involved in prob. 8 and check your results.

10. Write 4, 9, 18, 12 in a system with base 3, i.e., using the digits 0, 1, 2. In this system the first few integers are 0, 1, 2, 10, 11, 12, 20, 21, 22, 100.

11. Make up an addition and multiplication table for the base 3. Translate into this system and carry through the following operations: (a) 28 + 4. (b) 18 + 36 + 13. (c) 8·19. (d) 45·72.

12. Write 8, 24, 18, 12 in the system with base 4. (The first few integers in this system are 0, 1, 2, 3, 10, 11, 12, 13, 20, ····.)

13. When we use a base b, we need digits to represent each integer less than b. When $b \leq 10$, we can use the usual decimal digits. For $b > 10$, we must invent new ones. For example, if we wish to use the base 12 (the **duodecimal system**) we might use the letters d and e to represent the two numbers following 9. Then the first twelve integers would be represented by 1, 2, 3, 4, 5, 6, 7, 8, 9, d, e, 10. We have 13 = 11(12), 25 = 21(12), 23 = 1e(12), and 144 = 100(12). This system has been called "counting by dozens," and there is a group of people who think it should be adopted in place of our decimal system. Translate the following numbers into the duodecimal system: (a) 3. (b) 18. (c) 36. (d) 38. (e) 90. (f) 100. (g) 144. (h) 143. (i) 149. (j) 200. (k) 50. (l) 1000. [*Answers.* (a) 3. (c) 30. (e) 76. (g) 100. (i) 105. (k) 42.]

14. So far we have dealt only with integers. Other numbers, when written in decimal form, involve negative powers of the base, the power corresponding to each digit being indicated by its position. In terminating decimals these negative powers may always be eliminated by multiplication and division by a power of the base. Thus 0.1(2) = $\frac{1}{2}$, 0.01(2) = $\frac{1}{4}$, 0.001(2) = $\frac{1}{8}$, etc. Find the values to base 10 of the following binary decimals: (a) 1.01. (b) 101.001. (c) 10.101. (d) 0.111. (e) 0.0101. (f) 1.0001. (g) 0.11101. (h) 10101. [*Answers.* (a) $\frac{5}{4}$. (c) $2\frac{1}{8}$. (e) $\frac{5}{16}$. (g) $2\frac{9}{32}$.]

15. The duodecimal system has advantages for expressing fractions because many that give infinite decimals in the 10 system give terminating decimals in the 12 system. Illustrate this by translating the following numbers in duodecimal notation to the decimal system: (a) 0.1. (b) 0.2. (c) 0.3. (d) 0.4. (e) 0.8. (f) 0.9. (g) 0.d. (h) 0.e. (i) 0.11. (j) 0.15. (k) 0.25. [*Answers.* (a) $\frac{1}{12}$. (c) $\frac{1}{4}$. (e) $\frac{2}{3}$. (g) $\frac{5}{6}$.]

16. We have not considered division as yet. However, the algorithm for it consists essentially in making trials, checking by multiplication, and rewriting the dividend as a product of the divisor and a quotient plus a remainder that gets smaller as the work continues. The justification of the procedure is contained in the sort of reasoning we employed in 12.1. The same technique is applicable when any base is used, using of course the appropriate addition and multiplication tables for that base. Compute the following quotients, where the numbers are given in binary notation. Check by translating data and results to the decimal system:

(a) 1010/101 (c) 11101/1010 (e) 100111/111
(b) 110/100 (d) 10111/1000 (f) 101000/1101

†**17.** Evaluate to the base 10 the following infinite repeating binary decimals:

(a) 0.1111 $\cdots$ (c) 0.011011011 $\cdots$ (e) 1101011.$\dot{1}0\dot{1}$
(b) 1.010101 $\cdots$ (d) 10.1010 $\cdots$ (f) 0.$\dot{0}\dot{1}\dot{0}\dot{1}\dot{1}$

18. The base 60 is used in writing angles in degrees, minutes, and seconds, and in writing time in hours, minutes, and seconds. Thus 1 hour = 60 minutes = 60^2 seconds. Explain the usual rules for handling such numbers in terms of the following examples, and indicate the essential similarity to the techniques of decimal arithmetic:

(a) 6° 35′ 47″ + 2° 50′ 18″ (c) (5° 20′ 39″)/2
(b) 2° 18′ 40″ − 1° 36′ 50″ (d) 8.2(18° 51′ 8″)

19. Sometimes a number is expressed as a polynomial involving units equal to an arbitrary number of basic units, rather than powers of a single base. Examples are yards, feet, and inches or pounds, shillings, and pence. Thus 2 yards, 7 feet, 3 inches = [2(36) + 7(12) + 3] inches. Thus for yards, feet, and inches we deal with polynomials in 36, 12, and 1 and always keep the coefficients of each number smaller than the number of times it is contained in the next larger one. Thus if units equal to a and b basic units are used, where $a > b$, a number is expressed as $xa + yb + z$ basic units with $z < b$, $y < a/b$. The decimal system is the special case given by $a = b^2 = 10^2$. Find a and b for: (a) Tons, pounds, and ounces. (b) Hours, minutes, and seconds. (c) Gallons, quarts, and pints. (e) Pounds, shillings, and pence.

20. Carry out the following computations and explain them in terms of the comments in prob. 19:

(a) 11 pounds + (5 pounds, 1 ounce) + 9 ounces
(b) 13 hours, 50 minutes, 40 seconds + 30 minutes, 29 seconds
(c) 2(3 yards, 2 feet, 8 inches)
(d) 0.5(2 yards, 1 foot, 9 inches)
(e) (1 hour, 37′, 14.5″)(4.2)
(f) 4(2 gallons, 3 quarts, 1 pint)
(g) (3 feet, 6 inches)(2 feet, 4 inches)
(h) (4 yards, 2 feet, 1 inch)^2
(i) (2 inches)(5 yards, 3 inches)
(j) (16 feet, 4 inches)/(11 inches)

†**21.** Prove that in the decimal system a number is divisible by 9 if and only if the sum of its digits is divisible by 9.

†**22.** Prove the same property for divisibility by 3.

†**23.** The "excess of nines" in a number is the remainder when the number is divided by 9. Show that: (a) The excess of nines in a number is the excess in the sum of its digits. (b) The excess of nines in any sum is the excess in the sum of the excesses of the summands. (c) The excess in a product is the excess in the product of the excesses of the factors. (These propositions justify the checking procedure known as "casting out nines.")

Algebraic Functions

We say that y is an **algebraic function** of x if y is defined by an algebraic equation in which the coefficients are polynomials in x. Thus "y is an algebraic function of x" means that y and x are related by an equation of the form:

$$\text{(1)} \quad P_n(x)y^n + P_{n-1}(x)y^{n-1} + \cdots + P_3(x)y^3$$
$$+ P_2(x)y^2 + P_1(x)y + P_0(x) = 0$$

where the P's are polynomials in x and $P_n(x) \not\equiv 0$. We have already studied a number of algebraic functions. All polynomials are algebraic functions, since $y = P(x)$ results from setting $n = 1$, $P_1(x) = 1$, and $P_0(x) = -P(x)$ in (1). The linear equation $Ax + By + C = 0$ and the equation of a conic section are other examples of equations that determine y as an algebraic function of x. On the other hand, the circular, exponential, and logarithmic functions are not algebraic. There is no equation of the type (1) whose solution is $y = \sin x$. Functions that are not algebraic are called **transcendental**.

In this chapter we consider a few additional types of algebraic functions. When (1) is of the first degree in y, we have $y = -P_0/P_1$, that is, y is the quotient of two polynomials in x. Such a function is called a **rational function**. Polynomials are special cases of rational functions. (Why?) Functions that are not rational are called **irrational**. They include some irrational algebraic functions and all the transcendental functions. A type of irrational function that appears often is a function involving fractional powers of x or a polynomial in x under a radical. A simple type is one of the form $y = \sqrt{R(x)}$, where $R(x)$ is a rational function that is not a perfect square. It is a solution of $y^2 - R(x) = 0$.

The foregoing classification of functions is similar to the classification of numbers. Polynomials play a role among functions similar to that played by integers among numbers. Thus rational numbers are quotients of integers, and rational functions are quotients of polynomials. Irrational numbers are all other real numbers, and irrational functions all other functions. Likewise, although we have not previously mentioned this, an **algebraic number** is a real number that is the root of an algebraic equation with integral coefficients. Other real numbers are called **transcendental**. Irrational numbers include some algebraic numbers and all non-algebraic numbers. Examples of transcendentals are π and e.

13.1 Graphing rational functions

In this section we are concerned with functions that can be put in the form

$$(1) \qquad\qquad y = P(x)/Q(x)$$

where $P(x)$ and $Q(x)$ are polynomials in x. To graph such a function, we begin as usual by considering certain key features of the curve such as intercepts, symmetry, asymptotes, and bend points. We discuss the general method and illustrate it with the function

$$(2) \qquad\qquad y = (x - 1)(x + 4)/(x + 1)$$

To find the *intercepts* we substitute $x = 0$ and then $y = 0$. In (2), $x = 0$ yields $y = -4$. To find the values for which $y = 0$, we note that a fraction is zero if and only if its numerator is zero and its denominator is not zero. In (2), $y = 0$ implies either $x = 1$ or $x = -4$. Hence the intercepts are $(0, -4)$, $(1, 0)$, $(-4, 0)$. We plot these points in Fig. 13.1. Looking at (1), we see that for $x = 0$ there is just one y given by $P(0)/Q(0)$. Hence a rational function has just one y-intercept unless $Q(0) = 0$, in which case it will have none. (Why?) Also (1) indicates that the x-intercepts are given by the real roots of $P(x) = 0$ that are not also roots of $Q(x) = 0$.

In order to test for *symmetry* we consider the effect of replacing x by $-x$, y by $-y$, and both x and y by $-x$ and $-y$. If the equation remains unchanged after these substitutions, the graph is symmetric with respect to the y-axis, the x-axis, and the origin, respectively. (Why?) It is clear from the form of (1) that the equation is always changed if y is replaced by $-y$, but the other two types of symmetry are possible for rational functions. For example, $y = 1/x$ is a rational function whose graph is symmetric with respect to the origin, and the graph of $y = 1/(x^2 + 1)$ is

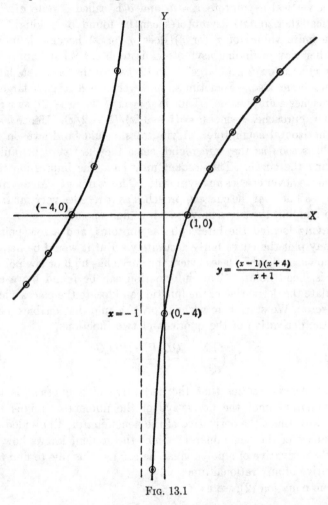

$$y = \frac{(x-1)(x+4)}{x+1}$$

(−4, 0)

(1, 0)

$x = -1$

(0, −4)

FIG. 13.1

symmetric with respect to the y-axis. Equation (2) is not symmetric with respect to the axes or the origin.

Asymptotes play an important part in graphing rational functions. It is clear from (2) that when $x = -1$, y is undefined (why?), and when x approaches -1, y gets very large. Thus the curve approaches the line $x = -1$ as y gets very large in absolute value. We call the line a vertical asymptote. In general, for any value a of x for which $Q(a) = 0$ and $P(a) \neq 0$, the curve (1) has a vertical asymptote $x = a$, and a is called a **pole** of P/Q. Vertical (horizontal) asymptotes can be found by asking: "For what finite values of x (or y) does y (or x) become infinite?" Another way of finding asymptotes is to ask: "What happens to y (or x) as x (or y) gets large?" In (2) we see that as x gets large, y gets large also. Sometimes, however, when x gets large, y approaches a finite value. Thus in $y = (x - 1)/(x - 2)$, as x gets large y approaches 1, since $y = (1 - 1/x)/(1 - 2/x)$. Hence $y = 1$ is a horizontal asymptote. Asymptotes should be drawn on the graph as soon as they are found, since they are very helpful in plotting the curve. The student may have the impression that the curve never crosses an asymptote. This is not so. An **asymptote** is a line that the curve approaches as it recedes very far from the origin, but the curve may cross it at finite points.

Having located the intercepts, asymptotes, and a few points, we may plot the curve fairly accurately. But it would be helpful to know exactly the places where the curve has high or low points, that is, bend points. This information can be found if we can calculate the derivative of the function and locate the places where it is zero. We state here without proof a rule that enables us to find the derivative of the quotient of two functions:

*13.1.1
$$D_x\left(\frac{P}{Q}\right) = \frac{QD_xP - PD_xQ}{Q^2}$$

This theorem states that the derivative of a quotient is the denominator times the derivative of the numerator minus the numerator times the derivative of the denominator, all divided by the square of the denominator. Since the student knows how to find the derivative of a polynomial, he can use this rule to find the derivative of any rational function.

From equation (2), we have:

(3) $y = (x^2 + 3x - 4)/(x + 1)$

(4) $D_xy = \dfrac{(x + 1)(2x + 3) - (x^2 + 3x - 4)(1)}{(x + 1)^2}$

(5) $= (x^2 + 2x + 7)/(x + 1)^2$

Since the numerator has no real roots (why?), there are no bend points.

We can now plot the entire curve with assurance. Since y changes sign at $x = 1$ and $x = -4$ and goes "off to infinity" at $x = -1$, a few points should be found to the left of $x = -4$, to the right of $x = 1$, and in each of the intervals -4 to -1 and -1 to 1. These points and the final curve are shown in Fig. 13.1. It appears to be a hyperbola, and this can be verified from equation (2). (How?)

It may sometimes be of interest to find the values of the variable for which a given rational function is positive or negative. This requires the solution of an inequality of the form $P(x)/Q(x) > 0$. The solution may be found graphically by plotting the left member and observing where it is above the x-axis, but it may also be found by algebra alone. We know that a fraction is positive if and only if the numerator and denominator have the same sign. Hence the solution is given by those x's that satisfy $P(x) > 0$ *and* $Q(x) > 0$ and those that satisfy $P(x) < 0$ *and* $Q(x) < 0$. In the case of the function given by (2), $y > 0$ when

$$(6) \qquad (x-1)(x+4) > 0 \quad \text{and} \quad x + 1 > 0$$

or when

$$(7) \qquad (x-1)(x+4) < 0 \quad \text{and} \quad x + 1 < 0$$

The first inequality in (6) holds if $x - 1 > 0$ and $x + 4 > 0$ or if $x - 1 < 0$ and $x + 4 < 0$. The first inequality in (7) holds if $x - 1 > 0$ and $x + 4 < 0$ or if $x - 1 < 0$ and $x + 4 > 0$. Thus we have four cases to consider:

(6′) $x - 1 > 0$, $x + 4 > 0$, and $x + 1 > 0$,

$$\text{i.e., } x > 1, x > -4, x > -1$$

(6″) $x - 1 < 0$, $x + 4 < 0$, and $x + 1 > 0$,

$$\text{i.e., } x < 1, x < -4, x > -1$$

(7′) $x - 1 > 0$, $x + 4 < 0$, and $x + 1 < 0$,

$$\text{i.e., } x > 1, x < -4, x < -1$$

(7″) $x - 1 < 0$, $x + 4 > 0$, and $x + 1 < 0$,

$$\text{i.e., } x < 1, x > -4, x < -1$$

Now (6′) is satisfied if $x > 1$ (why?), (6″) and (7′) are impossible (why?), and (7″) is satisfied if $-4 < x < -1$. Hence the values of x that satisfy $(x - 1)(x + 4)/(x + 1) > 0$ are those in the intervals $x > 1$ and $-4 < x < -1$. These are just the regions where the curve is above the x-axis in Fig. 13.1. The computations in this procedure are easy, but the student must be careful to consider all cases in which the function has the desired sign. In this connection, it is helpful to recall that a product is positive or negative according as the number of negative factors is even or odd.

EXERCISE 13.1

1. Graph the function $y = (x - 2)(x + 2)/(x - 1)$, following the procedure of the section. Graph it also by the methods of Chapter 11.

2. Why is there no point on the graph of $y = P(x)/Q(x)$ where $Q(x) = 0$? Are the functions $y = x(x - 1)/x$ and $y = x - 1$ the same? [*Answer.* They are not the same, because the first one is undefined when $x = 0$. Thus the first yields the straight line $y = x - 1$ with the exception of the point $(0, -1)$.]

3. Graph the following, in each case finding the intercepts, symmetry, asymptotes, and bend points.

(a) $x^2y = 1$

(b) $y = x(1 + x^2)^{-1}$

(c) $y = x(1 - x^2)^{-1}$

(d) $y = (x - 1)x^{-2}$

(e) $2x = y(x - 1)(x + 3)$

(f) $(x + 2)y = x^2(x - 2)$

(g) $y = (x^2 + 1)(x^2 - 1)^{-1}$

(h) $y = (x^2 + x + 1)(x^2 + 1)^{-1}$

(i) $y = 2x^{-3}$

(j) $x^2y = x^2 + x - 6$

(k) $y(x^2 - 1) = x^3$

(l) $y = (x^2 - 1)(x + 2)(x + 3)$

(m) $(x - 1)^2(x + 3)y = x + 1$

(n) $y(1 - x^2)(x^2 + 1)^2 = 1$

(o) $y(x + 4) = x^2 - 1$

4. Solve the following inequalities, showing the region in the x,y-plane.

(a) $(x - 1)(x + 2) < 0$

(b) $x(x + 1)^2 > 0$

(c) $x(x + 1)(x - 3) < 0$

(d) $x(x - 1)^{-1} > 0$

(e) $x^2(x - 2)(x + 3)^{-1} \leq 0$

(f) $x(x + 1)(x - 4)^{-1} \leq 0$

(g) $x(x - 3)(x^2 - 1)^{-1} \geq 0$

(h) $(x + 1)(x - 2) > (x - 3)(x + 2)$

Answers. (a) $-2 < x < 1$. (c) $x < -1$ or $0 < x < 3$. (e) $-3 < x \leq 2$.

5. Graph the functions that appear in prob. 4.

6. Why is it incorrect to try to solve $P(x)/Q(x) > 0$ by rewriting it as $P(x) > Q(x)$ and then $P(x) - Q(x) > 0$? Is the solution of $P(x)Q(x) > 0$ the same as that of $P(x)/Q(x) > 0$?

7. Graph the following equation, which appears in spectroscopy:

$$\lambda = 3646n^2(n^2 - 4)^{-1}$$

8. Find the cubic polynomial passing through $(0, 0)$, $(5, 0)$, $(-3, -1)$, $(2, 4)$.

9. Find the quartic polynomial passing through $(0, 1)$, $(2, 0)$, $(-3, 0)$, $(6, 3)$, $(-2, 1)$.

10. Show that a polynomial $y = A_n x^n + \cdots + A_1 x + A_0$ has the same sign as the first term $A_n x^n$ for all x greater than some positive number N and all x less than some negative number M. (*Suggestion.* Write y as $A_n x^n$ times a rational function.)

13.2 Manipulation of fractions

We have been using the definitions and identities of 3.7 and 3.11 to manipulate fractional expressions. In this section we deal more systematically with fractions involving polynomials.

If the numerator and denominator of a fraction have no polynomial factor in common, it is said to be in **lowest terms.** A fraction may be put in lowest terms by factoring numerator and denominator as far as possible and then using the identity $ca/cb \equiv a/b$. Thus

$$(1) \quad \frac{(x^2 - 1)(x^2 + 2x - 8)}{(x^2 - 3x + 2)(x^2 + 1)} \equiv \frac{(x - 1)(x + 1)(x - 2)(x + 4)}{(x - 1)(x - 2)(x^2 + 1)}$$

$$(2) \qquad\qquad\qquad \equiv \frac{(x + 1)(x + 4)}{x^2 + 1}$$

A fraction is called **proper** or **improper** according as the numerator is or is not of lower degree than the denominator. An improper fraction may be rewritten as a polynomial plus a proper fraction. The process may be described as dividing the denominator into the numerator to find a quotient and remainder. (Compare Ex. 12.8.16.) The algorithm for doing this is the same as for long division of one number by another or a polynomial by a binomial (see 12.1). Both numerator and denominator are arranged in descending powers of the variable. The terms in the quotient are obtained by dividing the first term in the denominator into the term of highest degree in the numerator or the remainder after previous steps. We illustrate:

$$
(3) \qquad 3x^2 - x + 1 \overline{\smash{\big)}\
\begin{array}{l}
\ 2x\ +\ (7/3) \\
\ 6x^3 + 5x^2 - 2x + 8 \\
\ 6x^3 - 2x^2 + 2x \\
\hline
\ 7x^2 - 4x + 8 \\
\ 7x^2 - (7/3)x + (7/3) \\
\hline
\ (-5/3)x + (17/3)
\end{array}
}
$$

(4) $6x^3 + 5x^2 - 2x + 8 \equiv (3x^2 - x + 1)\,[2x + (7/3)]$

$$+ (-5x + 17)/3$$

If the remainder is zero, we say that the numerator is **divisible** by the denominator or is a **multiple** of the denominator and the denominator is a **factor** of the numerator.   If a polynomial is not divisible by any other polynomial (except a constant), it is said to be **prime** or irreducible.   Whether a polynomial is prime or not depends upon the limitations placed on the coefficients in its possible factors.   Thus $x^2 - 2$ is prime if we limit ourselves to integral coefficients, but it factors into $(x - \sqrt{2})(x + \sqrt{2})$ if we permit any real coefficients.   In this section we limit ourselves to integral coefficients.

It frequently happens that it is necessary to simplify a **compound fraction,** that is, one that contains fractions in its numerator or denominator.   This can be done by simplifying numerator and denominator separately and then "inverting," but the student should be alert for opportunities to avoid some of these manipulations by appropriate multiplications of the numerator and denominator of the entire fraction.   (Compare 3.11.)   For example:

(5)
$$\frac{\dfrac{x-2}{x-1} + \dfrac{1}{x^2-1}}{\dfrac{x^2+x-1}{x+1}} \equiv \frac{(x-2)(x+1)+1}{(x-1)(x^2+x-1)}$$

(6)
$$\equiv \frac{x^2-x-1}{(x-1)(x^2+x-1)}$$

Here we have multiplied both numerator and denominator by $x^2 - 1 = (x - 1)(x + 1)$.

Multiplication and division of rational functions involve nothing new.   We use *3.11.4 and *3.11.8 to rewrite the product or quotient as a single fraction and simplify by the methods already discussed.

In adding fractions, it would be possible to use *3.11.3 mechanically and simplify, but the most efficient procedure is to rewrite over a common denominator that is as simple as possible.   (Compare Ex. 3.11.10.)   Thus consider:

(7)
$$\frac{1}{x+1} + \frac{3}{x^3} + \frac{1}{x(x+1)}$$

The expression $x^3(x + 1)$ is divisible by each of the denominators. Hence, if we multiply the entire expression by $x^3(x + 1)$, we eliminate all denominators. If we do this and then divide by $x^3(x + 1)$, we have:

(8)
$$\frac{x^3 + 3(x + 1) + x^2}{x^3(x + 1)} \equiv \frac{x^2(x + 1) + 3(x + 1)}{x^3(x + 1)}$$

(9)
$$\equiv \frac{x^2 + 3}{x^3}$$

We call $x^3(x + 1)$ the least common denominator. It is the "smallest" expression that could be used in this way, in the sense that it is a multiple of each of the denominators and that it is a factor of any expression that is a multiple of each one. This is so because any expression that is a multiple of all denominators must have each one as a factor, that is, it must contain x^3 and $x + 1$ and hence $x^3(x + 1)$ as factors.

Let us make these ideas more precise. We call one polynomial a **common multiple** of several polynomials if it is a multiple of each one. The **least common multiple** (l.c.m.) of several polynomials is the common multiple that is a divisor of every other common multiple. For example, the numbers 48 and 24 are both common multiples of 3, 8, and 6, but 24 is the l.c.m. since any common multiple of 3, 8, and 6 must have 24 as a divisor. In order to see this, we factor the numbers into prime factors: $3, 2^3$, and $2 \cdot 3$. Now any common multiple must contain $3, 2^3$, and $2 \cdot 3$ as factors. Hence it must contain 3 as a factor and 2 as a factor three times. Hence it must contain $3 \cdot 2^3 = 24$ as a factor. This suggests the following rule for finding the l.c.m.:

To find the l.c.m. of several polynomials, factor each into its prime factors. The l.c.m. is the product of all different prime factors, each factor being taken a number of times equal to the largest number of times it appears in any one polynomial.

Thus, to find the l.c.m. of $x^2 - 3x - 10$, $(x - 5)^2$ and $x^2(x^2 - 4x - 5)$, we note that

(10)
$$x^2 - 3x - 10 = (x - 5)(x + 2)$$

(11)
$$x^2(x^2 - 4x - 5) = x^2(x - 5)(x + 1)$$

Then the l.c.m. is $x^2(x - 5)^2(x + 1)(x + 2)$. If we wished to add fractions whose denominators were given by these three expressions, we would multiply all terms by this l.c.m. and write the result over it. The student may prefer to think of this step as rewriting each fraction with the l.c.m. as denominator or as multiplying each numerator by those factors of the l.c.m. that do not appear in its denominator. The l.c.m. of the denominators of several fractions is called the **least common denominator** (l.c.d.).

We have spoken only of polynomials here, but the methods of this section apply to any fractions, although the particular techniques for factoring and division might not be effective. Frequently, expressions that are not polynomials in the variable that appears may be considered as polynomials in some other variable. For example, $2 \sin^2 x + 3 \sin x - 4$ is not a polynomial in x, but it is a polynomial in $\sin x$. Such expressions can, of course, be treated as polynomials in the appropriate variable.

EXERCISE 13.2

1. In the following expressions, perform the indicated operations and express in the form $P(x)/Q(x)$ in lowest terms:

(a) $\left[\dfrac{x + y}{x - y} - \dfrac{x - y}{x + y} \right](x + y)^{-1}$

(b) $[(a - x)^{-1} + (a + x)^{-2}]^{-1}(a^2 - x^2)^{-1}$

(c) $\dfrac{1}{2x} + \dfrac{3x}{x - 1} + \dfrac{x^2}{(x - 1)^2}$

(d) $(x + 1)(x^2 + 3x + 2)^{-1} + x^2(x^2 - 3x + 2)^{-1}$

(e) $\dfrac{1}{x - 1} - \dfrac{1}{x + 1}$

(f) $\dfrac{x^5 - xy^4 - y^4 + x^4}{x^2 + 2xy + y^2}$

(g) $\dfrac{p^2 - 1}{(1 + px)^2 - (p + x)^2}$

(h) $\left[x + \dfrac{xy}{x - y} \right] \left[x - \dfrac{xy}{x + y} \right]^{-1}$

(i) $\dfrac{4}{x - 1} + \dfrac{2}{3(x - 1)^2} + \dfrac{1}{x + 2}$

(j) $\dfrac{1}{x + \dfrac{1}{x + \dfrac{1}{x}}}$

$$(k)\ \frac{x-a}{3} + \frac{\dfrac{1}{x-a} - \dfrac{1}{2}}{\dfrac{1}{x-a}}$$

$$(l)\ 2x - 3 + \frac{x-11}{4x+1}$$

$$(m)\ \frac{ax}{x^2-4} - \frac{b}{x+2} + 1$$

Answers. (a) $4xy/(x+y)^2(x-y)$. (c) $(8x^3 - 5x^2 - 2x + 1)/2x(x-1)^2$.

(e) $2/(x^2-1)$. (g) $(x^2-1)^{-1}$. (i) $\dfrac{15x^2 + 8x - 17}{3(x-1)^2(x+2)}$. (k) $(6+a-x)/6$.

(m) $\dfrac{x^2 + (a-b)x + 2b - 4}{x^2 - 4}$.

2. Rewrite the following in the form $P(x) + R(x)$, where $P(x)$ is a polynomial and $R(x)$ is a proper rational fraction:

(a) $(2x^5 + x^3 + x^2 - 3)/(x^2 - 1)$
(b) $(x^4 + 10x^3 - 11x^2 + 2x)/(x^2 + x + 1)$
(c) $(15x^4 - x + 3x^2 + 2)/(1 - 2x^2)$
(d) $(2x^6 + x^3 - 3x^4 + 2x - 18)/(2x^2 - 3)$
(e) $(1 + x^7 - 3x^2)/(1 - 2x^2)$
(f) $(x^9 + y^9)/(x^3 + y^3)$
(g) $(6x^3 - 4x^2 + x - 3)/(x^2 - x + 6)$
(h) $(x^8 + 3x^6 - x^4 + 14x^2 - 2)/(x^2 + 2)$
(i) $(5x^6 + 3x^5 + x - 1)/(x^3 - x + 10)$
(j) $(8x^7 + 5)/(2x^3 - 4x^4 + 3x)$
(k) $(x^3 + 2x^2y + y^2)/(x - y)$
(l) $3x^7/(x^2 - 5x + 7)$

Answers. (a) $2x^3 + 3x + 1 + (4x - 2)/(x^2 - 1)$. (c) $15x + (3x^2 + 44x + 2)/(x^3 - 3)$. (e) $-(\tfrac{1}{8})(4x^5 + 2x^3 + x - 12) + (x - 4)/8(1 - 2x^2)$. (g) $6x + 2 - (33x + 15)/(x^2 - x + 6)$.

3. Simplify:

$$(a)\ \frac{\dfrac{cr}{16}(1-r) + cr^2(1-r)^2}{r^2(1-r^2)}$$

$$(b)\ \frac{(a+b+c)^2 - (a-b)^2}{(a+c+d)^2}$$

$$(c)\ \frac{\dfrac{x}{3}[(x+y)^{-1} - (x-y)^{-1}]}{(x^2-y^2)(x^2y + xy^2)^{-1}}$$

(d) $1/x + 1/x^2 + 1/x^3 + \cdots + 1/x^n$

(e) $\dfrac{(1+x)^2}{(1-x^{-2})(1-x)^{-1}}$

(f) $\left[\dfrac{3a^4+48+24a^2}{32-2a^4}\right]\left[\dfrac{3a-6}{2+a}\right]$

(g) $\dfrac{1}{2b}\left[\dfrac{2x}{x^2+\dfrac{a}{b}}\right]$

(h) $\dfrac{\dfrac{1}{c}\left[\sqrt{x^2+c^2}-x\dfrac{x}{\sqrt{x^2+c^2}}\right](x^2+c^2)^{-1}}{\sqrt{1-\dfrac{x^2}{x^2+c^2}}}$

(i) $\dfrac{1}{b^2}\left[\dfrac{b}{a+bx}-\dfrac{ab}{(a+bx)^2}\right]$

(j) $\dfrac{1}{b^3}\left[(a+bx)b-2ab+\dfrac{a^2b}{a+bx}\right]$

(k) $-a^{-1}\left[\dfrac{x}{a+bx}\right]\left[\dfrac{bx-(a+bx)}{x^2}\right]$

(l) $\dfrac{b}{b'}+\left[\dfrac{ab'-a'b}{b'^2}\right]\left[\dfrac{b'}{a'+b'x}\right]$

(m) $\left[\dfrac{ab}{b(a+bx)}-\dfrac{a'b'}{b'(a'+b'x)}\right](ab'-a'b)^{-1}$

(n) $\left[\dfrac{a+bx}{a'+b'x}\right]\left[\dfrac{(a+bx)b'-(a'+b'x)b}{(ab'-a'b)(a+bx)^2}\right]$

Answers. (a) $c(1+16r-16r^2)/16r(r+1)$. (c) $-2x^2y^2/3(x^2-y^2)(x-y)$. (e) $-x^2(x+1)$. (g) $x/(bx^2+a)$. (i) $x(a+bx)^{-2}$. (k) $1/x(a+bx)$. (m) $x/(a+bx)(a'+b'x)$.

4. For each expression $f(x)$ find and simplify $\dfrac{f(x+\Delta x)-f(x)}{\Delta x}$:

(a) $f(x)=1/x$ (d) $f(x)=(x^2-1)(x+3)$
(b) $f(x)=(x+2)/(x+1)$ (e) $f(x)=(2x+1)/(x^2+1)$
(c) $f(x)=1/x^2$ (f) $f(x)=1/(x+3)^3$

Answers. (a) $-1/x(x+\Delta x)$. (c) $[-2x-(\Delta x)]/[(x+\Delta x)^2x^2]$.

5. Find the l.c.m. of:

(a) $x^3(x^2+2x+1)$, $x(x^2-1)$, x^3-1, $2x$, $6x^2$
(b) x^4-1, $(x-1)^3$, ab^3, a^2-ab
(c) $e^x\sin x$, e^{2x}, $1-\cos^2 x$
(d) x^2+x-6, $5+x-2x^2$, x^2+6x+9

Answers. (a) $6x^3(x^3-1)(x+1)^2$. (c) $e^{2x}\sin^2 x$.

6. Equations in which the unknown appears in a denominator are called **fractional equations.** If the equation can be put in the form $P(x)/Q(x) = 0$, the solutions are the roots of $P(x) = 0$ that are not also roots of $Q(x) = 0$. For example, to solve

$$(12) \qquad \frac{3}{x-1} - \frac{1}{x} = 2$$

we proceed as follows:

$$(13) \qquad \frac{3}{x-1} - \frac{1}{x} - 2 = 0$$

$$(14) \qquad \frac{3x - (x-1) - 2x(x-1)}{x(x-1)} = 0$$

$$(15) \qquad \frac{2x^2 - 4x - 1}{x(x-1)} = 0$$

The roots of the numerator are $x = 1 \pm 0.5\sqrt{6}$, and they are also the roots of the original equation, since the left members of (13) and (15) are identical. Some authors advise dealing with (12) by multiplying both sides by $x(x-1)$ to "clear of fractions." This usually leads to the same results as the above method, but it may introduce **extraneous roots,** that is, values that satisfy the final equation but not the original. Solve and check the following equations:

(a) $\dfrac{1}{x} - \dfrac{1}{x-1} = 1$

(b) $\dfrac{1}{x^2} - \dfrac{2a}{x} + a = 0$

(c) $C = (xE)\Big/\Big(R + \dfrac{rx^2}{n}\Big)$ (Solve for x and for n.)

(d) $\dfrac{x-b}{3} = \dfrac{x^2}{x-a}$

(e) $p - p' = (h-1)q/(h+g)(p+1)$ (Solve for h.)

(f) $p' = (e-1-h)/(h+g)$ (Solve for h.)

(g) $\dfrac{x+1}{x-2} + \dfrac{x^2+2x+2}{2x^2-x-6} = 2$

(h) $\dfrac{1}{x-1} + \dfrac{1}{x-2} = \dfrac{2}{x-3}$

7. Find the zeros (values of x) of some of the functions that appear in probs. 1 and 2 and check.

8. The **highest common factor** (h.c.f.) of several polynomials is the common factor that is divisible by every common factor. Thus the h.c.f. of $18 = 2 \cdot 3^2$ and $12 = 3 \cdot 2^2$ is $2 \cdot 3 = 6$. It is the product of all the different prime factors of the polynomials, each being taken a number of times equal to the smallest number of times it appears in any one polynomial. Find the h.c.f. of:

(a) 6, 8, 12

(b) 18, 12, 36

(c) $x^2 - y^2$, $x^4 - y^4$, $(x + y)^2$

(d) $x^2 - xy$, $x^3 - y^3$, $x + y$

(e) $64x^2(y - c)^2$, $40(cx - xy)$

(f) $x^{10} - x^5$, $x^4 - 2x^2y + y^2$

Answers. (a) 2. (c) $x + y$. (e) $8x(y - c)$.

9. It is a known theorem that the product of the l.c.m. and the h.c.f. of two polynomials is equal to the product of the two. Verify this for several examples of your own construction.

10. Prove that, if $2s = a + b + c$, then

$$\frac{abc}{(s - a)(s - b)(s - c)} = \frac{s}{s - a} + \frac{s}{s - b} + \frac{s}{s - c} - 1$$

11. From $\quad B = \dfrac{U}{\dfrac{1}{s_1} + i}\left(\dfrac{\dfrac{c}{s_2} + e + ci}{u}\right) - \dfrac{E}{\dfrac{1}{s_1} + i}, \quad \dfrac{1}{s_1} + i = \dfrac{1}{a_1}, \quad$ and

$\dfrac{1}{s_2} + i = \dfrac{1}{a_2}$, show that

$$B = Ua_1\left(\frac{\dfrac{c}{a_2} + e}{u} - \frac{E}{U}\right)$$

†12. The student is familiar with the process of adding several fractions to get a single fraction. The inverse process of rewriting a fraction as the sum of fractions is called **expanding in partial fractions.** It is useful in calculus. The purpose is to rewrite the fraction in terms of others with as simple denominators as possible. We need consider only proper fractions (why?), and we assume that the denominator has been factored into linear and quadratic factors. Then the procedure is to set the given fraction equal to a sum of fractions with unknown coefficients in the numerators and then evaluate these coefficients so as to make the two expressions identical. For example, we write:

(16) $$\frac{2x + 3}{(x - 1)(x + 2)} \equiv \frac{A}{x - 1} + \frac{B}{x + 2}$$

Then A and B must be chosen to make:

(17) $$2x + 3 \equiv A(x + 2) + B(x - 1)$$

Letting $x = 1$, we find $A = 5/3$. Similarly $x = -2$ yields $B = 1/3$. We could get the same result by collecting the right member around powers of x and setting corresponding coefficients equal. (See Ex. 12.2.8.) Thus

(18) $$2x + 3 \equiv (A + B)x + (2A - B)$$

Then $A + B = 2$ and $2A - B = 3$ yield the same A and B.

The question remains: What denominators should be considered and what unknown parameters should be inserted? When the denominator is factored into linear and quadratic factors, the following rule works: For each factor of the denominator include a partial fraction with that factor as denominator. If a factor appears to the nth power, insert a partial fraction for each power of this factor from 1 to n. Numerators associated with linear factors are constants, and those associated with quadratic factors are linear. Thus to expand:

$$(19) \qquad \frac{f(x)}{(x-1)(2x+3)(x+2)^3(x^2+2x-1)(x^2-4x-2)^2}$$

we set it equal to

$$(20) \qquad \frac{A}{x-1} + \frac{B}{2x+3} + \frac{C}{x+2} + \frac{D}{(x+2)^2} + \frac{E}{(x+2)^3}$$
$$+ \frac{Fx+G}{x^2+2x-1} + \frac{Hx+I}{x^2-4x-2} + \frac{Jx+K}{(x^2-4x-2)^2}$$

and evaluate the constants. Of course in a particular problem, some constants may be zero, but the method consists in inserting all possible partial fractions. If too few are considered, it will turn out that the constants cannot be evaluated so as to make the two expressions identical. Expand in partial fractions and check by addition.

(a) $\dfrac{2-x}{(x-1)(x-2)}$

(b) $\dfrac{1}{x(x+1)}$

(c) $\dfrac{3x-2}{x(x-1)^2}$

(d) $\dfrac{x^2-2}{x(x+1)(x-1)^2}$

(e) $\dfrac{3x^2-x+1}{(x^2+1)(x+2)}$

(f) $\dfrac{1}{(x^2-3)^2(x+3)^2}$

(g) $\dfrac{2x^2}{(x^3+1)(x^3-1)}$

(h) $\dfrac{3x^3-6x^2+2x-8}{x^2-3x-18}$

(i) $\dfrac{x^4}{x^2-1}$

(j) $\dfrac{x^4+2x+1}{(x^2+2)(x-1)}$

Note. In parts h, i, and j begin by rewriting as a polynomial plus a proper fraction.

13.3 Graphing irrational functions

The irrational functions are a very large class, comprising all functions that are not rational. We consider only rather simple examples, especially those of the form $y = \sqrt{R(x)}$, where $R(x)$ is a rational function.

To graph an irrational function we proceed very much as for rational functions. Intercepts are found as usual. If the equation is given in the form $y = \sqrt{R(x)}$, then there is no symmetry with respect to the x-axis, since the radical means only the positive square root. However, the graph of $y^2 - R(x) = 0$ is symmetric with respect to the x-axis and is equivalent to the graphs of $y = \sqrt{R(x)}$ and $y = -\sqrt{R(x)}$. The asymptotes are found as for rational functions. The bend points of $y = \sqrt{R(x)}$ are at the same values of x as those of $y = R(x)$, since the two functions achieve maxima and minima for the same values of x. In addition, it is helpful to consider **excluded values** of the variables, that is, those values of x that do not yield real values of y. Since any value of x that makes $R(x)$ negative makes $y = \sqrt{R(x)}$ imaginary, to find the excluded values of x, we solve the inequality $R(x) < 0$. (See 13.1 and Ex. 13.1.4.)

We illustrate by discussing and sketching the equation

$$(1) \qquad (x - 1)y^2 - x^2 - 2x = 0$$

Evidently this determines y as an algebraic function of x. (Why?) Solving for y and x, we find:

$$(2) \qquad y = \pm \sqrt{\frac{x(x + 2)}{(x - 1)}}$$

$$(3) \qquad x = \frac{y^2 - 2 \pm \sqrt{y^4 - 8y^2 + 4}}{2}$$

(a) *Intercepts.* When $x = 0$, $y = 0$. When $y = 0$, we have from either (2) or (3) that $x = 0$ or $x = -2$. Hence the intercepts are $(0, 0)$ and $(-2, 0)$.

(b) *Symmetry.* From (1), (2), or (3) it is evident that the curve is symmetric with respect to the x-axis.

(c) *Asymptotes.* From (2) we see that y approaches infinity as x approaches 1. From the same equation it appears that y gets large as x gets large. Equation (3) shows that there are no finite values of y for which x is undefined. Hence the only vertical or horizontal asymptote is $x = 1$.

(d) *Bend Points.* Using *13.1.1, we find that the derivative of $x(x + 2)/(x - 1)$ is $(x^2 - 2x - 2)/(x - 1)^2$. This is zero if $x = 1 \pm \sqrt{3}$. Hence these *may* yield bend points.

(e) *Excluded Values.* We leave it to the student to show that the solution of $x(x + 2)/(x - 1) < 0$ consists of those values of x for which $x < -2$ or $0 < x < 1$. Also $y^4 - 8y^2 + 4 < 0$ when $\sqrt{4 - 2\sqrt{3}} < y < \sqrt{4 + 2\sqrt{3}}$ or $-\sqrt{4 + 2\sqrt{3}} < y < -\sqrt{4 - 2\sqrt{3}}$. Hence these regions are excluded. Note that the boundary values of y are just those at the possible bend points.

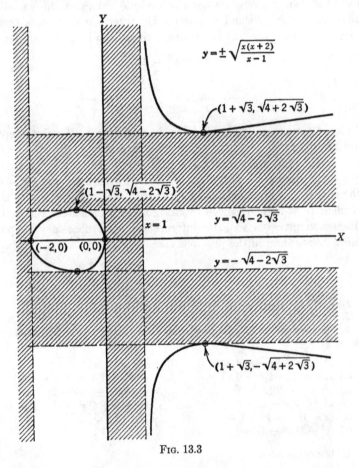

$$y = \pm \sqrt{\frac{x(x+2)}{x-1}}$$

$(1 + \sqrt{3}, \sqrt{4 + 2\sqrt{3}})$

$(1 - \sqrt{3}, \sqrt{4 - 2\sqrt{3}})$

$y = \sqrt{4 - 2\sqrt{3}}$

$x = 1$

$(-2, 0) \quad (0, 0)$

$y = -\sqrt{4 - 2\sqrt{3}}$

$(1 + \sqrt{3}, -\sqrt{4 + 2\sqrt{3}})$

Fig. 13.3

In Fig. 13.3 we shade the excluded regions, plot the intercepts, draw the asymptotes, and place horizontal tangents at the possible bend points. We do not need to plot a large number of points to be sure of the shape of the graph. For example, we know that the curve diverges from the x-axis as x gets large, because we have found all points where the tangent is horizontal, and we can see that y gets large as x gets large because x appears to the second power in the numerator in (2).

In the equation just considered, the solutions for x and y involved only square roots. In other cases, higher roots may appear. Whenever the root is even, we have to exclude values of the variable that make the radicand negative. However, for odd roots there exist real values for positive or negative radicand. For example,

(4) $$x^3 + y^3 = 1$$

yields

(5) $$y = \sqrt[3]{1 - x^3} \quad \text{and} \quad x = \sqrt[3]{1 - y^3}$$

Evidently there are no excluded values of x or y. The intercepts are $(1, 0)$ and $(0, 1)$. The curve is not symmetric with respect to the origin, and there are no vertical or horizontal asymptotes. Since the derivative of $1 - x^3$ is $-3x^2$ and is zero when $x = 0$, the curve has a horizontal tangent when $x = 0$. We leave it to the student to sketch the curve by using this information. Since the discussion gives us so little information, it is necessary to plot more points than in the previous case.

EXERCISE 13.3

1. Verify the statements in the discussion of (1).

2. Graph (4).

3. For each of the following equations write a systematic discussion and sketch:

(a) $x^2 + 3x - (x - 3)y^2 = 0$

(b) $x(x^2 - 1) - 2y^2 = 0$

(c) $y^2 = (x - 2)(x + 1)^2$

(d) $x^2y^2 - 3(x - 1) = 0$

(e) $(x + 1)y^3 + 3x - 1 = 0$

(f) $(y^2 - 1)(x^2 - 1) = x$

(g) $x^2y^3 = 1$

(h) $y^2(x^2 - 1)(x + 2) - x = 0$

(i) $y^2 = x(x - 1)(x - 2)$

(j) $(x^2 - x + 1)y^2 = x^2 + x + 1$

(k) $\sigma = 1/\sqrt{n(n - 1)}$

(l) $(x^2 - 1)y^3 = x$

4. Sketch the following:

(a) $x^3 - y^3 = 1$

(b) $x^4 + y^4 = 1$

(c) $x^{10} + y^{10} = 1$

(d) $x^5 + y^5 = 1$

(e) $x^{0.5} + y^{0.5} = 1$

(f) $x^{1/3} + y^{1/3} = 1$

(g) $x^{2/3} + y^{2/3} = 1$

(h) $y = x^{0.5} + x$

5. Graham's law states that rates of diffusion of gases are inversely proportional to the square roots of their densities. Letting y stand for the diffusion rate and x for the density, write the formula and graph it.

13.4 Manipulation of radicals

We have been dealing throughout the book with expressions of the form $\sqrt[n]{F(x)}$, usually with $n = 2$. As a rule only the simplest manipulations have been necessary and could easily be carried out by reference to the definitions and theorems of 5.3 and 6.10. The student will recall that the theorems of those sections depend upon the assumption that the numbers involved are real and that the radicals stand for the principal value of the root. The case where complex numbers appear was discussed in Chapter 10. Here we shall consider only the principal value of an expression of the form $\sqrt[n]{F(x)}$. Then our radicals are single-valued functions, and the methods of 5.3 and 6.10 apply. Thus *5.3.4 and *5.3.3 justify

$$(1) \qquad \sqrt{(x+1)^2(x+2)} \equiv \sqrt{(x+1)^2}\sqrt{x+2}$$

$$(2) \qquad \equiv |x+1|\sqrt{x+2}$$

and *6.10.3 and *6.10.5 justify

$$(3) \qquad \sqrt[3]{\frac{x^5(x-y)^2(x^2-y^2)}{(x+y)(2x+1)}} \equiv \sqrt[3]{\frac{x^3(x-y)^3(2x+1)^2x^2}{(2x+1)^3}}$$

$$(4) \qquad \equiv \frac{x(x-y)\sqrt[3]{x^2(2x+1)^2}}{(2x+1)}$$

$$(5) \qquad \equiv x^{5/3}(2x+1)^{-1/3}(x-y)$$

The student should use fractional and negative exponents when convenient. There is no hard and fast rule for deciding what the "simplest form" is. Multiplication, division, raising to a power, and root extraction involving radicals can be handled by means of the principles of 6.10. However, it is usually better to use fractional exponents as we did in section 6.11, especially if radicals of different orders are involved.

If we have an expression of the form

$$(6) \qquad F(x)\sqrt[n]{A(x)} + G(x)\sqrt[n]{A(x)}$$

we may use the distributive law to "add" the two radicals in order to get $[F(x) + G(x)]\sqrt[n]{A(x)}$. Since this possibility depends on the radicals being identical, the problem of adding radicals sometimes involves rewriting them so that they involve a radical as a common factor. For example,

(7) $\sqrt{2x} + \sqrt{x} + \sqrt{16x^3} \equiv \sqrt{2}\sqrt{x} + \sqrt{x} + 4x\sqrt{x}$

(8) $\equiv (\sqrt{2} + 1 + 4x)\sqrt{x}$

We see from this that the coefficient of the common radical may itself involve a radical. Thus

(9) $\sqrt{x^2 - y^2} + \sqrt{x + y} = (1 + \sqrt{x - y})(\sqrt{x + y})$

Of course, which of these two ways of writing (9) is more convenient depends on the situation. Radicals may always be added by rewriting each term in the sum as some function times the l.c.m. of the radicals, but this may not be an improvement. For example,

(10) $\sqrt{x} + \sqrt{y} + \sqrt{xy} = \sqrt{xy}\left[\dfrac{1}{\sqrt{x}} + \dfrac{1}{\sqrt{y}} + 1\right]$

but the right member is hardly "simpler." Better than either would be $x^{1/2} + x^{1/2}y^{1/2} + y^{1/2}$.

Equations that involve an unknown under a radical are called **irrational equations.** Consider

(11) $\sqrt{x} - 2\sqrt{x - 1} = 0$

The usual procedure is to write it as

(12) $\sqrt{x} = 2\sqrt{x - 1}$

and then square both sides to get

(13) $x = 4(x - 1)$

(14) $3x = 4$ and $x = 4/3$

This can be shown to be a solution by substituting in the left member of (12). Now consider

(15) $\sqrt{x} + 2\sqrt{x - 1} = 0$

Following the same procedure, we find

(16) $\sqrt{x} = -2\sqrt{x - 1}$

(17) $x = 4(x - 1)$ and $x = 4/3$

as before. But $x = 4/3$ is not a solution of (15)! As a matter of fact there is no solution to (15), because the sum of two positive numbers is positive, and it is impossible for both $\sqrt{x}$ and $\sqrt{x-1}$ to be zero. (Why?)

How does it happen that we picked up an extraneous root? It happened because $a = b \longrightarrow a^2 = b^2$, but not conversely. We squared both sides of (16) and then found an x that satisfied the result. But it does not follow that this x satisfies the original equation before squaring. Let us look at it another way. Squaring both sides of (16) is the same as multiplying both sides of (15) by $\sqrt{x} - 2\sqrt{x-1}$. Thus, if we have $a + b = 0$, multiplication by $a - b$ yields $a^2 - b^2 = 0$ or $a^2 = b^2$. Thus the result of squaring is the same as

(18) $\qquad (\sqrt{x} - 2\sqrt{x-1})(\sqrt{x} + 2\sqrt{x-1}) = 0$

Now the root we found in (17) is a root of this equation, but it is a zero of the factor by which we multiplied and not of the left member of the original. Thus squaring both sides of an equation is equivalent to multiplying by a function of the unknown, and the zeros of this function will appear as roots of the result, possibly extraneous. For this reason it is essential to check the roots found by multiplying by a function of the unknown.

EXERCISE 13.4

1. Simplify the following expressions, using fractional exponents where convenient:

(a) $\sqrt{(x^2 + x - 2)(x^2 - 3x + 2)}$

(b) $\sqrt{x^6 - x^5 - 4x^4 - 3x^3}$

(c) $\sqrt{(x+1)(x-1)^{-1}}$

(d) $\sqrt[3]{x(x^2-1)/(x^2-x)(a-b)}$

(e) $(\sqrt[3]{ab^2})(\sqrt{ac^3b})$

(f) $\sqrt[6]{8a^3(a-x)^{-3}(a+x)^{-2}}$

(g) $\sqrt{x^2 - a^2}/\sqrt[3]{(x-a)/(x+a)}$

(h) $\dfrac{1}{\sqrt{x+2} - \sqrt{x-2}}$

(i) $\dfrac{1}{\sqrt{x-1} + \sqrt{x+1}}$

(j) $\dfrac{1}{\sqrt{x}} + \dfrac{1}{\sqrt{2x}} + \sqrt{x}$

(k) $\sqrt{x^2 - a^2} + \sqrt{x - a}$

(l) $\dfrac{x + \sqrt{x^2-1}}{x - \sqrt{x^2-1}} - \dfrac{x - \sqrt{x^2-1}}{x + \sqrt{x^2-1}}$

(m) $\sqrt[3]{\dfrac{(a^2-b^2)(a+b)^2c^2}{a^3-b^3}}$

(n) $\sqrt[4]{\sqrt[3]{(x-y)^2(x^4-y^4)}}$

2. Discuss the function $y = \sqrt{x^2 + 2x + 1}$. When is $y = x + 1$ and when is $y = -(x + 1)$? Graph the function.

3. Simplify:

(a) $3\sqrt{18} - 7\sqrt{8} - \sqrt{72}$

(b) $2\sqrt{18} - 7\sqrt{30}$

(c) $\dfrac{x(x + 1)^{0.5}(x^2 + 1)^{-0.5} - 0.5(x + 1)^{-0.5}(x^2 + 1)^{0.5}}{(x + 1)}$

(d) $\dfrac{2x(x^2 + 2)^{0.5} - x^3(x^2 + 1)^{-0.5}}{x^2 + 2}$

(e) $x^2(a^2 - x^2)^{-0.5} + (a^2 - x^2)^{0.5} + a(1 - a^{-2}x^2)^{-0.5}$

(f) $[2 + 0.5(2 + x^2)^{-0.5}(2x)]/2x\sqrt{2 + x^2}$

4. Solve the following equations for x:

(a) $\sqrt{x + 1} - 2\sqrt{x} = 0$

(b) $x + 1 = (x + 7)^{0.5}$

(c) $\sqrt{2x - 4} - \sqrt{x + 5} = 1$

(d) $x - 4 = \sqrt{x} + 1$

(e) $(x - 1)^{-0.5} + x^{0.5} = 0$

(f) $\sqrt{x^2 - 3x - 1} = 2x + 7$

(g) $\sqrt{2 + x^2} = 4 - x$

(h) $\sqrt{3 - x^2} = 2 - x$

(i) $\sqrt{3 - x^2} = x - 2$

(j) $x + \sqrt{x} = \sqrt{2 + x^2}$

5. Describe and graph the following functions:

(a) $y^2 = \sin x$

(b) $y = \sqrt{x} + \sqrt{x - 1}$

(c) $y^2 = \sin x + \cos x$

(d) $y^3 = (x - 1)x^{-2}(x + 2)^{-1}$

(e) $x^2(y - 1) = (y + 2)^3$

(f) $y = \dfrac{(x + 1)^{0.5} - 2x^{0.5}}{x - 3}$

6. If two sources of light of intensities a and b are at a distance d apart, the position on the line between them where their intensities are equal is given by $ax^{-2} = b(d - x)^{-2}$. Show that $x = d\sqrt{a}\,(\sqrt{a} \pm \sqrt{b})^{-1}$.

7. The following is the equation of the Lorenz transformation, which appears in the theory of relativity. Solve it for v.

$$t' = \frac{t - \left(\dfrac{v}{c^2}\right)x}{\sqrt{1 - \dfrac{v^2}{c^2}}}$$

8. Eliminate radicals from the denominators of the following:

(a) $\dfrac{1}{1 - \sqrt{3} + \sqrt{2}}$

(b) $\dfrac{x}{1 + \sqrt{x} + \sqrt{y}}$

(c) $\dfrac{x - y}{1 + \sqrt{5} - \sqrt{x + y}}$

(d) $\dfrac{1}{\sqrt[3]{2} - 1}$

(e) $\dfrac{2}{1 + x^{\frac{1}{3}} + x^{\frac{2}{3}}}$

(f) $\dfrac{x}{x^{\frac{1}{3}} - y^{\frac{1}{3}}}$

9. A function $f(x, y)$ is called **homogeneous of degree** n if $f(tx, ty) = t^n f(x, y)$ for any t. Show that the following are homogeneous and find their degrees:

(a) $\dfrac{x^2 + y^2}{x^2 - y^2}$

(b) $x^{0.5}y + y^{2/3}x^{5/6}$

(c) $\dfrac{x + y}{x - y}$

(d) $\dfrac{\sqrt{x + y}}{\sqrt{x - y}} - \sqrt{\dfrac{x}{y}}$

(e) $x^3 - 2x^2y + 3y^3$

(f) $\log_{10} 2^{xy}$

10. Find $\sqrt{2 + \sqrt{3}}$. (*Suggestion.* Let $x + y\sqrt{3} = \sqrt{2 + \sqrt{3}}$.)

11. Find $\sqrt{4 - 3i}$.

12. According to the relativity theory, the mass M of a body is given by $M = m\left(1 - \dfrac{v^2}{c^2}\right)^{-0.5}$, where m is the mass of the body at rest, v its velocity, and c the velocity of light. (a) Solve for v. (b) What happens to M as v approaches c?

13. For each $f(x)$, find $[f(x + h) - f(x)]/h$ and simplify by eliminating radicals that appear in denominators.

(a) $f(x) = \sqrt{x}$

(b) $f(x) = x^{1/3}$

(c) $f(x) = x^{-0.5}$

(d) $f(x) = (x^2 + 1)^{0.5}$

(e) $f(x) = (1 - x)^{-0.5}$

(f) $f(x) = x^{2/3}$

14. Derive the first equation below from the other two. (*Note.* The derivation arises in quantum mechanics.)

(a) $n - n' = \dfrac{hnn'}{mc^2}(1 - \cos\theta)$

(b) $hn = hn' + mc^2\left[\dfrac{1}{\sqrt{1 - \dfrac{v^2}{c^2}}} - 1\right]$

(c) $\dfrac{m^2v^2}{1 - \dfrac{v^2}{c^2}} = \dfrac{h^2n^2}{c^2} + \dfrac{h^2n'^2}{c^2} - 2\dfrac{hn}{c}\dfrac{hn'}{c}\cos\theta$

13.5 Summary of curve plotting

The locus of an equation $F(x, y) = 0$ is the set of points whose coordinates satisfy this equation. This set of points may be a single continuous curve, for example, when $F = Ax + By + C$. It may consist of several branches, for example, when $F = (x - 1)y^2 - x^2 - 2x$. (See 13.3.) It may consist in an isolated point, for example, when $F = x^2 + y^2$, or it may be an empty set, for example, when $F = x^2 + y^2 + 1$. As a rule it will include an infinite number of points, and it is obviously impossible to plot them all. The crudest method of curve plotting consists in simply plotting many points and then linking them up. But no matter how many points we plot we cannot be certain what happens between them, unless we investigate the locus as a whole. This consideration we have called the *discussion* of the curve. The discussion should consist in a systematic consideration of certain key features of the equation and its locus. As each piece of information is found, it is indicated on the graph. Then, with the help of a few well-chosen points, the locus can be sketched with assurance. Of course such a sketch will not be exactly accurate, but that is not its purpose. If we wish to know the value of y corresponding to a given value of x, we may find it by algebraic means from the equation or from a graph made with great care. The primary purpose of the sketch is to help us grasp the nature of the relationship between x and y, and for this purpose the location of key points and "the big picture" are the important things.

The following is a check list for the discussion of a curve. The order need not be preserved. We should first consider those items that are most "obvious," since they may help with more difficult points or even eliminate the need for considering them.

1. *Familiar Loci.* If the equation is of a familiar locus, such as a straight line, power function, or a conic, it should be put in a standard form and graphed by the special methods appropriate. Thus

$$(1) \qquad y = \frac{x - 1}{x - 2}$$

is a hyperbola since it can be written

$$(2) \qquad xy - x - 2y + 1 = 0$$

or

$$(3) \qquad (x - 2)(y - 1) = 1$$

Hence the curve is a rectangular hyperbola with asymptotes $x = 2$ and $y = 1$.

2. *Degeneracy.* It may happen that the equation represents two or more distinct curves. If $F(x, y) = 0$ is of the form $f(x, y)g(x, y) = 0$, its locus is said to be **degenerate** and consists of $f(x, y) = 0$ and $g(x, y) = 0$. This is true because a point lies on the locus $fg = 0$ if and only if its coordinates satisfy $f = 0$ or $g = 0$ or both.

Thus the equation

(4) $$x^4 - y^4 + x^2 + y^2 = 0$$

can be rewritten as

(5) $$(x^2 + y^2)(x^2 - y^2 + 1) = 0$$

and its locus consists in the loci of

(6) $$x^2 + y^2 = 0 \quad \text{and} \quad x^2 - y^2 + 1 = 0$$

Hence the locus consists of the point $(0, 0)$ and the rectangular hyperbola given by the second equation.

3. *Transformation of Coordinates.* The equation may sometimes be greatly simplified by transforming to some other coordinate system. The possibility of shifting from rectangular to polar coordinates or from polar to rectangular coordinates should be considered. An obvious example is

(7) $$\sqrt{x^2 + y^2} = \arctan(y/x)$$

This is rather unmanageable in rectangular coordinates, but it becomes

(8) $$r = \theta$$

in polar coordinates and is then easy to plot. We have seen how appropriate translations and rotations are the key to plotting conic sections. They may often be used in other cases. For example,

(9) $$x^2 - (y + 3)^3 = 1$$

becomes

(10) $$y' = \sqrt[3]{x^2 - 1}$$

after translating the origin to $(0, -3)$.

4. *Intersections with Key Lines.* The intercepts are the intersections of the curve with the axes. They are often easy to find. When we find the y's corresponding to $x = a$, we are finding the intersections of the curve and the line $x = a$. Similarly, we find the intersections with the line $y = b$ by substituting $y = b$ in the

equation of the locus. But it may also be helpful to find the intersections of the curve with other lines, in particular lines of the form $y = mx$. Consider, for example, the equation

$$(11) \qquad\qquad x^3 + y^3 - 3xy = 0$$

It is easy to see that its locus crosses the axes only at the origin, but other points are hard to find by substituting values of x or y. (Why?) However, if we try to find the intersections of the curve with lines through the origin given by $y = mx$, we find

$$(12) \qquad\qquad x^3 + (mx)^3 - 3x(mx) = 0$$

$$(13) \qquad\qquad x = 3m/(m^3 + 1) \qquad y = 3m^2/(m^3 + 1)$$

By substituting various values of m we can easily find points on the curve. We can see also that as m approaches -1, both x and y approach infinity. This suggests that the line $y = -x$ or some line parallel to it might be an asymptote. The equations (13) give the locus in parametric form. Thus it is sometimes convenient to introduce a parameter.

5. *Symmetry.* Symmetry with respect to an axis or a center was defined in Ex. 5.1.10 and Ex. 6.1.5. We have seen that a curve is symmetric with respect to the x-axis (y-axis) if and only if its equation remains unchanged when y is replaced by $-y$ (x replaced by $-x$). Also a curve is symmetric with respect to the origin if and only if its equation remains unchanged when x and y are replaced by $-x$ and $-y$ at the same time. Of course a curve may be symmetric with respect to some other lines or points. This may be tested by translating and rotating so that an axis becomes parallel to the suspected line of symmetry or so that the origin coincides with the possible center of symmetry. However, certain symmetries may be detected without this. If the equation is unchanged when x and y are interchanged, the curve is symmetric with respect to the line $y = x$. Similarly, if the equation remains unchanged when x is replaced by $-y$ and y by $-x$, the curve is symmetric with respect to the line $y = -x$. Also, if substituting $x = a + b$ gives the same y as substituting $x = a - b$, the curve is symmetric with respect to the vertical line $x = a$. Similar remarks apply to symmetry with respect to a horizontal line. We leave it to the student to verify these statements and to recall the means of testing for symmetry in polar coordinates.

6. *Behavior at Infinity.* "At infinity" is just a picturesque way
of saying "when x or y gets very large." Here we are interested
in the questions: "What happens to x when y gets large?" "What
happens to y when x gets large?" If one variable approaches
infinity as the other approaches a finite value, we have a horizontal
or vertical asymptote. Our general definition of an **asymptote**
is a line that is approached by a point on the curve as x or y
approaches infinity. We may think of it as a line that is "tangent
to the curve at infinity." Sometimes we can conjecture the
existence of an asymptote and then prove our guess as we did
for the hyperbola. This may sometimes be done by rotating so
that one axis becomes parallel to the suspected asymptote. For
example, under a rotation through $-45°$, (11) becomes

$$(14) \qquad 2y'^3 + 6x'^2y' + 3\sqrt{2}x'^2 - 3\sqrt{2}y'^2 = 0$$

$$(15) \qquad x'^2 = \frac{y'^2(3\sqrt{2} - 2y')}{3\sqrt{2} + 6y'}$$

Equation (15) shows that the line $y' = -1/\sqrt{2}$ in the new system
is an asymptote.

7. *Excluded Values.* Since we are considering only pairs of real
coordinates (a branch of mathematics called functions of a complex
variable considers complex pairs), we must exclude values of either
variable that make the other imaginary. The easiest way to locate
these excluded values is to solve (if possible) the equation for each
variable. We then have $y = G(x)$ and $x = H(y)$. If G or H
involves even roots, we exclude those values of the variable that
make the radicands negative.

8. *Slope of the Curve.* By the slope of the curve at a point, we
mean the slope of the line tangent to the curve at that point. The
slope of the line tangent to $y = f(x)$ at a point (x_0, y_0) is given by
the derivative of y with respect to x evaluated at x_0, that is,
$[D_x f(x)]_{x=x_0}$. We have seen how to find this for various functions
of x. Where the slope is zero the curve has a horizontal tangent
and may have either a bend point (maximum or minimum) or a
horizontal inflection. The slope tells us the direction and steep-
ness of the curve. If it is positive, the curve is rising as x increases;
if it is negative, the curve is falling. In calculus these ideas are
fully developed and used extensively in curve plotting and many
other applications.

We shall illustrate the ideas of this section by giving a complete discussion of equation (11). The equation is not that of a familiar locus, and it does not factor in any obvious way. The only intercept is $(0, 0)$. Substitution of numerical values for x or y would yield a cubic equation, which would be difficult to solve. There is no symmetry with respect to the axes or the origin, but we note that x and y may be interchanged without affecting the equation.

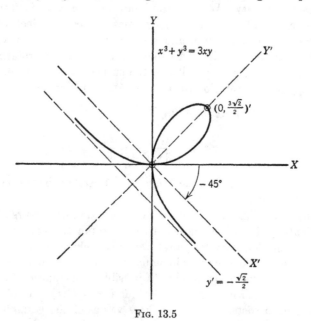

FIG. 13.5

Hence the curve is symmetric with respect to the line $y = x$. This suggests rotating the axes through $-45°$ so that the curve will be symmetric with respect to the new y-axis. When we do so, we find equation (15). We see that its locus is symmetric with respect to the y'-axis and has $y' = -1/\sqrt{2}$ as an asymptote. It meets the y'-axis at $y' = 0$ and $y' = 3/\sqrt{2}$. Values of $y' > 3/\sqrt{2}$ and $y' < -1/\sqrt{2}$ are excluded. (Why?) We may use (15) to find a few points on the curve, or we may make use of the parametric equations in the original coordinates given by (13). The curve, sketched in Fig. 13.5, is called the folium of Descartes.

EXERCISE 13.5

1. Illustrate each of the statements in the section by reference to curves previously plotted.

2. Justify statements about symmetry that were not proved in the section.

3. What is the equation of the asymptote in Fig. 13.5 in the original coordinates?

4. Discuss fully and plot each of the following equations. Organize your discussion under the eight headings dealt with in the section, stating negative as well as positive findings and justifying each statement.

(a) $xy = \sqrt{1 - x^2}$

(b) $y = x^2\sqrt{4 - x^2}$

(c) $xy = y^3 - 2y^2 + y + 10$

(d) $y^4 - (x + 1)(x - 2) = 0$

(e) $x^4 - y^4 = 1$

(f) $x^3 - y^3 = 0$

(g) $x^3 + 2y^3 - 3xy = 0$

(h) $x^3y = x^3 - 2$

5. A method of graphing called addition of ordinates is sometimes useful when the function can be expressed in the form of the sum of other functions. The method consists in plotting each part and then adding ordinates on the graph. Use this method to plot: (a) $y = x + x^2$. (b) $y = 1/x + x^2$. (c) $y = e^x - (1 + x)$.

6. Discuss and sketch the following, proceeding in what appears to be the most efficient way. Write your discussion systematically, but not necessarily in the order considered in the section.

(a) $x^2y = x^2 + 3x + 1$

(b) $y = 0.5(x^2 + 4)^{0.5} - (x + 1)$

(c) $x^3(y^2 + 2y - 1) = 1$

(d) $x^4 + y^4 - xy = 0$

(e) $y^2 - 3xy = x^3$

(f) $y = 3 \pm (x - 2)\sqrt{2}$

(g) $y = x^{3/2} + x^2$

(h) $(x^2 + y^2)^2 - 2(x^2 - y^2) = 2$

(i) $(y - 2)(y + 1)(y^2 + 4) - x = 0$

(j) $x^4 = y^4$

(k) $x^3y - x^2 - xy - y^2 + xy^2 + 1 = 0$

(l) $(x^2 + y^2)^3 = 4x^2y^2$

7. Discuss the appearance of the curves $y = x^2$ and $y = x^4$ at and near the origin. Where do they cross? Make a careful drawing of both curves on the same graph.

8. The following types of curves have aesthetic and geometrical properties that have caused them to be given special names. Choose numerical values of the parameters and plot.

(a) $y^2 = x^3/(2a - x)$ (Cissoid of Diocles)

(b) $yx^2 = a^2(a - y)$ (Witch of Agnesi)

(c) $y^2 = x^2(a - x)/(a + x)$ (Strophoid)

(d) $xy = cx^3 + dx^2 + ex + f$ (Trident of Newton)

(e) $x^2y + aby - a^2x = 0$ $(ab > 0)$ (Serpentine)

(f) $(x^2 + y^2)^2 = a^2(x^2 - y^2)$ (Lemniscate)

(g) $(x^2 + y^2 + 2ax)^2 = b^2(x^2 + y^2)$ (Limaçon)

(h) $(x - b)^2(x^2 + y^2) - ax^2 = 0$ (Conchoid)

(For additional information about these and other special curves, see the article *Curves, Special,* in the *Encyclopaedia Britannica.*)

9. The following functions are transcendental. However, the methods of this section apply. Discuss and sketch each one.

(a) $y = |x|$

(b) $y = |x^3|$

(†c) $y = x^{1/x}$

(d) $y = x + |x|$

(e) $y = x + |x| + |x + 2| + (x + 2)$

(f) $y = 0.5 \log_e \left(\dfrac{1 + x}{1 - x} \right)$

(g) $y = x \sin x^2$

(h) $y = \dfrac{\sin^2 x}{x}$

(i) $y = \sin x + |\sin x|$

(j) $y = (1 + e^{-x})^{-1}$

(k) $\sin x = \cos y$

(l) $y = x + \sin x$

(m) $y = x + e^{-x}$

(n) $y = \log_{10} [(x + 1)(x^2 - 1)^{-1}]$

(o) $y = \log_{10} (1 - x)^2$

(p) $\log_{10} y = 2x^2 + x + 1$

(q) $y = \sin (1/x)$

(r) $y = x \sin (1/x)$

(s) $y = x^2 \sin x$

(t) $y = \dfrac{\sin x}{x^2}$

10. We define $[x]$ as the greatest integer less than or equal to x. Plot the following:

(a) $y = [x]$

(b) $y = [|x|]$

(c) $y = [x^2]$

(d) $y = [x^3]$

(e) $y = x + [x]$

(f) $y = x^2 + [x]$

11. Solve simultaneously and sketch:

(a) $x^3 - y^3 = 9$
$x - y = 3$

(b) $x^4 - y^4 = 10$
$x + y = 2$

(c) $x^4 + y^4 = 100$
$x + y = 2$

(d) $2y = e^x + e^{-x}$
$x^2 + y^2 = 1$

(*Note.* You may have to use successive approximations.)

12. Plot $x^a y^b = 1$ for various values of a and b. Summarize your results.

13. Graph examples of the following curves given in parametric form:

(a) $x = a(\theta - \sin \theta)$, $y = a(1 - \cos \theta)$. This is the path of a point on the circumference of a rolling wheel of radius a. It is called a **cycloid.**

(b) $x = a\theta - b \sin \theta$, $y = a - b \cos \theta$.   This is the path of a point fixed at a distance b from the center of a rolling wheel of radius a.   It is called a **trochoid** or a **prolate** or **curtate cycloid**, according as $b > a$ or $b < a$.

(c) $x = (a \pm b) \cos \theta \mp b \cos \dfrac{a \pm b}{b} \theta$, $y = (a \pm b) \sin \theta - b \sin \dfrac{a \pm b}{b} \theta$.

This is the path of a point on the circumference of a circle of radius b rolling on a circle of radius a.   If it rolls outside, the upper signs are appropriate and it is called an **epicycloid**.   If it rolls inside, the lower signs apply and it is called a **hypocycloid**.   (*Suggestion.* Try cases where a is a multiple of b and also those in which it is not.)

14. Show that $x^{2/3} + y^{2/3} = a^{2/3}$ is a hypocycloid.

15. The curve $y = ae^{bx^2}$ passes through the points $(0, 3)$ and $(1, 2)$.   Find a and b.

16. Van der Waals' equation states that $\left(P + \dfrac{a}{V^2}\right)(V - b) = c$, where P is the pressure and V the volume of a gas.   Graph P as a function of V by choosing convenient values of the constants a, b, c.   Compare the graph with that of $PV = c$ for large P and for large V.

14

Functions of
Two Variables

We have so far concentrated our attention on functions of one variable, although we have sometimes used a function of two variables to define a function of one variable. Thus $x^2 + y^2 - 1 = 0$ defines y as a function of x, but the left member of the equation is a function of two variables, x and y. In fact, the locus of the equation is just the set of all pairs (x, y) for which this function is zero.

We say that a variable z is a **function of two variables** x and y, if for every pair of values (x, y) there is a corresponding value of z. As for functions of one variable, the values of the variables need not be numbers. We call x and y the **independent variables** and z the **dependent variable**. Other terms such as "single-valued" and "multiple-valued" are carried over from functions of a single variable. Evidently three variables are involved in a function of two variables. Accordingly, plane analytic geometry is not sufficient to represent these functions. In the first section of this chapter we introduce the fundamental ideas of analytic geometry of three dimensions. In the following sections we discuss the simplest types of functions of two variables and the simplest geometric figures in three dimensions.

14.1 Coordinates in three dimensions

In order to represent a function of two variables we want to be able to indicate a correspondence between pairs of values (x, y) and corresponding values z. It would seem natural to represent the pairs in a familiar plane coordinate system and to represent the corresponding z's by means of a third axis drawn perpendicular to the xy-plane at the origin. The two essentially different ways in which this third axis may be drawn relative to the xy-plane are indicated in Fig. 14.1.1. *We shall use the left-handed system.*

In Fig. 14.1.1 we have three **coordinate axes.** Taken in pairs, they determine three **coordinate planes,** called the xy-plane, the xz-plane, and the yz-plane. These planes are mutually perpendicular, that is, each is perpendicular to the other two. They divide the space into eight regions called **octants.** Consider any point P in the space. Drop a perpendicular PQ from it to the xy-plane. Then the x- and y-coordinates of P are just the coordinates of Q in the xy-plane. The z-coordinate is the length QP with a plus or minus sign according to whether the point is above or

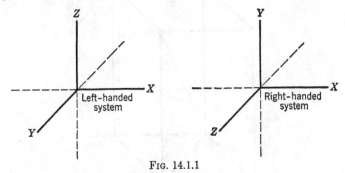

Fig. 14.1.1

below the xy-plane. Thus for each point we have a unique ordered triple of real numbers (x, y, z). Conversely, given any ordered triple (x, y, z), we can locate a unique point at a directed distance z above or below the point (x, y) in the xy-plane. Thus there is a one-to-one correspondence between points in space and ordered triples of numbers.

Given the coordinates (x, y, z), we arrive at the same point if we locate (x, z) in the xz-plane and then erect the perpendicular y or if we locate (y, z) in the yz-plane and then erect the perpendicular x. If we draw planes through the point parallel to the coordinate planes we have a rectangular parallelopiped (box) whose edges are of lengths $|x|$, $|y|$, and $|z|$. (See Fig. 14.1.2.) We can arrive at the point P via any one of six different paths by starting at the origin and moving along edges parallel to each of the axes in succession. Two such paths are shown in the figure. The coordinates of a point may be described as the directed distances of the point from the coordinate planes. Note that each coordinate is the directed distance from the plane of the *other* two coordinates. For example, the x-coordinate is the directed distance from the yz-plane.

A point P determines a vector OP with initial point at the origin. We call it the **radius vector**. In order to find the length of this vector, which is just the distance of the point from the

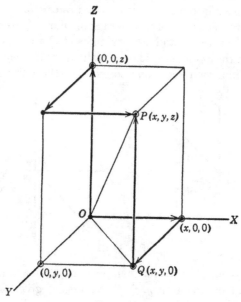

Fig. 14.1.2

origin, we note that, in Fig. 14.1.2, $x^2 + y^2 = \overline{OQ}^2$. (Why?) Also we have $\overline{OQ}^2 + \overline{QP}^2 = \overline{OP}^2$. (Why?) Hence

*14.1.1 The distance from the origin to a point (x, y, z) is given by

$$d = \sqrt{x^2 + y^2 + z^2}$$

The foregoing result is seen to be an obvious generalization of *8.1.2. The student will find that many ideas from plane analytic geometry carry over to solid analytic geometry by merely taking into account the additional coordinate. Thus we have:

*14.1.2 The distance between two points (x_1, y_1, z_1) and (x_2, y_2, z_2) is given by

$$d = \sqrt{(x_2 - x_1)^2 + (y_2 - y_1)^2 + (z_2 - z_1)^2}$$

This is easy to prove by drawing planes through both points parallel to the coordinate planes as in Fig. 14.1.3. The desired distance is the diagonal of the rectangular box whose edges are given by the differences of the coordinates.

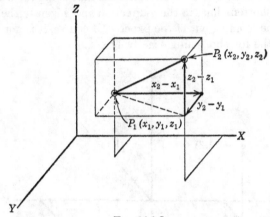

FIG. 14.1.3

It is very valuable to be able to visualize three-dimensional figures. Three-dimensional models are helpful in developing this ability, and the student should make them when he can. More frequently the three-dimensional situation is represented by a diagram on a plane surface. We adopt certain conventions for drawing, so that our diagram suggests the figure represented. The x- and z-axes are drawn as though the plane of the paper were the xz-plane. The y-axis is drawn in the plane of the paper at an angle of $135°$ with the x-axis, so that this angle of $135°$ represents an angle of $90°$ in space. (Fig. 14.1.4.)

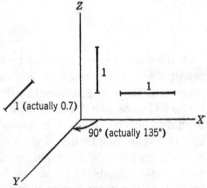

FIG. 14.1.4. Drawing conventions.

Lines that are parallel in space are drawn parallel on the diagram. Hence lines that are parallel to the axes are drawn parallel to OX,

OY, and OZ, respectively. For example, the vectors that represent the coordinates of a point are drawn parallel to the axes in Fig. 14.1.2. Distances along lines parallel to the x- and z-axes are measured with the same unit. The unit of measure along the y-axis is taken as $1/\sqrt{2} \doteq 0.7$ times the unit along the other axes. This foreshortens lines in the y-direction and suggests the appearance of lines coming out of the paper. Of course, it is not required to make the foreshortening exactly $1/\sqrt{2}$.

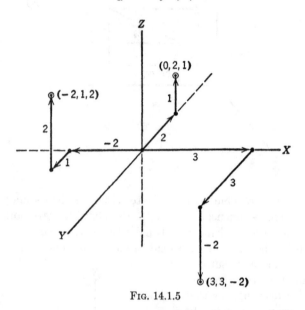

Fig. 14.1.5

In drawings made according to these conventions there is no perspective. For example, a segment of length 2 parallel to the x-axis will have length 2 in the drawing no matter where it is located. Hence, in order to indicate position in depth, it is desirable to show additional lines. The position of a point can be indicated by drawing its coordinates (Figs. 14.1.2 and 14.1.5). The position of a segment can be indicated by showing the coordinates of its end points. Figure 14.1.6 shows how three points in the same position on the paper can represent quite different points in space. In order to indicate a surface, we draw the representations of key points and curves on the surface. Particularly suggestive are the curves (called **traces**) in which the surface meets other surfaces,

especially the axis planes or planes parallel to them. In Fig.
14.1.7a we indicate a plane parallel to the xz-plane by drawing its
intersection with the yz-plane, the xy-plane, and two planes

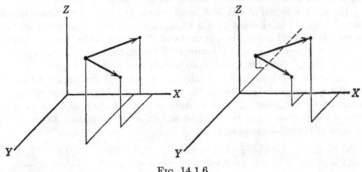

FIG. 14.1.6

parallel to the yz-plane. In Fig. 14.1.7b we suggest one-eighth of
a sphere by drawing the circles in which it intersects the axis
planes, a plane through the z-axis, and a plane parallel to the
xy-plane.

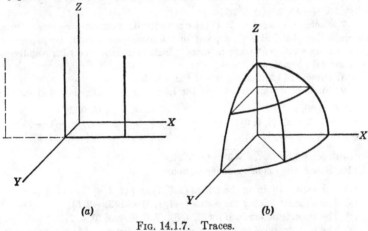

(a)   (b)

FIG. 14.1.7. Traces.

The student should practice making drawings according to these
conventions. It is essential to make these sketches large and
simple. Avoid shading and other devices for making a representa-
tional "picture." Draw only a part of the configuration if drawing
more would make the figure complicated.

EXERCISE 14.1

1. Make a model of a left-handed coordinate system of stiff paper. Label the axes and indicate the coordinates of several points.

†*2.* Make a model of a right-handed system and convince yourself that it cannot be made to coincide with a left-handed system. Thus there are two essentially different orientations of the axes. Is the same thing true in a plane? Work out an explanation of the terms "right-handed" and "left-handed" in terms of the relative positions of your thumbs and first two fingers.

3. The octants are not generally numbered, but the octant where all coordinates are positive is called the **first octant.** We may describe other octants by stating their positions above or below the xy-plane, in front or behind the xz-plane, and to the right or left of the yz-plane. Describe the position of the following points:

(a) $(5, 3, 1)$	(d) $(-1, 2, -6)$	(g) $(2, 1, -3)$	(j) $(-2, 1, 0)$
(b) $(0, 1, 2)$	(e) $(-3, -1, -4)$	(h) $(-3, -1, 4)$	(k) $(0, 0, -7)$
(c) $(1, -1, 5)$	(f) $(8, -1, -7)$	(i) $(8, 0, -3)$	(l) $(-3, 0, 0)$

4. Draw six large sketches showing the different paths from the origin to a point in the first quadrant, where each path is made up of three coordinate vectors.

5. Draw large sketches showing each of the points in prob. 3.

6. Find the distance from the origin of the points in prob. 3. [*Answers.* (a) $\sqrt{35}$. (c) $3\sqrt{3}$. (e) $\sqrt{26}$.]

7. Sketch the point $(3, 2, 1)$ together with its coordinate vectors. Draw several other sketches with a point in the same position on the paper but representing a different point in space. Include at least one with negative y and one with negative z.

8. Prove *14.1.2, making use of Fig. 14.1.3.

9. Find the distances between the following pairs of points and sketch:

(a) $(1, 3, 6), (2, 1, 5)$	(d) $(0, \pi, -3), (5, 0, 2)$
(b) $(-1, 3, -3), (1, 0, 0)$	(e) $(-1, -5, -1), (2, -1, -2)$
(c) $(-3, 2, 4), (0, 1, -1)$	(f) $(-3, 0, \sqrt{3}), (-2, -1, 0)$

Answers. (a) $\sqrt{6}$. (c) $\sqrt{35}$. (e) $\sqrt{26}$.

10. Sketch the following configurations:

(a) The segment joining the points $(1, 3, 1)$ and $(4, 1, 6)$.

(b) The segment joining the points $(-1, 1, 2)$ and $(3, -2, 5)$.

(c) The triangle determined by $(3, 4, 0)$, $(2, 0, 3)$, and $(0, 4, 2)$.

(d) The plane parallel to the zy-plane and at a distance 3 to the right of it.

(e) The plane parallel to the xy-plane and 4 units above it.

(†*f*) The plane through the points $(2, 0, 0)$, $(0, 2, 0)$, and $(0, 0, 2)$.

(g) Three segments of length 4, parallel to the x-axis, 2 units above the xy-plane and at a distance of 1, 3, and 5 units respectively in front of the xz-plane.

(†*h*) A line from the origin, making equal angles with the positive directions on the axes.

(*i*) A sphere of radius 2 with center at the origin. (Draw only the portion in the first octant.)

(*j*) Two segments of length 5 in each of the coordinate planes parallel to the axes in those planes.

(*k*) Three squares of sides 2, one in each of the axis planes with edges parallel to the axes in that plane.

(†*l*) Three circles of radius 2, one in the *xz*-plane, with center at (3, 0, 3), the second in the *yz*-plane with center at (0, 3, 3), and the third in the *xy*-plane with center at (3, 3, 0). (Note the distortion of the circles in the *yz*- and *xy*-planes. Begin by drawing diameters of the circles parallel to the axes.)

†11. Show that:

*14.1.3   The point that divides the segment from (x_1, y_1, z_1) to (x_2, y_2, z_2) in the ratio r_1/r_2 is given by

$$x = \frac{r_1 x_2 + r_2 x_1}{r_1 + r_2} \qquad y = \frac{r_1 y_2 + r_2 y_1}{r_1 + r_2} \qquad z = \frac{r_1 z_2 + r_2 z_1}{r_1 + r_2}$$

Suggestion. Compare Ex. 4.6.11.

12. For each pair of points in prob. 9, find the point that divides the segment in each of the following ratios: (*a*) 2/3. (*b*) 1. (*c*) −1. (*d*) 0. (*e*) 10. (*f*) −1/2. (*g*) −10.

13. Write formulas for the midpoint of a segment joining two given points.

14. Find the distance of the point (3, 2, 4) from each of the coordinate axes. (*Suggestion.* Make a drawing and recall that the distance of a point from a line is the length of the perpendicular from the point to the line.)

15. Find formulas for the distance of any point (x, y, z) from each of the coordinate axes.

14.2   Direction in space

In order to indicate the direction of a directed line or vector in a plane it was sufficient to indicate a single angle, which we called the direction angle. In 9.3 and 9.7 we defined it as the angle through which the positive *x*-axis or a line parallel to it had to be rotated to coincide with the vector. The angle α is the direction angle of the vectors shown in Fig. 14.2.1. Now we could also indicate the direction of a plane vector by giving the angle through which the positive *y*-axis would have to be rotated to coincide with the vector. This angle is labeled β in Fig. 14.2.1. Evidently $\alpha - \beta = 90°$ and $\sin \alpha = \cos \beta$ for any position of the vector. (Why?) The two angles α and β are called the **direction angles** of the vector. We note that

*14.2.1 $\cos^2 \alpha + \cos^2 \beta = 1$

The cosines of the direction angles, cos α and cos β, are called the **direction cosines** of the vector. If they are known, the angles α and β can be found. (How? Why would just one direction cosine be insufficient?) The use of the direction angles and

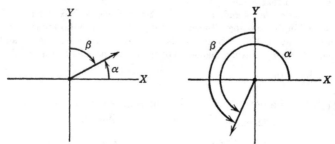

Fig. 14.2.1.   Direction angles.

direction cosines has the advantage of making familiar formulas more symmetric in appearance. For example, the normal form of the equation of a straight line becomes:

$$(1) \qquad\qquad x \cos α + y \cos β = p$$

where α and β are the direction angles of the normal vector from the origin.

In terms of these direction cosines we can also derive a very simple symmetric expression for the angle between two lines. Let the two lines have direction angles α, β and α', β'. In Fig. 14.2.2 we show the two lines intersecting at the origin. (If the given lines do not both pass through the origin we may take lines parallel to them which do.) Let (x, y) and (x', y') be any two points on the lines. Then by the law of cosines we have:

$$(2) \qquad\qquad d^2 = r^2 + r'^2 - 2rr' \cos θ$$

where d, r, and r' are the distances indicated in the figure. Solving this equation for cos θ and replacing the squares by their values in terms of the coordinates of the points, we find:

$$(3) \qquad \cos θ = \frac{xx' + yy'}{rr'} = \left(\frac{x}{r}\right)\left(\frac{x'}{r'}\right) + \left(\frac{y}{r}\right)\left(\frac{y'}{r'}\right)$$

But

$$(4) \qquad \begin{array}{lll} x = r \cos α & \quad & x' = r' \cos α' \\ & \text{and} & \qquad\qquad \text{(Why?)} \\ y = r \cos β & & y' = r' \cos β' \end{array}$$

Hence

*14.2.2 The angle between two plane vectors with direction
angles α, β, and α', β' is given by

$$\cos \theta = \cos \alpha \cos \alpha' + \cos \beta \cos \beta'$$

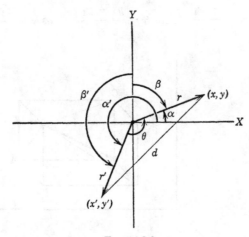

Fig. 14.2.2

The foregoing ideas of direction angles and direction cosines are
readily generalized to three dimensions.   In Fig. 14.2.3 we show a
vector OP with initial point
at the origin.   Let α, β, and
γ be the angles through
which the positive x-, y-
and z-axes would have to
be rotated to coincide with
this vector. These angles
are in the planes XOP,
YOP, and ZOP, respec-
tively. We call them the
direction angles of the
vector.

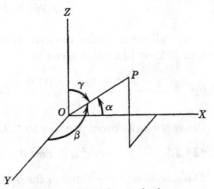

Fig. 14.2.3.   Direction angles in space.

In order to deal with a
vector with initial point not
at the origin, we replace it by a parallel radius vector or draw
through the initial point of the vector lines parallel to the

coordinate axes (Fig. 14.2.4). Then the direction angles of the vector are the angles α, β, and γ between the vector and these directed lines. If we let (x_1, y_1, z_1) and (x_2, y_2, z_2) be the coordinates of the initial and terminal points and r the length of the vector, we have in the triangles P_1AP_2, P_1BP_2, and P_1CP_2

$$(5) \quad \cos \alpha = \frac{x_2 - x_1}{r} \quad \cos \beta = \frac{y_2 - y_1}{r} \quad \cos \gamma = \frac{z_2 - z_1}{r}$$

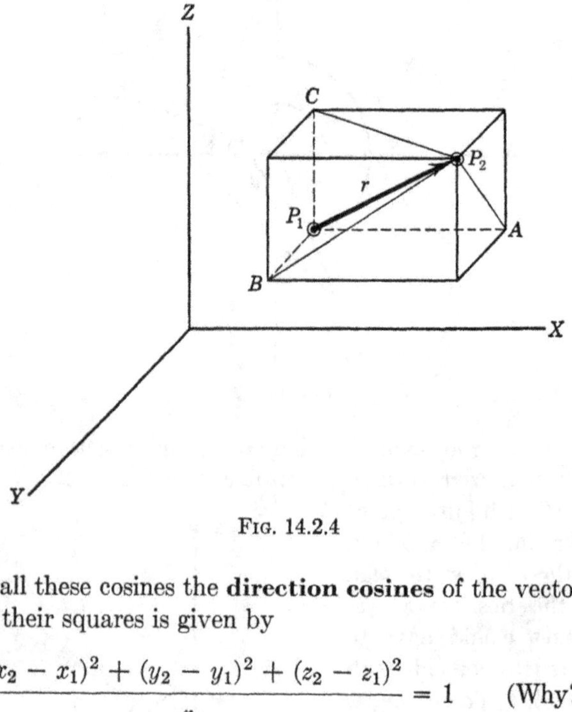

Fig. 14.2.4

We call these cosines the **direction cosines** of the vector. The sum of their squares is given by

$$(6) \quad \frac{(x_2 - x_1)^2 + (y_2 - y_1)^2 + (z_2 - z_1)^2}{r^2} = 1 \quad \text{(Why?)}$$

Hence we have corresponding to *14.2.1:

*14.2.3 $\qquad \cos^2 \alpha + \cos^2 \beta + \cos^2 \gamma = 1$

The case of a vector through the origin is the special case in which the initial point is replaced by $(0, 0, 0)$.

The direction of a vector in space is uniquely determined by its direction cosines. As in the plane, if we reverse the direction of the vector, we change the direction angles by 180° and reverse the sign of each of the direction cosines. For an undirected line there are two possible sets of direction angles and cosines according to which direction on the line we choose. We speak of either set as being the direction cosines and direction angles of the line.

We now develop a formula corresponding to *14.2.2. If two vectors do not have the same initial point, we define the angle between them to be the angle between two radius vectors with the same direction angles. In Fig. 14.2.5 we show two radius

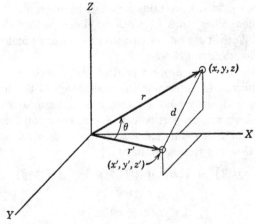

FIG. 14.2.5

vectors of lengths r and r' with terminal points (x, y, z) and (x', y', z') and direction angles α, β, γ and α', β', γ'. If d is the distance between the terminal points and θ the angle between the two vectors, we have equation (2) as in the plane case. However, now

$$r^2 = x^2 + y^2 + z^2$$

(7) $$r'^2 = x'^2 + y'^2 + z'^2$$

$$d^2 = (x - x')^2 + (y - y')^2 + (z - z')^2$$

Solving (2) for $\cos \theta$, substituting these expressions, and simplifying, we find:

(8) $$\cos \theta = \frac{xx' + yy' + zz'}{rr'} = \left(\frac{x}{r}\right)\left(\frac{x'}{r'}\right) + \left(\frac{y}{r}\right)\left(\frac{y'}{r'}\right) + \left(\frac{z}{r}\right)\left(\frac{z'}{r'}\right)$$

But from (5), each of these fractions is the corresponding direction cosine. Hence

*14.2.4 The angle between two vectors with direction angles α, β, γ and α', β', γ' is given by

$$\cos \theta = \cos \alpha \cos \alpha' + \cos \beta \cos \beta' + \cos \gamma \cos \gamma'$$

This formula gives the smallest angle between the positive directions of the vectors. If we wish the angle between two un-directed lines in space, we take the angle between any two vectors lying on these lines, that is, we may choose either direction on each line. We will get one of two supplementary angles according to which directions we choose.

Any three numbers proportional to the direction cosines of a line are sufficient to determine the orientation of the line, that is, its direction cosines except for a factor ± 1. Suppose we are given A, B, C, which are proportional to $\cos \alpha$, $\cos \beta$, and $\cos \gamma$. This means that there is a constant k such that $(\cos \alpha)/A = (\cos \beta)/B = (\cos \gamma)/C = k$. Hence we have

(9) $(kA)^2 + (kB)^2 + (kC)^2 = 1$ (Why?)

and hence

(10) $k = \pm(A^2 + B^2 + C^2)^{-\frac{1}{2}}$

With this value of k we may find the direction cosines as $\cos \alpha = kA$, $\cos \beta = kB$, and $\cos \gamma = kC$. Either the plus or the minus sign must be used for all three. The two signs give the direction cosines of vectors lying on the same line in opposite directions. Numbers that are proportional to the direction cosines of a line are called **direction numbers.** We see that a set of direction numbers determines the orientation of a line.

EXERCISE 14.2

Direction Angles in the Plane

1. In each of the following problems, one of the direction angles and a point on the line are given. Draw the line and find the other direction angle.

(a) $\alpha = 30°$, $(0, 0)$ (f) $\beta = 186°$, $(2, 1)$
(b) $\beta = 45°$, $(0, 0)$ (g) $\beta = 82°$, $(-1, 4)$
(c) $\beta = -45°$, $(0, 0)$ (h) $\alpha = 260°$, $(-2, -5)$
(d) $\alpha = 210°$, $(0, 0)$ (i) $\beta = -10°$, $(3, -2)$
(e) $\alpha = 300°$, $(0, 0)$

2. The following pairs of numbers are direction cosines of certain radius vectors. Find the direction angles.

(a) $0.5, 0.5\sqrt{3}$ (c) $-0.77, -0.64$ (e) $-0.98, 0.21$
(b) $0.5\sqrt{2}, -0.5\sqrt{2}$ (d) $0, 1$ (f) $-1, 0$

3. Why is *14.2.1 true? Justify (1). Carry through the algebra to get from (2) to (3). Why is (4) correct?

4. For two lines we have $\alpha = 31°$ and $\alpha' = 264°$. Find the angle between them from *14.2.2. Do the same for $\alpha = 211°$ and $\alpha' = 264°$. Draw a sketch for each situation and indicate the angles.

5. Find the direction cosines of the line $y = 2x + 1$. (Note that there are two possible answers.)

6. A set of numbers proportional to the direction cosines of a line in the plane is called a set of **direction numbers.** Derive a formula corresponding to (10) from which you can find the direction cosines if a set of direction numbers is known. Find the direction cosines of the lines with the following direction numbers: (a) 1, 2. (b) -3, 1. (c) 0, 1. (d) $-1, -1$. (e) 1, 0.

7. Show that the equation of a straight line may be written in the following forms:

$$(11) \qquad \frac{x - x_1}{\cos \alpha} = \frac{y - y_1}{\cos \beta}$$

$$(12) \qquad \frac{x - x_1}{A} = \frac{y - y_1}{B}$$

$$(13) \qquad \frac{x - x_1}{x_2 - x_1} = \frac{y - y_1}{y_2 - y_1}$$

$$(14) \qquad x = s \cos \alpha + x_1$$
$$y = s \cos \beta + y_1$$

where (x_1, y_1), (x_2, y_2) are any two points on the line, α and β are the direction angles, A and B are a set of direction numbers, and s is the distance measured along the line in the positive direction from (x_1, y_1) to (x, y).

8. Show that in the equation of a straight line the coefficients of x and y are a set of direction numbers of the normal to the line.

9. Show that a necessary and sufficient condition for two lines to be perpendicular in the plane is that

$$\cos \alpha \cos \alpha' + \cos \beta \cos \beta' = 0$$

Show that this condition is equivalent to *9.2.3.

10. Show that a necessary and sufficient condition that two lines in a plane be parallel is that

$$\cos \alpha \cos \alpha' + \cos \beta \cos \beta' = \pm 1$$

Show that this is equivalent to the equality of their slopes.

11. Show that *14.2.2 is consistent with *9.2.2. (Note that *14.2.2 does not indicate the sign of the angle, but does indicate without ambiguity the angles between two directed lines, whereas *9.2.2 gives the sign of the angle from the first to the second line but does not distinguish between the two supplementary angles formed by two intersecting lines. The slope does not distinguish between vectors in exactly opposite directions.)

Directions in Space

12. In each of the following, two points are given. Find the direction cosines of a vector from the first to the second. Sketch the lines and determine the direction angles in each case.

(a) $(0, 0, 0)$, $(2, 1, 4)$ (e) $(-2, 3, 1)$, $(5, 5, 6)$

(b) $(0, 0, 0)$, $(1, 1, 1)$ (f) $(-1, -1, -1)$, $(4, -5, 2)$

(c) $(0, 0, 0)$, $(-3, -1, 4)$ (g) $(0, 9, -2)$, $(5, 0, 1)$

(d) $(1, 8, 3)$, $(4, 2, 7)$ (h) $(1, 3, 7)$, $(2, 18, 0)$

Answers. (a) $2/\sqrt{21}$, $1/\sqrt{21}$, $4/\sqrt{21}$, $\alpha = 64° 7'$, $\beta = 77° 24'$, $\gamma = 29° 13'$. (e) $\alpha = 37° 34'$, $\beta = 76° 55'$, $\gamma = 54° 31'$.

13. Find the direction cosines of each of the axes.

14. Verify *14.2.3 for several lines in prob. 12.

15. Find the direction cosines of the lines with the following direction numbers:

(a) $1, 1, 1$ (c) $0, 1, 1$ (e) $2, -1, 4$ (g) $1, -4, 0$

(b) $-1, 3, -5$ (d) $2, -3, -2$ (f) $-2, -2, -2$ (h) $\sqrt{2}, 1, \sqrt{3}$

16. Find the angles between the pairs of lines with the following direction numbers. Find both possibilities.

(a) $(4, 1, 1)$, $(2, 1, -1)$ (c) $(1, 1, 1)$, $(-1, 2, 1)$

(b) $(-2, -1, 3)$, $(0, 1, 0)$ (d) $(1, 2, -2)$, $(0, 1, 1)$

17. Rewrite *14.2.4 in terms of direction numbers.

18. Write necessary and sufficient conditions for two lines to be perpendicular in terms of their direction angles and in terms of their direction numbers.

19. Write necessary and sufficient conditions for two lines to be parallel in terms of their direction angles and in terms of their direction numbers.

20. Given $\alpha = \beta = \gamma$, find them.

21. Given $\alpha = 20°$, $\cos \beta = -1$. Find γ.

22. Given $\alpha = 30°$, $\beta = 60°$. Find γ.

23. Can two of the direction angles be given arbitrarily? Justify your answer. Given α, what restriction is imposed on β? on γ?

24. Show that a necessary and sufficient condition for three lines to be mutually perpendicular is $\Delta^2 = 1$, where

$$\Delta = \begin{vmatrix} A_1 & B_1 & C_1 \\ A_2 & B_2 & C_2 \\ A_3 & B_3 & C_3 \end{vmatrix}$$

and the letters are direction numbers of the lines.

14.3 The plane

A linear function in one variable or a linear equation in two variables represents a straight line in plane analytic geometry. A linear function in two variables

$$(1) \qquad z = mx + ny + b$$

or a linear equation in three variables

$$(2) \qquad Ax + By + Cz + D = 0 \qquad \text{(Standard Form)}$$

represents a plane in solid analytic geometry. It is easy to see this in simple cases. Thus the equations $x = 0$, $y = 0$, $z = 0$ are each of the form (2) and obviously are the equations of the yz-, xz-, and xy-planes, respectively. Similarly, $x = c$, $y = d$, and $z = e$ are, respectively, the equations of planes parallel to the coordinate planes. We now wish to prove that:

*14.3.1 Every equation of the form (2) is the equation of a plane, and every plane has an equation of the form (2).

We begin by considering a plane and its normal from the origin, that is, the vector from the origin perpendicular to the plane (Fig. 14.3). Let p be the length of this vector OP and α, β, and γ its direction angles. Now consider any point $Q(x, y, z)$ on the plane. Let α', β', and γ' be the direction angles of OQ and let θ be the angle POQ. Since OP is normal to the plane, it is perpendicular to PQ. (Why?) Hence in the right triangle OPQ, $\cos \theta = p/OQ$. Also

$$(3) \quad \cos \alpha' = x/OQ, \quad \cos \beta' = y/OQ, \quad \text{and} \quad \cos \gamma' = z/OQ$$

Hence for every point on the plane and only for such points we have, from *14.2.4,

$$(4) \quad p/OQ = (x/OQ) \cos \alpha + (y/OQ) \cos \beta + (z/OQ) \cos \gamma$$

Our argument holds no matter what the position of the plane unless it passes through the origin. In that case $p = 0$ and the triangle POQ does not exist. However, the necessary and sufficient condition that a point lie in such a plane is that the vector from the origin to the point be perpendicular to the normal. (Why? Note that the normal may be taken in either of two directions.) But the condition for this perpendicularity is just (4) with $p = 0$.

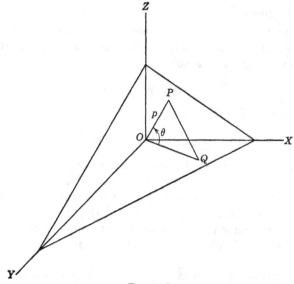

FIG. 14.3

(Why? See Ex. 14.2.18.) Hence (4) holds in all cases, and we have:

*14.3.2 The equation of the plane whose normal from the origin has length p and direction angles α, β, γ is

(5) $x \cos \alpha + y \cos \beta + z \cos \gamma = p$ (Normal Form)

This result proves the second part of *14.3.1.

In order to prove the first part of *14.3.1 we consider the equation (2). We shall find the direction cosines of the line that has A, B, C as its direction numbers. From 14.2(10) these direction cosines are kA, kB, kC, where $k = \pm(A^2 + B^2 + C^2)^{-\frac{1}{2}}$. Rewriting (2) by multiplying both sides by k and transposing the constant term, we find

(6) $kAx + kBy + kCz = -kD$

Finally, we choose the sign of k so that the right member of (6) is positive. Then we see that (6) is just in the form (5), that is, the coefficients of x, y, and z are direction cosines and the right member is a positive number. Hence (6) must be the equation of a plane with normal of length $-kD$ and direction cosines kA, kB, and kC, where the sign of k has been properly chosen. This result not only completes the proof of *14.3.1 but tells us how to put an equation of the form (2) in the form (5). We see that the coefficients of the variables in a linear equation are a set of direction numbers of the normal to the corresponding plane.

We leave it to the student to prove the following theorems:

*14.3.3 The equation of a plane through three points (x_1, y_1, z_1), (x_2, y_2, z_2), (x_3, y_3, z_3) is given by

$$\begin{vmatrix} x & y & z & 1 \\ x_1 & y_1 & z_1 & 1 \\ x_2 & y_2 & z_2 & 1 \\ x_3 & y_3 & z_3 & 1 \end{vmatrix} = 0 \qquad \text{(Three-Point Form)}$$

*14.3.4 The equation of the plane through the point (x_0, y_0, z_0) and whose normal has direction numbers A, B, C is given by

$$A(x - x_0) + B(y - y_0) + C(z - z_0) = 0$$

$$\text{(Point Direction Form)}$$

*14.3.5 The equation of the plane with intercepts $(a, 0, 0)$, $(0, b, 0)$, and $(0, 0, c)$ is

$$\frac{x}{a} + \frac{y}{b} + \frac{z}{c} = 1 \qquad \text{(Intercept Form)}$$

It is sometimes of interest to talk about the **angle between two planes.** This is defined as the angle between the normals to the two planes. Hence we can use *14.2.4 to find the angle between two planes since we can easily find the direction cosines of the normals from the coefficients of the equations. In particular,

*14.3.6 The angle between two planes

$$Ax + By + Cz + D = 0$$

$$A'x + B'y + C'z + D' = 0$$

is given by

$$\cos \theta = \frac{AA' + BB' + CC'}{(A^2 + B^2 + C^2)^{1/2}(A'^2 + B'^2 + C'^2)^{1/2}}$$

EXERCISE 14.3

1. Explain why the equations $x = a$, $y = b$, and $z = c$ represent planes.

2. Find the equations of the following planes:

(a) Having a normal with direction numbers 1, 1, 1 and length 2.

(b) Through $(0, 1, 5)$, $(2, -1, 4)$, and $(-1, -1, 6)$.

(c) Having intercepts $(0, 0, 2)$, $(3, 0, 0)$, and $(0, 4, 0)$.

(d) Passing through the point $(1, 5, -1)$ and having a normal with equal direction angles.

(e) Passing through the origin and perpendicular to a line with direction numbers 1, 2, 3.

3. An effective way to sketch a plane is to draw its traces in the axis planes. The easiest way to do this is to find the intercepts of the plane (by setting two variables at a time equal to zero) and join them by straight lines. In this way sketch the following planes:

(a) $x + y + z = 3$

(b) $x + 2y + 3z = 6$

(c) $x - y + 2z = 8$

(d) $x + 3y = 4z - 24$

(e) $\dfrac{x}{2} + \dfrac{y}{5} + \dfrac{z}{-1} = 1$

(f) $z + 5y = 2$

(g) $2y - 3z = 12$

(h) $3x - 4z = 12$

(i) $3z - 2 = 0$

(j) $2x - y - 3z = 6$

(k) $x + y + z = -5$

(l) $y = 2x + 1$

4. Sketch the planes in prob. 2.

5. Put each of the equations in prob. 3 in normal form; find the direction angles and length of the normal.

6. Put each of the equations in prob. 3 in intercept form.

7. Find the equation of the plane through $(1, -1, 3)$, $(5, 1, 4)$, and $(-6, 3, -5)$. Put the equation in the standard form, normal form, and intercept form. Find its intercepts, length of normal, and direction angles of normal. Sketch.

8. Do the same for the plane through the points $(0, 0, 1)$, $(0, 3, 0)$, and $(-4, 0, 0)$.

9. Prove *14.3.6. Show that a necessary and sufficient condition for two planes to be perpendicular is that $AA' + BB' + CC' = 0$, where A, B, C and A', B', C' are the coefficients of x, y, z in their equations.

10. Show that a necessary and sufficient condition for two planes to be parallel is that the coefficients of the variables are proportional. What happens if all the coefficients are proportional?

11. Find the equation of the plane tangent to a sphere with center at the origin and radius 3 at the point $(2, 1, -2)$.

†*12.* Find the equations of the planes passing through $(10, 2, 6)$ and $(8, 6, 1)$ and tangent to a sphere of radius 5 with center at the origin.

13. Find the equation of the plane through the point $(2, 5, -9)$ and parallel to the plane $3x - 2y + 7z = 4$.

†14. Show that:

*14.3.7 The distance from the plane

$$x \cos \alpha + y \cos \beta + z \cos \gamma = p$$

to the point (x_1, y_1, z_1) is given by

$$d = x_1 \cos \alpha + y_1 \cos \beta + z_1 \cos \gamma - p$$

Suggestion. Draw a plane parallel to the given plane through the given point and write its equation in normal form. Then the distance from plane to point is the same as from plane to plane.

15. Find the distance from the following planes to the indicated points:

> (a) $x + y + z = 0$ to $(1, 1, 4)$
> (b) $4x - 2y + z = -2$ to $(-1, -3, -2)$
> (c) $3y - x - 4z + 12 = 0$ to $(1, -3, 0)$
> (d) $4x - 3y - 2z = 1$ to $(1, 1, 1)$
> (e) $3x - 15y - 2z = 4$ to $(-1, -3, 0)$
> (f) $x + 2y - 5z + 7 = 0$ to $(-8, 3, -14)$

16. Find the distance from each plane in prob. 3 to $(1, -6, 2)$.

17. Put the equations $x = 5$, $z = -4$, and $y = 3$ in normal form and find the lengths and direction angles of the normals.

18. Find the equations of the planes perpendicular to the line joining $(-1, 1, 3)$ and $(8, 1, 2)$ and passing through: (a) $(2, 2, 1)$. (b) The origin. (c) $(-5, 4, -3)$.

†19. In 4.10 we indicated how to solve three linear equations in three unknowns. Since each equation represents a plane, what is the interpretation of the solution? Draw a sketch of one or more of the problems in Ex. 4.10.

†20. What is the geometric interpretation of three linear equations that have no solution? An infinite number of solutions?

21. Find the locus of points equidistant from the end points of the segment $(3, 9, -1)$, $(2, -1, -3)$. Show that the locus is a plane bisecting the segment and perpendicular to it.

22. Generalize the results of prob. 21 to any segment and prove your statement.

14.4 The straight line in space

In 14.2(5) we found that the direction cosines of a line are given by the differences of the corresponding coordinates of two of its points divided by the distance between them. Any two points on the same line would yield the same direction cosines or their negatives, depending upon the order in which the points are taken. This follows because the derivation of 14.2(5) was quite independent of the choice of the two points. Hence if (x_1, y_1, z_1) is a fixed point on a line and α, β, γ are its direction angles, then, for any point (x, y, z) on the line, and only for such points

(1) $\qquad \dfrac{x - x_1}{\cos \alpha} = r, \qquad \dfrac{y - y_1}{\cos \beta} = r, \qquad \dfrac{z - z_1}{\cos \gamma} = r$

where r is the distance between the two points. It follows that the equations of the line are

(2) $\qquad \dfrac{x - x_1}{\cos \alpha} = \dfrac{y - y_1}{\cos \beta} = \dfrac{z - z_1}{\cos \gamma} \qquad$ (Symmetric Form)

There are only two independent equations in (2), since the equality of any two pairs of terms implies the equality of the third. Since two planes meet in a line, it is not surprising to find that it takes two linear equations to represent a line in space. Each of the three planes in (2) is parallel to one of the axes (why?) and is called a **projecting plane** because it meets one of the coordinate planes in the projection of the line.

Evidently (2) would be unaffected if we multiplied all the denominators by any constant. By doing so we could rewrite it with the direction cosines replaced by any set of direction numbers of the line. Hence another form of the equations of a line is:

(3) $\qquad \dfrac{x - x_1}{A} = \dfrac{y - y_1}{B} = \dfrac{z - z_1}{C} \qquad$ (Point Direction Form)

where A, B, C is a set of direction numbers of the line. But from these equations we see that the differences of the coordinates of any two points on a line are direction numbers of the line. (Why?) Hence a third form of (2) is:

(4) $\qquad \dfrac{x - x_1}{x_2 - x_1} = \dfrac{y - y_1}{y_2 - y_1} = \dfrac{z - z_1}{z_2 - z_1} \qquad$ (Two-Point Form)

Finally, it is easy to see that the line may be given parametrically by any set of linear functions in the parameter, for example,

(5) $\qquad x = At + x_0 \qquad y = Bt + y_0 \qquad z = Ct + z_0$

$\qquad\qquad\qquad\qquad\qquad\qquad\qquad\qquad\qquad$ (Parametric Form)

This is so because if we eliminate t from these three equations we get equations of the form (3). Moreover, A, B, C are direction numbers of the line. In particular, if the parameter is the distances measured along the line from the point (x_0, y_0, z_0) in the direction α, β, γ, these equations become

(6) $x = s \cos \alpha + x_0,$ $y = s \cos \beta + y_0,$ $z = s \cos \gamma + z_0$

The foregoing forms of the equations of the straight line are the most convenient because they supply us immediately with direction

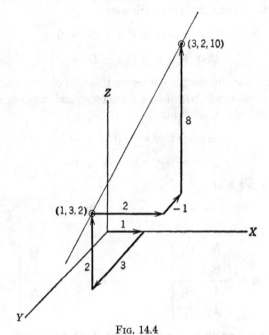

FIG. 14.4

numbers and a point on the line. A second point can always be found by using the fact that the differences of the coordinates of any two points on the line are a set of direction numbers. This means that we can start at any point on the line and find another by going parallel to the axes distances equal to any set of direction numbers. Thus to plot the line given by

(7) $x = 2t + 1,$ $y = -t + 3,$ $z = 8t + 2$

or

(8) $$\frac{x - 1}{2} = \frac{y - 3}{-1} = \frac{z - 2}{8}$$

we locate the point $(1, 3, 2)$ on the line (see Fig. 14.4) and then locate another point by going directed distances 2, -1, 8. We could of course locate another point by substituting an arbitrary value of t in (7) or by substituting some value for one variable in (8).

It may happen that we are given the equations of a line by means of two linear equations not in a convenient symmetric or parametric form. Thus any two equations

$$(9) \quad \begin{aligned} A_1x + B_1y + C_1z + D_1 &= 0 \\ A_2x + B_2y + C_2z + D_2 &= 0 \end{aligned}$$

determine a line, unless the corresponding planes are parallel. We put these equations in parametric form by considering z as a parameter and calling it t. Solving

$$(10) \quad \begin{aligned} A_1x + B_1y &= -(C_1t + D_1) \\ A_2x + B_2y &= -(C_2t + D_2) \end{aligned}$$

for x and y, we find

$$(11) \qquad x = At + x_0, \qquad y = Bt + y_0$$

where

$$(12) \qquad A = \frac{\begin{vmatrix} B_1 & C_1 \\ B_2 & C_2 \end{vmatrix}}{\begin{vmatrix} A_1 & B_1 \\ A_2 & B_2 \end{vmatrix}} \qquad B = \frac{\begin{vmatrix} C_1 & A_1 \\ C_2 & A_2 \end{vmatrix}}{\begin{vmatrix} A_1 & B_1 \\ A_2 & B_2 \end{vmatrix}}$$

It follows from our discussion of (5) that A, B, 1 is a set or direction numbers of the line. Multiplying the set by the denominator in (12), we find

*14.4.1 Direction numbers of the line of intersection of (9) are given by

$$(13) \qquad \begin{vmatrix} B_1 & C_1 \\ B_2 & C_2 \end{vmatrix} \qquad \begin{vmatrix} C_1 & A_1 \\ C_2 & A_2 \end{vmatrix} \qquad \begin{vmatrix} A_1 & B_1 \\ A_2 & B_2 \end{vmatrix}$$

The student should note that each determinant contains the coefficients of the variables *not* corresponding to the direction number sought. Also the letters involved come from the set $\{A, B, C\}$ by omitting A, B, C in succession and taking the remaining letters always in the order $ABCA$. Once a set of direction numbers has been found, the equations of the line may be written in any one of the symmetric or parametric forms after finding the coordinates of any point. But $(x_0, y_0, 0)$, where x_0 and y_0 are the quantities that emerge in (11), is one point on the line, so that the procedure (9)–(12) supplies us with all the information desired.

EXERCISE 14.4

1. Use the two-point form to write the equations of each of the lines in Ex. 14.2.12. Write also in the parametric form.

Answers. (a) $\dfrac{x}{2} = \dfrac{y}{1} = \dfrac{z}{4}$, $x = 2t$, $y = t$, $z = 4t$. (e) $\dfrac{x+2}{7} = \dfrac{y-3}{2} = \dfrac{z-1}{5}$, $x = -2 + 7t$, $y = 3 + 2t$, $z = 1 + 5t$.

2. A line is perpendicular to the plane $2x - 3y + 7z - 1 = 0$ and passes through the point $(-1, 2, 5)$. Find its equations.

3. Same as prob. 2, with $(-1, 2, 5)$ replaced by: (a) The origin. (b) $(-1, -1, 0)$. (c) $(2, -8, 4)$.

4. Find the equations of the lines through the following points and having the given direction numbers:

(a) $(0, 1, 5)$; $1, 1, 1$

(b) $(-3, 2, 1)$; $2, -1, 3$

(c) $(-1, -3, -5)$; $-3, -4, 5$

(d) $(-3, 0, -2)$; $2, -1, 3$

(e) $(4, -r, s)$; $-3, 2a, 3a$

(f) (a, b, c); a^2, ab, b^2

Answers. (a) $\dfrac{x}{1} = \dfrac{y-1}{1} = \dfrac{z-5}{1}$. (c) $\dfrac{x+1}{3} = \dfrac{y+3}{4} = \dfrac{z+5}{-5}$.

5. Write the equations of the lines with the following direction angles and passing through the indicated points:

(a) $\alpha = 60°$, $\beta = 40°$, $(1, 3, -3)$

(b) $\alpha = 100°$, $\beta = 10°$, $(0, 0, 0)$

6. Sketch the lines given by

(a) $x = t$, $y = t$, $z = t$

(b) $x = 3t - 1$, $y = t + 2$, $z = t - 1$

(c) $x - 3 = y - 1 = z + 3$

(d) $\dfrac{x+1}{2} = \dfrac{y-3}{4} = \dfrac{x+5}{-1}$

7. Use *14.4.1 to find direction numbers of the lines determined by

(a) $x - 3y + z = 2$
    $-x + 2y - 3z + 2 = 0$

(b) $x + y - z - 1 = 0$
    $-x + 7y - 2z = 3$

(c) $x - y = 0$
    $4y + 2x - z = 0$

(d) $x = 2$
    $y = 0$

(e) $ax + by + cz = d$
    $a'x + b'y + c'z = d'$

(f) $z = x - y + 2$
    $z = x - 3y - 1$

8. Carry through the procedure (9)–(11) for the examples in prob. **7** and so find the equations of the lines in parametric form.

9. Find and sketch the projecting planes of each line in prob. 6. (*Suggestion*. Each may be found by eliminating one variable between the two equations. It is not necessary to put in symmetric form.)

10. Solve (10) to verify (12) and evaluate x_0 and y_0. What happens if the planes (9) are parallel?

†11. Derive *14.4.1 by making use of the fact that the line determined by (9) must lie in each plane and hence be perpendicular to the normal to each.

12. In order to find where a line meets a given plane, we solve the equations of the line simultaneously with those of the plane. Unless the line is parallel to the plane, this will result in three simultaneous equations in three unknowns. If the line is given parametrically we substitute the parametric expressions in the equation of the plane and solve for the value of the parameter. Find where the following lines meet the indicated planes:

(*a*) $x = 2t$, $y = t - 1$, $z = 3t + 4$, and the axis planes
(*b*) $3x + y - z = 2$, $4z - y + x - 1 = 0$, and the axis planes
(*c*) $x = t + 1$, $y = 3t$, $z = 2t$, and $3x - y + 4z = 5$
(*d*) $x + 2y - z = 4$, $-x - 3y - 2z = 2$, and $x + y = 3$
(*e*) $x + 3y + z = 4$, $y + 2x + z = 0$, and $7x - 2y + 10 = z$
(*f*) $x = 1 - t$, $y = 3t + 2$, $z = t - 5$, and $4x - 2z + y = 7$

13. Find the equations of the line through the origin and parallel to the line given by $x - y + 2z - 1 = 0$ and $2x - z = 1$.

14. Can the equations of a line parallel to one of the coordinate axes be put in the symmetric form or one of those derived from it? What forms appear to be always possible? Write the equations of: (*a*) The x-axis. (*b*) The y-axis. (*c*) The z-axis. (*d*) A line parallel to the xy-plane and making equal angles with the positive x-axis and y-axis. (*e*) A line in the xz-plane bisecting the angle at the origin.

15. Find the line through the point $(1, -3, 2)$ perpendicular to the line $x = 3t - 1$, $y = t - 2$, $z = 2t + 3$, and to the line $2x - y + z = 0$, $z = 1 - 3y + 2$

14.5 Loci in space

Any equation of the form

$$(1) \qquad\qquad F(x, y, z) = 0$$

represents a locus that is just the set of points whose coordinates satisfy (1). Here $F(x, y, z)$ stands for a function of the point (x, y, z) or of x, y, and z, and the locus is the set of points for which it is zero. Of course the locus may consist of a single point; for example, the locus of

$$(2) \qquad\qquad x^2 + y^2 + z^2 = 0$$

is the single point $(0, 0, 0)$. Or the locus may contain no points at all; for example, the locus of

$$(3) \qquad\qquad x^2 + y^2 + z^2 = -1$$

is a null (empty) set of points, because there are no points whose coordinates satisfy (3). We have seen that when F is linear, the locus is a plane. Of course, F may represent any function and hence a tremendous variety of surfaces. For example, the equation

$$(4) \qquad\qquad x^2 + y^2 + z^2 = R^2$$

is the equation of a sphere with center at the origin and radius R, because the coordinates of a point satisfy this equation if and only if it is at a distance R from the origin. More generally,

$$(5) \qquad\qquad (x - h)^2 + (y - k)^2 + (z - m)^2 = R^2$$

is the equation of a sphere of radius R and center (h, k, m).

It is not possible to give a simple rule from which we can know the nature of the surface represented by the equation (1). Just as in plane analytic geometry, it is necessary to study the nature of each surface. If (1) is solvable for one of the variables, so that it can be rewritten in the form

$$(6) \qquad\qquad z = f(x, y)$$

we can find as many points as we wish by substituting values of x and y. But even more than in the plane case, the unsystematic plotting of points does not help us very much. The discussion of a surface has much in common with the discussion of a plane curve. It is certainly useful to find the intercepts, that is, the points at which the surface cuts the axes. This can be done by letting two variables at a time be zero. Consideration of symmetry may be helpful also. The surface is symmetric with respect to the xy-plane, for example, if the equation remains unchanged when z is replaced by $-z$. It is symmetric with respect to the origin if the equation is unchanged when all variables are replaced by their negatives. It may also be helpful to consider excluded values of the variables and the behavior of the surface when one or more of the variables is large.

A new and very important technique appears in plotting surfaces. This is the method of studying a surface by considering its traces in various planes. It is particularly revealing to consider the traces of a surface in the coordinate planes and planes parallel to them.

We illustrate these ideas by discussing

(7) $$\frac{x^2}{a^2} + \frac{y^2}{b^2} + \frac{z^2}{c^2} = 1$$

We note that the intercepts are $(\pm a, 0, 0)$, $(0, \pm b, 0)$, and $(0, 0, \pm c)$. Since the locus is symmetric to all the coordinate planes and to

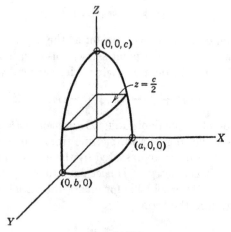

FIG. 14.5.1

the origin, we need consider it only in the first octant. The equation can be solved for any one of the variables, for example,

(8) $$z = \pm c \sqrt{1 - \frac{x^2}{a^2} - \frac{y^2}{b^2}}$$

From this it follows that $|x| < a$ and $|y| < b$. (Why?) Similarly, we could show that $|z| < c$. (How?) Hence the whole surface is within a rectangular box of sides $2a$, $2b$, $2c$ parallel to the axes. To find the trace in the xy-plane, we let $z = 0$ in (7). Then the equation of the trace is

(9) $$\frac{x^2}{a^2} + \frac{y^2}{b^2} = 1$$

which we recognize as an ellipse with center at the origin. Similarly, the locus meets each of the other coordinate planes in an ellipse. Moreover, the student can easily verify that any plane parallel to a coordinate plane meets the surface in an ellipse. Thus the plane $z = c/2$ meets the surface in the ellipse whose equation in the plane $z = c/2$ is

$$(10) \qquad \frac{x^2}{a^2} + \frac{y^2}{b^2} = 1 - \frac{(c/2)^2}{c^2} = 3/4$$

These traces give a fairly good idea of the locus (Fig. 14.5.1). The surface is called an **ellipsoid**.

Surfaces for which $F(x, y, z)$ is a quadratic polynomial are called **quadric surfaces**. They correspond to conics in the plane, and their sections by any planes are conics. (Why?) Some of the simplest types are included in the exercise.

A surface may also be given parametrically by three equations involving *two* parameters. Consider, for example:

$$(11) \qquad \begin{aligned} x &= a\sqrt{1 - s^2}\, \cos t \\ y &= b\sqrt{1 - s^2}\, \sin t \\ z &= \pm cs \end{aligned}$$

If we square the first two equations, add, and replace s by $\pm z/c$, we find just equation (7). Hence (11) represents an ellipsoid.

EXERCISE 14.5

1. State rules by which to test symmetry with respect to the xz-plane and the yz-plane.

2. Show that there are no points on the surface (4) outside the cube given by $|x| < R$, $|y| < R$, and $|z| < R$.

3. Sketch the surface $\dfrac{x^2}{4} + \dfrac{y^2}{9} + z^2 = 1$.

4. Sketch the following surfaces by finding their intercepts, symmetry, and traces in the coordinate planes:

(a) $4x^2 + 16y^2 + 25z^2 = 400$

(b) $x^2 - y^2 + z^2 = 1$

(c) $\dfrac{x^2}{4} + \dfrac{y^2}{9} - \dfrac{z^2}{16} = 1$

(d) $3x^2 - y = z$

(e) $\dfrac{x^2}{9} - \dfrac{y^2}{4} - \dfrac{z^2}{16} = 1$

(f) $x^2 + y^2 = z$

(g) $\dfrac{x^2}{4} + \dfrac{z^2}{9} = y$

(h) $z = xy$

(i) $xyz = 1$

(j) $x = 3s + t$, $y = s - t + 1$, $z = s + t$

(k) $x = t$, $y = t^2$, $z = t^3$

(†l) $x = \sin t$, $y = \sin t$, $z = t$

5. Sketch the locus given by the parametric equations $x = s^2 + t^2$, $y = 2t$, $z = 3s$. (*Suggestion.* Eliminate s and t.)

6. Show that any set of linear parametric equations in two parameters, that is, one of the form

(12) $x = as + bt + c, \; y = a's + b't + c', \; z = a''s + b''t + c''$

represents a plane.

7. If one of the variables is missing from the equations $F(x, y, z) = 0$, the surface will have the same trace in every plane perpendicular to the axis of the missing variable. (Why?) The surface is therefore a **cylinder,** which is defined as a surface traced by a straight line, called the **genetrix,** moving parallel to a fixed line and always passing through a fixed curve, called the **directrix.** In this case the surface is traced by a line moving parallel to the axis of the missing variable, and any trace in a plane perpendicular to this axis may be considered as a directrix. For example, the surface $x^2 + y^2 = 25$ meets every plane $z = $ constant in a circle of radius 5 and center on the z-axis. It is a circular cylinder with the z-axis as axis of symmetry. Identify and sketch the following cylinders:

(a) $y = x^2$ (e) $x^2 - y^2 = 1$
(b) $x^2 + y^2 = 4$ (f) $y = 2x$
(c) $z^2 + y^2 = 1$ (g) $y = x^3$
(d) $x^2 + z^2 = 9$ (h) $(x - 1)^2 + (y - 2)^2 = 25$

8. A **surface of revolution** is one formed by rotating a curve about an axis. The equation of a surface of revolution whose axis is one of the coordi-

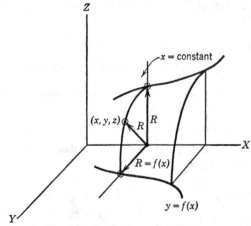

Fig. 14.5.2. Surface of revolution.

nate axes is easy to write or to recognize. Consider a curve $y = f(x)$ in the xy-plane and let it rotate about the x-axis. (See Fig. 14.5.2.) Its trace in a plane $x = $ constant is a circle (why?) of radius $R = f(x)$. At any point on this circle, and only at such points, we have $y^2 + z^2 = [f(x)]^2$. Since the argument holds for any x, this is the equation of the surface. Similarly, surfaces of revolution about the y-axis and z-axis may be of the form $x^2 + z^2 = [f(y)]^2$ and $x^2 + y^2 = [f(z)]^2$. Find the equations of the following surfaces formed by rotating the given curve about the indicated axis. Sketch.

> (a) $y = 2x$ about the x-axis
> (b) $y = 2x$ about the y-axis
> (c) $z = 3x^2$ about the x-axis
> (d) $z = 3x^2$ about the z-axis
> (e) $y = \sin x$ about the x-axis
> (f) $y = e^x$ about the x-axis
> (†g) $x^2 - 2y^2 = 3$ about the x-axis
> (†h) $(x - 2)^2 + (y - 3)^2 = 1$ about the y-axis

9. Describe and sketch the following surfaces of revolution:

(a) $x^2 + y^2 + 2z^2 = 2$ (e) $(x^2 + y^2)^2 = z$

(b) $x^2 + z^2 = 2y^4$ (f) $y^2 + z^2 = (2x^2 + x)^2$

(c) $x^2 + z^2 = y^2$ (g) $(\sqrt{x^2 + y^2} - 2)^2 + z^2 = 1$

(d) $x^2 + y^2 = z$ (h) $(x^2 + y^2)^2 + x^2 + y^2 = 1 - z$

Note in parts g and h that the equation is not in the exact form mentioned in prob. 8, where we assumed that the equation of the curve was in the form $y = f(x)$. If it is in the form $F(x, y) = 0$, the surface of revolution about the x-axis is given by replacing y by $\sqrt{y^2 + z^2}$. Similar remarks apply when the y-axis or z-axis is the axis of revolution.

†10. A curve in space may be represented by two equations in three variables, e.g., $F(x, y, z) = 0$ and $G(x, y, z) = 0$. The locus is the set of points that satisfy both equations. A curve may also be represented by three parametric equations in one parameter. The locus is then the set of points determined by all values of the parameter. If the parameter is eliminated, two equations in x, y, and z are found, which represent surfaces upon which the curve lies. Sketch the following curves in space.

(a) $x = 2 \cos t,\ y = 2 \sin t,\ z = 3$ (g) $x = t,\ y = 2t,\ z = \sin t$

(b) $x = 3 \cos t,\ y = 4 \sin t,\ z = 2$ (h) $x = t^2,\ y = 2t,\ z = 2$

(c) $x = \cos t,\ y = \sin t,\ z = t$ (i) $x = t \cos t,\ y = t \sin t,\ z = \sin t$

(d) $x^2 + y^2 = z,\ x^2 + y^2 = 3$ (j) $x^2 + y^2 = 1,\ z = 3$

(e) $x = t \cos t,\ y = t \sin t,\ z = t$ (k) $x^2 + y^2 + z^2 = 1,\ y = 2x$

(f) $x = t,\ y = t^2,\ z = t^3$ (l) $xy = 1,\ z = 2x$

11. Instead of making a three-dimensional sketch of a surface according to the conventions of this chapter, we could represent it by means of contour lines. This amounts to finding intersections of the surface with planes $z =$ constant and then drawing the curves in the xy-plane. Each such curve is a **contour line,** and it has the property that all points above it on the surface have the same z. The method is used in making topographic maps (where z is the altitude), in economics, and in other sciences. If the values of z are equally spaced, the contour map of a surface gives a very good idea of its shape. Where there are more than one z for a given (x, y), a map must be drawn separately for each. Sketch contour maps of the following surfaces, taking equally spaced values of z and labeling each contour curve.

(a) $3x + 4y + 2z = 12$

(b) $x^2 + y^2 + z^2 = 25$

(c) $x^2 + y^2 = z$

(d) $x^2 + 4y^2 + 9z^2 = 36$

(e) $x^2 - 4y^2 = z$

(f) $x = yz$

(g) $xy = z$

(h) $z = \sqrt{xy}$

12. Graph $f(x,y)$, where

(a) $f(x, y) = x + y$ when x and y are positive integers
$= 0$ otherwise

(b) $f(x, y) = 2$ for x and y integers
$= 0$ otherwise

14.6 Cylindrical and spherical coordinates

The rectangular coordinates of a point in space may be considered as the coordinates (x, y) of a point in the xy-plane and the

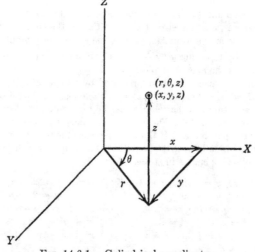

FIG. 14.6.1. Cylindrical coordinates.

directed distance z from this point. If the point in the xy-plane is indicated by its polar coordinates (r, θ), the point in space is determined by (r, θ, z), as indicated in Fig. 14.6.1. Such coordinates are called **cylindrical**. If $r \geq 0$ and $0 \leq \theta < 2\pi$, there is a one-one correspondence between points and coordinates, except that the origin has coordinates $(0, \theta, 0)$, where θ is arbitrary.

Another possible coordinate system is like the one that we use to locate positions on the earth. In Fig. 14.6.2 the coordinate ρ (read "rho") is the length of the radius vector from the origin. It is similar to the distance from the center of the earth and gives the same information as altitude. The angle θ is measured from the xz-plane to the plane defined by the z-axis and the radius vector. It is similar to longitude. The angle ϕ is measured from the

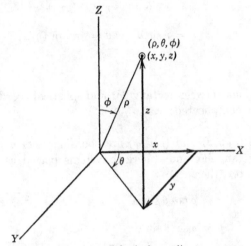

Fig. 14.6.2. Spherical coordinates.

positive z-axis to the radius vector. It corresponds to the colatitude (the complement of latitude), since latitude is measured from the equator (the xy-plane) instead of from the pole (the z-axis). The coordinates (ρ, θ, ϕ) are called **spherical**. If $\rho \geq 0$, $0 \leq \theta < 2\pi$, and $0 \leq \phi \leq \pi$, there is a one-one correspondence between points and coordinates, except that the origin has the coordinates $(0, \theta, \phi)$ where θ and ϕ are arbitrary.

Cylindrical coordinates are useful to represent loci that are symmetric with respect to the z-axis. Thus the equation of a circular cylinder with radius c and axis on the z-axis is $r = c$ in cylindrical coordinates. Spherical coordinates are useful to represent surfaces symmetric with respect to the origin. Thus the equation of a sphere with center at the origin and radius R is just $\rho = R$ in spherical coordinates. These examples suggest the origin of .he names.

The relations between rectangular and cylindrical coordinates follow at once from those between rectangular and polar coordinates:

*14.6.1 If (x, y, z) and (r, θ, z) are coordinates of a point in rectangular and cylindrical systems placed as in Fig. 14.6.1, then

$$x = r \cos \theta \qquad r = \sqrt{x^2 + y^2}$$

$$y = r \sin \theta \qquad \theta = \arctan (y/x)$$

$$z = z \qquad z = z$$

The relations between rectangular and spherical coordinates are a little more complicated:

*14.6.2 If (x, y, z) and (ρ, θ, ϕ) are coordinates of a point in rectangular and spherical systems placed as in Fig. 14.6.2, then

$$x = \rho \cos \theta \sin \phi \qquad \rho = \sqrt{x^2 + y^2 + z^2}$$

$$y = \rho \sin \theta \sin \phi \qquad \theta = \arctan (y/x)$$

$$z = \rho \cos \phi \qquad \phi = \arccos (z/\rho)$$

EXERCISE 14.6

1. Sketch each of the following points given in cylindrical coordinates and find its rectangular and spherical coordinates.

(a) $(2, 25°, 3)$ (c) $(1, \pi, -3)$ (e) $(0, 3, 4)$
(b) $(2, 110°, -1)$ (d) $(5, 1.8\pi, 3)$ (f) $(-5, 120°, 3)$

Answers. (a) $(1.813, 0.8452, 3)$; $(3.606, 25°, 33° 43')$.

2. Sketch each of the following points given in spherical coordinates and find its rectangular and cylindrical coordinates.

(a) $(0, 1, 3)$ (c) $(1, 80°, 100°)$ (e) $(1, 1, 0)$
(b) $(1, 0, 30°)$ (d) $(5, 3\pi/2, 60°)$ (f) $(-2, 200°, 30°)$

Answers. (a) $(0, 0, 0)$; $(0, \theta, 0)$. (c) $(0.1710, 0.9698, -0.1736)$; $(0.9847, 80°, -0.1736)$.

3. Sketch each of the following points given in rectangular coordinates and find its cylindrical and spherical coordinates.

(a) $(1, 3, -2)$ (c) $(-3, -1, 0)$ (e) $(0, 3, 0)$

(b) $(0, 8, 4)$ (d) $(-2, 1, -4)$ (f) $(2, -6, 5)$

Answers. (a) $(\sqrt{10}, 70° 34', -2)$; $(\sqrt{14}, 71° 34', 122° 19')$.

4. Find the formulas for cylindrical in terms of spherical coordinates and for spherical in terms of cylindrical. (*Answers.* $r = \rho \sin \phi, \theta = \theta, a = \rho \cos \phi$; $\rho = \sqrt{r^2 + z^2}, \theta = \theta, \phi = \arctan (r/z)$.

5. Justify *14.6.1 and *14.6.2.

6. Sketch the following surfaces whose equations are given in cylindrical coordinates. Transform to rectangular and to spherical coordinates.

(a) $r = 3$ (e) $r(1 - \cos \theta) = 1$ (i) $\theta = z$

(b) $r = z - 2$ (f) $r = z$ (j) $r = \cos 2\theta$

(c) $\theta = 1$ (g) $z = -2$ (k) $r = \theta$

(d) $r \cos \theta = 1$ (h) $r = \cos z$ (l) $r = z\theta$

7. Sketch the following surfaces whose equations are given in spherical coordinates. Transform to rectangular and to cylindrical coordinates.

(a) $\rho = 3$ (d) $\rho \cos \phi = 3$ (g) $\rho = \theta$

(b) $\theta = 1$ (e) $\theta = -45°$ (h) $\rho = -2$

(c) $\phi = 30°$ (f) $\rho = 2 \cos \phi$ (i) $\phi = \theta$

8. Transform to cylindrical and spherical coordinates the equations of the surfaces and curves of Exs. 14.5.7, 14.5.9, 14.5.11.

9. What are the spherical, cylindrical, and rectangular coordinates of a ship at latitude $43° 20'$ and west longitude $25° 15'$? (Take the z-axis through the poles and the positive x-axis through the meridian of zero longitude.)

10. Find in the three different coordinate systems the equations of the following:

(a) A sphere with center at the origin and radius 2.

(b) A plane through the z-axis and meeting the xy-plane in the line $y = x$.

(c) A right circular cylinder with radius 1 and axis the z-axis.

(d) A right circular cylinder with radius 2 and axis the y-axis.

(e) A right circular cone with axis the z-axis and with generating line making an angle of $45°$ with this axis.

(f) A sphere having radius 2 and center with rectangular coordinates $(1, 1, 1)$.

(g) A surface formed by rotating $b^2x^2 + a^2y^2 = a^2b^2$ about the x-axis.

11. Describe the curve given in cylindrical coordinates by $r = 3$, $z = \theta$. Sketch.

12. Describe and sketch the curve given in spherical coordinates by

$$\rho = 3, \qquad \phi = \theta$$

TABLE I. POWERS AND ROOTS 599

SQUARES AND CUBES   SQUARE ROOTS AND CUBE ROOTS

No.	Square	Cube	Square Root	Cube Root	No.	Square	Cube	Square Root	Cube Root
1	1	1	1.000	1.000	51	2,601	132,651	7.141	3.708
2	4	8	1.414	1.260	52	2,704	140,608	7.211	3.733
3	9	27	1.732	1.442	53	2,809	148,877	7.280	3.756
4	16	64	2.000	1.587	54	2,916	157,464	7.348	3.780
5	25	125	2.236	1.710	55	3,025	166,375	7.416	3.803
6	36	216	2.449	1.817	56	3,136	175,616	7.483	3.826
7	49	343	2.646	1.913	57	3,249	185,193	7.550	3.849
8	64	512	2.828	2.000	58	3,364	195,112	7.616	3.871
9	81	729	3.000	2.080	59	3,481	205,379	7.681	3.893
10	100	1,000	3.162	2.154	60	3,600	216,000	7.746	3.915
11	121	1,331	3.317	2.224	61	3,721	226,981	7.810	3.936
12	144	1,728	3.464	2.289	62	3,844	238,328	7.874	3.958
13	169	2,197	3.606	2.351	63	3,969	250,047	7.937	3.979
14	196	2,744	3.742	2.410	64	4,096	262,144	8.000	4.000
15	225	3,375	3.873	2.466	65	4,225	274,625	8.062	4.021
16	256	4,096	4.000	2.520	66	4,356	287,496	8.124	4.041
17	289	4,913	4.123	2.571	67	4,489	300,763	8.185	4.062
18	324	5,832	4.243	2.621	68	4,624	314,432	8.246	4.082
19	361	6,859	4.359	2.668	69	4,761	328,509	8.307	4.102
20	400	8,000	4.472	2.714	70	4,900	343,000	8.367	4.121
21	441	9,261	4.583	2.759	71	5,041	357,911	8.426	4.141
22	484	10,648	4.690	2.802	72	5,184	373,248	8.485	4.160
23	529	12,167	4.796	2.844	73	5,329	389,017	8.544	4.179
24	576	13,824	4.899	2.884	74	5,476	405,224	8.602	4.198
25	625	15,625	5.000	2.924	75	5,625	421,875	8.660	4.217
26	676	17,576	5.099	2.962	76	5,776	438,976	8.718	4.236
27	729	19,683	5.196	3.000	77	5,929	456,533	8.775	4.254
28	784	21,952	5.292	3.037	78	6,084	474,552	8.832	4.273
29	841	24,389	5.385	3.072	79	6,241	493,039	8.888	4.291
30	900	27,000	5.477	3.107	80	6,400	512,000	8.944	4.309
31	961	29,791	5.568	3.141	81	6,561	531,441	9.000	4.327
32	1,024	32,768	5.657	3.175	82	6,724	551,368	9.055	4.344
33	1,089	35,937	5.745	3.208	83	6,889	571,787	9.110	4.362
34	1,156	39,304	5.831	3.240	84	7,056	592,704	9.165	4.380
35	1,225	42,875	5.916	3.271	85	7,225	614,125	9.220	4.397
36	1,296	46,656	6.000	3.302	86	7,396	636,056	9.274	4.414
37	1,369	50,653	6.083	3.332	87	7,569	658,503	9.327	4.431
38	1,444	54,872	6.164	3.362	88	7,744	681,472	9.381	4.448
39	1,521	59,319	6.245	3.391	89	7,921	704,969	9.434	4.465
40	1,600	64,000	6.325	3.420	90	8,100	729,000	9.487	4.481
41	1,681	68,921	6.403	3.448	91	8,281	753,571	9.539	4.498
42	1,764	74,088	6.481	3.476	92	8,464	778,688	9.592	4.514
43	1,849	79,507	6.557	3.503	93	8,649	804,357	9.644	4.531
44	1,936	85,184	6.633	3.530	94	8,836	830,584	9.695	4.547
45	2,025	91,125	6.708	3.557	95	9,025	857,375	9.747	4.563
46	2,116	97,336	6.782	3.583	96	9,216	884,736	9.798	4.579
47	2,209	103,823	6.856	3.609	97	9,409	912,673	9.849	4.595
48	2,304	110,592	6.928	3.634	98	9,604	941,192	9.899	4.610
49	2,401	117,649	7.000	3.659	99	9,801	970,299	9.950	4.626
50	2,500	125,000	7.071	3.684	100	10,000	1,000,000	10.000	4.642

From "General Mathematics," by Currier, Watson, and Frame. Reprinted by permission of The Macmillan Company, publishers.

	0	1	2	3	4	5	6	7	8	9
10	0000	0043	0086	0128	0170	0212	0253	0294	0334	0374
11	0414	0453	0492	0531	0569	0607	0645	0682	0719	0755
12	0792	0828	0864	0899	0934	0969	1004	1038	1072	1106
13	1139	1173	1206	1239	1271	1303	1335	1367	1399	1430
14	1461	1492	1523	1553	1584	1614	1644	1673	1703	1732
15	1761	1790	1818	1847	1875	1903	1931	1959	1987	2014
16	2041	2068	2095	2122	2148	2175	2201	2227	2253	2279
17	2304	2330	2355	2380	2405	2430	2455	2480	2504	2529
18	2553	2577	2601	2625	2648	2672	2695	2718	2742	2765
19	2788	2810	2833	2856	2878	2900	2923	2945	2967	2989
20	3010	3032	3054	3075	3096	3118	3139	3160	3181	3201
21	3222	3243	3263	3284	3304	3324	3345	3365	3385	3404
22	3424	3444	3464	3483	3502	3522	3541	3560	3579	3598
23	3617	3636	3655	3674	3692	3711	3729	3747	3766	3784
24	3802	3820	3838	3856	3874	3892	3909	3927	3945	3962
25	3979	3997	4014	4031	4048	4065	4082	4099	4116	4133
26	4150	4166	4183	4200	4216	4232	4249	4265	4281	4298
27	4314	4330	4346	4362	4378	4393	4409	4425	4440	4456
28	4472	4487	4502	4518	4533	4548	4564	4579	4594	4609
29	4624	4639	4654	4669	4683	4698	4713	4728	4742	4757
30	4771	4786	4800	4814	4829	4843	4857	4871	4886	4900
31	4914	4928	4942	4955	4969	4983	4997	5011	5024	5038
32	5051	5065	5079	5092	5105	5119	5132	5145	5159	5172
33	5185	5198	5211	5224	5237	5250	5263	5276	5289	5302
34	5315	5328	5340	5353	5366	5378	5391	5403	5416	5428
35	5441	5453	5465	5478	5490	5502	5514	5527	5539	5551
36	5563	5575	5587	5599	5611	5623	5635	5647	5658	5670
37	5682	5694	5705	5717	5729	5740	5752	5763	5775	5786
38	5798	5809	5821	5832	5843	5855	5866	5877	5888	5899
39	5911	5922	5933	5944	5955	5966	5977	5988	5999	6010
40	6021	6031	6042	6053	6064	6075	6085	6096	6107	6117
41	6128	6138	6149	6160	6170	6180	6191	6201	6212	6222
42	6232	6243	6253	6263	6274	6284	6294	6304	6314	6325
43	6335	6345	6355	6365	6375	6385	6395	6405	6415	6425
44	6435	6444	6454	6464	6474	6484	6493	6503	6513	6522
45	6532	6542	6551	6561	6571	6580	6590	6599	6609	6618
46	6628	6637	6646	6656	6665	6675	6684	6693	6702	6712
47	6721	6730	6739	6749	6758	6767	6776	6785	6794	6803
48	6812	6821	6830	6839	6848	6857	6866	6875	6884	6893
49	6902	6911	6920	6928	6937	6946	6955	6964	6972	6981
50	6990	6998	7007	7016	7024	7033	7042	7050	7059	7067
51	7076	7084	7093	7101	7110	7118	7126	7135	7143	7152
52	7160	7168	7177	7185	7193	7202	7210	7218	7226	7235
53	7243	7251	7259	7267	7275	7284	7292	7300	7308	7316
54	7324	7332	7340	7348	7356	7364	7372	7380	7388	7396

From "College Algebra and Trigonometry," by F. H. Miller. Publisher: John Wiley and Sons (1945).

TABLE II. COMMON LOGARITHMS 601

	0	1	2	3	4	5	6	7	8	9
55	7404	7412	7419	7427	7435	7443	7451	7459	7466	7474
56	7482	7490	7497	7505	7513	7520	7528	7536	7543	7551
57	7559	7566	7574	7582	7589	7597	7604	7612	7619	7627
58	7634	7642	7649	7657	7664	7672	7679	7686	7694	7701
59	7709	7716	7723	7731	7738	7745	7752	7760	7767	7774
60	7782	7789	7796	7803	7810	7818	7825	7832	7839	7846
61	7853	7860	7868	7875	7882	7889	7896	7903	7910	7917
62	7924	7931	7938	7945	7952	7959	7966	7973	7980	7987
63	7993	8000	8007	8014	8021	8028	8035	8041	8048	8055
64	8062	8069	8075	8082	8089	8096	8102	8109	8116	8122
65	8129	8136	8142	8149	8156	8162	8169	8176	8182	8189
66	8195	8202	8209	8215	8222	8228	8235	8241	8248	8254
67	8261	8267	8274	8280	8287	8293	8299	8306	8312	8319
68	8325	8331	8338	8344	8351	8357	8363	8370	8376	8382
69	8388	8395	8401	8407	8414	8420	8426	8432	8439	8445
70	8451	8457	8463	8470	8476	8482	8488	8494	8500	8506
71	8513	8519	8525	8531	8537	8543	8549	8555	8561	8567
72	8573	8579	8585	8591	8597	8603	8609	8615	8621	8627
73	8633	8639	8645	8651	8657	8663	8669	8675	8681	8686
74	8692	8698	8704	8710	8716	8722	8727	8733	8739	8745
75	8751	8756	8762	8768	8774	8779	8785	8791	8797	8802
76	8808	8814	8820	8825	8831	8837	8842	8848	8854	8859
77	8865	8871	8876	8882	8887	8893	8899	8904	8910	8915
78	8921	8927	8932	8938	8943	8949	8954	8960	8965	8971
79	8976	8982	8987	8993	8998	9004	9009	9015	9020	9025
80	9031	9036	9042	9047	9053	9058	9063	9069	9074	9079
81	9085	9090	9096	9101	9106	9112	9117	9122	9128	9133
82	9138	9143	9149	9154	9159	9165	9170	9175	9180	9186
83	9191	9196	9201	9206	9212	9217	9222	9227	9232	9238
84	9243	9248	9253	9258	9263	9269	9274	9279	9284	9289
85	9294	9299	9304	9309	9315	9320	9325	9330	9335	9340
86	9345	9350	9355	9360	9365	9370	9375	9380	9385	9390
87	9395	9400	9405	9410	9415	9420	9425	9430	9435	9440
88	9445	9450	9455	9460	9465	9469	9474	9479	9484	9489
89	9494	9499	9504	9509	9513	9518	9523	9528	9533	9538
90	9542	9547	9552	9557	9562	9566	9571	9576	9581	9586
91	9590	9595	9600	9605	9609	9614	9619	9624	9628	9633
92	9638	9643	9647	9652	9657	9661	9666	9671	9675	9680
93	9685	9689	9694	9699	9703	9708	9713	9717	9722	9727
94	9731	9736	9741	9745	9750	9754	9759	9763	9768	9773
95	9777	9782	9786	9791	9795	9800	9805	9809	9814	9818
96	9823	9827	9832	9836	9841	9845	9850	9854	9859	9863
97	9868	9872	9877	9881	9886	9890	9894	9899	9903	9908
98	9912	9917	9921	9926	9930	9934	9939	9943	9948	9952
99	9956	9961	9965	9969	9974	9978	9983	9987	9991	9996

Base e = 2.71828...

NOTE. $\log_e 10N = \log_e N + \log_e 10$

$\log_e \dfrac{N}{10} = \log_e N - \log_e 10$

$\log_e 10 = 2.30259$

Examples: $\log_e 35 = \log_e 3.5 + \log_e 10$
$= 1.25276 + 2.30259 = 3.55535$

$\log_e .35 = \log_e 3.5 - \log_e 10$
$= 1.25276 - 2.30259 = 8.95017 - 10$

N	0	1	2	3	4	5	6	7	8	9
1.0	0.0 0000	0995	1980	2956	3922	4879	5827	6766	7696	8618
1.1	9531	*0436	*1333	*2222	*3103	*3976	*4842	*5700	*6551	*7395
1.2	0.1 8232	9062	9885	*0701	*1511	*2314	*3111	*3902	*4686	*5464
1.3	0.2 6236	7003	7763	8518	9267	*0010	*0748	*1481	*2208	*2930
1.4	0.3 3647	4359	5066	5767	6464	7156	7844	8526	9204	9878
1.5	0.4 0547	1211	1871	2527	3178	3825	4469	5108	5742	6373
1.6	7000	7623	8243	8858	9470	*0078	*0682	*1282	*1879	*2473
1.7	0.5 3063	3649	4232	4812	5389	5962	6531	7098	7661	8222
1.8	8779	9333	9884	*0432	*0977	*1519	*2058	*2594	*3127	*3658
1.9	0.6 4185	4710	5233	5752	6269	6783	7294	7803	8310	8813
2.0	9315	9813	*0310	*0804	*1295	*1784	*2271	*2755	*3237	*3716
2.1	0.7 4194	4669	5142	5612	6081	6547	7011	7473	7932	8390
2.2	8846	9299	9751	*0200	*0648	*1093	*1536	*1978	*2418	*2855
2.3	0.8 3291	3725	4157	4587	5015	5442	5866	6289	6710	7129
2.4	7547	7963	8377	8789	9200	9609	*0016	*0422	*0826	*1228
2.5	0.9 1629	2028	2426	2822	3216	3609	4001	4391	4779	5166
2.6	5551	5935	6317	6698	7078	7456	7833	8208	8582	8954
2.7	9325	9695	*0063	*0430	*0796	*1160	*1523	*1885	*2245	*2604
2.8	1.0 2962	3318	3674	4028	4380	4732	5082	5431	5779	6126
2.9	6471	6815	7158	7500	7841	8181	8519	8856	9192	9527
3.0	9861	*0194	*0526	*0856	*1186	*1514	*1841	*2168	*2493	*2817
3.1	1.1 3140	3462	3783	4103	4422	4740	5057	5373	5688	6002
3.2	6315	6627	6938	7248	7557	7865	8173	8479	8784	9089
3.3	9392	9695	9996	*0297	*0597	*0896	*1194	*1491	*1788	*2083
3.4	1.2 2378	2671	2964	3256	3547	3837	4127	4415	4703	4990
3.5	5276	5562	5846	6130	6413	6695	6976	7257	7536	7815
3.6	8093	8371	8647	8923	9198	9473	9746	*0019	*0291	*0563
3.7	1.3 0833	1103	1372	1641	1909	2176	2442	2708	2972	3237
3.8	3500	3763	4025	4286	4547	4807	5067	5325	5584	5841
3.9	6098	6354	6609	6864	7118	7372	7624	7877	8128	8379
4.0	8629	8879	9128	9377	9624	9872	*0118	*0364	*0610	*0854
4.1	1.4 1099	1342	1585	1828	2070	2311	2552	2792	3031	3270
4.2	3508	3746	3984	4220	4456	4692	4927	5161	5395	5629
4.3	5862	6094	6326	6557	6787	7018	7247	7476	7705	7933
4.4	8160	8387	8614	8840	9065	9290	9515	9739	9962	*0185
4.5	1.5 0408	0630	0851	1072	1293	1513	1732	1951	2170	2388
4.6	2606	2823	3039	3256	3471	3687	3902	4116	4330	4543
4.7	4756	4969	5181	5393	5604	5814	6025	6235	6444	6653
4.8	6862	7070	7277	7485	7691	7898	8104	8309	8515	8719
4.9	8924	9127	9331	9534	9737	9939	*0141	*0342	*0543	*0744
5.0	1.6 0944	1144	1343	1542	1741	1939	2137	2334	2531	2728
N	0	1	2	3	4	5	6	7	8	9

TABLE III. NATURAL LOGARITHMS OF NUMBERS 603

N	0	1	2	3	4	5	6	7	8	9
5.0	1.6 0944	1144	1343	1542	1741	1939	2137	2334	2531	2728
5.1	2924	3120	3315	3511	3705	3900	4094	4287	4481	4673
5.2	4866	5058	5250	5441	5632	5823	6013	6203	6393	6582
5.3	6771	6959	7147	7335	7523	7710	7896	8083	8269	8455
5.4	8640	8825	9010	9194	9378	9562	9745	9928	*0111	*0293
5.5	1.7 0475	0656	0838	1019	1199	1380	1560	1740	1919	2098
5.6	2277	2455	2633	2811	2988	3166	3342	3519	3695	3871
5.7	4047	4222	4397	4572	4746	4920	5094	5267	5440	5613
5.8	5786	5958	6130	6302	6473	6644	6815	6985	7156	7326
5.9	7495	7665	7834	8002	8171	8339	8507	8675	8842	9009
6.0	9176	9342	9509	9675	9840	*0006	*0171	*0336	*0500	*0665
6.1	1.8 0829	0993	1156	1319	1482	1645	1808	1970	2132	2294
6.2	2455	2616	2777	2938	3098	3258	3418	3578	3737	3896
6.3	4055	4214	4372	4530	4688	4845	5003	5160	5317	5473
6.4	5630	5786	5942	6097	6253	6408	6563	6718	6872	7026
6.5	7180	7334	7487	7641	7794	7947	8099	8251	8403	8555
6.6	8707	8858	9010	9160	9311	9462	9612	9762	9912	*0061
6.7	1.9 0211	0360	0509	0658	0806	0954	1102	1250	1398	1545
6.8	1692	1839	1986	2132	2279	2425	2571	2716	2862	3007
6.9	3152	3297	3442	3586	3730	3874	4018	4162	4305	4448
7.0	4591	4734	4876	5019	5161	5303	5445	5586	5727	5869
7.1	6009	6150	6291	6431	6571	6711	6851	6991	7130	7269
7.2	7408	7547	7685	7824	7962	8100	8238	8376	8513	8650
7.3	8787	8924	9061	9198	9334	9470	9606	9742	9877	*0013
7.4	2.0 0148	0283	0418	0553	0687	0821	0956	1089	1223	1357
7.5	1490	1624	1757	1890	2022	2155	2287	2419	2551	2683
7.6	2815	2946	3078	3209	3340	3471	3601	3732	3862	3992
7.7	4122	4252	4381	4511	4640	4769	4898	5027	5156	5284
7.8	5412	5540	5668	5796	5924	6051	6179	6306	6433	6560
7.9	6686	6813	6939	7065	7191	7317	7443	7568	7694	7819
8.0	7944	8069	8194	8318	8443	8567	8691	8815	8939	9063
8.1	9186	9310	9433	9556	9679	9802	9924	*0047	*0169	*0291
8.2	2.1 0413	0535	0657	0779	0900	1021	1142	1263	1384	1505
8.3	1626	1746	1866	1986	2106	2226	2346	2465	2585	2704
8.4	2823	2942	3061	3180	3298	3417	3535	3653	3771	3889
8.5	4007	4124	4242	4359	4476	4593	4710	4827	4943	5060
8.6	5176	5292	5409	5524	5640	5756	5871	5987	6102	6217
8.7	6332	6447	6562	6677	6791	6905	7020	7134	7248	7361
8.8	7475	7589	7702	7816	7929	8042	8155	8267	8380	8493
8.9	8605	8717	8830	8942	9054	9165	9277	9389	9500	9611
9.0	9722	9834	9944	*0055	*0166	*0276	*0387	*0497	*0607	*0717
9.1	2.2 0827	0937	1047	1157	1266	1375	1485	1594	1703	1812
9.2	1920	2029	2138	2246	2354	2462	2570	2678	2786	2894
9.3	3001	3109	3216	3324	3431	3538	3645	3751	3858	3965
9.4	4071	4177	4284	4390	4496	4601	4707	4813	4918	5024
9.5	5129	5234	5339	5444	5549	5654	5759	5863	5968	6072
9.6	6176	6280	6384	6488	6592	6696	6799	6903	7006	7109
9.7	7213	7316	7419	7521	7624	7727	7829	7932	8034	8136
9.8	8238	8340	8442	8544	8646	8747	8849	8950	9051	9152
9.9	9253	9354	9455	9556	9657	9757	9858	9958	*0058	*0158
10.0	2.3 0259	0358	0458	0558	0658	0757	0857	0956	1055	1154
N	0	1	2	3	4	5	6	7	8	9

[Characteristics of Logarithms omitted—determine by the usual rule from the value]

Radians	Degrees	Sine Value	Sine Log₁₀	Tangent Value	Tangent Log₁₀	Cotangent Value	Cotangent Log₁₀	Cosine Value	Cosine Log₁₀		
.0000	0° 00′	.0000	———	.0000	———	———	———	1.0000	.0000	90° 00′	1.5708
.0029	10	.0029	.4637	.0029	.4637	343.77	.5363	1.0000	.0000	50	1.5679
.0058	20	.0058	.7648	.0058	.7648	171.89	.2352	1.0000	.0000	40	1.5650
.0087	30	.0087	.9408	.0087	.9409	114.59	.0591	1.0000	.0000	30	1.5621
.0116	40	.0116	.0658	.0116	.0658	85.940	.9342	.9999	.0000	20	1.5592
.0145	50	.0145	.1627	.0145	.1627	68.750	.8373	.9999	.0000	10	1.5563
.0175	1° 00′	.0175	.2419	.0175	.2419	57.290	.7581	.9998	.9999	89° 00′	1.5533
.0204	10	.0204	.3088	.0204	.3089	49.104	.6911	.9998	.9999	50	1.5504
.0233	20	.0233	.3668	.0233	.3669	42.964	.6331	.9997	.9999	40	1.5475
.0262	30	.0262	.4179	.0262	.4181	38.188	.5819	.9997	.9999	30	1.5446
.0291	40	.0291	.4637	.0291	.4638	34.368	.5362	.9996	.9998	20	1.5417
.0320	50	.0320	.5050	.0320	.5053	31.242	.4947	.9995	.9998	10	1.5388
.0349	2° 00′	.0349	.5428	.0349	.5431	28.636	.4569	.9994	.9997	88° 00′	1.5359
.0378	10	.0378	.5776	.0378	.5779	26.432	.4221	.9993	.9997	50	1.5330
.0407	20	.0407	.6097	.0407	.6101	24.542	.3899	.9992	.9996	40	1.5301
.0436	30	.0436	.6397	.0437	.6401	22.904	.3599	.9990	.9996	30	1.5272
.0465	40	.0465	.6677	.0466	.6682	21.470	.3318	.9989	.9995	20	1.5243
.0495	50	.0494	.6940	.0495	.6945	20.206	.3055	.9988	.9995	10	1.5213
.0524	3° 00′	.0523	.7188	.0524	.7194	19.081	.2806	.9986	.9994	87° 00′	1.5184
.0553	10	.0552	.7423	.0553	.7429	18.075	.2571	.9985	.9993	50	1.5155
.0582	20	.0581	.7645	.0582	.7652	17.169	.2348	.9983	.9993	40	1.5126
.0611	30	.0610	.7857	.0612	.7865	16.350	.2135	.9981	.9992	30	1.5097
.0640	40	.0640	.8059	.0641	.8067	15.605	.1933	.9980	.9991	20	1.5068
.0669	50	.0669	.8251	.0670	.8261	14.924	.1739	.9978	.9990	10	1.5039
.0698	4° 00′	.0698	.8436	.0699	.8446	14.301	.1554	.9976	.9989	86° 00′	1.5010
.0727	10	.0727	.8613	.0729	.8624	13.727	.1376	.9974	.9989	50	1.4981
.0756	20	.0756	.8783	.0758	.8795	13.197	.1205	.9971	.9988	40	1.4952
.0785	30	.0785	.8946	.0787	.8960	12.706	.1040	.9969	.9987	30	1.4923
.0814	40	.0814	.9104	.0816	.9118	12.251	.0882	.9967	.9986	20	1.4893
.0844	50	.0843	.9256	.0846	.9272	11.826	.0728	.9964	.9985	10	1.4864
.0873	5° 00′	.0872	.9403	.0875	.9420	11.430	.0580	.9962	.9983	85° 00′	1.4835
.0902	10	.0901	.9545	.0904	.9563	11.059	.0437	.9959	.9982	50	1.4806
.0931	20	.0929	.9682	.0934	.9701	10.712	.0299	.9957	.9981	40	1.4777
.0960	30	.0958	.9816	.0963	.9836	10.385	.0164	.9954	.9980	30	1.4748
.0989	40	.0987	.9945	.0992	.9966	10.078	.0034	.9951	.9979	20	1.4719
.1018	50	.1016	.0070	.1022	.0093	9.7882	.9907	.9948	.9977	10	1.4690
.1047	6° 00′	.1045	.0192	.1051	.0216	9.5144	.9784	.9945	.9976	84° 00′	1.4661
.1076	10	.1074	.0311	.1080	.0336	9.2553	.9664	.9942	.9975	50	1.4632
.1105	20	.1103	.0426	.1110	.0453	9.0098	.9547	.9939	.9973	40	1.4603
.1134	30	.1132	.0539	.1139	.0567	8.7769	.9433	.9936	.9972	30	1.4573
.1164	40	.1161	.0648	.1169	.0678	8.5555	.9322	.9932	.9971	20	1.4544
.1193	50	.1190	.0755	.1198	.0786	8.3450	.9214	.9929	.9969	10	1.4515
.1222	7° 00′	.1219	.0859	.1228	.0891	8.1443	.9109	.9925	.9968	83° 00′	1.4486
.1251	10	.1248	.0961	.1257	.0995	7.9530	.9005	.9922	.9966	50	1.4457
.1280	20	.1276	.1060	.1287	.1096	7.7704	.8904	.9918	.9964	40	1.4428
.1309	30	.1305	.1157	.1317	.1194	7.5958	.8806	.9914	.9963	30	1.4399
.1338	40	.1334	.1252	.1346	.1291	7.4287	.8709	.9911	.9961	20	1.4370
.1367	50	.1363	.1345	.1376	.1385	7.2687	.8615	.9907	.9959	10	1.4341
.1396	8° 00′	.1392	.1436	.1405	.1478	7.1154	.8522	.9903	.9958	82° 00′	1.4312
.1425	10	.1421	.1525	.1435	.1569	6.9682	.8431	.9899	.9956	50	1.4283
.1454	20	.1449	.1612	.1465	.1658	6.8269	.8342	.9894	.9954	40	1.4254
.1484	30	.1478	.1697	.1495	.1745	6.6912	.8255	.9890	.9952	30	1.4224
.1513	40	.1507	.1781	.1524	.1831	6.5606	.8169	.9886	.9950	20	1.4195
.1542	50	.1536	.1863	.1554	.1915	6.4348	.8085	.9881	.9948	10	1.4166
.1571	9° 00′	.1564	.1943	.1584	.1997	6.3138	.8003	.9877	.9946	81° 00′	1.4137
		Value Cosine	Log₁₀	Value Cotangent	Log₁₀	Value Tangent	Log₁₀	Value Sine	Log₁₀	Degrees	Radians

From "Logarithmic and Trigonometric Tables," by Earle Raymond Hedrick. Reprinted by permission of The Macmillan Company, publishers.

TABLE IV. FOUR-PLACE TRIGONOMETRIC FUNCTIONS 605

[Characteristics of Logarithms omitted—determine by the usual rule from the value]

RADIANS	DEGREES	SINE Value	SINE Log₁₀	TANGENT Value	TANGENT Log₁₀	COTANGENT Value	COTANGENT Log₁₀	COSINE Value	COSINE Log₁₀		
.1571	9° 00′	.1564	.1943	.1584	.1997	6.3138	.8003	.9877	.9946	81° 00′	1.4137
.1600	10	.1593	.2022	.1614	.2078	6.1970	.7922	.9872	.9944	50	1.4108
.1629	20	.1622	.2100	.1644	.2158	6.0844	.7842	.9868	.9942	40	1.4079
.1658	30	.1650	.2176	.1673	.2236	5.9758	.7764	.9863	.9940	30	1.4050
.1687	40	.1679	.2251	.1703	.2313	5.8708	.7687	.9858	.9938	20	1.4021
.1716	50	.1708	.2324	.1733	.2389	5.7694	.7611	.9853	.9936	10	1.3992
.1745	10° 00′	.1736	.2397	.1763	.2463	5.6713	.7537	.9848	.9934	80° 00′	1.3963
.1774	10	.1765	.2468	.1793	.2536	5.5764	.7464	.9843	.9931	50	1.3934
.1804	20	.1794	.2538	.1823	.2609	5.4845	.7391	.9838	.9929	40	1.3904
.1833	30	.1822	.2606	.1853	.2680	5.3955	.7320	.9833	.9927	30	1.3875
.1862	40	.1851	.2674	.1883	.2750	5.3093	.7250	.9827	.9924	20	1.3846
.1891	50	.1880	.2740	.1914	.2819	5.2257	.7181	.9822	.9922	10	1.3817
.1920	11° 00′	.1908	.2806	.1944	.2887	5.1446	.7113	.9816	.9919	79° 00′	1.3788
.1949	10	.1937	.2870	.1974	.2953	5.0658	.7047	.9811	.9917	50	1.3759
.1978	20	.1965	.2934	.2004	.3020	4.9894	.6980	.9805	.9914	40	1.3730
.2007	30	.1994	.2997	.2035	.3085	4.9152	.6915	.9799	.9912	30	1.3701
.2036	40	.2022	.3058	.2065	.3149	4.8430	.6851	.9793	.9909	20	1.3672
.2065	50	.2051	.3119	.2095	.3212	4.7729	.6788	.9787	.9907	10	1.3643
.2094	12° 00′	.2079	.3179	.2126	.3275	4.7046	.6725	.9781	.9904	78° 00′	1.3614
.2123	10	.2108	.3238	.2156	.3336	4.6382	.6664	.9775	.9901	50	1.3584
.2153	20	.2136	.3296	.2186	.3397	4.5736	.6603	.9769	.9899	40	1.3555
.2182	30	.2164	.3353	.2217	.3458	4.5107	.6542	.9763	.9896	30	1.3526
.2211	40	.2193	.3410	.2247	.3517	4.4494	.6483	.9757	.9893	20	1.3497
.2240	50	.2221	.3466	.2278	.3576	4.3897	.6424	.9750	.9890	10	1.3468
.2269	13° 00′	.2250	.3521	.2309	.3634	4.3315	.6366	.9744	.9887	77° 00′	1.3439
.2298	10	.2278	.3575	.2339	.3691	4.2747	.6309	.9737	.9884	50	1.3410
.2327	20	.2306	.3629	.2370	.3748	4.2193	.6252	.9730	.9881	40	1.3381
.2356	30	.2334	.3682	.2401	.3804	4.1653	.6196	.9724	.9878	30	1.3352
.2385	40	.2363	.3734	.2432	.3859	4.1126	.6141	.9717	.9875	20	1.3323
.2414	50	.2391	.3786	.2462	.3914	4.0611	.6086	.9710	.9872	10	1.3294
.2443	14° 00′	.2419	.3837	.2493	.3968	4.0108	.6032	.9703	.9869	76° 00′	1.3265
.2473	10	.2447	.3887	.2524	.4021	3.9617	.5979	.9696	.9866	50	1.3235
.2502	20	.2476	.3937	.2555	.4074	3.9136	.5926	.9689	.9863	40	1.3206
.2531	30	.2504	.3986	.2586	.4127	3.8667	.5873	.9681	.9859	30	1.3177
.2560	40	.2532	.4035	.2617	.4178	3.8208	.5822	.9674	.9856	20	1.3148
.2589	50	.2560	.4083	.2648	.4230	3.7760	.5770	.9667	.9853	10	1.3119
.2618	15° 00′	.2588	.4130	.2679	.4281	3.7321	.5719	.9659	.9849	75° 00′	1.3090
.2647	10	.2616	.4177	.2711	.4331	3.6891	.5669	.9652	.9846	50	1.3061
.2676	20	.2644	.4223	.2742	.4381	3.6470	.5619	.9644	.9843	40	1.3032
.2705	30	.2672	.4269	.2773	.4430	3.6059	.5570	.9636	.9839	30	1.3003
.2734	40	.2700	.4314	.2805	.4479	3.5656	.5521	.9628	.9836	20	1.2974
.2763	50	.2728	.4359	.2836	.4527	3.5261	.5473	.9621	.9832	10	1.2945
.2793	16° 00′	.2756	.4403	.2867	.4575	3.4874	.5425	.9613	.9828	74° 00′	1.2915
.2822	10	.2784	.4447	.2899	.4622	3.4495	.5378	.9605	.9825	50	1.2886
.2851	20	.2812	.4491	.2931	.4669	3.4124	.5331	.9596	.9821	40	1.2857
.2880	30	.2840	.4533	.2962	.4716	3.3759	.5284	.9588	.9817	30	1.2828
.2909	40	.2868	.4576	.2994	.4762	3.3402	.5238	.9580	.9814	20	1.2799
.2938	50	.2896	.4618	.3026	.4808	3.3052	.5192	.9572	.9810	10	1.2770
.2967	17° 00′	.2924	.4659	.3057	.4853	3.2709	.5147	.9563	.9806	73° 00′	1.2741
.2996	10	.2952	.4700	.3089	.4898	3.2371	.5102	.9555	.9802	50	1.2712
.3025	20	.2979	.4741	.3121	.4943	3.2041	.5057	.9546	.9798	40	1.2683
.3054	30	.3007	.4781	.3153	.4987	3.1716	.5013	.9537	.9794	30	1.2654
.3083	40	.3035	.4821	.3185	.5031	3.1397	.4969	.9528	.9790	20	1.2625
.3113	50	.3062	.4861	.3217	.5075	3.1084	.4925	.9520	.9786	10	1.2595
.3142	18° 00′	.3090	.4900	.3249	.5118	3.0777	.4882	.9511	.9782	72° 00′	1.2566
		Value COSINE	Log₁₀	Value COTANGENT	Log₁₀	Value TANGENT	Log₁₀	Value SINE	Log₁₀	DEGREES	RADIANS

From "Logarithmic and Trigonometric Tables," by Earle Raymond Hedrick. Reprinted by permission of The Macmillan Company, publishers.

[Characteristics of Logarithms omitted—determine by the usual rule from the value]

Radians	Degrees	Sine Value	Sine Log₁₀	Tangent Value	Tangent Log₁₀	Cotangent Value	Cotangent Log₁₀	Cosine Value	Cosine Log₁₀		
.3142	18° 00′	.3090	.4900	.3249	.5118	3.0777	.4882	.9511	.9782	72° 00′	1.2566
.3171	10	.3118	.4939	.3281	.5161	3.0475	.4839	.9502	.9778	50	1.2537
.3200	20	.3145	.4977	.3314	.5203	3.0178	.4797	.9492	.9774	40	1.2508
.3229	30	.3173	.5015	.3346	.5245	2.9887	.4755	.9483	.9770	30	1.2479
.3258	40	.3201	.5052	.3378	.5287	2.9600	.4713	.9474	.9765	20	1.2450
.3287	50	.3228	.5090	.3411	.5329	2.9319	.4671	.9465	.9761	10	1.2421
.3316	19° 00′	.3256	.5126	.3443	.5370	2.9042	.4630	.9455	.9757	71° 00′	1.2392
.3345	10	.3283	.5163	.3476	.5411	2.8770	.4589	.9446	.9752	50	1.2363
.3374	20	.3311	.5199	.3508	.5451	2.8502	.4549	.9436	.9748	40	1.2334
.3403	30	.3338	.5235	.3541	.5491	2.8239	.4509	.9426	.9743	30	1.2305
.3432	40	.3365	.5270	.3574	.5531	2.7980	.4469	.9417	.9739	20	1.2275
.3462	50	.3393	.5306	.3607	.5571	2.7725	.4429	.9407	.9734	10	1.2246
.3491	20° 00′	.3420	.5341	.3640	.5611	2.7475	.4389	.9397	.9730	70° 00′	1.2217
.3520	10	.3448	.5375	.3673	.5650	2.7228	.4350	.9387	.9725	50	1.2188
.3549	20	.3475	.5409	.3706	.5689	2.6985	.4311	.9377	.9721	40	1.2159
.3578	30	.3502	.5443	.3739	.5727	2.6746	.4273	.9367	.9716	30	1.2130
.3607	40	.3529	.5477	.3772	.5766	2.6511	.4234	.9356	.9711	20	1.2101
.3636	50	.3557	.5510	.3805	.5804	2.6279	.4196	.9346	.9706	10	1.2072
.3665	21° 00′	.3584	.5543	.3839	.5842	2.6051	.4158	.9336	.9702	69° 00′	1.2043
.3694	10	.3611	.5576	.3872	.5879	2.5826	.4121	.9325	.9697	50	1.2014
.3723	20	.3638	.5609	.3906	.5917	2.5605	.4083	.9315	.9692	40	1.1985
.3752	30	.3665	.5641	.3939	.5954	2.5386	.4046	.9304	.9687	30	1.1956
.3782	40	.3692	.5673	.3973	.5991	2.5172	.4009	.9293	.9682	20	1.1926
.3811	50	.3719	.5704	.4006	.6028	2.4960	.3972	.9283	.9677	10	1.1897
.3840	22° 00′	.3746	.5736	.4040	.6064	2.4751	.3936	.9272	.9672	68° 00′	1.1868
.3869	10	.3773	.5767	.4074	.6100	2.4545	.3900	.9261	.9667	50	1.1839
.3898	20	.3800	.5798	.4108	.6136	2.4342	.3864	.9250	.9661	40	1.1810
.3927	30	.3827	.5828	.4142	.6172	2.4142	.3828	.9239	.9656	30	1.1781
.3956	40	.3854	.5859	.4176	.6208	2.3945	.3792	.9228	.9651	20	1.1752
.3985	50	.3881	.5889	.4210	.6243	2.3750	.3757	.9216	.9646	10	1.1723
.4014	23° 00′	.3907	.5919	.4245	.6279	2.3559	.3721	.9205	.9640	67° 00′	1.1694
.4043	10	.3934	.5948	.4279	.6314	2.3369	.3686	.9194	.9635	50	1.1665
.4072	20	.3961	.5978	.4314	.6348	2.3183	.3652	.9182	.9629	40	1.1636
.4102	30	.3987	.6007	.4348	.6383	2.2998	.3617	.9171	.9624	30	1.1606
.4131	40	.4014	.6036	.4383	.6417	2.2817	.3583	.9159	.9618	20	1.1577
.4160	50	.4041	.6065	.4417	.6452	2.2637	.3548	.9147	.9613	10	1.1548
.4189	24° 00′	.4067	.6093	.4452	.6486	2.2460	.3514	.9135	.9607	66° 00′	1.1519
.4218	10	.4094	.6121	.4487	.6520	2.2286	.3480	.9124	.9602	50	1.1490
.4247	20	.4120	.6149	.4522	.6553	2.2113	.3447	.9112	.9596	40	1.1461
.4276	30	.4147	.6177	.4557	.6587	2.1943	.3413	.9100	.9590	30	1.1432
.4305	40	.4173	.6205	.4592	.6620	2.1775	.3380	.9088	.9584	20	1.1403
.4334	50	.4200	.6232	.4628	.6654	2.1609	.3346	.9075	.9579	10	1.1374
.4363	25° 00′	.4226	.6259	.4663	.6687	2.1445	.3313	.9063	.9573	65° 00′	1.1345
.4392	10	.4253	.6286	.4699	.6720	2.1283	.3280	.9051	.9567	50	1.1316
.4422	20	.4279	.6313	.4734	.6752	2.1123	.3248	.9038	.9561	40	1.1286
.4451	30	.4305	.6340	.4770	.6785	2.0965	.3215	.9026	.9555	30	1.1257
.4480	40	.4331	.6366	.4806	.6817	2.0809	.3183	.9013	.9549	20	1.1228
.4509	50	.4358	.6392	.4841	.6850	2.0655	.3150	.9001	.9543	10	1.1199
.4538	26° 00′	.4384	.6418	.4877	.6882	2.0503	.3118	.8988	.9537	64° 00′	1.1170
.4567	10	.4410	.6444	.4913	.6914	2.0353	.3086	.8975	.9530	50	1.1141
.4596	20	.4436	.6470	.4950	.6946	2.0204	.3054	.8962	.9524	40	1.1112
.4625	30	.4462	.6495	.4986	.6977	2.0057	.3023	.8949	.9518	30	1.1083
.4654	40	.4488	.6521	.5022	.7009	1.9912	.2991	.8936	.9512	20	1.1054
.4683	50	.4514	.6546	.5059	.7040	1.9768	.2960	.8923	.9505	10	1.1025
.4712	27° 00′	.4540	.6570	.5095	.7072	1.9626	.2928	.8910	.9499	63° 00′	1.0996
		Value Log₁₀ COSINE		Value Log₁₀ COTANGENT		Value Log₁₀ TANGENT		Value Log₁₀ SINE		DEGREES	RADIANS

From "Logarithmic and Trigonometric Tables," by Earle Raymond Hedrick. Reprinted by permission of The Macmillan Company, publishers.

TABLE IV. FOUR-PLACE TRIGONOMETRIC FUNCTIONS 607

[Characteristics of Logarithms omitted—determine by the usual rule from the value]

Radians	Degrees	Sine Value	Sine Log₁₀	Tangent Value	Tangent Log₁₀	Cotangent Value	Cotangent Log₁₀	Cosine Value	Cosine Log₁₀	Degrees	Radians
.4712	27° 00′	.4540	.6570	.5095	.7072	1.9626	.2928	.8910	.9499	63° 00′	1.0996
.4741	10	.4566	.6595	.5132	.7103	1.9486	.2897	.8897	.9492	50	1.0966
.4771	20	.4592	.6620	.5169	.7134	1.9347	.2866	.8884	.9486	40	1.0937
.4800	30	.4617	.6644	.5206	.7165	1.9210	.2835	.8870	.9479	30	1.0908
.4829	40	.4643	.6668	.5243	.7196	1.9074	.2804	.8857	.9473	20	1.0879
.4858	50	.4669	.6692	.5280	.7226	1.8940	.2774	.8843	.9466	10	1.0850
.4887	28° 00′	.4695	.6716	.5317	.7257	1.8807	.2743	.8829	.9459	62° 00′	1.0821
.4916	10	.4720	.6740	.5354	.7287	1.8676	.2713	.8816	.9453	50	1.0792
.4945	20	.4746	.6763	.5392	.7317	1.8546	.2683	.8802	.9446	40	1.0763
.4974	30	.4772	.6787	.5430	.7348	1.8418	.2652	.8788	.9439	30	1.0734
.5003	40	.4797	.6810	.5467	.7378	1.8291	.2622	.8774	.9432	20	1.0705
.5032	50	.4823	.6833	.5505	.7408	1.8165	.2592	.8760	.9425	10	1.0676
.5061	29° 00′	.4848	.6856	.5543	.7438	1.8040	.2562	.8746	.9418	61° 00′	1.0647
.5091	10	.4874	.6878	.5581	.7467	1.7917	.2533	.8732	.9411	50	1.0617
.5120	20	.4899	.6901	.5619	.7497	1.7796	.2503	.8718	.9404	40	1.0588
.5149	30	.4924	.6923	.5658	.7526	1.7675	.2474	.8704	.9397	30	1.0559
.5178	40	.4950	.6946	.5696	.7556	1.7556	.2444	.8689	.9390	20	1.0530
.5207	50	.4975	.6968	.5735	.7585	1.7437	.2415	.8675	.9383	10	1.0501
.5236	30° 00′	.5000	.6990	.5774	.7614	1.7321	.2386	.8660	.9375	60° 00′	1.0472
.5265	10	.5025	.7012	.5812	.7644	1.7205	.2356	.8646	.9368	50	1.0443
.5294	20	.5050	.7033	.5851	.7673	1.7090	.2327	.8631	.9361	40	1.0414
.5323	30	.5075	.7055	.5890	.7701	1.6977	.2299	.8616	.9353	30	1.0385
.5352	40	.5100	.7076	.5930	.7730	1.6864	.2270	.8601	.9346	20	1.0356
.5381	50	.5125	.7097	.5969	.7759	1.6753	.2241	.8587	.9338	10	1.0327
.5411	31° 00′	.5150	.7118	.6009	.7788	1.6643	.2212	.8572	.9331	59° 00′	1.0297
.5440	10	.5175	.7139	.6048	.7816	1.6534	.2184	.8557	.9323	50	1.0268
.5469	20	.5200	.7160	.6088	.7845	1.6426	.2155	.8542	.9315	40	1.0239
.5498	30	.5225	.7181	.6128	.7873	1.6319	.2127	.8526	.9308	30	1.0210
.5527	40	.5250	.7201	.6168	.7902	1.6212	.2098	.8511	.9300	20	1.0181
.5556	50	.5275	.7222	.6208	.7930	1.6107	.2070	.8496	.9292	10	1.0152
.5585	32° 00′	.5299	.7242	.6249	.7958	1.6003	.2042	.8480	.9284	58° 00′	1.0123
.5614	10	.5324	.7262	.6289	.7986	1.5900	.2014	.8465	.9276	50	1.0094
.5643	20	.5348	.7282	.6330	.8014	1.5798	.1986	.8450	.9268	40	1.0065
.5672	30	.5373	.7302	.6371	.8042	1.5697	.1958	.8434	.9260	30	1.0036
.5701	40	.5398	.7322	.6412	.8070	1.5597	.1930	.8418	.9252	20	1.0007
.5730	50	.5422	.7342	.6453	.8097	1.5497	.1903	.8403	.9244	10	.9977
.5760	33° 00′	.5446	.7361	.6494	.8125	1.5399	.1875	.8387	.9236	57° 00′	.9948
.5789	10	.5471	.7380	.6536	.8153	1.5301	.1847	.8371	.9228	50	.9919
.5818	20	.5495	.7400	.6577	.8180	1.5204	.1820	.8355	.9219	40	.9890
.5847	30	.5519	.7419	.6619	.8208	1.5108	.1792	.8339	.9211	30	.9861
.5876	40	.5544	.7438	.6661	.8235	1.5013	.1765	.8323	.9203	20	.9832
.5905	50	.5568	.7457	.6703	.8263	1.4919	.1737	.8307	.9194	10	.9803
.5934	34° 00′	.5592	.7476	.6745	.8290	1.4826	.1710	.8290	.9186	56° 00′	.9774
.5963	10	.5616	.7494	.6787	.8317	1.4733	.1683	.8274	.9177	50	.9745
.5992	20	.5640	.7513	.6830	.8344	1.4641	.1656	.8258	.9169	40	.9716
.6021	30	.5664	.7531	.6873	.8371	1.4550	.1629	.8241	.9160	30	.9687
.6050	40	.5688	.7550	.6916	.8398	1.4460	.1602	.8225	.9151	20	.9657
.6080	50	.5712	.7568	.6959	.8425	1.4370	.1575	.8208	.9142	10	.9628
.6109	35° 00′	.5736	.7586	.7002	.8452	1.4281	.1548	.8192	.9134	55° 00′	.9599
.6138	10	.5760	.7604	.7046	.8479	1.4193	.1521	.8175	.9125	50	.9570
.6167	20	.5783	.7622	.7089	.8506	1.4106	.1494	.8158	.9116	40	.9541
.6196	30	.5807	.7640	.7133	.8533	1.4019	.1467	.8141	.9107	30	.9512
.6225	40	.5831	.7657	.7177	.8559	1.3934	.1441	.8124	.9098	20	.9483
.6254	50	.5854	.7675	.7221	.8586	1.3848	.1414	.8107	.9089	10	.9454
.6283	36° 00′	.5878	.7692	.7265	.8613	1.3764	.1387	.8090	.9080	54° 00′	.9425
		Value Log₁₀ Cosine		Value Log₁₀ Cotangent		Value Log₁₀ Tangent		Value Log₁₀ Sine		Degrees	Radians

From "Logarithmic and Trigonometric Tables," by Earle Raymond Hedrick. Reprinted by permission of The Macmillan Company, publishers.

608 TABLE IV. FOUR-PLACE TRIGONOMETRIC FUNCTIONS

[Characteristics of Logarithms omitted—determine by the usual rule from the value]

Radians	Degrees	Sine Value	Sine Log₁₀	Tangent Value	Tangent Log₁₀	Cotangent Value	Cotangent Log₁₀	Cosine Value	Cosine Log₁₀		
.6283	36° 00′	.5878	.7692	.7265	.8613	1.3764	.1387	.8090	.9080	54° 00′	.9425
.6312	10	.5901	.7710	.7310	.8639	1.3680	.1361	.8073	.9070	50	.9396
.6341	20	.5925	.7727	.7355	.8666	1.3597	.1334	.8056	.9061	40	.9367
.6370	30	.5948	.7744	.7400	.8692	1.3514	.1308	.8039	.9052	30	.9338
.6400	40	.5972	.7761	.7445	.8718	1.3432	.1282	.8021	.9042	20	.9308
.6429	50	.5995	.7778	.7490	.8745	1.3351	.1255	.8004	.9033	10	.9279
.6458	37° 00′	.6018	.7795	.7536	.8771	1.3270	.1229	.7986	.9023	53° 00′	.9250
.6487	10	.6041	.7811	.7581	.8797	1.3190	.1203	.7969	.9014	50	.9221
.6516	ʹ20	.6065	.7828	.7627	.8824	1.3111	.1176	.7951	.9004	40	.9192
.6545	30	.6088	.7844	.7673	.8850	1.3032	.1150	.7934	.8995	30	.9163
.6574	40	.6111	.7861	.7720	.8876	1.2954	.1124	.7916	.8985	20	.9134
.6603	50	.6134	.7877	.7766	.8902	1.2876	.1098	.7898	.8975	10	.9105
.6632	38° 00′	.6157	.7893	.7813	.8928	1.2799	.1072	.7880	.8965	52° 00′	.9076
.6661	10	.6180	.7910	.7860	.8954	1.2723	.1046	.7862	.8955	50	.9047
.6690	20	.6202	.7926	.7907	.8980	1.2647	.1020	.7844	.8945	40	.9018
.6720	30	.6225	.7941	.7954	.9006	1.2572	.0994	.7826	.8935	30	.8988
.6749	40	.6248	.7957	.8002	.9032	1.2497	.0968	.7808	.8925	20	.8959
.6778	50	.6271	.7973	.8050	.9058	1.2423	.0942	.7790	.8915	10	.8930
.6807	39° 00′	.6293	.7989	.8098	.9084	1.2349	.0916	.7771	.8905	51° 00′	.8901
.6836	10	.6316	.8004	.8146	.9110	1.2276	.0890	.7753	.8895	50	.8872
.6865	20	.6338	.8020	.8195	.9135	1.2203	.0865	.7735	.8884	40	.8843
.6894	30	.6361	.8035	.8243	.9161	1.2131	.0839	.7716	.8874	30	.8814
.6923	40	.6383	.8050	.8292	.9187	1.2059	.0813	.7698	.8864	20	.8785
.6952	50	.6406	.8066	.8342	.9212	1.1988	.0788	.7679	.8853	10	.8756
.6981	40° 00′	.6428	.8081	.8391	.9238	1.1918	.0762	.7660	.8843	50° 00′	.8727
.7010	10	.6450	.8096	.8441	.9264	1.1847	.0736	.7642	.8832	50	.8698
.7039	20	.6472	.8111	.8491	.9289	1.1778	.0711	.7623	.8821	40	.8668
.7069	30	.6494	.8125	.8541	.9315	1.1708	.0685	.7604	.8810	30	.8639
.7098	40	.6517	.8140	.8591	.9341	1.1640	.0659	.7585	.8800	20	.8610
.7127	50	.6539	.8155	.8642	.9366	1.1571	.0634	.7566	.8789	10	.8581
.7156	41° 00′	.6561	.8169	.8693	.9392	1.1504	.0608	.7547	.8778	49° 00′	.8552
.7185	10	.6583	.8184	.8744	.9417	1.1436	.0583	.7528	.8767	50	.8523
.7214	20	.6604	.8198	.8796	.9443	1.1369	.0557	.7509	.8756	40	.8494
.7243	30	.6626	.8213	.8847	.9468	1.1303	.0532	.7490	.8745	30	.8465
.7272	40	.6648	.8227	.8899	.9494	1.1237	.0506	.7470	.8733	20	.8436
.7301	50	.6670	.8241	.8952	.9519	1.1171	.0481	.7451	.8722	10	.8407
.7330	42° 00′	.6691	.8255	.9004	.9544	1.1106	.0456	.7431	.8711	48° 00′	.8378
.7359	10	.6713	.8269	.9057	.9570	1.1041	.0430	.7412	.8699	50	.8348
.7389	20	.6734	.8283	.9110	.9595	1.0977	.0405	.7392	.8688	40	.8319
.7418	30	.6756	.8297	.9163	.9621	1.0913	.0379	.7373	.8676	30	.8290
.7447	40	.6777	.8311	.9217	.9646	1.0850	.0354	.7353	.8665	20	.8261
.7476	50	.6799	.8324	.9271	.9671	1.0786	.0329	.7333	.8653	10	.8232
.7505	43° 00′	.6820	.8338	.9325	.9697	1.0724	.0303	.7314	.8641	47° 00′	.8203
.7534	10	.6841	.8351	.9380	.9722	1.0661	.0278	.7294	.8629	50	.8174
.7563	20	.6862	.8365	.9435	.9747	1.0599	.0253	.7274	.8618	40	.8145
.7592	30	.6884	.8378	.9490	.9772	1.0538	.0228	.7254	.8606	30	.8116
.7621	40	.6905	.8391	.9545	.9798	1.0477	.0202	.7234	.8594	20	.8087
.7650	50	.6926	.8405	.9601	.9823	1.0416	.0177	.7214	.8582	10	.8058
.7679	44° 00′	.6947	.8418	.9657	.9848	1.0355	.0152	.7193	.8569	46° 00′	.8029
.7709	10	.6967	.8431	.9713	.9874	1.0295	.0126	.7173	.8557	50	.7999
.7738	20	.6988	.8444	.9770	.9899	1.0235	.0101	.7153	.8545	40	.7970
.7767	30	.7009	.8457	.9827	.9924	1.0176	.0076	.7133	.8532	30	.7941
.7796	40	.7030	.8469	.9884	.9949	1.0117	.0051	.7112	.8520	20	.7912
.7825	50	.7050	.8482	.9942	.9975	1.0058	.0025	.7092	.8507	10	.7883
.7854	45° 00′	.7071	.8495	1.0000	.0000	1.0000	.0000	.7071	.8495	45° 00′	.7854
		Value Log₁₀ Cosine		Value Log₁₀ Cotangent		Value Log₁₀ Tangent		Value Log₁₀ Sine		Degrees	Radians

From "Logarithmic and Trigonometric Tables," by Earle Raymond Hedrick. Reprinted by permission of The Macmillan Company, publishers.

Index

In this index the numbers refer to chapters, sections, theorems, definitions, and exercises, and are indicated according to the numbering system of the book. References to definitions are in bold-face type. A single number refers to the introductory paragraphs of a chapter; double numbers refer to sections; triple numbers refer to exercises; triple numbers preceded by an asterisk refer to theorems, formal definitions, laws, axioms. For example:

9 refers to Chapter 9 (introductory paragraph).

6.5 refers to Section 5 of Chapter 6.

6.5.26 refers to Exercise 26, Section 5, Chapter 6.

*2.5.1 refers to Theorem 1, Section 5, Chapter 2.

11.4(12) refers to equation (12), Section 4, Chapter 11.

609

Circular functions, inverse, 8.9
  line values, 8.4.17
  reciprocal relations, *8.4.3
  signs, 8.4.8
  of special angles, 8.4, 8.4.9
  of sum, 8.7
Circular measure, *see* Radians
Circular reasoning, 1.2, 2.6
Circular sector, *8.3.5, 8.3.13, 8.3.17, 8.3.18
Circular segment, 8.11.22
Circumcenter, 9.5.12, 9.5.20–21
Circumference, 4.1.13, 8.2.13
  see also Arc length
Circumscribed circle, 8.1.11, 8.11.25, 11.1.6l
Cissoid of Diocles, 13.5.8
Closed interval, 4.12
Closed system, 10, 10.5.14
Coefficient, 3.5, 3.5.13
  of absorption, 6.8.17
  binomial, 6.5
  leading, 12
  of volume expansion, 11.5.7
Cofunctions, 8.6, 8.6.9
Coincident lines, 4.9
Coincident roots, 5.6, 12.2
Colatitude, 14.6
Collinear points, 4.7.14, 9.5.21
Cologarithm, 7.4
Column of determinant, 4.10
Column subscript, 4.11.15
Combination, 6.4, 6.5.14
Combinatorial analysis, 6.4
Combinatorial product, 3.5.18–21, 3.8.10
Common chord of circles, *see* Radical axis
Common denominator, 3.11.10, 13.2
Common difference, 4.5
Common factor, 3.6, 13.2.8–9
Common logarithms, 7.3
Common multiple, 13.2
  see also Least common multiple
Common ratio, 6.6
Commutative law, for actions, 3.13.23
  of addition, *3.4.2, 3.4.16, 3.10.3, 10.2, 10.2.6

Commutative law, of multiplication, *3.5.3, 3.5.21, 6.4.12
  for operations, 3.13.24
  for sets, 3.4.15, 3.4.19, 3.5.20
  for transformations, 9.1.12, 9.1.19
  see also Field, Integral domain, Laws
Comparing numbers, 3.12.18
Complementary angle, 2.2.5, 8.6
Completing the square, 5.5, 5.6, 5.6.4, 5.6.5, 8.2.15, 11.1–11.3
Complex numbers, 3.13, 10, 10.2
  absolute value, 10.4
  addition, *10.2.4
  amplitude, 10.4
  argument, 10.4
  conjugate, 10.2.12, 10.2.15, 12.2.12, 12.2.13
  division, *10.2.7, *10.4.3
  equality, *10.2.1–2
  as exponents, 10.4.12
  as field, 10.2, 10.2.16
  geometric interpretation, 10.3
  modulus, 10.4
  multiplication, *10.2.5, *10.4.2
  negative, *10.2.8
  as ordered pairs, 10.3
  polar form, 10.4
  reciprocal, *10.2.10, *10.4.4
  roots of, 10.5
  scalar multiplication, 10.2.11, 10.2.15, 10.3
  subtraction, *10.2.6
  trigonometric form, 10.4
  and vectors, 10.3
Component, imaginary, 10.2
  real, 10.2
  of vector, 9.7
Composite function, 4.3.14–15, 12.5, 12.5.6–10
  derivative of, *12.5.3
Composition of proportion, 3.11.17
Compound fraction, 13.2
Compound interest law, *6.6.4
  see also Interest
Compression ratio, 6.8.7
Computations, *see* Arithmetic, Numerical methods

6.2.22–23, 6.3.16–20, 6.5.27,
6.9.14, 6.11.19, 7.2.32, 7.5.17,
8.2.11, 8.2.17, 8.4.11–12, 8.5.10,
8.8.18, 8.9.19, 8.11.28–29, 9.5.22,
9.6.13, 9.8.8, 12.2.9, 12.7.9–10,
13.1, 13.1.4–5
absolute, **4.12**
conditional, **4.12**
linear, 4.12
Inequality, 2.4.1, 3.3, 3.4, 3.8, 3.9,
3.12, 3.12.5
Abel's, 6.5.27
graph of, **4.12**
for infinite decimals, 3.13.14
monotonic laws, *3.4.4–8, *3.5.5,
*3.5.7, *3.10.14–15, 3.13.17
triangular, 8.1.12
Infinite decimal, 3.12.17, 3.13,
3.13.14, 6.7, 12.8.17
Infinite sets, 2.7, 2.7.4, 3.1, 3.1.8,
3.12–3.13
Infinity, 6.7–6.8, 8.4, 8.8, 13.1, 13.5
Inflexion, **12.5**
Informal proof, 2.3
Initial side of angle, **8.3**
Inscribed angle, 2.1.4g
Inscribed circle, 8.11.4, 11.1.6l
Instantaneous acceleration, rate,
speed, *see* Exact acceleration,
Exact rate of change, Exact
speed
Integers, 2.2.7k, 2.3.1h, **3.10**, 3.10.45a
negative, **3.9**
positive, 3.1–3.7, 4.5
Integral, definite, **12.7**
indefinite, **12.6**
Integral calculus, 12.6–12.7
fundamental theorem of, *12.7.1
Integral domain, 3.10.45, 10.2.18
Integrand, **12.7**
Integration, 12.6–12.7
Intercept, normal, **9.3**
Intercept form, of line, **4.8.9**
of plane, *14.3.5
Intercepts, 4.3, 9.8, 13.1, 13.3, 13.5
of ellipse, 11.1
of hyperbola, 11.3
in polar coordinates, 9.8

Intercepts, of surface, 14.5
Interest, compound, 4.4.7, 6.6.10–11,
6.9.12, 7.3.11, 7.3.16, 7.4.15–16,
7.4.32
continuous, 7.1.12
discount factor, **6.9.12**, **7.4.32**
effective rate, **7.3.16**
force of, **7.1.12**
nominal rate, **7.3.16**
perpetuity, 6.7.6
present value, **6.6.10**
principal, **6.6.10**
simple, 4.1.13i, 4.3.11, 4.5.10–13
see also Economics
Interior angles, of polygon, 6.2
of triangle, 2.1.4, 2.7.5a
Interpolation, **5.2**, 5.2.6–13, 5.3,
6.8.26, 6.10.6, 7.3, 8.10, 8.12,
12.4
Intersections, of curves, 5.7.12, 5.7.14,
9.4, 9.4.12–14, 9.8.14, 11.1.9–15,
11.3.7, 13.5
of lines, 4.9–4.10
of planes, 14.4.1, 14.4.12
see also Simultaneous equations
Intervals, **4.12**
Invariance, 4.7.15, 9.1.5, 9.1.13, 11.5
Invariants of conic, **11.5**
Inverse, 3.7, 5.3, 6.10, 8.9, 9.1.22,
12.6
Inverse circular functions, **8.9**
Inverse functions, 4.10.11, **6.10**,
6.10.22–31, 6.11.25, 7, 7.2, 7.2.14,
7.2.17, 8.9.23–24
Inverse operations, 3.7, 3.11, 5.3
Inverse square law, of gravitation,
6.9.6
of light intensity, 6.8.6
Inverse transformations, **9.1.22**
Inverse variation, **6.8**
Inversely proportional, **6.8**, 13.3.5
Inversion of proportion, **3.11.17**
Inversions, **4.11.19**
Inverting, 3.11, 13.2
Investment, *see* Interest
Irrational equations, 6.11.10, 13.4,
13.4.4
Irrational functions, **13**, 13.3

Irrational numbers, **3.13**, 7.3, 7.3.15, 13
Irrational roots, 12.4
Irreducible polynomials, **13.2**
Isosceles triangle, **2.2.7f**, 2.3, 2.3.1k, 2.3.3, 2.4

Joint variation, 4.1.17, 6.8.8, 6.8.18

Kendall and Smith, 3.14.29
Kepler's third law, **6.10.17**

Language, 1.4, 2.7.7
Lateral area of cone, 2.1.4p, **2.2.7j**
Latitude, 14.6, 14.6.9
Latus rectum, of ellipse, **11.1.28–29**
  of hyperbola, **11.3.20**
  of parabola, **11.2.10**
Law, of cosines, *8.11.2
  dichotomy, *3.3.1
  Graham's, 13.3.5
  of growth, see Growth
  Kepler's third, **6.10.17**
  Newton's, of gravitation, **6.9.6**
  of sines, *8.11.1
  Stefan's, **6.1.21**
  of tangents, *8.11.3
  trichotomy, *3.3.5
Laws, of algebra, 3, 3.10.45, 3.11, 3.11.23–24, 3.13.21, 10.2, 10.2.6;
  see also Associative law, Cancellation law, Commutative law, Distributive law, Field, Integral domain
  of exponents, *6.3.3–8, 6.9, 6.11
  of logarithms, 7.2
  monotonic, see Monotonic law
  Newton's, of motion, 1.3, 2.6.1e, 4.1.13j
  of vectors, 9.7.22
Leading coefficient, **12**
Least common denominator, 3.11.10, **13.2**
Least common multiple, 2.2.9j, **13.2**, 13.2.9
Left-handed system, Fig. 14.1.1, 14.1.1–2
Left member, **3.3**

Lemniscate, **9.8.11**, 13.5.8f
Length, see Distance
  arc, see Arc length
  of vector, **9.7**, 9.7.24
Less, 3.3, 4.4.9
Less than, **3.12.18**, 4.4.9
Light, 4.1.15, 4.4.7, 5.1.4, 5.7.17, 6.8.6, 6.8.16, 7.4.19, 11.1.19, 11.2.21, 13.4.6
Light year, **7.4.19**
Limaçon, **9.8.9**, 13.5.8
Limit, 5.4, 6.7, 6.8.24, 6.8.27, 11.6, 12.7
Limits, of integration, **12.7**
  lower, **12.7**
  of summation, **6.5**
  upper, **12.7**
Line, 2.2, 4.2, 4.7
  directed, **3.2**
  direction angles, **9.3**, **14.2**
  equations of, 4.3, 4.7–4.9, 9.1, 9.3, 14.2.6, 14.4
  half, **8.3**
  inclination of, **9.2**
  intercept form, **4.8.9**
  normal form, 9.3, 14.2(1)
  parametric form, *9.6.1, 9.6.4, 14.2.7, 14.4
  point-slope form, 4.8(7)
  in polar coordinates, 9.8, 9.8.15
  secant, 2.2.8s, 5.2
  in space, 2.1.1k, 14.2, 14.4
  symmetric form, 14.2.7, 14.4(2)
  tangent, see Tangent
  line segment, 2.7, 9.5.9
  division of, 4.6.11–18, 14.1.11–13
  length of, see Distance
  midpoint, 4.6.12, 14.1.13
  perpendicular bisector, see Perpendicular bisector
Line values of circular functions, 8.4.17
Linear equations, **4.4**, 4.5.15, 4.8, 4.9–4.10, 4.10.9, 5.6.10, 14.3
Linear functions, **4**, 4.3, 4.5.21, 4.10.11, *5.4.5, 14.2, 14.3
Linear inequalities, 4.12

9.6.17, 9.7, 9.7.15, 11.5.6, 13.3.5,
13.4.12, 13.4.14
see also Astronomy, Freely falling
body, Physics, Rotating body
Median, of trapezoid, 9.5.7
of triangle, 2.1.4i, 2.2.6b
Medicine, *see* Human biology
Member, of equation, **3.3**
left and right, **3.3**
of set, **2.7**, 2.7.11
Mental arithmetic, 3.4.3, 3.4.9, 3.5.2,
3.5.3–4, 3.6.7
Meters, 6.8.9d
Method, of false position, 12.4.9
scientific, 1.3, 6.2
Micron, 6.8.13
Midpoint, 3.12.7, 4.6.12–13, 14.1.13
Mil, **8.3.26–29**
Miles, 6.8.9
Military problems, 4.1.16, 4.1.18,
4.4.17, 4.4.19, 4.9.10, 6.4.13,
6.4.28–29, 6.8.18, 8.10.5, 8.10.8,
9.6.17, 9.7.12
Millimeters, 6.8.9
Minimum, **5.5.11**
see also Maxima and minima
Minor, in determinant, **4.11, 4.11.15**
Minor axis, **11.1**
Minuend, **3.7**
Minus sign, 3.7, 3.10
Minutes of angle, 8.3.7
Mixed numbers, **3.11.11**
Mixtures, 1.2, 4.4.20
see also Solutions
Modulus, of complex number, **10.4**
of elasticity, **5.2.18**
Moisture content, 3.12.28
Mole, 3.14.26
Mollweide's equations, 8.12.9
Monomial, **3.4**
Monotonic law, of addition, *3.4.4
of multiplication, *3.5.5, *3.10.14
see also Inequality
More than, **3.12.18**
Motion, *see* Freely falling body, Mechanics, Rotating body
harmonic, 8.8.14
Multinomial, **3.4**

Multiple, **3.5, 13.2**
common, **13.2**
least common, 2.2.9j, **13.2**, 13.2.9
Multiple roots, 5.6, **12.2**
Multiple-valued, **5.3**, 12.6, 14
Multiples of nine, 6.2.26, 12.8.21
Multiplication, 6.2.13, 12.8
associative law, *3.5.4, 3.5.20–21,
6.4.12
commutative law, *3.5.3, 3.5.21,
6.4.12
of complex numbers, *10.2.5,
*10.4.2
distributive law, 3.6, 3.6.13, 3.6.15,
6.2.12
of fractions, 3.11, 13.2
of integers, **3.10**
monotonic law, *3.5.5, *3.10.14
of positive integers, **3.5, 3.5.19**
of rationals, *3.11.4
scalar, **9.7.20**, 10.2.11, 10.2.15, 10.3
of sets, *see* Logical product, Combinatorial product
of transformations, 9.1.19–20
by zero, *3.8.3

n-valued, **5.3**
National income, 5.2.10, 5.2.15
National Mathematics Magazine,
6.2.15
Natural logarithms, 7.5
Natural numbers, *see* Positive integers
Navigation, 4.1.2, 5.7.22, 8.10.7, 9.6,
9.6.1, 9.6.7–9, 9.7, 9.7.10–14,
14.6.9
Necessary condition, **2.1**
Necessary and sufficient, **2.5**, 2.5.6–7,
2.7.7f
Negative (algebra), angles, 8.3
bases, 6.11, 7.2.7
of complex numbers, *10.2.8
exponents, **6.9**
integers, **3.9**
of numbers, 3.9, 3.12, *10.2.8
of rationals, **3.12**
of vectors, **9.7.18**